Signals and Communication T

Series Editors

Emre Celebi, Department of Computer Science, University of Central Arkansas, Conway, AR, USA

Jingdong Chen, Northwestern Polytechnical University, Xi'an, China

E. S. Gopi, Department of Electronics and Communication Engineering, National Institute of Technology, Tiruchirappalli, Tamil Nadu, India

Amy Neustein, Linguistic Technology Systems, Fort Lee, NJ, USA

H. Vincent Poor, Department of Electrical Engineering, Princeton University, Princeton, NJ, USA

Antonio Liotta, University of Bolzano, Bolzano, Italy

Mario Di Mauro, University of Salerno, Salerno, Italy

This series is devoted to fundamentals and applications of modern methods of signal processing and cutting-edge communication technologies. The main topics are information and signal theory, acoustical signal processing, image processing and multimedia systems, mobile and wireless communications, and computer and communication networks. Volumes in the series address researchers in academia and industrial R&D departments. The series is application-oriented. The level of presentation of each individual volume, however, depends on the subject and can range from practical to scientific.

Indexing: All books in "Signals and Communication Technology" are indexed by Scopus and zbMATH

For general information about this book series, comments or suggestions, please contact Mary James at mary.james@springer.com or Ramesh Nath Premnath at ramesh.premnath@springer.com.

Anupam Biswas • Emile Wennekes •
Alicja Wieczorkowska • Rabul Hussain Laskar
Editors

Advances in Speech and Music Technology

Computational Aspects and Applications

≙ Springer

Editors
Anupam Biswas
Department of Computer Science & Engineering
National Institute of Technology Silchar
Cachar, Assam, India

Emile Wennekes
Department of Media and Culture Studies
Utrecht University
Utrecht, Utrecht, The Netherlands

Alicja Wieczorkowska
Multimedia Department
Polish-Japanese Academy of Information Technology
Warsaw, Poland

Rabul Hussain Laskar
Department of Electronics & Communication Engineering
National Institute of Technology Silchar
Cachar, India

ISSN 1860-4862 ISSN 1860-4870 (electronic)
Signals and Communication Technology
ISBN 978-3-031-18446-8 ISBN 978-3-031-18444-4 (eBook)
https://doi.org/10.1007/978-3-031-18444-4

© The Editor(s) (if applicable) and The Author(s), under exclusive license to Springer Nature Switzerland AG 2023

This work is subject to copyright. All rights are solely and exclusively licensed by the Publisher, whether the whole or part of the material is concerned, specifically the rights of translation, reprinting, reuse of illustrations, recitation, broadcasting, reproduction on microfilms or in any other physical way, and transmission or information storage and retrieval, electronic adaptation, computer software, or by similar or dissimilar methodology now known or hereafter developed.
The use of general descriptive names, registered names, trademarks, service marks, etc. in this publication does not imply, even in the absence of a specific statement, that such names are exempt from the relevant protective laws and regulations and therefore free for general use.
The publisher, the authors, and the editors are safe to assume that the advice and information in this book are believed to be true and accurate at the date of publication. Neither the publisher nor the authors or the editors give a warranty, expressed or implied, with respect to the material contained herein or for any errors or omissions that may have been made. The publisher remains neutral with regard to jurisdictional claims in published maps and institutional affiliations.

This Springer imprint is published by the registered company Springer Nature Switzerland AG
The registered company address is: Gewerbestrasse 11, 6330 Cham, Switzerland

Preface

Speech and music are two prominent research areas in the domain of audio signal processing. With recent advancements in speech and music technology, the area has grown tremendously, bringing together the interdisciplinary researchers of computer science, musicology, and speech analysis. The language we speak propagates as sound waves through various media and allows communication between, or entertainment for us, humans. Music we hear or create can be perceived in different aspects as rhythm, melody, harmony, timbre, or mood. The multifaceted nature of speech or music information requires algorithms, systems using sophisticated signal processing, and machine learning techniques to optimally extract useful information. This book provides both profound technological knowledge and a comprehensive treatment of essential and innovative topics in speech and music processing.

Recent computational developments have opened up several avenues to further explore the domains of speech and music. A profound understanding of both speech and music in terms of perception, emotion, mood, gesture, and cognition is in the forefront, and many researchers are working in these domains. In this digital age, overwhelming data have been generated across the world that require efficient processing for better maintenance and retrieval. Machine learning and artificial intelligence are best suited for these computational tasks.

The book comprises four parts. The first part covers state of the art in computational aspects of speech and music. The second part covers machine learning techniques applied in various music information retrieval tasks. The third part comprises chapters dealing with perception, health, and emotion involving music. The last part includes several case studies.

Audio technology, covering speech, music, and other signals, is a very broad domain. Part I contains five review chapters, presenting state of the art in selected aspects of speech and music research, namely automatic speaker recognition, music composition based on artificial intelligence, music recommendation systems, and investigations on Indian classical music, which is very different from Western music that most of us are used to.

Chapter "A Comprehensive Review on Speaker Recognition", written by Banala Saritha, Mohammad Azharuddin Laskar, and Rabul Hussain Laskar, offers a comprehensive review on speaker recognition techniques, mainly focusing on text-dependent methods, where predefined text is used in the identification process. The authors review feature extraction techniques applied often as pre-processing, and then present various models that can be trained for speaker identification, with a special section devoted to deep learning. Measures that can be applied to assess the speaker recognition quality are also briefly discussed.

Chapter "Music Composition with Deep Learning: A Review", authored by Carlos Hernandez-Olivan and Jose R. Beltran, presents a review of music composition techniques, based on deep learning. Artificial intelligence has been applied to music composition since the previous millennium, as briefly reviewed in this chapter. Obviously, deep neural networks are also applied for this purpose, and these techniques are presented in this chapter. Next, the authors delve into the details of the music composition process, including musical form and style, melody, harmony, and instrumentation. Evaluation metrics are also provided in this chapter. Finally, the authors pose and answer interesting questions regarding automatic music composition: how creative it is, what network architectures perform best, how much data is needed for training, etc. Possible directions of future works in this area conclude this chapter.

Chapters "Music Recommendation Systems: Overview and Challenges" and "Music Recommender Systems: A Review Centered on Biases" describe music recommendation systems. Chapter "Music Recommendation Systems: Overview and Challenges", written by Yesid Ospitia-Medina, Sandra Baldassarri, Cecilia Sanz, and José Ramón Beltrán, offers a general overview of such systems, whereas chapter Music Recommender Systems: A Review Centered on Biases, by Makarand Velankar and Parag Kulkarni, presents a review focusing on biases. Chapter "Music Recommendation Systems: Overview and Challenges" presents very broadly content-based approach to music recommendation systems, as well as collaborative and approach, and the hybrid approach to creating such systems. Context-aware recommendation systems, which result in better recommendations, are also briefly presented in this chapter. The authors discuss business aspects of music recommendation systems as well. A special section is devoted to user profiling and psychological aspects, as the type of music the users want to listen to depends on their mood and emotional state. The chapter is concluded with current challenges and trends in music recommendation.

Chapter "Music Recommender Systems: A Review Centered on Biases" presents an overview of biases in music recommendation systems. The authors set off with presenting research questions that are the basis of research in this area, and thus answering them can introduce biases. These research questions include the main characteristics of music recommender systems (approaches to the creation of such systems are presented in the chapter), and how new songs are introduced. The authors review what are the main biases in such systems, and the relationships between the biases and both recommendation strategies and music datasets used. Biases are classified into three categories, namely pre-existing, technical, and emerging biases, detected in use of the system. Works on biases, as well as general works on music recommendation systems, are reviewed here. The authors discuss on how biases impact these systems, and also propose guidelines for handling biases in such systems.

Chapter "Computational Approaches for Indian Classical Music: A Comprehensive Review" presents a review of research on computational techniques applied in Indian classical music, by Yeshwant Singh and Anupam Biswas. This traditional music has roots in singing swaras. Nowadays, it is divided into Hindustani music, with ragas (raags), mainly practiced in northern India, and Carnatic music in southern part of the country. Microtones called shruti are specific to Indian classical music, and make it very different from the Western music. The authors review papers on tonic identification in classical Indian music, including feature extraction and distribution, and melody processing, with segmentation, similarity analysis, and melody representation. The automatic recognition of ragas is also covered in this chapter. The authors also describe datasets of classical Indian music, and evaluation metrics for the research on this music. Before concluding the chapter, the authors present open challenges in this interesting research area.

Machine learning is helpful in understanding and learning from data, identifying patterns, and making decisions with minimal human interaction. This is why machine learning for audio signal processing has attracted attention recently for its applications in both speech and music processing, presented in the five chapters of Part II. Two chapters are focused on speech and multimodal audio signal processing, and three on music, including instruments, raags, shruti, and emotion recognition from music.

Chapter "A Study on Effectiveness of Deep Neural Networks for Speech Signal Enhancement in Comparison with Wiener Filtering Technique" by Vijaya Kumar Padarti, Gnana Sai Polavarapu, Madhurima Madiraju, V. V. Naga Sai Nuthalapati, Vinay Babu Thota, and V.D. Subramanyam Veeravalli explores speech signal enhancement with deep learning and Wiener filtering techniques. The speech signal in general is highly susceptible to various noises. Therefore, speech denoising is essential to produce noise-free speech signals from noisy recordings, thus improving the perceived speech quality and increasing its intelligibility. Common approach is to remove high frequency components from the original signal, but it leads to removal of parts of the original signal, resulting in undesirable quality degradation. In this chapter, Wiener filtering and neural networks are compared as tools for speech signal enhancement. The output signal quality is assessed in terms of signal to noise ratio (SNR) and peak signal to noise ratio (PSNR). Advanced MATLAB toolboxes such as Deep Learning toolbox, Audio toolbox, and Signal Processing toolbox are utilized for the analysis.

Chapter "Video Soundtrack Evaluation with Machine Learning: Data Availability, Feature Extraction, and Classification" by Georgios Touros and Theodoros Giannakopoulos evaluates multimodal signals using machine learning techniques, with a combined analysis of both video and audio data, in order to find satisfactory accompaniment music for video content. The availability of data, feature extraction, and classification are discussed in this chapter. Creating or choosing music that accompanies visual content, i.e. video soundtracks, is an artistic task that is usually taken up by dedicated professionals, namely a composer and music supervisor, to have the musical content that best accentuates each scene. In this chapter, a method is proposed for collecting and combining relevant data from three modalities: audio, video, and symbolic representations of music, in an end-to-end classification pipeline. A comprehensive multimodal feature library is described, together with a database that has been obtained by applying the proposed method on a small dataset representing movie scenes. Furthermore, a classifier that aims to discriminate between real and fake examples of video soundtracks from movies has been implemented. This chapter also presents potential research directions and possible improvements in the investigated area.

Chapter "Deep Learning Approach to Joint Identification of Instrument Pitch and Raga for Indian Classical Music" by Ashwini Bhat, Karrthik G. K., Vishal Mahesh, and Vijaya Krishna A. explores deep learning approaches for joint identification of instruments, pitch, and ragas in Indian classical music. The concept of raag and shruti is fundamental in Indian classical music, so their identification, although difficult, is crucial for the analysis of a very complex Indian classical music. The chapter offers a comprehensive comparison of Convolution Neural Network (CNN), Recurrent Neural Network (RNN), and XGboost as tools to achieve the goal. Three feature sets have been created for each task at hand, three models trained, and next a combined RNN model created, yielding approximately 97% accuracy.

Chapter "Comparison of Convolutional Neural Networks and K-Nearest Neighbors for Music Instrument Recognition" by Dhivya S and Prabu Mohandas analyses convolutional neural networks and k-nearest neighbours (k-NN) for identifying instruments from music. Music instrument recognition is one of the main tasks of music information retrieval, as it can enhance the performance of other tasks like automatic music transcription, music genre identification, and source separation. Identification of instruments from the recording is a challenging task in the case of polyphonic music, but it is feasible in the monophonic case. Temporal, spectral, and perceptual features are used for identifying instruments. The chapter compares a convolutional neural network architecture and k-nearest neighbour classifier to identify the musical instrument from monophonic music. Mel-spectrogram representation is used to extract features for the neural network model, and mel-frequency cepstral coefficients are the basis for the k-NN classification. The models were trained on the London Philharmonic dataset consisting of six classes of musical instruments, yielding up to 99% accuracy.

Chapter "Emotion Recognition in Music Using Deep Neural Networks" written by Angelos Geroulanos and Theodoros Giannakopoulos deals with the emotion recognition in music, using deep learning techniques. Although accessing music content online is easy nowadays, and streaming platforms provide automatic recommendations to the users, the suggested list often does not match the current emotional state of the listener; even the classification of emotions poses difficulty, due to the lack of universal definitions. In this chapter, the task of music emotion recognition is investigated using deep neural networks, and adversarial architectures are applied for music data augmentation. Traditional classifiers such as support vector machines, k-NN, random forests, and trees have also been applied, using hand-crafted features representing the audio signals. Mel scale spectrograms were used as a basis to create inputs to the deep convolutional networks. Six architectures (AlexNet, VGG16bn, Inception v3, DenseNet121, SqueezeNet1.0, and ResNeXt101-32x8d) with an equal number of ImageNet pre-trained models were applied in transfer learning. The classification was evaluated for the recognition of valence, energy, tension, and emotions (anger, fear, happy, sad, and tender).

In the era of deep learning, speech and music signal processing offers unprecedented opportunities to transform the healthcare industry. In addition, the quality of the perceived speech and music signals, both for normal and hard of hearing people, is one of the most important requirements of the end users. Music can help deal with stress, anxiety, and various emotions, and influence activity-related brain plasticity. Part III comprises five chapters that explore the potential use of speech and music technology for our well-being. The first three chapters focus on music processing for the hearing impaired, as well as on music therapy addressed to relieve anxiety in diabetic patients and stress in the era of pandemic. The fourth chapter sheds light on the plasticity of the brain when learning music, and the fifth chapter is focused on expressing emotions in speech automatically generated from text.

Chapter "Music to Ears in Hearing Impaired: Signal Processing Advancements in Hearing Amplification Devices" by Kavassery Venkateswaran Nisha, Neelamegarajan Devi, and Sampath Sridhar explores music perception in the hearing impaired, using hearing aids and cochlear implants. The hearing aids improve the auditory perception of speech sounds, using various signal processing techniques. However, the music perception is usually not improved, as hearing aids do not compensate the non-linear response of human cochlea, a pre-requisite for music perception. The limited input dynamic range and higher crest ratio in analogue-to-digital converters of hearing aids fall short of processing live music. The cochlear implants were developed to improve speech perception rather than music perception, and they have limitations for music perception in terms of encoding fine structure information in music. The electrode array that is surgically implanted results in difficulty in perceiving pitch and higher harmonics of musical sounds. This chapter provides elaborate discussion on the advancements in signal processing techniques in hearing amplification devices such as hearing aids and cochlear implants that can address their drawbacks.

Chapter "Music Therapy: A Best Way to Solve Anxiety and Depression in Diabetes Mellitus Patients" by Anchana P. Belmon and Jeraldin Auxillia evaluates the potential of music therapy as an alternative solution towards the anxiety and depression in diabetic patients. There are pharmacological and non-pharmacological treatments available to deal with anxiety and depression. Music therapy along with relaxation and patients training is the main non pharmacological method. The effect of music in human body is unbelievable. There are two types of music therapy, namely passive and active music therapy. In this chapter, the effectiveness of music therapy in 50 diabetic patients has been assessed using Beck Anxiety Inventory and Beck Depression Inventory, reporting 0.67 reliability. The anxiety and depression measures were assessed in pre-evaluation, post-evaluation, and follow-up stages.

The statistical analysis suggests that music is an effective tool to accelerate the recovery of patients.

Chapter "Music and Stress During Covid-19 Lockdown: Influence of Locus of Control and Coping Styles on Musical Preferences" by Junmoni Borgohain, Rashmi Ranjan Behera, Chirashree Srabani Rath, and Priyadarshi Patnaik explores music as one of the effective strategies to enhance well-being during the lockdown. It tries to analyse the relation between stress during Covid-19 lockdown and preferences towards various types of music as a remedial tool. Music helps to reduce stress, but the ways people deal with stress are influenced by individual traits and people's musical tastes. The reported study was conducted on 138 Indian participants, representing various age, social, and demographic groups. Several quantitative measures (scaled from 1 to 5) such as BRIEF-COPE inventory, Perceived Stress Scale, the Cantril scale, and Brief-COPE Inventory are used for parametric representation of various activities performed by the subjects, and statistical analysis applied for data analysis. This study has observed several patterns in music preference during lockdown period. The study shows how music can be used as a tool for socio-emotional management during stressful times, and it can be helpful for machine learning experts to develop music-recommendation systems.

Chapter "Biophysics of Brain Plasticity and Its Correlation to Music Learning" authored by Sandipan Talukdar and Subhendu Ghosh explores the correlation of brain plasticity with learning music, based on experimental evidence. Brain plasticity is one of the key mechanisms of learning new things through growth and reorganization of neural networks in the brain. Human brains can change both structurally and functionally, which is a basis of the remarkable capacity of the brain to learn and memorize or unlearn and forget. The plasticity of the brain manifests itself at the level of synapses, single neurons, and networks. Music learning involves all of these mechanisms of brain plasticity, which requires intensive brain activities at different regions, whether it is simply listening to a music pattern, or performing, or even imaging music. The chapter investigates the possibility of any correlation between the biological changes induced in the brain and the sound wave during music perception and learning. Biophysical mechanisms involved in brain plasticity at the level of synapses and single neurons are considered for experimentation. The ways in which this plasticity is involved in music learning are discussed in this chapter, taking into account the experimental evidence.

Chapter "Analyzing Emotional Speech and Text: A Special Focus on Bengali Language" by Krishanu Majumder and Dipankar Das deals with the development of a text-to-speech (TTS) system in Bengali language and incorporates as much naturalistic features as possible using deep learning technique. The existing multilingual state-of-the-art TTS systems that produce speech for given text have several limitations. Most of them lack naturalness and sound artificial. Also, very limited work has been carried out on regional languages like Bengali, and no standard database is available to carry out the research work. This has motivated the authors to collect a database in Bengali language, with different emotions, for developing TTS engine. TTS systems are generally trained on a single language, and the possibility of training a TTS on multiple languages has also been explored. The chapter explores the possibility of including the contextual emotional aspects in the synthesized speech to enhance its quality. Another contribution of this chapter is to develop a bilingual TTS in Bengali and English languages. The objectives of the chapter have been validated in several experiments.

The concluding part of this volume comprises six chapters addressing an equal amount of case studies. They span from research addressing the duplication of audio material to Dutch song structures, and from album covers to measurements of the tabla's timbre. Musical influence on visual aesthetics as well as the study on emotions in audio-visual domain complete the picture of this part's contents.

Chapter "Duplicate Detection for Digital Audio Archive Management: Two Case Studies", by Joren Six, Federica Bressan, and Koen Renders, presents research aimed at identifying duplicate audio material in large digital music archives. The recent and rapid developments of Music Information Retrieval (MIR) have yet to be exploited by digital music archive management, but there is promising potential for this technology to aid such tasks as duplicate management. This research comprises two cases studies to explore the effectiveness of MIR for this task. The first case study is based on a VRT shellac disc archive at the Belgian broadcasting institute. Based on 15,243 digitized discs (out of about 100,000 total), the study attempts to determine the amount of unique versus duplicate audio material. The results show difficulties in discriminating between a near exact noisy duplicate and a translated version of a song with the same orchestral backing, when based on duplicate detection only. The second case study uses an archive of tapes from the Institute for Psychoacoustic and Electronic Music (IPEM). This study had the benefit of the digitized archive, first in 2001 and then in 2016. The results showed that in this case, MIR was highly effective at correctly identifying tracks and assigning meta data. This chapter concludes with a deeper dive into the recent Panako system for acoustic fingerprinting (i.e. the technology for identifying same or similar audio data in the database), to show its virtues.

Yke Paul Schotanus shows in chapter "How a Song's Section Order Affects Both 'Refrein' Perception and the Song's Perceived Meaning" how digital restructuring of song sections influences the lyrical meaning, as well as our understanding of the song's structure. When the section order of a song is manipulated, the listeners' understanding of a song is primarily based on how and where they perceive the chorus (*refrein* in Dutch), and/or the leading refrain line. A listening experiment was conducted, involving 111 listeners and two songs. Each participant listened to one of three different versions of the same Dutch cabaret song. The experiment showed that section order affects refrain perception and (semantic) meaning of Western pop songs. Manipulating musical properties such as pitch, timing, phrasing, or section order shows that popular music is more complex than thus far presumed; "the refrain of a song cannot be detected on the basis of strict formal properties", he concludes.

The objective of chapter "Musical Influence on Visual Aesthetics: An Exploration on Intermediality from Psychological, Semiotic, and Fractal Approach" (by Archi Banerjee, Pinaki Gayen, Shankha Sanyal, Sayan Nag, Junmoni Borgohain, Souparno Roy, Priyadarshi Patnaik, and Dipak Ghosh) is to determine the degree to which music and visuals interact in terms of human psychological perception.

Auditory and visual senses are not isolated aspects of psychological perception; rather, they are entwined in a complex process known as intermediality. Furthermore, the senses are not always equal in their impact on each other in a multimodal experience, as some senses may dominate others. This study attempts to investigate the relationship between auditory and visual senses to discover which is more dominant in influencing the total emotional outcome for a certain audience. To this end, abstract paintings have been used (chosen because of the lack of semantic dominance and the presence of pure, basic visual elements – lines, colours, shapes, orientation, etc.) and short piano clips of different tempo and complexity. Forty-five non-artist participants are then exposed to various combinations of the music and art – both complimentary and contradictory – and asked to rate their response using predefined emotion labels. The results are then analysed using a detrended fluctuation analysis to determine the nature of association between music and visuals – indifferent, compatible, or incompatible. It is found that music has a more significant influence on the total emotional outcome. This study reveals that intermediality is scientifically quantifiable and merits additional research.

Chapter "Influence of Musical Acoustics on Graphic Design: An Exploration with Indian Classical Music Album Cover Design", by Pinaki Gayen, Archi Banerjee, Sankha Sanyal, Priyadarshi Patnaik, and Dipak Ghosh, analyses strategies for graphic design for Indian classical music album covers and options to determine new possible design strategies to move beyond status quo conventions. The study is conducted with 30 design students who are asked to draw their own designs upon hearing two types of musical examples – *Komal Rishav Asavari* and *Jaunpuri*, which had been rated as "sad music" and "happy music", respectively, by a previous 70-person experiment. The design students were split into two groups, and each given 1 hour to complete their own designs while listening to the music. The resulting designs were then analysed using semiotic analysis and fractal analysis (detrended fluctuation analysis) to identify patterns of intermediality. This semiotic analysis entailed analysing iconic or symbolic representation (direct association to objects) versus indexical representation (cause and effect relationships). The findings showed that album cover designs fell in three categories: direct mood or emotional representation using symbolic followed by indexical representation; visual imageries derived from indexical followed by iconic representational; and musical feature representation primarily relying on iconic representation. In summary, the study provides new design possibilities for Indian classical music album covers and offers a quantitative approach to establishing effective intermediality towards successful designs.

Shankha Sanyal, Sayan Nag, Archi Banerjee, Souparno Roy, Ranjan Sengupta, and Dipak Ghosh present in chapter "A Fractal Approach to Characterize Emotions in Audio and Visual Domain: A Study on Cross-Modal Interaction" their study about classifying the emotional cues of sound and visual stimuli solely from their source characteristics. The study uses as a sample data set a collection of six audio signals of 15 seconds each and six affective pictures, of which three belonged to positive and negative valence, respectively ("excited", "happy", "pleased", etc., versus "sad", "bored", "angry", etc.). Then, using detrended fluctuation analysis (DFA), the study calculates the long-range temporal correlations (the Hurst exponent) corresponding to the audio signals. The results of the DFA technique were then applied on the array of pixels corresponding to affective pictures of contrast emotions to obtain a single unique scaling exponent corresponding to each audio signal and three scaling exponents corresponding to red/green/blue (RGB) component in each of the images. Finally, detrended cross-correlation (DCCA) technique was used to calculate the degree of nonlinear correlation between the sample audio and visual clips. The results were next confirmed by a follow-up human response study based on the emotional Likert scale ratings. The study presents an original algorithm to automatically classify and compare emotional appraisal from cross-modal stimuli based on the amount of long-range temporal correlations between the auditory and visual stimulus.

The closing chapter, chapter "Inharmonic Frequency Analysis of Tabla Strokes in North Indian Classical Music", by Shambhavi Shivraj Shete and Saurabh Harish Deshmukh, features the tabla. It is one of the essential instruments in North Indian Classical Music (NICM) and is highly unique compared to Western drums for its timbre. The tabla's timbre is related to inharmonicity (i.e. its overtones departing from harmonic series, being multiple of the fundamental) which is due to a complex art involving the application of ink (Syahi) to the table drum surface. This study aims to create a set of standard measurements of the tabla's timbre as this could be useful for instrument makers and performers. This measurement process is accomplished in two steps. First, a recording session collects 10 samples of a tabla playing the 9 common strokes within NICM for a total of 90 audio samples. These samples are then processed by a fast Fourier transform function to extract a frequency spectrum and determine the fundamental. The results are then compiled and organized by stroke type with comments about which overtones are most defining and which aspects of the stroke technique are especially important towards affecting those overtones responses.

We cordially thank all the authors for their valuable contributions. We also thank the reviewers for their input and valuable suggestions, and the Utrecht University intern Ethan Borshansky, as well as Mariusz Kleć from the Polish-Japanese Academy of Information Technology for their editorial assistance.

Finally, we thank all the stakeholders who have contributed directly or indirectly to making this book a success.

Cachar, Assam, India	Anupam Biswas
Utrecht, The Netherlands	Emile Wennekes
Warsaw, Poland	Alicja Wieczorkowska
Cachar, India	Rabul Hussain Laskar
December 2022	

Contents

Part I State-of-the-Art

A Comprehensive Review on Speaker Recognition 3
Banala Saritha, Mohammad Azharuddin Laskar, and Rabul Hussain Laskar

Music Composition with Deep Learning: A Review 25
Carlos Hernandez-Olivan and José R. Beltrán

Music Recommendation Systems: Overview and Challenges 51
Makarand Velankar and Parag Kulkarni

Music Recommender Systems: A Review Centered on Biases 71
Yesid Ospitia-Medina, Sandra Baldassarri, Cecilia Sanz,
and José Ramón Beltrán

Computational Approaches for Indian Classical Music: A
Comprehensive Review 91
Yeshwant Singh and Anupam Biswas

Part II Machine Learning

A Study on Effectiveness of Deep Neural Networks for Speech
Signal Enhancement in Comparison with Wiener Filtering Technique.... 121
Vijay Kumar Padarti, Gnana Sai Polavarapu, Madhurima Madiraju,
V. V. Naga Sai Nuthalapati, Vinay Babu Thota,
and V. D. Subramanyam Veeravalli

Video Soundtrack Evaluation with Machine Learning: Data
Availability, Feature Extraction, and Classification 137
Georgios Touros and Theodoros Giannakopoulos

Deep Learning Approach to Joint Identification of Instrument
Pitch and Raga for Indian Classical Music.................................. 159
Ashwini Bhat, Karrthik Gopi Krishnan, Vishal Mahesh,
and Vijaya Krishna Ananthapadmanabha

Comparison of Convolutional Neural Networks and K-Nearest Neighbors for Music Instrument Recognition 175
S. Dhivya and Prabu Mohandas

Emotion Recognition in Music Using Deep Neural Networks 193
Angelos Geroulanos and Theodoros Giannakopoulos

Part III Perception, Health and Emotion

Music to Ears in Hearing Impaired: Signal Processing Advancements in Hearing Amplification Devices 217
Kavassery Venkateswaran Nisha, Neelamegarajan Devi, and Sampath Sridhar

Music Therapy: A Best Way to Solve Anxiety and Depression in Diabetes Mellitus Patients 237
Anchana P. Belmon and Jeraldin Auxillia

Music and Stress During COVID-19 Lockdown: Influence of Locus of Control and Coping Styles on Musical Preferences 249
Junmoni Borgohain, Rashmi Ranjan Behera, Chirashree Srabani Rath, and Priyadarshi Patnaik

Biophysics of Brain Plasticity and Its Correlation to Music Learning 269
Sandipan Talukdar and Subhendu Ghosh

Analyzing Emotional Speech and Text: A Special Focus on Bengali Language 283
Krishanu Majumder and Dipankar Das

Part IV Case Studies

Duplicate Detection for for Digital Audio Archive Management: Two Case Studies 311
Joren Six, Federica Bressan, and Koen Renders

How a Song's Section Order Affects Both 'Refrein' Perception and the Song's Perceived Meaning 331
Yke Paul Schotanus

Musical Influence on Visual Aesthetics: An Exploration on Intermediality from Psychological, Semiotic, and Fractal Approach 353
Archi Banerjee, Pinaki Gayen, Shankha Sanyal, Sayan Nag, Junmoni Borgohain, Souparno Roy, Priyadarshi Patnaik, and Dipak Ghosh

Influence of Musical Acoustics on Graphic Design: An Exploration with Indian Classical Music Album Cover Design 379
Pinaki Gayen, Archi Banerjee, Shankha Sanyal, Priyadarshi Patnaik, and Dipak Ghosh

A Fractal Approach to Characterize Emotions in Audio and Visual Domain: A Study on Cross-Modal Interaction 397
Shankha Sanyal, Archi Banerjee, Sayan Nag, Souparno Roy, Ranjan Sengupta, and Dipak Ghosh

Inharmonic Frequency Analysis of Tabla Strokes in North Indian Classical Music 415
Shambhavi Shivraj Shete and Saurabh Harish Deshmukh

Index 441

Part I
State-of-the-Art

A Comprehensive Review on Speaker Recognition

Banala Saritha ⓘ, Mohammad Azharuddin Laskar ⓘ, and Rabul Hussain Laskar

1 Introduction

Speech is the universal mode of human communication. In addition to exchanging thoughts and ideas, speech is considered to be useful for extracting a lot of other information like language identity, gender, age, emotion, cognitive behavior, and speaker identity. One of the goals in speech technology is to make human-machine interaction as natural as possible, with systems like intelligent assistants, e.g., Apple Siri, Cortana, and Google Now. Speaker recognition also has a huge scope in this line of products. Every human has a unique speech production system [1]. The unique characteristics of the speech production system help to find a speaker's identity based on his or her speech signal. The task of recognizing the identity of a person from speech signal is called speaker recognition. It may be classified into speaker identification and speaker verification. The process of identifying an unknown speaker from a set of known speakers is speaker identification, while authentication of an unknown speaker claiming a person's identity already registered with the system is called speaker verification [2]. Speaker recognition finds numerous applications across different fields like biometrics, forensics, and access control systems [3]. Further, speaker recognition is classified into text-dependent and text-independent tasks based on whether the test subject is required to use a particular fixed utterance or is free to utter any valid text for recognition purposes [4]. Research on speaker recognition has been carried out since the 1960s [5]. Significant advancements have been made in this field over the recent decades where various aspects like features, modeling techniques, and scoring have been explored. The advances in deep learning and machine learning techniques

---

B. Saritha · M. A. Laskar (✉) · R. H. Laskar (✉)
Department of Electronics and Communication Engineering, National Institute of Technology Silchar, Silchar, India
e-mail: rhlaskar@ece.nits.ac.in

© The Author(s), under exclusive license to Springer Nature Switzerland AG 2023
A. Biswas et al. (eds.), *Advances in Speech and Music Technology*, Signals and Communication Technology, https://doi.org/10.1007/978-3-031-18444-4_1

have helped to promote speaker recognition and develop renewed interest among researchers in this field. Owing to its ease of use and higher accuracy, text-dependent speaker verification has been one of the focus areas. It plays a vital role in fraud prevention and access control. This chapter presents a comprehensive review of the techniques and methods employed for speaker recognition, with emphasis on text-dependent speaker verification.

The chapter's organization is as follows: Sect. 2 describes the basic structure of speaker recognition system. Section 3 presents a review of feature extraction techniques with an emphasis on the Mel-frequency cepstral coefficient (MFCC) feature extraction method. Speaker modeling involving classical techniques is discussed in Sect. 4. Advancements in speaker recognition with deep learning are discussed in Sect. 5. It also describes the performance metric for speaker recognition. The last section concludes the chapter.

2 Basic Overview of a Speaker Recognition System

Figure 1 represents the basic block diagram of the speaker verification system. The design of the speaker verification system mainly consists of two modules, namely, frontend and backend. Frontend takes the speech as an input signal and extracts the features. Generally, features are the more convenient representation of a given speech signal. It is represented in the form of a set of vectors and is termed as acoustic features.

The acoustic features are fed to the backend, which consists of a pre-designed speaker model along with classification and decision-making modules. The model based on these features is then compared with that of the registered speakers' models to determine the match between speakers. In text-dependent speaker verification (TDSV), the systems need to model both the text and the speaker characteristics. Figure 2 presents the block diagram representation of a text-dependent speaker verification system.

The speech signal input is first pre-processed using pre-emphasis filter followed by windowing and voice activity detection. Feature extraction is carried out using the voiced speech frames. These features are further modeled using techniques like Gaussian mixture model (GMM), identity vector (i-vector), or neural network and

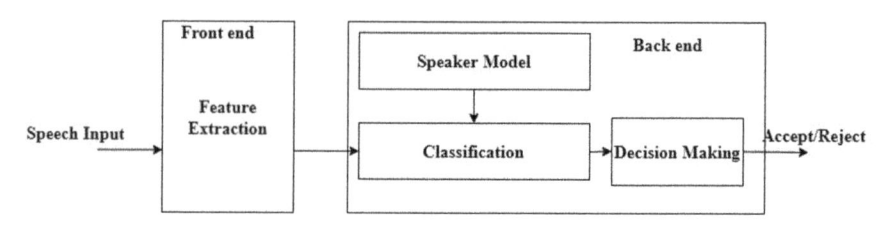

Fig. 1 Basic block diagram of a speaker verification system

Fig. 2 Text-dependent speaker verification system block diagram representation

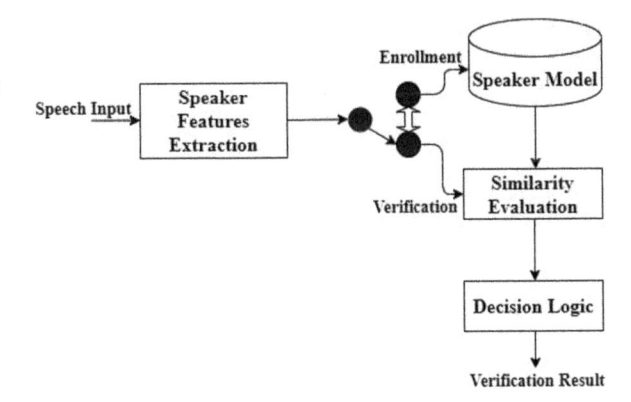

used for enrollment and verification. In the enrollment phase, speech utterance of adequate duration is taken and subjected to the said feature extraction and modeling modules to obtain the speaker model. Generally, a background or development set data is also required in conjunction with the enrolment data for training. During verification, the test utterance undergoes similar transformations and is then compared with the model corresponding to the claimed speaker identity. The comparison results in a score that helps to decide whether to accept or reject the claim [6].

3 Review on Feature Extraction

Every speaker has a unique speech production system. The process of capturing vocal characteristics is called feature extraction. Features can be classified into two types, namely, behavior-based (learned) features and physiological features. Behavior-based features include prosodic, spectro-temporal, and high-level features. Rhythm, energy, duration, pitch, and temporal features constitute the prosodic and spectro-temporal features. Phones, accents, idiolect, semantics, and pronunciation are the high-level features [7]. Figure 3 shows a classification of feature characteristics.

The physiological features are representative of the vocal tract length, dimension, and vocal fold size. Short-term spectral feature representations are commonly used to characterize these speaker-specific attributes. Some of the commonly used spectral features include Mel-frequency cepstral coefficients (MFCCs), gammatone feature, gammatone frequency cepstral coefficients (GFCCs), relative spectral-perceptual linear prediction (RASTA-PLP), Hilbert envelope of gammatone filter bank, and mean Hilbert envelope coefficients (MHECs) [8]. Of these features, MFCCs are the most widely used spectral features in the state-of-the-art speaker identification and verification systems.

Fig. 3 Classification of feature characteristics [7]

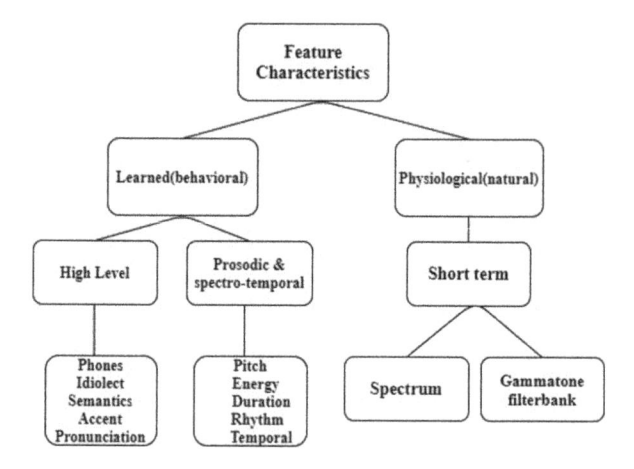

3.1 MFCC's Extraction Method

MFCCs are based on human auditory perception. Auditory perception is nonlinear and can be approximated using linear functions. The Mel scale is roughly linear at low frequencies and logarithmic at higher frequencies [9]. For a given frequency f in Hz, corresponding Mel scale frequency can be determined by the following formula:

$$mel(f) = 2595 * \log_{10}(1 + f/700) \quad (1)$$

Figure 4 represents the commonly followed feature extraction process. Speech signals are non-stationary in nature, i.e., the spectral content of the signals varies with time. Hence, in order to process a speech signal, it is divided into short (overlapping) temporal segments called frames, 10–30 ms long, as speech signal is quasi-stationary over a very short time frame, and short-time Fourier transform can be applied to analyze this signal. Further, to reduce the artifacts due to sudden signal truncations at the boundaries of the frames, windowing is done. Generally, the Hamming windowing (2) is applied to all frames to obtain smooth boundaries and to minimize the spectral distortions.

$$w(n) = 0.54 - 0.46\cos(2\pi n/M - 1); 0 \leq n \leq M \quad (2)$$

where M is the number of samples in the frame.

The short segments can be assumed to be stationary and are used for short-term frequency analysis. To obtain MFCCs, the windowed frames are subjected to fast Fourier transform (FFT) followed by the Mel filterbank, as shown in Fig. 5. The Mel spectrum is then represented on log scale before performing discrete cosine transform (DCT) to obtain the MFCC features. Usually, the first 13 coefficients, C0, C1, ..., C12, are considered. The coefficients are then normalized

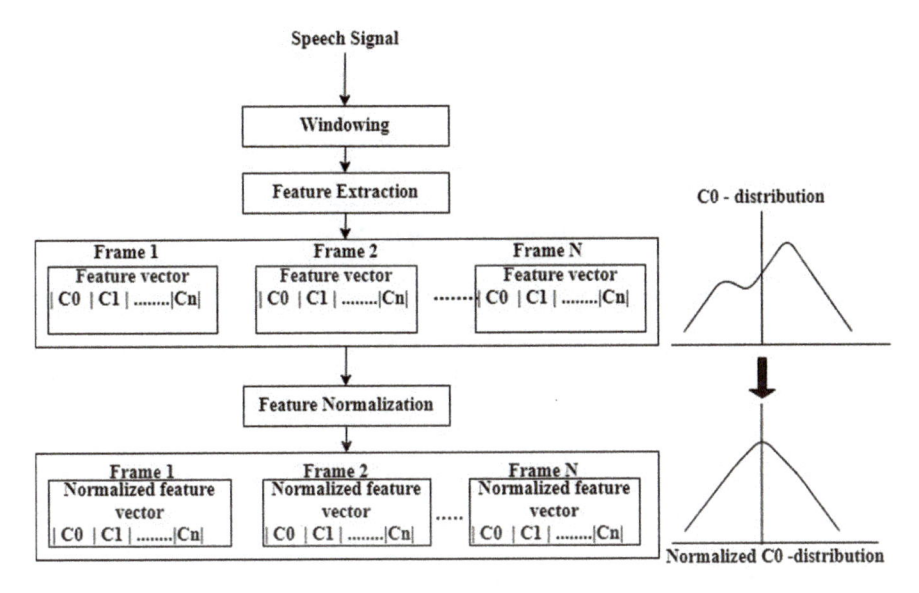

Fig. 4 Process of extracting feature vectors

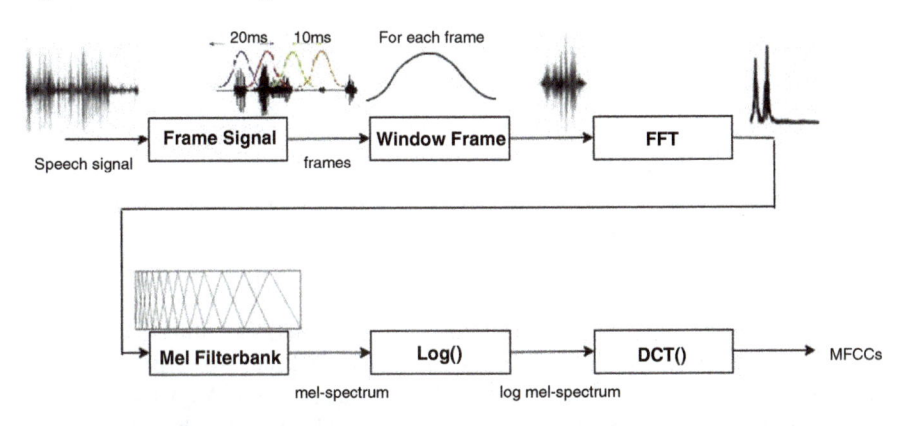

Fig. 5 MFCC feature extraction process

using techniques like cepstral mean subtraction (CMS), relative spectral filtering (RASTRA), and feature warping techniques. Once the features are normalized, the difference between C0 coefficients of a frame and its subsequent frame is calculated to obtain the delta parameter d0. Similarly, d1, d2, . . . , dn are obtained from C1, C2, . . . , Cn coefficients, respectively, as shown in Fig. 6. These are known as delta features. In the same way, acceleration or double-delta features are obtained by using difference of delta features [10]. The 13 MFCCs, the 13 delta features, and the 13 double-delta features are concatenated to obtain 39-dimensional feature vector for every frame.

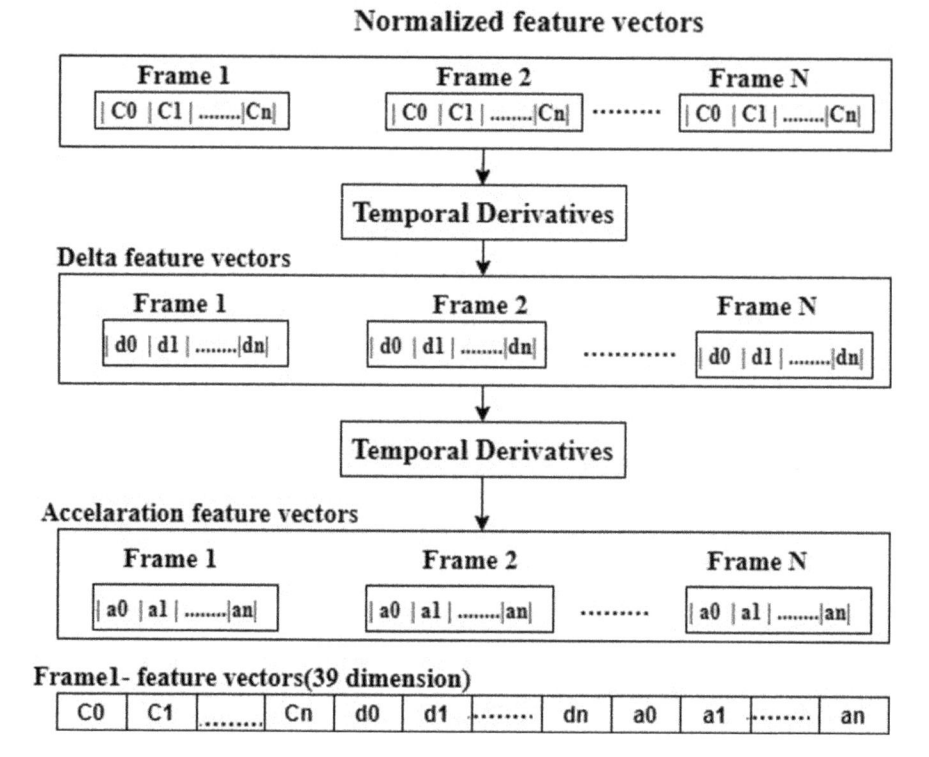

Fig. 6 Delta and acceleration features

4 Speaker Modeling

Once the acoustic features are extracted, speaker models are trained on it. The traditional speaker modeling techniques are categorized into two types. They are template models and stochastic models. Vector quantization (VQ) and dynamic time warping (DTW) approaches are the most popular template-based modeling techniques [11]. For text-dependent speaker verification, DTW template matching is a widely used technique. The acoustic feature sequence is obtained from the enrollment utterance and stored as a speaker-phrase template model. During the testing phase, the feature sequence corresponding to the test utterance is compared with the speaker-phrase template model using the DTW algorithm. DTW helps to time-align the two sequences and gives a similarity score, which is then used for the decision-making process. Another popular system for text-dependent speaker verification is realized using the VQ technique. In this method, the vector space consists of feature vectors and is mapped to a finite number of regions in the vector space. These regions are formed as clusters and represented by centroids. The set of centroids that represents the entire vector space is known as a codebook. Hence, the speaker-phrase models are prepared in terms of codebooks. This technique allows more flexibility in modeling the speaker-phrase model.

The stochastic models make use of probability theory. The most popular stochastic models are the Gaussian mixture model-universal background model (GMM-UBM) and hidden Markov model (HMM).

GMM-UBM is also a commonly used method for text-dependent speaker verification [12]. A UBM is built to represent a world model. Using the maximum a posteriori (MAP) adaptation technique, speaker-phrase-specific GMM is built from UBM using class-specific data [13]. The log-likelihood ratio is used to make the decision of whether to accept or reject the speaker-phrase subject. A number of HMM-based methods have also been proposed for text-dependent speaker verification. Such models are good at modeling the temporal information in the utterance and help to provide improved results. An unsupervised HMM-UBM and temporal GMM-UBM have also been proposed for TDSV [14]. In the case of HMM-UBM-based method, speaker-specific HMM is built through MAP adaptation from speaker-independent HMM-UBM trained in an unsupervised manner without using any transcription. In the temporal GMM-UBM approach, however, the temporal information is incorporated by computing the transition probability among the GMM-UBM mixture components using the speaker-specific training data. The HMM-UBM-based method is found to outperform the other systems by virtue of more effective modeling of the temporal information. Hierarchical multi-layer acoustic model (HiLAM) is an example of an HMM-based speaker-phrase model. It is a hierarchical acoustic model that adapts a text-independent, speaker-dependent GMM from UBM and then adapts the different HMM states from the mid-level model.

4.1 Gaussian Mixture Model-Universal Background Model

GMM-UBM model as shown in Fig. 7 is a popularly used method for text-dependent speaker verification. When all features are populated in large dimensional space, clusters are formed. Each cluster can be represented by a Gaussian distribution specified by mean and variance parameters. The overall data may be represented by a mixture of such Gaussian distributions, known as GMM, which is defined by a set of mean, variance, and weight parameters. Universal background model is a general speaker-independent model. It is used to obtain speaker-dependent GMMs with adaptation of mean, variance, and weight using target-specific data [15].

A training set is required to build up the UBM model, alongside a target set for which we are designing the system and a test set to evaluate the performance of the system.

A representation of GMM-based speaker model is given in Fig. 8. For a speaker model X1, each mixture ranging from 1 to 1024 has a 1×1 weight vector, 39×1 mean matrix, and 39×1 covariance matrix. Similarly, for developing a system for speaker models 2 to M, we have to store 1024 weight vectors and 39×1024 mean and covariance matrices.

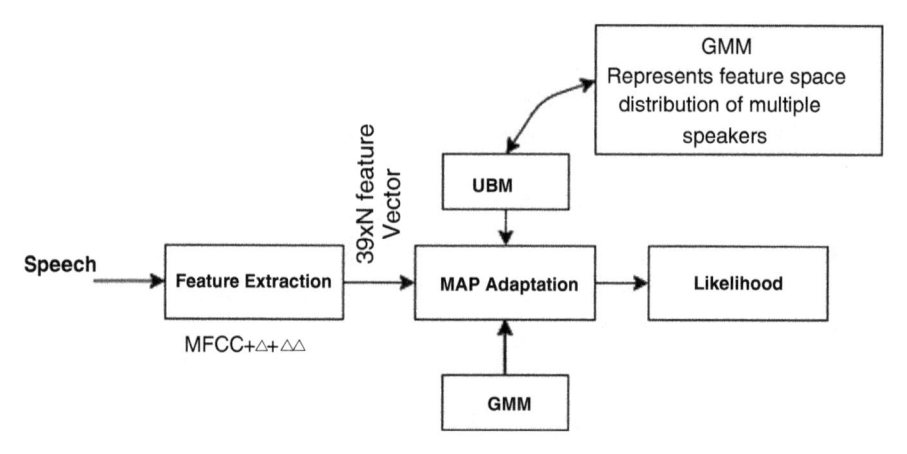

Fig. 7 A basic system of GMM-UBM model

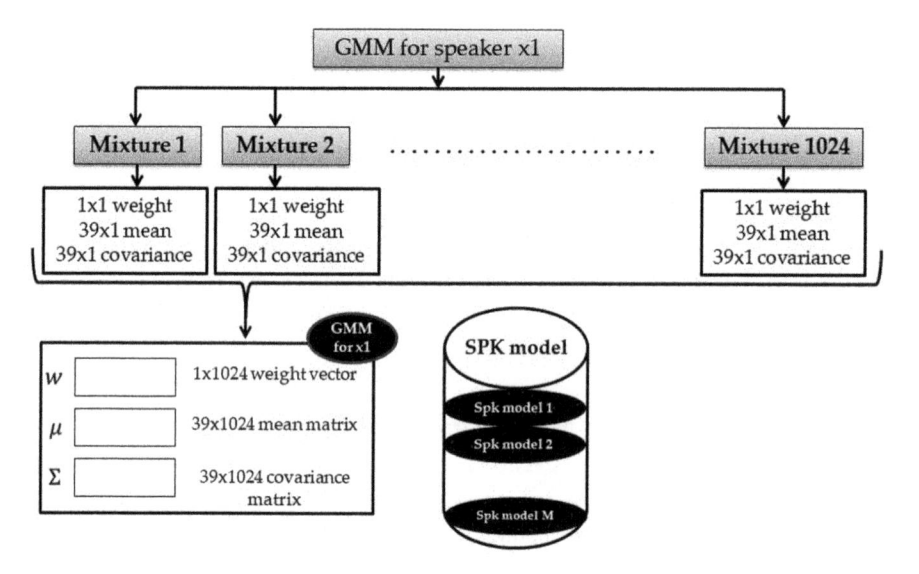

Fig. 8 Gaussian mixture model for speaker

When a speaker claims the identity of a registered speaker, the speaker verification system first extracts the features and compares them with the speaker model (GMM), determines the level of the match based on the log-likelihood ratio with a predefined threshold, and makes a decision whether to accept or reject the claimed speaker [16]. This process is shown in Fig. 9. The problem with GMM-UBM is that loads of vectors and matrices need to be stored. A single vector concept called a supervector S is introduced to represent a speaker to overcome the abovementioned difficulty.

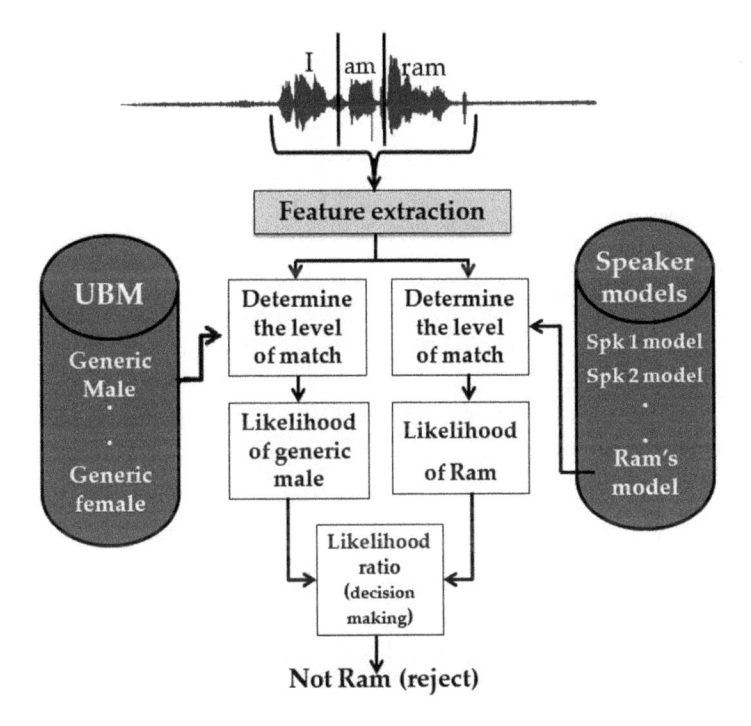

Fig. 9 Representation of universal background model

4.2 Supervector

Supervector is formed by concatenating all individual mixture means, scaled by the corresponding weights and covariances resulting in 39*1024-dimensional feature vector shown in Fig. 10. Each GMM (speaker model) can be represented by a supervector. Speech of different durations can be represented by supervectors of fixed size. Model normalization is typically carried out on a supervector when the speaker model is built. Supervectors are further modeled using various techniques available like nuisance attribute projection (NAP), joint factor analysis (JFA), i-vector, within-class covariance (WCCN), linear discriminant analysis, and probabilistic linear discriminant analysis (PLDA) [17].

Joint factor analysis is one of the popular techniques used to model GMM supervectors. The supervector is assumed to embed speaker information, channel information, some residual component, and speaker-independent components [18]. Accordingly, the supervector S is decomposed into different components given by

$$S = m + Vy + Ux + Dz \qquad (3)$$

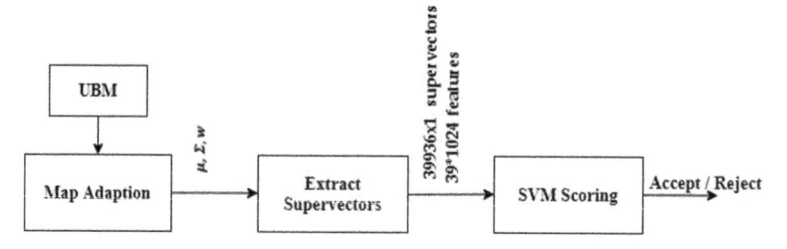

Fig. 10 The process of supervector formation

where m is the speaker-independent component, Vy represents the speaker information, Ux represents the channel information, and Dz represents the speaker-dependent residual component.

4.3 i-vector

In this technique, factor analysis is performed considering a common subspace for channel and speaker information as research indicated that relevant speaker information is also present in the channel feature obtained in JFA [19]. A lower-dimensional identity vector (w) is used to represent speaker model. The model may be represented as follows:

$$S = m + Tw \qquad (4)$$

where S is the GMM supervector, m is the UBM supervector mean, T represents the total variability matrix, and w indicates the standard normal distributed vector.

w may be represented as $p(w \| X) = N(\phi, L - 1)$

where w represents all the speaker information, and it follows a normal distribution with mean ϕ and variance $(L$ -1).

4.4 Trends in Speaker Recognition

Table 1 attempts to report the trends and progress in the field of speaker recognition. Techniques like NAP and WCCN have been used to normalize channel effects and session variability. Also, PLDA has been a popular backend model for many systems.

Table 1 Trends in speaker recognition with adopted techniques

S. no.	Year	Techniques adopted
1	2005	Supervector + NAP + support vector machine (SVM) scoring
2	2007	Supervector + JFA + y (reduced dimension vector) + WCNN + SVM scoring
3	2007	i-vector + WCNN + LDA (reduced dimension i-vector) + cosine distance scoring
4	2009	i-vector + PLDA (divides into channel and speaker space/channel compensation) + cosine distance scoring
5	2010	i-vector replaced with DNN + cosine distance/PLDA as decision-making

5 Deep Learning Methods for Speaker Recognition

The significant advancements in deep learning and machine learning techniques developed renewed interest among researchers in speaker recognition. The different deep learning architectures have received impetus from the availability of increased data and high computational power and have resulted in state-of-the-art systems. The DTW framework has also been implemented using deep neural network (DNN) posteriors extracted from the DNN-HMM automatic speech recognition (ASR) model [20]. The system leverages the discriminative power of the DNN-based model and is able to achieve enhanced performance. The deep neural network framework has two approaches in speaker recognition. The leading approach is feature extraction with deep learning methods. Another approach is classification and decision-making using deep learning methods [21]. In the first approach, Mel-frequency cepstral coefficients or spectra are taken as inputs and used to train a DNN with speaker IDs as the target variable.

Speaker feature embeddings are then obtained from the last layers of the trained DNN. The second approach replaces the cosine distance and probabilistic linear discriminate analysis; a deep network can be used for classification and decision-making.

2014 d-vector In d-vector framework shown in Fig. 11, instead of MFCC, stacked filter bank energies are used as input, to train a DNN in a supervised way [22]. The averaged activation function outputs from the last hidden layer of the trained network are used as the d-vector [23]. The 13-dimensional perceptual linear prediction coefficients (PLP) and delta and double-delta features were used in the training phase. "OK Google" database was used for experimentation [24].

2015 j-vector The multi-task learning approach shown in Fig. 12 extends the "d-vector" concept and leads to the j-vector framework [25]. The network is trained to discriminate both speakers and text at the same time.

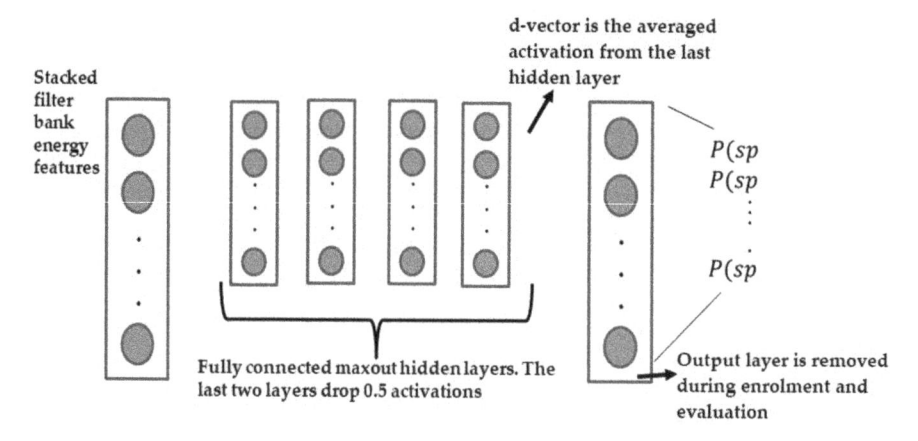

Fig. 11 d-vector framework in Variani et al. [23]

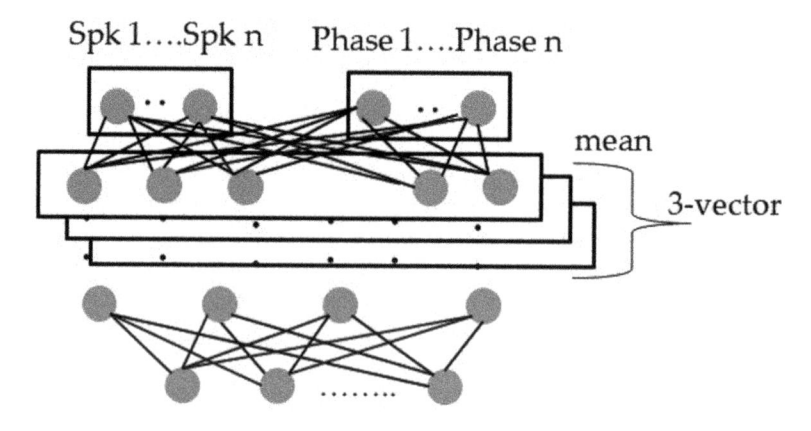

Fig. 12 Multi-task DNN in Chen et al. [25]

Like d-vectors, once the supervised training is finished, the output layer is discarded. Then joint feature vector called a j-vector is obtained from the last layer [26].

2018–2019 x-vector The authors (D. Snyder, D. Garcia-Romero) have proposed DNN embeddings in their work, called "x-vectors," to replace i-vectors for text-independent speaker verification [27, 28]. The main idea is to take variable length audio and to get a single vector representation for that entire audio. This single vector is capable of capturing speaker discriminating features. This architecture shown in Fig. 13 uses time delay neural network (TDNN) layers with a statistics

Fig. 13 x-vector DNN embedding architecture [26]

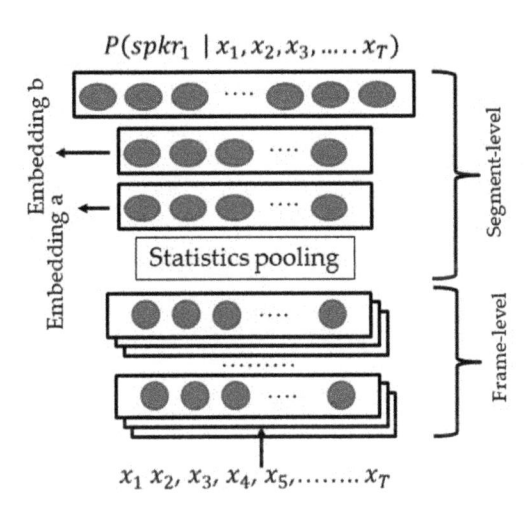

Fig. 14 TDNN framework [28]

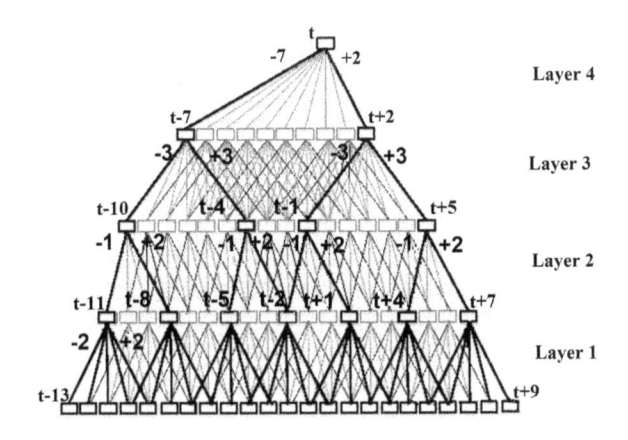

pooling layer. TDNN operates at frame level as shown in Fig. 14 and works better for a smaller amount of data. Speech input (utterances) is fed to the TDNN layer in frames (x1, x2..., xT), and it generates a sequence of frame-level features.

The statistics pooling layer determines the mean and standard deviation for the sequence of vectors. Concatenation of these two vectors is passed on as input to the next layer, which operates at the segment level. In the end, the softmax layer predicts the probability of speaker for a particular utterance. Additionally, data augmentation is done where noise and reverberation are added to original data with different SNR [29]. This makes the system more robust and improves accuracy compared to the i- and d-vector, particularly for short-duration utterances.

2018–2019 End-to-End System In the paper "End-to-end text-dependent speaker verification" by Heigold et al. [30], Google introduced DNN embedding-based speaker verification, which is one of state-of-the-art systems [30]. In this archi-

Fig. 15 End-to-end architecture used in [29]

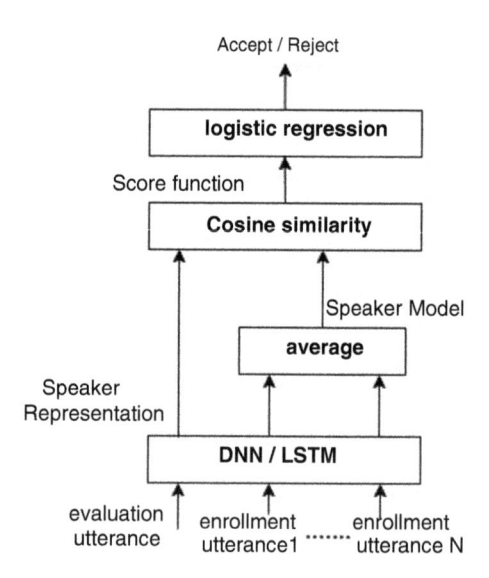

tecture, long short-term memory (LSTM) is used to process the "OK Google" kind of utterances. It gives speaker representations which are highly discriminative feature vectors. There are two inputs, namely, enrollment and evaluation utterances, applied to the network. As shown in Fig. 15, the network aggregates N vectors corresponding to N enrollment utterances to obtain a representation of the enrolled speaker. When the same speaker claims during the verification stage, it compares the generated vector with the previously stored vector from the enrollment data. Using cosine similarity, it compares the difference between vectors. If it is greater than the pre-determined threshold, the claimed speaker is accepted; otherwise, rejected. This end-to-end architecture performance on the "OK Google" database is similar to that of the "d-vector" technique.

2019–2020 Advancements in TDNN System A number of variants have been introduced to improve the performance of TDNN. They are factorized TDNN (F-TDNN), crossed TDNN (C-TDNN), densely connected TDNN (D-TDNN), extended TDNN (E-TDNN) [31], and emphasized channel attention and propagation and aggregation TDNN (ECAPA-TDNN). The current state-of-the-art TDNN-based speaker recognition is ECAPA-TDNN represented in Fig. 16. This architecture is proposed by Brecht Desplanques, Jenthe Thienpondt, and Kris Demuynck [32]. The complete architecture of squeeze-excitation (SE)-based Res2Net block (SE-Res2Block) of the ECAPA-TDNN is given in Fig. 17.

Fig. 16 Network topology of the ECAPA-TDNN [31]

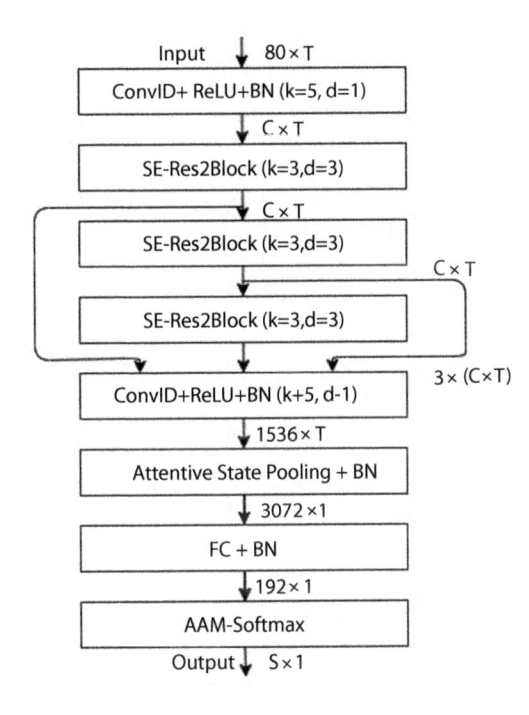

Fig. 17 The SE-Res2Block of the ECAPA-TDNN [31] architecture

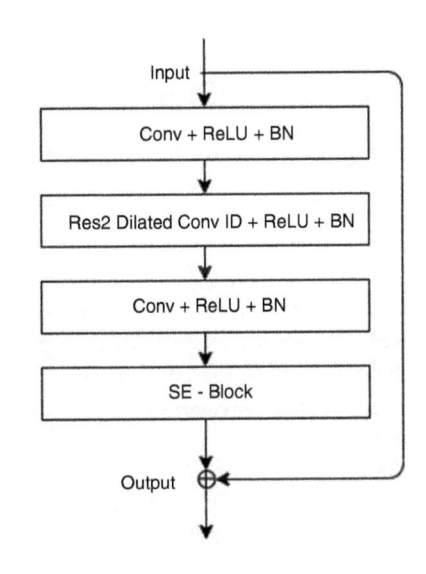

5.1 Deep Learning for Classification

Classification models with deep learning are presented in Table 2.

6 Performance Measure

To decide whether a speaker is accepted or rejected by the system, a threshold may require against which the score may generate.

	Correct decision	Miss
True speaker	Correct decision	Miss
Impostor	False acceptance	Correct decision

Correct Decision If a system correctly identifies the true speaker, it may be the "correct decision."

Miss If a system rejects the true speaker, it may be "miss."

False Acceptance If the system accepts the imposter, the system makes an error known as false acceptance.

Detection Error Trade-Off (DET) Curve
To plot the DET curve, first record the number of times the system rejected the true speaker (miss) and the number of times the imposter is accepted (false acceptance). Then express these two parameters in terms of percentage. Take false acceptance on the x-axis and miss rate on the y-axis for a particular threshold value $'\theta'$. We will get the point on two-dimensional space. By varying the $'\theta'$ value continuously, we will obtain a curve known as the detection error trade-off curve. As example, the dot point on the DET curve in Fig. 18 indicates the miss rate is very high and false acceptance is very low [48, 49]. This shows the system rejects true speakers in a good number of times and less number of imposters allowed. It is preferable in high-security systems like banking.

On the other end of the curve, the miss rate is meager, and false acceptance is very high. This shows the system is allowing imposters easily and not missing the true speaker. Such a scenario is useful in low-security applications [50].

The point at which the miss rate is equal to the false acceptance rate is the equal error rate (EER). Any system can be operated at this point. For example, let a system having EER of 1% indicates it has 1% of miss rate and 1% of false acceptance rate [51]. In Fig. 19, if the EER point is decaying toward the origin, it is a better system. If the EER point is rising toward the y=x line, it is not a better system.

Table 2 Classification models with deep learning

S. no.	Techniques adopted	Key concept	Merits/demerits
1	Variational autoencoder (VAE) [33–35]	VAE is used for voice conversion, speech, and speaker recognition, and it consists of stochastic neurons along with deterministic layers. The log-likelihood ratio scoring is used to discriminate between the same and different speakers	The performance of VAE is not superior to PLDA scoring
2	Multi-domain features [36]	Automatic speaker recognition (ASR) output is given as input to the speaker recognition evaluation (SRE) system and vice versa Extracted frame-level features are the inputs to ASR and SRE	For WSJ database, EERs are low compared to the i- vector
3	DNN in place of UBM [25]	DNN is used instead of UBMs. Universal deep belief networks are developed and are used as a backend. The target i-vector and imposter i-vector develop a vector such that it should have good discriminating properties. It is stored in a discriminative target model	This model not accomplished better performance compared to PLDA-based i-vector
4	Unlabeled data [37–39]	It introduces unlabeled samples continuously, and by taking a labeled corpus, it learns the DNN model	Proposed LSTM- and TDNN-based systems outperform the traditional methods
5	Hybrid framework [40]	Zero-order feature statistics are fed to a standard i-vector model through DNN. Also, speech is segmented into senones by training the DNN using the HMM-GMM model	A low equal error rate is achieved with this method
6	SincNet [41–45]	SincNet is a distinctive convolutional neural network (CNN) architecture that takes one-dimensional raw audio as input. A filter at the first layer of CNN acquires the knowledge of lower (fl) and higher (fu) cut-off frequencies. Also, the convolutional layer can adjust these frequencies before applying them to further standard layers	Fast convergence, improved accuracy, and computational efficiency. SincNet method outperforms the other DNN solutions in speaker verification
7	Far-field ResNet-BAM [46]	It is an easy and effective novel speaker embedding architecture with the ResNet-BAM method, in which the bottleneck attention module (BAM) with residual neural network (ResNet) is mixed. It focused on the smallest speech and domain mismatch where the frontend includes a data processing unit. The backend consists of speaker embedding with a DAT extractor	With the help of adversarial domain training along with gradient reversal layer provides domain mismatch
8	Bidirectional attention [47]	Proposed to unite CNN-based feature knowledge with a bidirectional attention method to attain improved performance with merely single enrollment speech	It is outperformed over the sequence-to-sequence and vector models

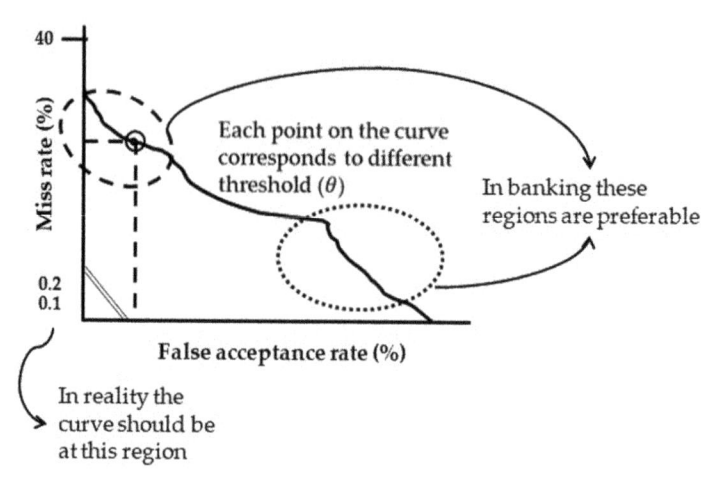

Fig. 18 Detection error trade-off (DET) curve

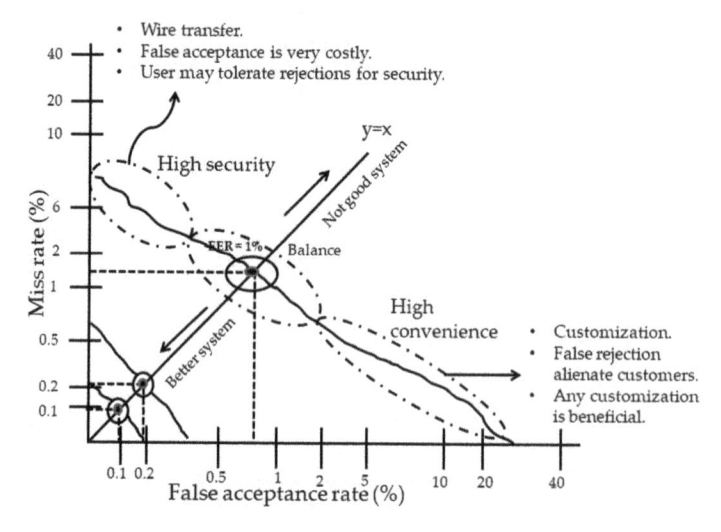

Fig. 19 Performance of system with EER on DET curve

7 Conclusion

This chapter attempts to present a comprehensive review of speaker recognition with more emphasis on text-dependent speaker verification. It discusses the most commonly used feature extraction and modeling techniques used for this task. It also surveys all the recent advancements proposed in the field of text-independent speaker recognition. The deep learning models have been found to outperform many classical techniques, and there exists a huge scope to further improve the performance of the systems with more advanced architectures and data augmentation techniques.

References

1. Kinnunen, T. and Li, H.Z. (2010). An Overview of Text-Independent Speaker Recognition: From Features to Super vectors. Speech Communication, 52, 12–40. https://doi.org/10.1016/j.specom.2009.08.009.
2. Hansen, John and Hasan, Taufiq. (2015). Speaker Recognition by Machines and Humans: A tutorial review. Signal Processing Magazine, IEEE. 32. 74–99. https://doi.org/10.1109/MSP.2015.2462851.
3. Todkar, S.P., Babar, S.S., Ambike, R.U., Suryakar, P.B., & Prasad, J.R. (2018). Speaker Recognition Techniques: A Review. 2018 3rd International Conference for Convergence in Technology (I2CT), 1–5.
4. Bimbot, F., Bonastre, J.F., Fredouille, C., Gravier, G., Magrin-Chagnolleau, I., Meignier, S., ... and Reynolds, D.A. (2004). A tutorial on text-independent speaker verification. EURASIP Journal on Advances in Signal Processing, 2004 (4), 1–22.
5. Pruzansky, S., Mathews, M., and Britner P.B. (1963). Talker-Recognition Procedure Based on Analysis of Variance. Journal of the Acoustical Society of America, 35, 1877–1877.
6. Nguyen, M.S., Vo, T. (2015). Vietnamese Voice Recognition for Home Automation using MFCC and DTW Techniques. 2015 International Conference on Advanced Computing and Applications (ACOMP), 150–156.
7. Tirumala, S.S., Shahamiri, S.R., Garhwal, A.S., Wang, R. Speaker identification features extraction methods: A systematic review, Expert Systems with Applications, Volume 90, 2017, Pages 250–271, ISSN 0957-4174, https://doi.org/10.1016/j.eswa.2017.08.015.
8. Islam, M.A., et al. A Robust Speaker Identification System Using the Responses from a Model of the Auditory Periphery. PloS one vol.11, 7e0158520.8 Jul.2016, https://doi.org/10.1371/journal.pone.0158520.
9. Sujiya, S., Chandra, E. (2017). A Review on Speaker Recognition. International journal of engineering and technology, 9, 1592–1598.
10. Alim, S.A., and Rashid, N.K.A. (December 12th 2018). Some Commonly Used Speech Feature Extraction Algorithms, From Natural to Artificial Intelligence - Algorithms and Applications, Ricardo Lopez-Ruiz, IntechOpen, https://doi.org/10.5772/intechopen.80419.
11. Brew, A., and Cunningham, P. (2010). Vector Quantization Mappings for Speaker Verification. 2010 20th International Conference on Pattern Recognition, 560–564.
12. Reynolds, D.A., Quatieri, T.F., and Dunn, R.B., "Speaker verification using adapted Gaussian mixture models", Digital Signal Processing, vol.10, no.1–3, pp. 19–41, 2000.
13. Larcher, A., Lee, K.A., Ma, B., Li, H., Text-dependent speaker verification: Classifiers, databases and RSR2015. Speech Communication, Elsevier: North-Holland, 2014.
14. Sarkar, A.K., and Tan, Z.H. "Text dependent speaker verification using un-supervised HMM-UBM and temporal GMM-UBM", in Seventeenth Annual Conference of the International Speech Communication Association (INTERSPEECH), pp. 425–429, 2016.
15. Zheng, R., Zhang, S., and Xu, B. (2004). Text-independent speaker identification using GMM-UBM and frame level likelihood normalization. 2004 International Symposium on Chinese Spoken Language Processing, 289–292.
16. Yin, S., Rose, R., and Kenny, p. "A Joint Factor Analysis Approach to Progressive Model Adaptation in Text-Independent Speaker Verification", in IEEE Transactions on Audio, Speech, and Language Processing, vol. 15, no. 7, pp. 1999–2010, Sept. 2007, https://doi.org/10.1109/TASL.2007.902410.
17. Campbell, W.M., Sturim, D.E., Reynolds, D.A. and Solomonoff, A. "SVM based speaker verification using a GMM supervector kernel and NAP variability compensation", in 2006 IEEE International Conference on Acoustics Speech and Signal Processing Proceedings (ICASSP), vol. 1, pp. I-I, IEEE, May 2006.
18. Kanagasundaram, A., Vogt, R., and Dean, D., Sridharan, S., Mason, M. (2011). i-vector Based Speaker Recognition on Short Utterances. Proceedings of the Annual Conference of the International Speech Communication Association, INTERSPEECH.

19. Li, W., Fu, T., Zhu, J. An improved i-vector extraction algorithm for speaker verification. J Audio Speech Music Proc. 2015, 18 (2015).https://doi.org/10.1186/s13636-015-0061-x
20. Dey, S., Motlicek, P., Madikeri, S., and Ferras, M., "Template-matching for text-dependent speaker verification", Speech Communication, vol.88, pp. 96–105, 2017.
21. Sztahó, D., Szaszák, G., Beke, A. (2019). Deep learning methods in speaker recognition: a review.
22. Bai, Z., and Zhang, X.-L. (2021). Speaker recognition based on deep learning: An overview. Neural Networks, 140, 65–99. https://doi.org/10.1016/j.neunet.2021.03.004.
23. Variani, E., Lei, X., McDermott, E., Moreno, I.L., and Gonzalez-Dominguez, J. "Deep neural networks for small footprint text-dependent speaker verification", 2014 IEEE International Conference on Acoustics, Speech and Signal Processing (ICASSP), 2014, pp. 4052–4056, https://doi.org/10.1109/ICASSP.2014.6854363.
24. Wan, L., Wang, Q., Papir, A., and Lopez-Moreno, I. (2018). Generalized End-to-End Loss for Speaker Verification. 2018 IEEE International Conference on Acoustics, Speech and Signal Processing (ICASSP), 4879–4883.
25. Chen, D., Mak, B., Leung, C., and Sivadas, S. "Joint acoustic modeling of triphones and trigraphemes by multi-task learning deep neural networks for low-resource speech recognition", 2014 IEEE International Conference on Acoustics, Speech and Signal Processing (ICASSP), 2014, pp. 5592–5596, https://doi.org/10.1109/ICASSP.2014.6854673.
26. Bai, Z., Zhang, X. (2021). Speaker recognition based on deep learning:An overview. Neural networks: the official journal of the International Neural Network Society, 140, 65–99.
27. Snyder, D., Garcia-Romero, D., Sell, G., Povey, D., Khudanpur, S. (2018). X-Vectors: Robust DNN Embeddings for Speaker Recognition. 2018 IEEE International Conference on Acoustics, Speech and Signal Processing (ICASSP), 5329–5333.
28. Fang, F., Wang, X., Yamagishi, J., Echizen, I., Todisco, M., Evans, N.,Bonastre, J. (2019). Speaker Anonymization Using X-vector and Neural Waveform Models. arXiv preprint arXiv: 1905.13561.
29. Snyder, D., Garcia-Romero, D., Povey, D., Khudanpur, S. (2017). Deep Neural Network Embeddings for Text-Independent Speaker Verification. INTERSPEECH.
30. Heigold, G., Moreno, I., Bengio, S., Shazeer, N.(2016). End-to-end text-dependent speaker verification.In: 2016 IEEE International Conference on Acoustics, Speech and Signal Processing (ICASSP). pp. 5115–5119.
31. Yu, Y., and Li, W. (2020). Densely Connected Time Delay Neural Network for Speaker Verification. INTERSPEECH.
32. Desplanques, B., Thienpondt, J., and Demuynck, K. (2020). ECAPA-TDNN: Emphasized Channel Attention, Propagation and Aggregation in TDNN Based Speaker Verification. INTERSPEECH.
33. Kingma, D.P., Welling, M. (2013). Auto-encoding variational bayes. arXiv preprint arXiv: 1312.6114.
34. Rezende, D.J., Mohamed, S., Wierstra, D. (2014). Stochastic Backpropagation and Approximate Inference in Deep Generative Models. Proceedings of the 31st International Conference on Machine Learning,in PMLR 32 (2), pp. 1278–1286.
35. Villalba, J., Brümmer, N., Dehak, N. (2017). Tied Variational Autoencoder Backends for i-Vector Speaker Recognition. In INTERSPEECH 2017, pp. 1004–1008.
36. Tang, Z., Li, L., Wang, D. (2016). Multi-task recurrent model for speech and speaker recognition. In 2016 Asia-Pacific Signal and Information Processing Association Annual Summit and Conference (APSIPA), pp. 1–4.
37. Marchi, E., Shum, S., Hwang, K., Kajarekar,S., Sigtia, S., Richards, H., Haynes, R.,Kim, Y.,Bridle J. (2018). Generalised discriminative transform via curriculum learning for speaker recognition. In: 2018 IEEE International Conference on Acoustics, Speech and Signal Processing, ICASSP 2018, pp. 5324–5328.
38. Ranjan, S., Hansen, J.H., Ranjan, S., Hansen, J.H. (2018). Curriculum learning based approaches for noise robust speaker recognition. IEEE/ACM Transactions on Audio, Speech and Language Processing (TASLP), 26 (1), pp. 197–210.

39. Zheng, S., Liu, G., Suo, H., Lei, Y. (2019). Auto encoder-based Semi-Supervised Curriculum Learning For Out-of-domain Speaker Verification. In: INTERSPEECH. 2019. pp. 4360–4364.
40. Lei, Y., Scheffer, N., Ferrer, L., McLaren, M. (2014). A novel scheme for speaker recognition using a phonetically-aware deep neural network. In: 2014 IEEE International Conference on Acoustics, Speech and Signal Processing (ICASSP). pp. 1695–1699.
41. Nagrani, A., Chung, J.S., Zisserman, A. (2017). Voxceleb: a large-scale speaker identification dataset. arXiv preprint arXiv:1706.08612.
42. Ravanelli, M., Bengio, Y. (2018). Speaker recognition from raw waveform with sincnet. In: 2018 IEEE Spoken Language Technology Workshop (SLT), pp. 1021–1028.
43. Hajavi, A., Etemad A. (2019). A Deep Neural Network for Short-Segment Speaker Recognition. In: Proc.Interspeech 2019, pp. 2878–2882.
44. Ravanelli, M., Bengio,Y. (2019). Learning speaker representations with mutual information. In: Proc.Interspeech 2019, pp. 1153–1157.
45. Salvati, D., Drioli, C., Foresti, G.L. (2019). End-to-End Speaker Identification in Noisy and Reverberant Environments Using Raw Waveform Convolutional Neural Networks. Proc. Interspeech 2019, pp. 4335–4339.
46. Zhang, Li, Wu, Jian, Xie, Lei. (2021). NPU Speaker Verification System for INTERSPEECH 2020 Far-Field Speaker Verification Challenge.
47. Fang, X., Gao, T., Zou, L., Ling, Z.: Bidirectional Attention for Text-Dependent Speaker Verification. Sensors. 20, 6784 (2020).
48. Doddington, G.R., Przybocki, M.A., Martin, A.F., and Reynolds, "The NIST speaker recognition evaluation–overview, methodology, systems, results, perspective", Speech Communication, vol.31, no.2–3, pp. 225–254, 2000.
49. Zeinali, H., Sameti, H., and Burget, L. "Text-dependent speaker verification based on i-vectors, Neural Networks and Hidden Markov Models", Computer Speech and Language, vol.46, pp. 53–71, 2017.
50. Bimbot, F., Bonastre, J F. Fredouille, C. et al. A Tutorial on Text-Independent Speaker Verification. EURASIP J.Adv.Signal Process.2004, 101962 (2004). https://doi.org/10.1155/S1110865704310024.
51. Cheng, J., and Wang, H. (2004). A method of estimating the equal error rate for automatic speaker verification. 2004 International Symposium on Chinese Spoken Language Processing, 285–288.

Music Composition with Deep Learning: A Review

Carlos Hernandez-Olivan and José R. Beltrán

1 Introduction

Music is generally defined as a succession of pitches or rhythms, or both, in some definite patterns [1]. Music composition (or generation) is the process of creating or writing a new piece of music. The music composition term can also refer to an original piece or work of music [1]. Music composition requires creativity. Chomsky defines creativity as "the unique human capacity to understand and produce an indefinitely large number of sentences in a language, most of which have never been encountered or spoken before" [9]. On the other hand, Root-Bernstein defines this concept as "creativity comes from finding the unexpected connections, from making use of skills, ideas, insights and analogies from disparate fields" [51]. Regarding music creativity, Gordon declares that there is not a clear definition of this concept. He stands that music creativity cannot be taught but the readiness for one to fulfill his potential for music creativity, that is, the audation vocabulary of tonal patterns or the varied and large rhythmic patterns [22]. This is a very important aspect that needs to be taken into account when designing or proposing an AI-based music composition algorithm. More specifically, music composition is an important topic in the music information retrieval (MIR) field. It comprises subtasks such as melody generation, multi-track or multi-instrument generation, style transfer, or harmonization. These aspects will be covered in this chapter from the point of view of the multitude of techniques that have flourished in recent years based on AI and DL.

C. Hernandez-Olivan (✉) · J. R. Beltrán (✉)
Department of Electronic Engineering and Communications, University of Zaragoza, Zaragoza, Spain
e-mail: carloshero@unizar.es; jrbelbla@unizar.es

© The Author(s), under exclusive license to Springer Nature Switzerland AG 2023
A. Biswas et al. (eds.), *Advances in Speech and Music Technology*, Signals and Communication Technology, https://doi.org/10.1007/978-3-031-18444-4_2

1.1 From Algorithmic Composition to Deep Learning

From the 1980s, the interest in computer-based music composition has never stop to grow. Some experiments came up in the early 1980s such as the Experiments in Musical Intelligence (EMI) [12] by David Cope from 1983 to 1989 or Analogiques A and B by Iannis Xenakis that follow the previous work from the author in 1963 [68]. Later in the 2000s, also David Cope proposed the combination of Markov chains with grammars for automatic music composition, and other relevant works such as Project1 (PR1) by Koening [2] were born. These techniques can be grouped in the field of algorithmic music composition which is a way of composing by means of formalizable methods [46, 58]. This type of composing consists of a controlled procedure which is based on mathematical instructions that must be followed in a fixed order. There are several methods inside the algorithmic composition such as Markov models, generative grammars, cellular automata, genetic algorithms, transition networks, or chaos theory [28]. Sometimes, these techniques and other probabilistic methods are combined with deep neural networks (NNs) in order to condition them or help them to better model music which is the case of DeepBach [25]. These models can generate and harmonize melodies in different styles, but the lack of generalizability capacity of these models and the rule-based definitions that must be done by hand make these methods less powerful and generalizable in comparison with DL-based models.

From the 1980s to the early 2000s, the first works which tried to model music with NNs were born [3, 17, 44]. In recent years, with the growing of deep learning (DL), lots of studies have tried to model music with deep NNs. DL models for music generation normally use NN architectures that are proven to perform well in other fields such as computer vision or natural language processing (NLP). There can also be used pre-trained models in these fields that can be used for music generation. This is called transfer learning [74]. Some NN techniques and architectures will be shown later in this chapter. Music composition today is taking input representations and NN's architectures from large-scale NLP applications, such as transformer-based models which are demonstrating very good performance in this task. This is due to the fact that music can be understood as a language in which every style or music genre has its own rules.

1.2 Neural Network Architectures for Music Composition with Deep Learning

First of all, we will provide an overview of the most widely used NN architectures that are providing the best results in the task of musical composition so far. The most used NN architectures in music composition task are generative models such as variational autoencoders (VAEs) or generative adversarial networks (GANs) and

NLP-based models such as long short-term memory (LSTM) or transformers. The following is an overview of these models.

1.2.1 Variational Autoencoders (VAEs)

The original VAE model [37] uses an encoder-decoder architecture to produce a latent space by reconstructing the input (see Fig. 1a). A latent space is a multidimensional space of compressed data in which the most similar elements are located closest to each other. In a VAE, the encoder approximates the posterior, and the decoder parameterizes the likelihood. The posterior and likelihood approximations are parametrized by a NN with λ and θ parameters for the encoder and decoder, respectively. The posterior inference is done by minimizing the Kullback-Leibler (KL) divergence between the encoder and approximate posterior, and the true posterior by maximizing the evidence lower bound (ELBO). The gradient is computed with the so-called *reparametrization* trick. There are variations of the original VAE model such as the β-VAE [27] which adds a penalty term β to the reconstruction loss in order to improve the latent space distribution. In Fig. 1a, we show the general VAE architecture. An example of a DL model for music composition based on a VAE is MusicVAE [50] which we describe in further sections in this chapter.

1.2.2 Generative Adversarial Networks (GANs)

GANs [21] are generative models composed by two NNs: the generator G and the discriminator D. The generator learns a distribution p_g over the input data. The training is done in order to let the discriminator maximize the probability of assigning the correct label to the training samples and the samples generated by the generator. This training idea can be understood as if D and G follow the two-player minimax game that Goodfellow et al. [21] described. In Fig. 1b, we show the general GAN architecture. The generator and the discriminator can be formed by different NN layers such as multi-layer perceptrons (MLPs) [52], LSTMs [30], or convolutional neural networks (CNNs) [19, 40].

1.2.3 Transformers

Transformers [61] are being currently used in NLP applications due to their well performance not only in NLP but also in computer vision models. Transformers can be used as auto-regressive models like the LSTMs which allow them to be used in generative tasks. The basic idea behind transformers is the attention mechanism. There are several variations of the original attention mechanism proposed by Vaswani et al. [61] that have been used in music composition [33]. The combination of the attention layer with feedforward layers leads to the formation of the encoder and decoder of the transformer which differs from purely autoencoder models that are also composed by the encoder and decoder. Transformers are trained with tokens which are structured representations of the inputs. In Fig. 1c, we show the general transformer architecture.

1.3 Challenges in Music Composition with Deep Learning

There are different points of view about the challenges perspective in music composition with DL that make ourselves ask questions related to the input representations and DL models that have been used in this field, the output quality of the actual state-of-the-art methods, or the way that researchers have measured the quality of the generated music. In this chapter, we ask ourselves the following questions that involve the composition process and output: Are the current DL models capable of generating music with a certain level of creativity? What is the best NN architecture to perform music composition with DL? Could end-to-end methods generate entire structured music pieces? Are the composed pieces with DL just an imitation of the inputs or can NNs generate new music in styles that are not present in the training data? Should NNs compose music by following the same logic and process as humans do? How much data do DL models for music generation need? Are current evaluation methods good enough to compare and measure the creativity of the composed music?

To answer these questions, we approach music composition or generation from the point of view of the process followed to obtain the final composition and the output of DL models, i.e., the comparison between the human composition process and the music generation process with DL and the artistic and creative characteristics presented by the generated music. We also analyze recent state-of-the-art models of music composition with DL to show the result provided by these models (*motifs*, complete compositions, etc.). Another important aspect analyzed is the input representation that these models use to generate music to understand if these representations are suitable for composing. This gives us some insights on how these models could be improved, if these NN architectures are powerful enough to compose new music with a certain level of creativity, and the directions and future work that should be done in music composition with DL.

1.4 Chapter Structure

In this chapter, we make an analysis of the symbolic music composition task from the composition process and the type of generated output perspectives. Therefore, we do not cover the performance or synthesis tasks. This chapter is structured as follows. Section 2 introduces a general view of the music composition process and music basic principles. In Sect. 3, we give an overview of state-of-the-art methods

Fig. 1 (**a**) VAE [37], (**b**) GAN [21], and (**c**) transformer general architecture. Reproduced from [61]

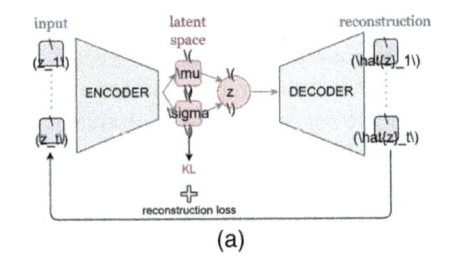

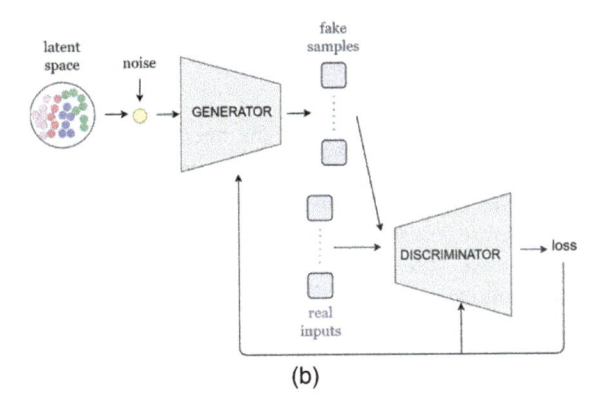

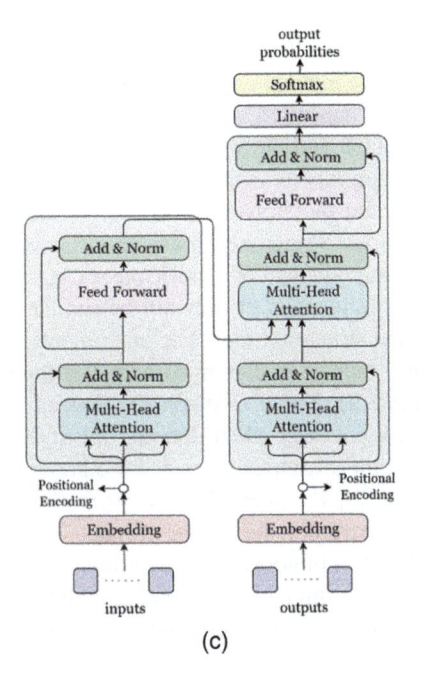

from the melodic composition perspective. We also examine how these models deal with the harmony and structure. In Sect. 4, we describe DL models that generate multi-track or multi-instrument music. In Sect. 5, we show different methods and metrics that are commonly used to evaluate the output of a music generation model. In Sect. 6, we describe the open questions in music generation field by analyzing the models described in the previous sections. Finally, in Sect. 7, we expose future work and challenges that are still being studied in the music generation with DL field.

2 The Music Composition Process

Much like written language, the music composition process is a complex process that depends on a large number of decisions [41]. In the music field, this process [11] depends on the music style we are working with. As an example, it is very common in Western classical music to start with a small unit of one or two bars called *motif* and develop it to compose a melody or music phrase, and in styles like pop or jazz, it is more common to take a harmonic progression and compose or improvise a melody ahead of it. In spite of the music style we are composing in, when a composer starts a piece of music, there is some basic melodic or harmonic idea behind it. From the Western classical music perspective, this idea (or *motif*) is developed by the composer to construct the melody or phrase that generates or follows a certain harmonic progression, and then these phrases are structured in sections. The melody can be constructed after the harmonic progression is set, or it can also be generated in the first place and then be harmonized. How the melody is constructed and the way it is harmonized are decisions made by the composer. Each section has its own purpose which means that it can be written in different tonalities and its phrases usually follow different harmonic progressions than the other sections. Sometimes, music pieces have a melodic part and an accompaniment part. The melodic part of a music piece can be played by different instruments whose frequency range may or may not be similar, and the harmonic part gives the piece a deep and structured feel. The instruments, which are not necessarily in the same frequency range, are combined with *Instrumentation* and *Orchestration* techniques (see Sect. 3.2). These elements are crucial in musical composition, and they are also important keys when defining the style or genre of a piece of music. Music has two dimensions, the time and the harmony dimensions. The time dimension is represented by the notes duration or rhythm which is the lowest level in this axis. In this dimension, notes can be grouped or measured in units called bars, which are ordered groups of notes. The other dimension, harmony, is related to the note values or pitch. If we think of an image, time dimension would be the horizontal axis and harmony dimension the vertical axis. Harmony does also have a temporal evolution, but this is not represented in music scores. There is a very common software-based music representation called piano-roll that follows this logic.

The music time dimension is structured in low-level units that are notes. Notes are grouped in bars that form *motifs*. In the time high-level dimension, we can find

Music Composition with Deep Learning: A Review

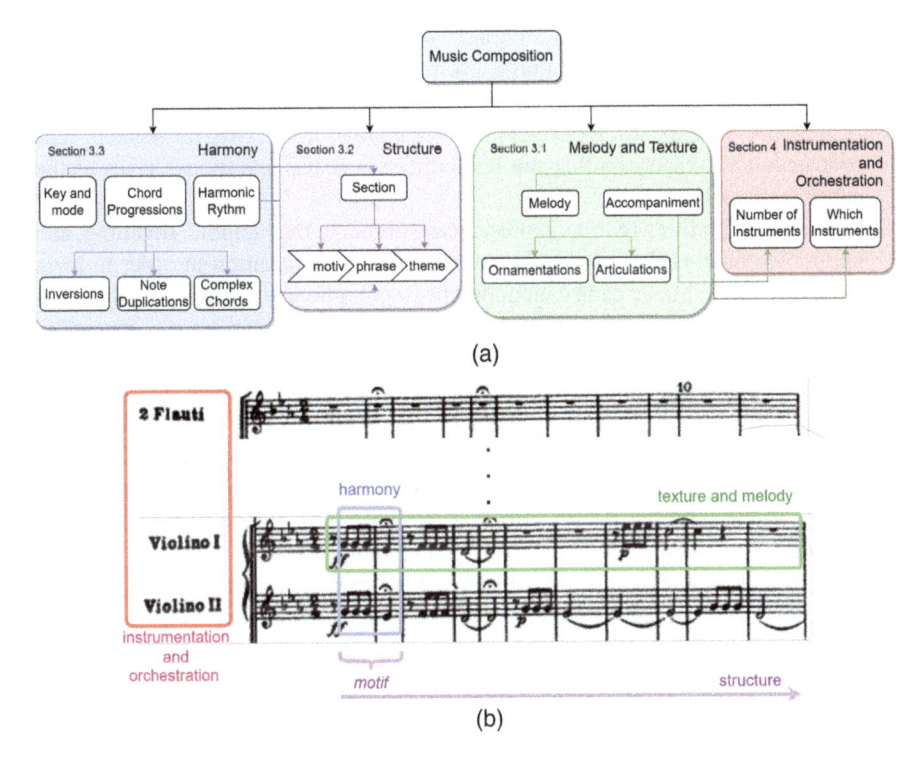

Fig. 2 (**a**) General music composition scheme and (**b**) an example of the beginning of Beethoven's fifth symphony with music levels or categories

sections which are composed by phrases that last eight or more bars (this depends on the style and composer). The lowest level in the harmony dimension is the note level. The superposition of notes played by different instruments creates chords. The sequence of chords is called harmonic progressions or chord progressions that are relevant to the composition, and they also have dependencies in the time dimension. Having said that, we can think about music as a complex language model that consists of short- and long-term relationships. These relationships extend in two dimensions, the time dimension which is related to music structure and the harmonic dimension which is related to the notes or pitches and chords, that is, the harmony.

From the symbolic music generation and analysis points of view, based on the ideas of Walton [63], some of the basic music principles or elements are (see Fig. 2):

- Harmony. It is the superposition of notes that form chords that compose a chord progression. The note level can be considered as the lowest level in harmony, and the next level that can be considered is the chord level. The highest level is the progression level which usually belongs to a certain tonality in tonal music.
- Music Form or Structure. It is the highest level that music presents, and it is related to the time dimension. The smallest part of a music piece is the motif

which is developed in a music phrase, and the combination of music phrases forms a section. Sections in music are ordered depending on the music style such as intro-verse-chorus-verse-outro for some pop songs (also represented as ABCBA) or exposition-development-recapitulation or ABA for sonatas. The concatenation of sections that can be in different scales and modes gives us the entire composition.
- Melody and Texture. Texture in music terms refers to the melodic, rhythmic, and harmonic contents that have to be combined in a composition in order to form the music piece. Music can be monophonic or polyphonic depending on the notes that are played at the same time step and homophonic or heterophonic depending on the melody, if it has or not accompaniment.
- Instrumentation and Orchestration. These are music techniques that take into account the number of instruments or tracks in a music piece. Whereas instrumentation is related to the combination of musical instruments which compose a music piece, orchestration refers to the assignment of melodies and accompaniment to the different instruments that compose a determined music piece. In recording or software-based music representation, instruments are organized as *tracks*. Each track contains the collection of notes played on a single instrument [18]. Therefore, we can call a piece with more than one instrument as multi-track which refers to the information that contains two or more tracks where each track is played by a single instrument. Each track can contain one note or multiple notes that sounds simultaneously, leading to monophonic tracks and polyphonic tracks, respectively.

Music categories are related between them. Harmony is related to the structure because a section is usually played in the same scale and mode. There are cadences between sections, and there can also be modulations which change the scale of the piece. Texture and instrumentation are related to timbral features, and their relationship is based on the fact that not all the instruments can play the same melodies. An example of that is when we have a melody with lots of ornamentation elements which cannot be played with determined instrument families (because of a fact of each instrument technique possibilities or a stylist reason).

Another important music attribute is the dynamics, but they are related to the performance rather than the composition itself, so we will not cover them in this chapter. In Fig. 2, we show the aspects of the music composition process that we cover in this chapter, and the relationships between categories and the sections of the chapter in which each topic is discussed are depicted.

3 Melody Generation

A melody is a sequence of notes with a certain rhythm ordered in an aesthetic way. Melodies can be monophonic or polyphonic. Monophonic refers to melodies in which only one note is played at a time step, whereas in polyphonic melodies, there

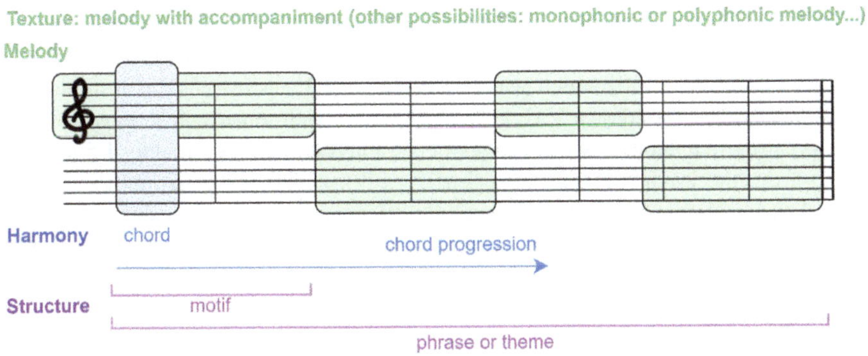

Music Composition with Deep Learning: A Review

Fig. 3 Scheme of an output-like score of melody generation models

is more than one note being played at the same time step. Melody generation is an important part of music composition, and it has been attempted with algorithmic composition and with several of the NN architectures that include generative models such as VAEs or GANs, recurrent neural networks (RNNs) used for auto-regression tasks such as LSTM, neural autoregressive distribution estimators (NADEs) [38], or current models used in natural language processing like transformers [61]. In Fig. 3, we show the scheme with the music basic principles of an output-like score of a melody generation model.

3.1 Deep Learning Models for Melody Generation: From Motifs to Melodic Phrases

Depending on the music genre of our domain, the human composition process usually begins with the creation of a *motif* or a chord progression that is then expanded to a phrase or melody. When it comes to DL methods for music generation, several models can generate short-term notes sequences. In 2016, the very first DL models attempted to generate short melodies with recurrent neural networks (RNNs) and semantic models such as unit selection [4]. These models worked for short sequences, so the interest to create entire melodies grew in parallel to the birth of new NNs. Derived from these first works and with the aim of creating longer sequences (or melodies), other models that combined NNs with probabilistic methods came up. An example of this is Google's Magenta Melody RNN models [62] released in 2016 and the Anticipation-RNN [24] and DeepBach [25] both published in 2017. DeepBach is currently considered one of the current state-of-the-art models for music generation because of its capacity to generate 4-voice chorales in the style of Bach.

However, these methods cannot generate new melodies with a high level of creativity from scratch. In order to improve the generation task, generative models were chosen by researchers to perform music composition. In fact, nowadays, one of the best-performing models to generate *motifs* or short melodies from 2 to 16 bars is MusicVAE[1] [50] which was published in 2018. MusicVAE is a model for music generation based on a VAE [37]. With this model, music can be generated by interpolating in a latent space. This model is trained with approximately 1.5 million songs from the Lakh MIDI Dataset (LMD)[2] [49], and it can generate polyphonic melodies for almost 3 instruments: melody, bass, and drums. After the creation of MusicVAE model along with the birth of new NN architectures in other fields, the necessity and availability of new DL-based models that can create longer melodies grew and this led to the birth of new transformer-based models for music generation. Examples of these models are the Music Transformer [33] in 2018, and models that use pre-trained transformers such as MuseNet in 2019 proposed by OpenAI [47] which uses the GPT-2 to generate music. These transformer-based models, such as Music Transformer, can generate longer melodies and continue a given sequence, but after a few bars or seconds, the melody ends up being a bit random, that is, there are notes and harmonies that do not follow the musical sense of the piece.

In order to overcome this problem and develop models that can generate longer sequences without losing the sense of the music generated in the previous bars or the main motifs, new models were born in 2020 and 2021 as combinations of VAEs, transformers, or other NNs or machine learning algorithms. Some examples of these models are the TransformerVAE [36] and PianoTree [66]. These models perform well even in polyphonic music, and they can generate music phrases. One of the latest released models to generate entire phrases is the model proposed in 2021 by Mittal et al. [42] which is based in denoising diffusion probabilistic models (DDPMs) [29] which are new generative models that generate high-quality samples by learning to invert a diffusion process from data to Gaussian noise. This model uses a MusicVAE 2-bar model to then train a diffusion model to capture the temporal relationships among the VAE latents z_k with $k = 32$ which are the 32 latent variables that allows to generate 64 bars (2 bars per latent). In spite that there can be generated longer polyphonic melodies, they do not follow a central motif so they tend to lose the sense of a certain direction.

3.2 Structure Awareness

As we mentioned in Sect. 1, music is a structured language. Once melodies have been created, they must be grouped into bigger sections (see Fig. 2) which play a fundamental role in a composition. These sections have different names that vary depending on the music style such as introduction, chorus, or verse for pop or trap genres and exposition, development, or recapitulation for classical sonatas. Sections

---

[1] https://magenta.tensorflow.org/music-vae, accessed August 2021.
[2] https://colinraffel.com/projects/lmd/, accessed August 2021.

can also be named with capital letters, and song structures can be expressed as ABAB, for example. Generating music with structure is one of the most difficult tasks in music composition with DL because structure means an aesthetical sense of rhythm, chord progressions, and melodies that are concatenated with bridges and cadences [39].

In DL, there have been models that have tried to generate structured music by imposing the high-level structure with the self-similarity constrains. An example of that is the model proposed by Lattner et al. in 2018 [39] which uses a convolutional restricted Boltzmann machine (C-RBM) to generate music and self-similarity constrain with a self-similarity matrix [45] to impose the structure of the piece as if it was a template. This method which imposes a structure template is similar to the composition process that a composer follows when composing music, and the resulting music pieces followed the imposed structure template. Although new DL models are trending to be end to end, and new studies about modeling music with structure are being released [7], there have not been DL models that are capable of generating structured music by themselves, that is, without the help of a template or high-level structure information that is passed to the NN.

3.3 Harmony and Melody Conditioning

Inside music composition with DL, there is a task that is the harmonization of a given melody which differs to the task of creating a polyphonic melody from scratch. On the one hand, if we analyze the harmony of a created melody from scratch with a DL model, we saw that music generated with DL is not well structured as it does not compose different sections and write aesthetic cadences or bridges between the sections in an end-to-end way yet. In spite of that, the harmony generated by transformer-based models that compose polyphonic melodies is coherent in the first bars of the generated pieces [33] because it follows a certain key. We have to emphasize here that these melodies are written for piano, which differs from multi-instrument music that presents added challenges such as generating appropriate melodies or accompaniments for each instrument or deciding which instruments make up the ensemble (see Sect. 4).

On the other hand, the task of melody harmonization consists of generating the harmony that accompanies a given melody. The accompaniment can be a chord accompaniment regardless of the instrument or the track where the chords are and multi-track accompaniment where the notes in each chord belong to a specific instrument. The first models for harmonization used HMM, but these models were improved by RNNs. Some models predicted chord functions [70], and other models match chord accompaniments for a given melody [69]. Regarding the generation of accompaniment with different tracks, there have been proposed GAN-based models which implement lead sheet arrangements. In 2018, a Multi-Instrument Co-Arrangement Model called MICA [72] and its improvement MSMICA in 2020 [73] were proposed to generate multi-track accompaniment. There is also a model called the Bach Doodle [32] which used Coconet [31] to generate accompaniments for a given melody in Bach style. The harmonic quality of these models improves the harmony generated by models that create polyphonic melodies from scratch because the model focus on the melodic content to perform the harmonization which represents a smaller challenge than generating an entire well-harmonized piece from scratch.

Several additional tasks remain within music generation with DL using conditioning, such as generating a melody given a chord progression, which is a way of composing that humans follow. These tasks have been addressed with variational autoencoders (VAEs) [57], generative adversarial networks or GAN-based models [26, 59],[3] and end-to-end models [72]. Other models perform the full process of composing, such as ChordAL [56]. This model generates chords, and then the obtained chord progression is sent to a melody generator, and the final output is sent to a music style processor. Models like BebopNet [26] generate a melody from jazz chords because this style presents additional challenges in the harmony context.

3.4 Genre Transformation with Style Transfer

In music, a style or genre is defined as a complex mixture of features ranging from music theory to sound design. These features include timbre, the composition process, and the instruments used in the music piece or the effects with which music is synthesized. Due to the fact that there are lots of music genres and the lack of datasets for some of those genres, it is common to use style transfer techniques to transform music in a determined style into other style by changing the pitch of existing notes or adding new instruments that fit in the style into which we want to transform the music.

In computer-based music composition, the most common technique for performing style transfer in music is to obtain an embedding of the style and to use this embedding or feature vector to generate new music. Style transfer [20] in NNs was introduced in 2016 by Gatys et al. with the idea of applying style features to an image from another image. One of the first studies that used style transfer for symbolic music generation was MIDI-VAE [5] in 2018. MIDI-VAE encodes the style in the latent space as a combination of pitch, dynamics, and instrument features to generate polyphonic music. Style transfer can also be achieved with transfer learning [74]. The first work that used transfer learning to perform style transfer was a recurrent VAE model for jazz which was proposed by Hung et al. [34] in 2019. Transfer learning is done by training the model on the source dataset and then finetuning the resulting model parameters on the target dataset that can be in a different style than the source dataset. This model showed that using transfer learning to transform a music piece in a determined style to another is a great solution because

---

[3] https://shunithaviv.github.io/bebopnet/, accessed August 2021.

it could not only be used to transform existing pieces in new genres but it could also be used to compose music from scratch in genres that are not present in the music composition datasets that are being used nowadays. An example of this could be the use of a NN trained with a large dataset for pop such as the Lakh MIDI Dataset (LMD) [49] and use of this pre-trained model to generate urban music through transfer learning.

Other music features such as harmony and texture (see Fig. 2) have been also used as style transfer features [65–67]. There have also been studied fusion genre models in which different styles are mixed to generate music in an unknown style [8].

4 Instrumentation and Orchestration

As we mentioned in Sect. 2, instrumentation and orchestration are fundamental elements in the musical genre being composed and may represent a characteristic signature of each composer by the use of specific instruments or the way their compositions are orchestrated. An example of that is the orchestration that Beethoven used in his symphonies that changed the way in which music was composed [23]. *Instrumentation* is the study of how to combine similar or different instruments in varying numbers in order to create an ensemble. *Orchestration* is the selection and combination of similarly or differently scored sections [54]. From that, we can relate instrumentation as the color of a piece and orchestration to the aesthetic aspect of the composition. Instrumentation and orchestration have a huge impact on the way we perceive music and so, to the emotional part of music, but, although they represent a fundamental part of the music, emotions are beyond the scope of this chapter.

4.1 From Polyphonic to Multi-Instrument Music Generation

In computer-based music composition, we can group instrumentation and orchestration concepts in multi-instrument or multi-track music. However, the DL-based models for multi-instrument generation do not work exactly with those concepts. Multi-instrument DL-based models generate polyphonic music for more than one instrument, but does the generated music follow a coherent harmonic progression? Is the resulting arrangement coherent in terms of instrumentation and orchestration, or do DL-based models just generate multi-instrument music not taking into account the color of each instrument or the arrangement? In Sect. 3, we showed that polyphonic music generation could compose music with a certain harmonic sense, but when facing multi-instrument music, one of the most important aspects to take into account is the color of the instruments and the ensemble. Deciding how many and which instruments are in the ensemble and how to divide the melody and accompaniment between them is not yet a solved problem in music generation with

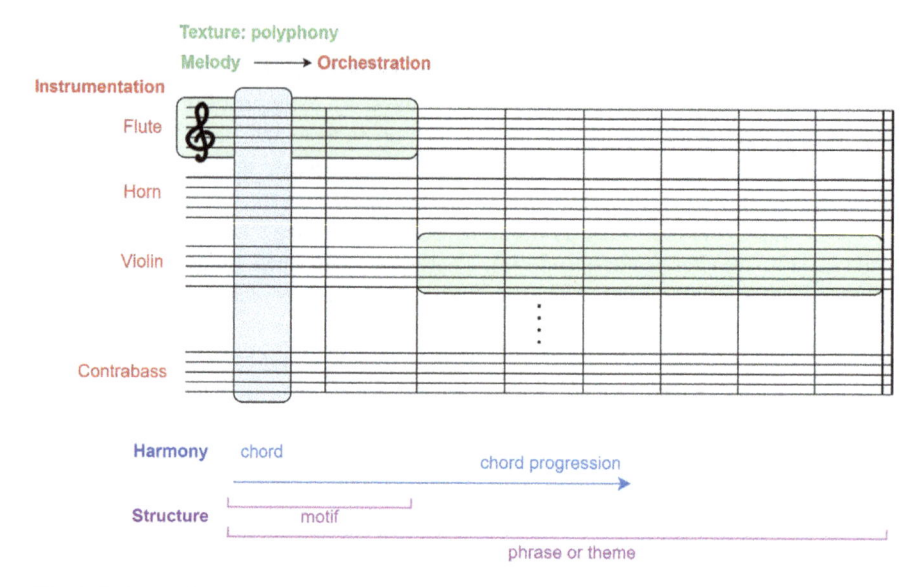

Fig. 4 Scheme of an output-like score of multi-instrument generation models

DL. In recent years, these challenges have been faced by building DL models that generate music from scratch that can be interactive models in which humans can select the instruments of the ensemble [18]. There are also models that allow to inpaint instruments or bars. We describe these models and answer to the exposed questions in Sect. 4.2. In Fig. 4, we show the scheme with the music basic principles of an output-like score of a multi-instrument generation model.

4.2 Multi-Instrument Generation from Scratch

The first models that could generate multi-track music have been proposed recently. Prior to multi-track music generation, some models generated the drums track for a given melody or chords. An example of those models is the model proposed in 2012 by Kang et al. [53] which accompanied a melody in a given scale with an automated drums generator. Later on, in 2017, Chu et al. [10] used a hierarchical RNN to generate pop music in which drums were present.

One of the most commonly used architectures in music generation is the generative models such as GANs and VAEs. The first considered and most well-known model for multi-track music generation is MuseGAN [15], presented in 2017. Then, more models followed the multi-instrument generation task [16, 71], and later in 2020, other models based on autoencoders such as MusAE [60] were released. The other big group of NN architectures that have been used recently to generate music are the transformers. The most well-known model for music

Music Composition with Deep Learning: A Review

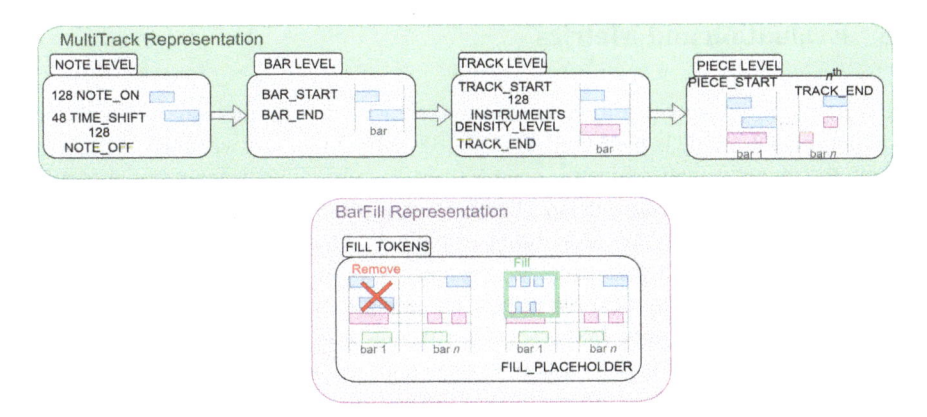

Fig. 5 MMM token representations. Reproduced from [18]

generation with transformers is the Music Transformer [33] for piano polyphonic music generation. In 2019, Donahue et al. [14] proposed LakhNES for multi-track music generation, and in 2020, Ens et al. proposed a conditional multi-track music generation model (MMM) [18] which is based on LakhNES and improves the token representation of this previous model by concatenating multiple tracks into a single sequence. This model uses a MultiInstrument and a BarFill representations which are represented in Fig. 5. In Fig. 5, we show the MultiInstrument representation which contains the tokens that the MMM model uses for generating music and the BarFill representation that is used to inpaint, that is, generating a bar or a couple of bars but maintaining the rest of the composition.

From the composition process perspective, these models do not orchestrate, but they create music from scratch or by inpainting. This means that these models do not choose the number of instruments and do not generate high-quality melodic or accompaniment content related to the instrument that is being selected. As an example, the MMM model generates melodic content for a predefined instrument which follows the timbral features of the instrument, but when inpainting or recreating a single instrument while keeping the other tracks, it is sometimes difficult to follow the key in which the other instruments are composed. This leads us to the conclusion that multi-instrumental models for music generation focus on end-to-end generation, but still do not work well when it comes to instrumentation or orchestration, as they still cannot decide the number of instruments in the generated piece of music. They generate music for the ensemble they were trained on, such as LakhNES [14], or they take predefined tracks to generate the content of each track [18]. More recent models, such as MMM, are opening up interactivity between human and artificial intelligence in terms of multi-instrument generation, which will allow better tracking of the human compositional process and thus improve the music generated with multiple instruments.

5 Evaluation and Metrics

Evaluation in music generation can be divided according to the way in which the output of a DL mode is measured. Ji et al. [35] differentiate between the evaluation from the objective and subjective perspectives. In music, it is necessary to measure the results from a subjective perspective because it is the type of evaluation that tells us how much creativity the model brings compared to human creativity. Sometimes, the objective evaluation that calculates the metrics of the results of a model can give us an idea of the quality of these results, but it is difficult to find a way to relate it to the concept of creativity. In this section, we show how the state-of-the-art models measure the quality of their results from objective and subjective points of view.

5.1 Objective Evaluation

The objective evaluation measures the performance of the model and the quality of its outputs using some numerical metrics. In music generation, there is the problem of comparing models trained for different purposes and models trained with different datasets, so we give a description of the most common metrics used in state-of-the-art models. Ji et al. [35] differentiate between model metrics and music metrics or descriptive statistics and other methods such as pattern repetition [64] or plagiarism detection [25].

When attempting to measure the performance of a model, the most used metrics depending on the DL model used to generate music are the loss, perplexity, BLEU score, precision (P), recall (R), or F-score (F_1). Normally, these metrics are used to compare different DL models that are built for the same purpose.

Loss is usually used to show the difference between the inputs and outputs of the model from a mathematical perspective, while, on the other hand, perplexity tells us the generalization capability that a model has, which is more related to how the model generates new music. As an example, the Music Transformer [33] uses the loss and the perplexity to compare the outputs between different transformer architectures in order to validate the model, TonicNet [48] only uses the loss for the same purpose, and MusicVAE [50] only uses a measure that indicates the quality of reconstruction that the model has, but it does not use any metric to compare between other DL music generation models.

Regarding metrics that are specifically related to music, that is, those that take into account musical descriptors, we can find that these metrics help to measure the quality of a composition. According to Ji et al. [35], these metrics can be grouped into four categories: pitch-related, rhythm-related, harmony-related, and style transfer-related. Pitch-related metrics [35], such as scale consistency, tone spam, ratio of empty bars, or number of pitch classes used, are metrics that measure pitch attributes in general. Rhythm-related metrics take into account the duration or pattern of the notes and are, for example, the rhythm variations, the number of concurrent three or four notes, or the duration of repeated pitches. The harmony-related metrics measure the chords entropy, distance, or coverage. These three metric categories are used by models like MuseGAN [15], C-RNN-GAN [43], or JazzGAN [59]. Finally, techniques related to style transfer help to understand how close or far the generation is from the desired style. These include, among others, the style fit or the content preservation [6].

5.2 Subjective Evaluation

The subjective perspective determines how the generated music is in terms of creativity and novelty, that is, to what extent the music generated can be considered art. There is no a way to define art, although art involves creativity and aesthetics. Sternberg and Kaufman [55] defined creativity as "the ability to make contributions that are both novel and appropriate to the task, often with an added component such as being quality, surprising, or useful." Creativity requires a deeper understanding of the nature and use of musical knowledge. According to Ji et al. [35], there is a lack of correlation between the quantitative evaluation of music quality and human judgement, which means that music generation models must be also evaluated from a subjective perspective which would give us insights of the creativity of the model. The most used method in subjective evaluations is the listening test which often consists of humans trying to differentiate between machine-generated or human-created music. This method is known as the Turing test which is performed to test DeepBach [25]. In this model, 1.272 people belonging to different musical experience groups took the test. This test showed that the more complex the model was, the better outputs it got. MusicVAE [50] also perform a listening test and the Kruskal-Wallis H test to validate the quality of the model, which conclude that the model performed better with the hierarchical decoder. MuseGAN [15] also conducted a listening test with 144 users divided into groups with different musical experience, but there were predefined questions that the users had to vote in a range from 1 to 5: the harmony complaisance, the rhythm, the structure, the coherence, and the overall rating.

Other listening methods require scoring the generated music; this is known as side-by-side rating [35]. It is also possible to ask some questions to the listeners about the creativity of the model or the naturalness of the generated piece, among other questions, depending on the generation goal of the model. One important thing to keep in mind in listening tests is the variability of the population that is chosen for the test (if listeners are music students with a basic knowledge of music theory, if they are amateurs and so they do not have any music knowledge, or if they are professional musicians). The listeners must have the same stimuli and also listen to the same pieces and have as reference (if it applicable) the same human-created pieces. Auditory fatigue must also be taken into account as there can be an induced bias in the listeners if they listen to similar samples for a long period of time.

Having said that, we can conclude that listening tests are indispensable when it comes to music generation because it gives a feedback of the quality of the model and they can also be a way to find the better NN architecture or DL model that is being studied.

6 Discussion

We have showed that music is a structured language model with temporal and harmonic short- and long-term relationships which requires a deep understanding of all of its insights in order to be modeled. This, in addition to the variety of genres and subgenres that exist in music and the large number of composing strategies that can be followed to compose a music piece, makes the field of music generation with DL a constantly growing and challenging field. Having described the music composition process and recent work in DL for music generation, we will now address the issues raised in Sect. 1.3.

Are the Current DL Models Capable of Generating Music with a Certain Level of Creativity? The first models that generated music with DL used RNNs such as LSTMs. These models could generate notes, but they failed when they generated long-term sequences. This was due to the fact that these NNs did not handle the long-term sequences that are required for music generation. In order to solve this problem and being able to generate short motifs by interpolating two existing motifs or sampling from a distribution, MusicVAE was created. But some questions arise from here: Do interpolations between existing motifs generate high-quality ones which have sense inside the same music piece? If we use MusicVAE for creating a short motif, we could get very good results, but if we use this kind of models to generate longer phrases or motifs that are similar to the inputs, these interpolations may output motifs that can be aesthetic, but sometimes they do not follow any rhythmic or notes direction (ascendent or descendent) pattern that the inputs have. Therefore, these interpolations usually cannot generate high-quality motifs because the model does not understand the rhythmic patterns and notes directions. In addition, chord progressions normally do have inversions, and there are rules in classical music or stylistic restrictions in pop, jazz, or urban music that determine how each chord is followed by another chord. If we analyze the generated polyphonic melodies of DL methods, there is a lack of quality in terms of harmonic content, because NNs which are trained to generate music could not understand all these intricacies that are present in the music language or because this information should be passed to the NN as part of the input, for example, as tokens.

What is the Best NN Architecture to Perform Music Composition with DL? The transformer architecture has been used with different attention mechanisms that allow longer sequences to be modeled. An example of this is the success of the MMM model [18] that used GPT-2 to generate multi-track music. Despite the fact that the model used a pre-trained transformer for text generation, it generates coherent music in terms of harmony and rhythm. Other architectures use generative networks such as GANs or VAEs and also a combination of these architectures with transformers. The power that brings these models is the possibility to extract high-level music attributes such as the style and low-level features that are organized in a latent space. This latent space is then used to interpolate between those features and attributes to generate new music based on existing pieces and music styles.

Analyzing the NN models and architectures that have been used in the past years to generate music with DL, there is not a specific NN architecture that performs better for this purpose because the best NN architecture that can be used to build a music generation model will depend on the output that we want to obtain. In spite of that, transformers and generative models are emerging as the best alternatives in this moment as the latest works in this domain demonstrate. A combination of both models is also a great option to perform music generation [36] although it depends on the output we want to generate, and sometimes the best solution comes from a combination of DL with probabilistic methods. Another aspect to take into account is that generally, music generation requires models with a large number of parameters and data. We can solve this problem by taking a pre-trained model as some state-of-the-art models which we have described in the previous sections and then performing fine-tuning to another NN architecture. Another option is having a pre-trained latent space which has been generated by training a big model with a huge dataset like MusicVAE proposes and then training a smaller NN with less data taking advantage of the pre-trained latent space in order to condition the style of the music compositions as MidiMe proposes [13].

Could End-to-End Methods Generate Entire Structured Music Pieces? As we described in Sect. 3.2, nowadays, there are structure template-based models that can generate structured music [39], but there is not yet an end-to-end method that can compose a structured music piece. The music composition process that a human composer follows is similar to this template-based method. In the near future, it would be likely that AI could compose structured music from scratch, but the question here is whether AI models for music generation will be used to compose entire music pieces from scratch or whether these models would be more useful as an aid to composers and thus as an interaction between humans and AI.

Are the Composed Pieces with DL Just an Imitation of the Inputs or Can NNs Generate New Music in Styles That Are Not Present in the Training Data? When training DL models, some information that is in the input which is passed to the NN can be present without any modification in the output. Even that, MusicVAE [50] and other DL models for music generation showed that new music can be composed without imitating existing music or committing plagiarism. Imitating the inputs could be a case of overfitting which is never the goal of a DL model. It should also be taken into account that it is very difficult to commit plagiarism in the generation of music due to the great variety of instruments, tones, rhythms, or chords that may be present in a piece of music.

Should NNs Compose Music by Following the Same Logic and Process as Humans Do? We showed that researchers started to build models that could generate polyphonic melodies but these melodies did not follow any direction after a few bars. When MusicVAE came out, it was possible to generate higher-quality motifs, and this encouraged new studies to generate melodies taking information of past time steps. New models such as diffusion models [42] are using these pre-trained models to generate longer sequences to let melodies follow patterns or directions. We also show that there are models that can generate a melody by conditioning it with chord progressions which is the way of composing music in styles like pop. Comparing the human way of composing to the DL models that are being used to generate music, we can see some similarities of both processes specially in auto-regressive models. Auto-regression consists of predicting the future values from past events. Some DL methods are auto-regressive, and the fact that new models are trying to generate longer sequences by taking information of past time steps resembles the human classical music composing process.

How Much Data do DL Models for Music Generation Need? This question can be answered partially if we look at the state-of-the-art models. MusicVAE uses the LMD [49] with 3.7 million melodies, 4.6 million drum patterns, and 116 thousand trios. Music Transformer instead uses only 1.100 piano pieces from the Piano-e-Competition to train the model. Other models such as the MMM takes the GPT-2 which is a pre-trained transformer with lots of text data. This leads us to affirm that DL models for music generation do need lots of data, specially when training generative models or transformers, but taking pre-trained models and performing transfer learning is also a good solution specially for music genres that are not represented in the actual datasets for symbolic music generation.

Are Current Evaluation Methods Good Enough to Compare and Measure the Creativity of the Composed Music? As we described in Sect. 5, there are two evaluation categories: the objective and the subjective evaluation. Objective evaluation metrics are similar between existing methods, but there is a lack of a general subjective evaluation method. The listening tests are the most used subjective evaluation method, but sometimes the Turing test which just asks to distinguish between a computer-based and a human composition is not enough to know all the characteristics of the compositions created by a NN. The solution to this problem would be to ask general questions related to the quality of the music features showed in Fig. 2 as MuseGAN proposes and use the same questions and the same rating method in the DL models to set a general subjective evaluation method.

7 Conclusions and Future Work

In this chapter, we have described the state of the art in DL music generation by giving an overview of the NN architectures that have been used for DL music generation and discussed the challenges that are still open in the use of deep NNs in music generation.

The use of DL architectures and techniques for the generation of music (as well as other artistic content) is a growing area of research. However, there are open challenges such as generating music with structure, analyzing the creativity of the generated music, and building interactivity models that could help composers. Future work should focus on better modeling the long-term relationships (in time and harmony axes) in order to generate well-structured and harmonized music that does not get loose after a few bars and inpainting or human-AI interaction which is a task with a growing interest in recent years. There is also a pending challenge that has to do with transfer learning or the conditioning of the generation of styles that allows not to be restricted only to the same authors and genres that are present in the publicly available datasets, such as the JSB Chorales Dataset or the Lakh MIDI Dataset, which makes most of the state-of-the-art works only focus on the same music styles. When it comes to multi-instrument generation, this task does not follow the human composing process, and it will be interesting to see new DL models that first compose a high-quality melodic content and then decide by themselves or with human's help the number of instruments of the music piece and be able to write music for each instrument attending to its timbral features. Further questions related to the directions that music generation with DL should focus on, that is, building end-to-end models that can generate high-creative music from scratch or interactive models in which composers could interact with the AI, are a task that the future will solve, although the trending of human-AI interaction is growing faster everyday.

There are more open questions in music composition with DL that are not in the scope of this chapter—questions like who owns the intellectual property of music generated with DL if the NNs are trained with copyrighted music. We suggest that this would be an important key in commercial applications. The main key here is to define what makes a composition different from others, and there are several features that play an important role here. As we mentioned in Sect. 1, these features include the composition itself, but also the timbre and effects used in creating the sound of the instruments. From the composition perspective which is the scope of our study, we can state that when generating music with DL, it is always likely to generate music that is similar to the inputs, and sometimes the music generated has patterns taken directly from the inputs, so further research would have to be done in this area from the point of view of music theory, intellectual property, and science to define what makes a composition different from others and how music generated with DL could be registered.

We hope that the analysis presented in this chapter will help to better understand the problems and possible solutions and thus may contribute to the overall research agenda of DL-based music generation.

Acknowledgments This research has been partially supported by the Spanish Ministry of Science, Innovation and Universities by the RTI2018-096986-B-C31 contract and the Government of Aragon by the AffectiveLab-T60-20R project.
We wish to thank Jürgen Schmidhuber for his suggestions.

References

1. https://www.copyright.gov/prereg/music.html (2019), accessed July 2021
2. https://koenigproject.nl/project-1/ (2019), accessed July 2021
3. Bharucha, J.J., Todd, P.M.: Modeling the perception of tonal structure with neural nets. Computer Music Journal **13**(4), 44–53 (1989)
4. Bretan, M., Weinberg, G., Heck, L.P.: A unit selection methodology for music generation using deep neural networks. In: Goel, A.K., Jordanous, A., Pease, A. (eds.) Proceedings of the Eighth International Conference on Computational Creativity, ICCC 2017, Atlanta, Georgia, USA, June 19-23, 2017. pp. 72–79. Association for Computational Creativity (ACC) (2017)
5. Brunner, G., Konrad, A., Wang, Y., Wattenhofer, R.: MIDI-VAE: modeling dynamics and instrumentation of music with applications to style transfer. In: Gómez, E., Hu, X., Humphrey, E., Benetos, E. (eds.) Proceedings of the 19th International Society for Music Information Retrieval Conference, ISMIR 2018, Paris, France, September 23-27, 2018. pp. 747–754 (2018)
6. Brunner, G., Wang, Y., Wattenhofer, R., Zhao, S.: Symbolic music genre transfer with cyclegan. In: Tsoukalas, L.H., Grégoire, É., Alamaniotis, M. (eds.) IEEE 30th International Conference on Tools with Artificial Intelligence, ICTAI 2018, 5-7 November 2018, Volos, Greece. pp. 786–793. IEEE (2018)
7. Chen, K., Zhang, W., Dubnov, S., Xia, G., Li, W.: The effect of explicit structure encoding of deep neural networks for symbolic music generation. In: 2019 International Workshop on Multilayer Music Representation and Processing (MMRP). pp. 77–84. IEEE (2019)
8. Chen, Z., Wu, C., Lu, Y., Lerch, A., Lu, C.: Learning to fuse music genres with generative adversarial dual learning. In: Raghavan, V., Aluru, S., Karypis, G., Miele, L., Wu, X. (eds.) 2017 IEEE International Conference on Data Mining, ICDM 2017, New Orleans, LA, USA, November 18-21, 2017. pp. 817–822. IEEE Computer Society (2017)
9. Chomsky, N.: Syntactic structures. De Gruyter Mouton (2009)
10. Chu, H., Urtasun, R., Fidler, S.: Song from PI: A musically plausible network for pop music generation. In: 5th International Conference on Learning Representations, ICLR 2017, Toulon, France, April 24-26, 2017, Workshop Track Proceedings. OpenReview.net (2017)
11. Collins, D.: A synthesis process model of creative thinking in music composition. Psychology of music **33**(2), 193–216 (2005)
12. Cope, D.: Experiments in musical intelligence (emi): Non-linear linguistic-based composition. Journal of New Music Research **18**(1-2), 117–139 (1989)
13. Dinculescu, M., Engel, J., Roberts, A. (eds.): MidiMe: Personalizing a MusicVAE model with user data (2019)
14. Donahue, C., Mao, H.H., Li, Y.E., Cottrell, G.W., McAuley, J.J.: Lakhnes: Improving multi-instrumental music generation with cross-domain pre-training. In: Flexer, A., Peeters, G., Urbano, J., Volk, A. (eds.) Proceedings of the 20th International Society for Music Information Retrieval Conference, ISMIR 2019, Delft, The Netherlands, November 4-8, 2019. pp. 685–692 (2019)

15. Dong, H., Hsiao, W., Yang, L., Yang, Y.: Musegan: Multi-track sequential generative adversarial networks for symbolic music generation and accompaniment. In: McIlraith, S.A., Weinberger, K.Q. (eds.) Proceedings of the Thirty-Second AAAI Conference on Artificial Intelligence, (AAAI-18), the 30th innovative Applications of Artificial Intelligence (IAAI-18), and the 8th AAAI Symposium on Educational Advances in Artificial Intelligence (EAAI-18), New Orleans, Louisiana, USA, February 2-7, 2018. pp. 34–41. AAAI Press (2018)
16. Dong, H., Yang, Y.: Convolutional generative adversarial networks with binary neurons for polyphonic music generation. In: Gómez, E., Hu, X., Humphrey, E., Benetos, E. (eds.) Proceedings of the 19th International Society for Music Information Retrieval Conference, ISMIR 2018, Paris, France, September 23-27, 2018. pp. 190–196 (2018)
17. Eck, D.: A network of relaxation oscillators that finds downbeats in rhythms. In: Dorffner, G., Bischof, H., Hornik, K. (eds.) Artificial Neural Networks - ICANN 2001, International Conference Vienna, Austria, August 21-25, 2001 Proceedings. Lecture Notes in Computer Science, vol. 2130, pp. 1239–1247. Springer (2001)
18. Ens, J., Pasquier, P.: Mmm: Exploring conditional multi-track music generation with the transformer. arXiv preprint arXiv:2008.06048 (2020)
19. Fukushima, K., Miyake, S.: Neocognitron: A new algorithm for pattern recognition tolerant of deformations and shifts in position. Pattern Recognit. **15**(6), 455–469 (1982)
20. Gatys, L.A., Ecker, A.S., Bethge, M.: Image style transfer using convolutional neural networks. In: 2016 IEEE Conference on Computer Vision and Pattern Recognition, CVPR 2016, Las Vegas, NV, USA, June 27-30, 2016. pp. 2414–2423. IEEE Computer Society (2016)
21. Goodfellow, I.J., Pouget-Abadie, J., Mirza, M., Xu, B., Warde-Farley, D., Ozair, S., Courville, A., Bengio, Y.: Generative adversarial nets. p. 2672–2680. NIPS'14, MIT Press, Cambridge, MA, USA (2014)
22. Gordon, E.E.: Audiation, music learning theory, music aptitude, and creativity. In: Suncoast Music Education Forum on Creativity. vol. 75, p. 81. ERIC (1989)
23. Grove, G.: Beethoven and his nine symphonies, vol. 334. Courier Corporation (1962)
24. Hadjeres, G., Nielsen, F.: Interactive music generation with positional constraints using anticipation-rnns. CoRR **abs/1709.06404** (2017)
25. Hadjeres, G., Pachet, F., Nielsen, F.: Deepbach: a steerable model for bach chorales generation. In: Precup, D., Teh, Y.W. (eds.) Proceedings of the 34th International Conference on Machine Learning, ICML 2017, Sydney, NSW, Australia, 6-11 August 2017. Proceedings of Machine Learning Research, vol. 70, pp. 1362–1371. PMLR (2017)
26. Hakimi, S.H., Bhonker, N., El-Yaniv, R.: Bebopnet: Deep neural models for personalized jazz improvisations. In: Proceedings of the 21st international society for music information retrieval conference, ismir (2020)
27. Higgins, I., Matthey, L., Pal, A., Burgess, C., Glorot, X., Botvinick, M., Mohamed, S., Lerchner, A.: beta-vae: Learning basic visual concepts with a constrained variational framework. In: 5th International Conference on Learning Representations, ICLR 2017, Toulon, France, April 24-26, 2017, Conference Track Proceedings. OpenReview.net (2017)
28. Hiller Jr, L.A., Isaacson, L.M.: Musical composition with a high speed digital computer. In: Audio Engineering Society Convention 9. Audio Engineering Society (1957)
29. Ho, J., Jain, A., Abbeel, P.: Denoising diffusion probabilistic models. In: Larochelle, H., Ranzato, M., Hadsell, R., Balcan, M., Lin, H. (eds.) Advances in Neural Information Processing Systems 33: Annual Conference on Neural Information Processing Systems 2020, NeurIPS 2020, December 6-12, 2020, virtual (2020)
30. Hochreiter, S., Schmidhuber, J.: Long short-term memory. Neural computation **9**(8), 1735–1780 (1997)
31. Huang, C.A., Cooijmans, T., Roberts, A., Courville, A.C., Eck, D.: Counterpoint by convolution. In: Cunningham, S.J., Duan, Z., Hu, X., Turnbull, D. (eds.) Proceedings of the 18th International Society for Music Information Retrieval Conference, ISMIR 2017, Suzhou, China, October 23-27, 2017. pp. 211–218 (2017)

32. Huang, C.A., Hawthorne, C., Roberts, A., Dinculescu, M., Wexler, J., Hong, L., Howcroft, J.: Approachable music composition with machine learning at scale. In: Flexer, A., Peeters, G., Urbano, J., Volk, A. (eds.) Proceedings of the 20th International Society for Music Information Retrieval Conference, ISMIR 2019, Delft, The Netherlands, November 4-8, 2019. pp. 793–800 (2019)
33. Huang, C.Z.A., Vaswani, A., Uszkoreit, J., Shazeer, N., Hawthorne, C., Dai, A.M., Hoffman, M.D., Eck, D.: Music transformer: Generating music with long-term structure. arXiv preprint arXiv:1809.04281 (2018)
34. Hung, H., Wang, C., Yang, Y., Wang, H.: Improving automatic jazz melody generation by transfer learning techniques. In: 2019 Asia-Pacific Signal and Information Processing Association Annual Summit and Conference, APSIPA ASC 2019, Lanzhou, China, November 18-21, 2019. pp. 339–346. IEEE (2019)
35. Ji, S., Luo, J., Yang, X.: A comprehensive survey on deep music generation: Multi-level representations, algorithms, evaluations, and future directions. CoRR **abs/2011.06801** (2020)
36. Jiang, J., Xia, G., Carlton, D.B., Anderson, C.N., Miyakawa, R.H.: Transformer VAE: A hierarchical model for structure-aware and interpretable music representation learning. In: 2020 IEEE International Conference on Acoustics, Speech and Signal Processing, ICASSP 2020, Barcelona, Spain, May 4–8, 2020. pp. 516–520. IEEE (2020)
37. Kingma, D.P., Welling, M.: Auto-encoding variational bayes. In: Bengio, Y., LeCun, Y. (eds.) 2nd International Conference on Learning Representations, ICLR 2014, Banff, AB, Canada, April 14-16, 2014, Conference Track Proceedings (2014)
38. Larochelle, H., Murray, I.: The neural autoregressive distribution estimator. In: Gordon, G.J., Dunson, D.B., Dudík, M. (eds.) Proceedings of the Fourteenth International Conference on Artificial Intelligence and Statistics, AISTATS 2011, Fort Lauderdale, USA, April 11-13, 2011. JMLR Proceedings, vol. 15, pp. 29–37. JMLR.org (2011)
39. Lattner, S., Grachten, M., Widmer, G.: Imposing higher-level structure in polyphonic music generation using convolutional restricted boltzmann machines and constraints. CoRR **abs/1612.04742** (2016)
40. LeCun, Y., Haffner, P., Bottou, L., Bengio, Y.: Object recognition with gradient-based learning. In: Forsyth, D.A., Mundy, J.L., Gesù, V.D., Cipolla, R. (eds.) Shape, Contour and Grouping in Computer Vision. Lecture Notes in Computer Science, vol. 1681, p. 319. Springer (1999)
41. Levi, R.G.: A field investigation of the composing processes used by second-grade children creating original language and music pieces. Ph.D. thesis, Case Western Reserve University (1991)
42. Mittal, G., Engel, J.H., Hawthorne, C., Simon, I.: Symbolic music generation with diffusion models. CoRR **abs/2103.16091** (2021)
43. Mogren, O.: C-RNN-GAN: continuous recurrent neural networks with adversarial training. CoRR **abs/1611.09904** (2016)
44. Mozer, M.C.: Neural network music composition by prediction: Exploring the benefits of psychoacoustic constraints and multi-scale processing. Connect. Sci. **6**(2-3), 247–280 (1994)
45. Müller, M.: Fundamentals of Music Processing - Audio, Analysis, Algorithms, Applications. Springer (2015)
46. Nierhaus, G.: Algorithmic composition: paradigms of automated music generation. Springer Science & Business Media (2009)
47. Payne, C.: Musenet, 2019. https://openai.com/blog/musenet (2019)
48. Peracha, O.: Improving polyphonic music models with feature-rich encoding. In: Cumming, J., Lee, J.H., McFee, B., Schedl, M., Devaney, J., McKay, C., Zangerle, E., de Reuse, T. (eds.) Proceedings of the 21th International Society for Music Information Retrieval Conference, ISMIR 2020, Montreal, Canada, October 11-16, 2020. pp. 169–175 (2020)
49. Raffel, C.: Learning-based methods for comparing sequences, with applications to audio-to-midi alignment and matching. Ph.D. thesis, Columbia University (2016)

50. Roberts, A., Engel, J.H., Raffel, C., Hawthorne, C., Eck, D.: A hierarchical latent vector model for learning long-term structure in music. In: Dy, J.G., Krause, A. (eds.) Proceedings of the 35th International Conference on Machine Learning, ICML 2018, Stockholmsmässan, Stockholm, Sweden, July 10-15, 2018. Proceedings of Machine Learning Research, vol. 80, pp. 4361–4370. PMLR (2018)
51. Root-Bernstein, R.S.: Music, creativity and scientific thinking. Leonardo **34**(1), 63–68 (2001)
52. Rosenblatt, F.: The perceptron: a probabilistic model for information storage and organization in the brain. Psychological review **65**(6), 386 (1958)
53. Semin Kang, S.Y.O., Kang, Y.M.: Automatic music generation and machine learning based evaluation. In: International Conference on Multimedia and Signal Processing. pp. 436–443 (2012)
54. Sevsay, E.: The cambridge guide to orchestration. Cambridge University Press (2013)
55. Sternberg, R.J., Kaufman, J.C.: The nature of human creativity. Cambridge University Press (2018)
56. Tan, H.H.: Chordal: A chord-based approach for music generation using bi-lstms. In: Grace, K., Cook, M., Ventura, D., Maher, M.L. (eds.) Proceedings of the Tenth International Conference on Computational Creativity, ICCC 2019, Charlotte, North Carolina, USA, June 17-21, 2019. pp. 364–365. Association for Computational Creativity (ACC) (2019)
57. Teng, Y., Zhao, A., Goudeseune, C.: Generating nontrivial melodies for music as a service. In: Cunningham, S.J., Duan, Z., Hu, X., Turnbull, D. (eds.) Proceedings of the 18th International Society for Music Information Retrieval Conference, ISMIR 2017, Suzhou, China, October 23-27, 2017. pp. 657–663 (2017)
58. Todd, P.M.: A connectionist approach to algorithmic composition. Computer Music Journal **13**(4), 27–43 (1989), http://www.jstor.org/stable/3679551
59. Trieu, N., Keller, R.: Jazzgan: Improvising with generative adversarial networks. In: MUME workshop (2018)
60. Valenti, A., Carta, A., Bacciu, D.: Learning a latent space of style-aware symbolic music representations by adversarial autoencoders. CoRR **abs/2001.05494** (2020)
61. Vaswani, A., Shazeer, N., Parmar, N., Uszkoreit, J., Jones, L., Gomez, A.N., Kaiser, L., Polosukhin, I.: Attention is all you need. In: Guyon, I., von Luxburg, U., Bengio, S., Wallach, H.M., Fergus, R., Vishwanathan, S.V.N., Garnett, R. (eds.) Advances in Neural Information Processing Systems 30: Annual Conference on Neural Information Processing Systems 2017, December 4-9, 2017, Long Beach, CA, USA. pp. 5998–6008 (2017)
62. Waite, E., et al.: Generating long-term structure in songs and stories. Web blog post. Magenta **15**(4) (2016)
63. Walton, C.W.: Basic Forms in Music. Alfred Music (2005)
64. Wang, C., Dubnov, S.: Guided music synthesis with variable markov oracle. In: Pasquier, P., Eigenfeldt, A., Bown, O. (eds.) Musical Metacreation, Papers from the 2014 AIIDE Workshop, October 4, 2014, Raleigh, NC, USA. AAAI Workshops, vol. WS-14-18. AAAI Press (2014)
65. Wang, Z., Wang, D., Zhang, Y., Xia, G.: Learning interpretable representation for controllable polyphonic music generation. CoRR **abs/2008.07122** (2020)
66. Wang, Z., Zhang, Y., Zhang, Y., Jiang, J., Yang, R., Zhao, J., Xia, G.: PIANOTREE VAE: structured representation learning for polyphonic music. CoRR **abs/2008.07118** (2020)
67. Wu, S., Yang, Y.: Musemorphose: Full-song and fine-grained music style transfer with just one transformer VAE. CoRR **abs/2105.04090** (2021)
68. Xenakēs, G.: Musiques formelles: nouveaux principes formels de composition musicale. Ed. Richard-Masse (1963)
69. Yang, W., Sun, P., Zhang, Y., Zhang, Y.: Clstms: A combination of two lstm models to generate chords accompaniment for symbolic melody. In: 2019 International Conference on High Performance Big Data and Intelligent Systems (HPBD&IS). pp. 176–180. IEEE (2019)
70. Yeh, Y., Hsiao, W., Fukayama, S., Kitahara, T., Genchel, B., Liu, H., Dong, H., Chen, Y., Leong, T., Yang, Y.: Automatic melody harmonization with triad chords: A comparative study. CoRR **abs/2001.02360** (2020)

71. Yu, L., Zhang, W., Wang, J., Yu, Y.: Seqgan: Sequence generative adversarial nets with policy gradient. In: Singh, S.P., Markovitch, S. (eds.) Proceedings of the Thirty-First AAAI Conference on Artificial Intelligence, February 4-9, 2017, San Francisco, California, USA. pp. 2852–2858. AAAI Press (2017)
72. Zhu, H., Liu, Q., Yuan, N.J., Qin, C., Li, J., Zhang, K., Zhou, G., Wei, F., Xu, Y., Chen, E.: Xiaoice band: A melody and arrangement generation framework for pop music. In: Guo, Y., Farooq, F. (eds.) Proceedings of the 24th ACM SIGKDD International Conference on Knowledge Discovery & Data Mining, KDD 2018, London, UK, August 19-23, 2018. pp. 2837–2846. ACM (2018)
73. Zhu, H., Liu, Q., Yuan, N.J., Zhang, K., Zhou, G., Chen, E.: Pop music generation: From melody to multi-style arrangement. ACM Trans. Knowl. Discov. Data **14**(5), 54:1–54:31 (2020)
74. Zhuang, F., Qi, Z., Duan, K., Xi, D., Zhu, Y., Zhu, H., Xiong, H., He, Q.: A comprehensive survey on transfer learning. Proc. IEEE **109**(1), 43–76 (2021)

Music Recommendation Systems: Overview and Challenges

Makarand Velankar and Parag Kulkarni

1 Introduction

The internet has revolutionized the way people interact and communicate. An enormous amount of data is available at the fingertips due to affordable technology. It includes audio, video, and text from various media, blogs, books, journals, and web pages. In music, a large number of songs are available on the internet. As a result, music listening habits have changed in recent years. The storage of songs on personal devices such as compact disks or hard disks is no longer relevant. Instead, online streaming services dominate the music industry, and music consumption patterns have shifted in recent years. Conventionally, the significant issues faced by music recommendation systems are the cold start problem and playlist generation. The cold start problem is recommending new users or songs when no user or song history is available. The playlist is a personalized list of favorite songs representing specific moods or artists or genres. Automatic playlist generation is another significant challenge for precisely deducing the purpose of the current playlist. The related problem is to rank songs in response to a user-selected metadata query.

As per one survey done in the year 2019 [1], it was observed that 27% of people use mobile phones for music listening. The trend to use smartphones for music listening was more among youth mainly in the age range of 18 to 24 years where 48% prefer to use a mobile device for enjoying music. Different music streaming apps have dominated the industry growth to cater for this changing need of users.

M. Velankar (✉)
MKSSS'S Cummins College of Engineering for women, Pune, India
e-mail: makarand.velankar@cumminscollege.in

P. Kulkarni (✉)
Iknowlation labs, Pune, India

In the survey done [2] in 2021, 443 million users had a paid subscription account, which is increased by 18.5% compared with the previous year.

A significant noticeable change in music production has been observed in recent years. Previously, only specific companies had recording set up used to produce the music. Now, anyone can create music and upload it on different websites or social media apps. With the massive amount of music available online, recommendation systems must adapt to this rapid growth in music data and online demand. Specific factors make the music recommendation system different from other recommendation systems [3]. It includes the duration of items, the magnitude of items, sequential consumption, repeated recommendations, consumption behavior, listening intent, occasion, context, and associated emotions. It is addressed using approaches such as content-based recommendation, hybridization, and cross-domain recommendations.

In a content-based approach, the automatic feature extraction for each item is done by extracting the feature vector. The feature vectors are created individually and independently, considering the cross-relation among the training data set items. Various approaches are used in recommendation systems. The collaborative and hybrid approaches are currently dominating the music recommendation systems, and researchers are also exploring different approaches. The scalability and commercial viability of these approaches will decide the future revolutions in the music recommendation systems. User orientation plays a crucial role in designing user interfaces, easing user interactions, and getting implicit and explicit feedback to improve the systems. User psychology and cognitive approaches are also being studied in order to design an efficient recommendation system. Machine learning approaches are explored to improve the performance for prediction, classification, and clustering [4]. Novel trends related to technology advancement and user perspectives, including popularity bias, provide new opportunities and challenges in the changing times.

The chapter is organized in the following manner. Overview of different approaches in recommendation systems and business aspects is covered in Sect. 2. Section 3 covers user orientation and related aspects of the recommendation. Current challenges and trends in music recommendation are covered in Sect. 4. Finally, Sect. 5 provides concluding remarks and future directions in the field of music recommendations.

2 Overview of Recommendation Systems

The recommendation systems can be broadly classified based on various parameters such as technology, domain, geographical coverage, items, and system users. The classification based on technology is as shown in Fig. 1. Content-based recommendation systems predominantly use content in the form of metadata of songs for the recommendation. User preferences are modelled using the history of user interactions and preferences. The demographic filtering approach recommends songs based on user location. Hybrid approaches attempt to build systems based on

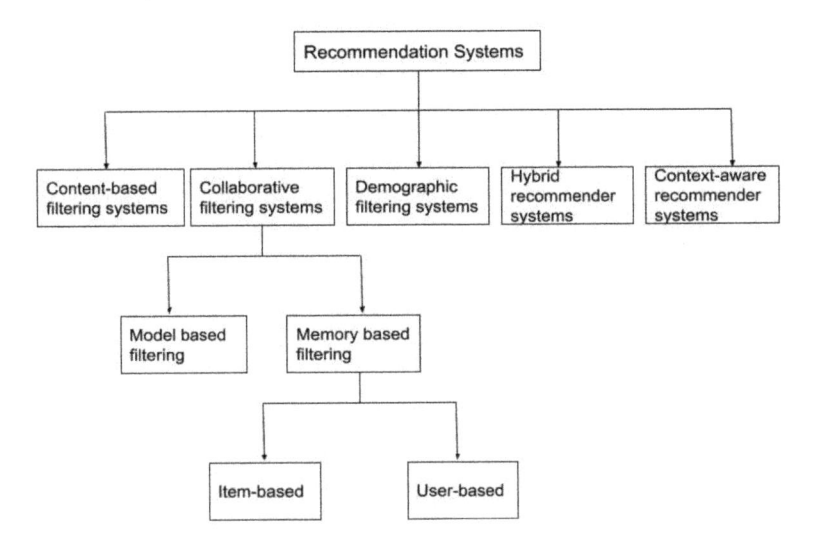

Fig. 1 Classification of recommendation systems

different approaches with benefits from them. Finally, the context-aware approach attempts to identify user context for an effective recommendation. The following subsections cover the popular approaches used today.

2.1 Content-Based Approach

The content-based technique recommends items by comparing their contents to a user profile as shown in Fig. 2. The content-based approach relies on item similarity based on identified features and recommends the items with similar features. K-means clustering and Monte Carlo sampling are some of the standard methods used to compute similarity. K-means clustering technique recommends songs based on very little data, and hence, the problem of cold start is solved partially. K-means clustering involves [5, 6] matching the similarity of cluster centroid attribute value and item object attribute values. Different cluster centroids are made based on the output of the K-means algorithm. New suggestions are provided by extracting the quantity of Mel-frequency cepstral coefficients (MFCC) features [7], clustering it, and thus obtaining the feature value of the item. Then, the metadata is put together with their feature value into the database. For recommendation, the feature value of the item is extracted. Next, the existing items in the database are used to compare the distance; the lesser the absolute value, the more significant the similarity of items.

Different researchers propose various systems based on a content-based approach. Zhong et al. [8] and Schedl [9] combined deep convolution neural networks (DCNN) with content-based music recommendation as MusicCNNs.

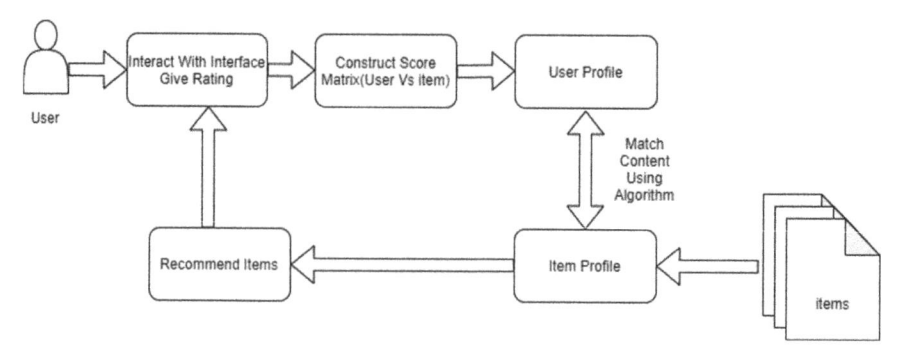

Fig. 2 Content-based filtering model

First, each song is split into 20 music segments and converted to images with the help of Fourier transformation to feed them easily into MusicCNNs. The set of audio descriptors are used to represent the content of each song [10–12]. Each song's acoustic features such as melody, tempo, loudness, rhythm, and timbre are analyzed for the recommendation

Current challenges in the content-based model are:

1. Terms automatically assigned to the items must be first extracted from them. Hence, a method must be chosen for this operation.
2. The term representations should compare items and user profiles meaningfully.
3. A learning algorithm learns the user profile based on seen items to make recommendations.

The inability of the systems to deal with items that have no ratings is a significant drawback according to [13, 14]. Various content-based approaches for feature identification in music analytics covered by Velankar et al. [4] provide more music features useful for the content-based approach. However, computational cost and scalability issues, considering the enormous speed of data increase, are significant challenges in effectively implementing a content-based approach.

2.2 Collaborative Approach

The collaborative filtering method produces automated predictions regarding users' interests by considering preferences and taste information from multiple users. The overall model is depicted in Fig. 3. Once a sufficient user history is generated, the system matures and provides recommendations based on similar user profiles. This approach assumes that users (considered as user clusters) with similar histories have similar song tastes.

The two crucial types of collaborative filtering are memory-based and model-based. Memory-based approach is further classified into item-item filtering and

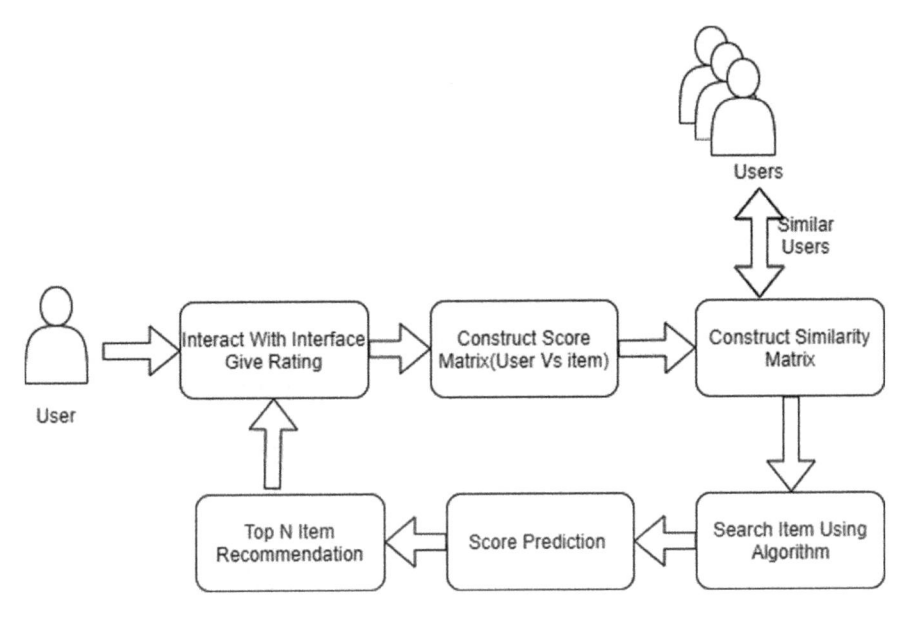

Fig. 3 Collaborative filtering model

user-item filtering [15, 16]. In the memory-based collaborative filtering method [17], user-item filtering predicts the result based on the previous rating collection. Grouping of users is done based on similar interests. Many user votes are used to find the nearest neighbor and thus produce a new item. Whereas, [18] suggests item-item filtering is independent of the number of items in the item inventory and the number of users. It targets finding items identical to those that the user bought instead of matching other identical users.

In model-based recommendations, the item attributes and users are represented as latent feature vectors. Then, based on predicted rating, items are recommended. Model-based collaborative filtering trains the system and models user's preferences using machine learning and data mining algorithms.

Sharma and Gera [19] and Sánchez-Moreno et al. [20] studied the problems faced by the personalized recommendation systems using collaborative filtering. The main problems faced are the prediction accuracy, cold start, and data sparsity problem. Another issue was that the user-item matrix would keep growing, so the computations could also increase exponentially. Cold start can be solved using deep multi-modal approach [21, 22]. For automatic playlist generation, a significant challenge is to deduce the purpose of the current playlist precisely. Another challenge is to rank songs in response to a user-selected metadata query to keep the user engaged for a longer time for the specific instance.

2.3 Hybrid Approach

The shortcomings of content-based as well as collaborative models are addressed in the hybrid model [23]. Furthermore, the recommendations based on hybrid models can adopt different weight strategies in different situations. As per findings in [24], the recommendation systems are majorly based on the two-dimension user vs item matrix.

Hybrid perspective is accomplished in the following ways:

1. Separately implement the content-based method as well as collaborative method and then merge their predictions.
2. Characteristics of the content-based method merged into a collaborative execution.
3. Collaborative characteristics merged into the execution of content-based. They built a consolidated model with a fusion of collaborative and content-based characteristics.

Figure 4 exhibits the general blocks in the hybrid model. Hybrid recommendation systems are becoming increasingly popular in various fields such as online news reading [25] and Netflix [26]. YouTube recommendation is of the top-N recommendations type. It makes use of the association rule mining technique, where algorithms such as personalized video ranker, top-N video ranker, and video-video similarity are used[27]. Furthermore, a systematic review of near-duplicate video retrieval techniques [28] provides new dimensions in video recommendation, which applies to music considering the uploads of the same song by different artists. Furthermore, context-based decision-making for emotion recognition for images

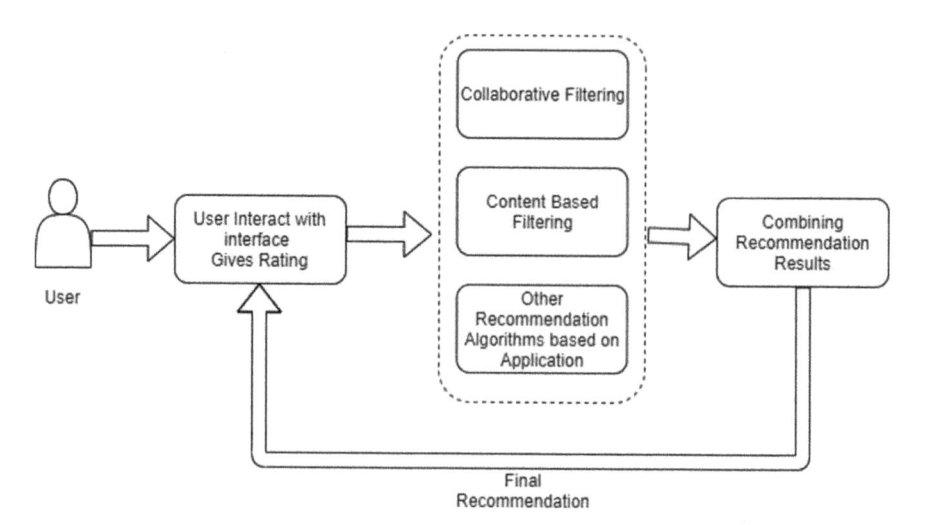

Fig. 4 Hybrid model

[29] can be extended further to music videos for identifying emotional cues in a hybrid approach. According to [30], the personalized TV recommendation system (RS) based on user perspective takes into account user activity, mood, demographic information, experience, and interest based on a hybrid technique. As the viewing information is gathered, the recommendation accuracy will keep on increasing.

Hybrid models are extensively used in music recommendation [31]. As per [32, 33], a music recommendation system, which gives emotion-aware recommendations and provides similarities between user information and music, is extracted. It can be achieved by a combination of the outputs from different approaches. Fessahaye et al. [34] proposed a hybrid algorithm using a deep learning classification model to result in error-free recommendations with real-time prediction. It scores every song based on hybridization and recommends k songs. Similarly, for Spotify recommendations, two visualization [35, 36] hybrid techniques radar chart and sliders are used.

2.4 Context-Aware Approach

The substantial rise in the data generated from automated and electronic devices has caused the need for intelligent applications and techniques, that can store information, access, process, and analyze it to maximize the benefit of the users. Context-aware recommendation systems (CARS) solve these data problems, which act as tools to help the information seeking process. Context-based approach explored by Khatavkar and Kulkarni [37], Kulkarni et al. [38], and Vidhate and Kulkarni [39] provided the utility of the approach for document retrieval and recommendation. CARS leverage contextual information as well as the two-dimensional search processes, which results in better recommendations [40]. According to [41], the dominant approach in CARS involves pre-filtering to integrate and represent the contextual information of the user. However, sizable research is needed to compare the results of various filtering methods. As a matter of fact, due to degrading performance, pre-filtering does not dominate other filtering approaches. Verbert et al. [42] suggests an example of a context-aware system for learning which takes into account the noise level and location of the user in order to recommend learning resources. Therefore, a contextual recommendation would suggest that learning activities examine users' knowledge of formerly learned topics. In this view, new challenges arise to capture and understand the context and exploit the information to create intelligent recommendations and consider the user's current needs. In conclusion, there is still no generalization in the algorithmic approaches being used in CARS. Partially, this can be because of the undeveloped knowledge of contextual preferences, nature of the data exploited in these studies, and lack of objective validation methods. The further investigation by Adomavicius and Tuzhilin [43] in this area identified challenges such as:

1. Building a generalized CARS framework
2. Developing standardized context-aware recommendation approaches

3. Including context identification and incorporation
4. Incorporating dimensionality reduction techniques
5. Standardizing user modelling
6. Bench-marking data sets and evaluation mechanisms.

2.5 Business Aspects

Recommendation systems have certainly played a crucial role in ensuring customer satisfaction and increasing revenue for online businesses. However, adopting a suitable architecture that can handle complex algorithms and result in new recommendations is essential for large-scale recommendation systems [44]. For giant retailers like Amazon, a decent recommendation algorithm is scalable over exceptionally massive client bases. Item lists require just sub-second handling time to generate online suggestions, can answer promptly to changes in a customer's information in real time, and thus make persuasive suggestions for clients irrespective of the number of buys as well as ratings.

Bauer et al. [45] suggests business in the music industry has a theory of long tail and short tail. Sales are pivoted on the hit songs, which form the head, and lesser-known items from a long tail behind it. Therefore, selling small numbers of the long tail is more profitable than substantial quantities of a small number of hits.

Business revenues and sales conversions rate can be significantly improved by incorporating personalized conversational recommendation agents [46]. Traditional conversation agents only make use of the present session information. Personalized dialogue systems use current as well as past user choices to optimize each user session. Information can be collected by asking questions, and deep reinforcement learning algorithms do learning. Understanding user behavior from historical data helps businesses overcome challenges like keeping appropriate stocks, managing website traffic, and providing effective deliveries during shopping festivals.

2.6 Summary

Business aspects dominate the current recommendation system, with revenues from advertisements as one of the significant sources of revenue. However, hybrid approaches are becoming increasingly popular, and technology advancements will lead to more mature and user-oriented recommendation systems in the coming years. User orientation is the need for a recommendation system; moreover, it plays a significant role in music recommendation due to enhancing or changing user mood.

3 User Orientation

The accuracy of recommendations majorly depends on the underlying filtering algorithm. The significant factors go beyond the quality of the algorithm when deciding the effectiveness of the recommendation system. Swearingen and Sinha [47] suggests a theory that from a user's point of view, an efficient system has logic that is a little transparent and directs the user to a new and not yet recommended item. It thus gives different ways to improve recommendations by considering or removing particular items. For practical recommendations, users can give more input to the system. Simply presenting the recommendation results is not sufficient. It is more of making it clear whether the item is a good choice or not for the user by giving a detailed description of the qualities of the recommended item. A variety of factors influence the effectiveness of music recommendations, including the user's specific intent, personality, how much novelty users seek, and the user's context. Initially, it should provide recommendations that will build the trust of the new users in the system [48]. Then, the familiar users may be provided diverse and novel suggestions. It is observed that the success of recommendation systems is more for low-risk items like movies, books, and music and less for high-risk fields such as cars. The cause for this is the reluctance of the user to trust recommendations they do not understand. Hence, detailed recommendations are required in high-risk domains. From [44] perspective, to increase the user's credibility over the system, explanations for recommendations like the predicted ratings and related items watched are provided. In music recommendation, the system should model the user profile and present needs for an effective recommendation.

3.1 User Profiling

In various phases of a customer session, their inclination toward a suggestion can change extensively. When a user logs in, the system has to know their inclination and preference. When the user begins centering on a specific set of items, the system can suggest comparative items. After a choice is made, a similar item proposal turns out to be less significant. For instance, the kind of music users listen to depends on their mood and emotions [49]. The general mood and the emotional aspect of a user can be considered at a particular time of day. In psychological studies, user personality is considered the main reason for variable user behavior and preferences. Strong and direct correlations between user models and user personality exist in recommendation systems. The Big Five Model helps to measure a user's personality quantitatively in terms of OCEAN known as Openness, Conscientiousness, Extraversion, Agreeableness, and Neuroticism are the main user personality traits. For generating group recommendations and the creation of diverse recommendations, personality has proved to help solve the cold start problem.

User profile modelling can be categorized into three domains—demographic, psychographic, and geographic. According to user listening experience modelling [50], user choices differ depending on their level of music expertise. It is observed that user psychology helps to build user profiles and helps determine recommendations.

3.2 Psychology and Cognitive Aspects

The data generated by social recommendation systems are heterogeneous, volatile, and massive in volume. Semantic technologies are used to overcome challenges faced by classical and social recommendation systems. For instance, personalized advertisements are recommended to users by dynamic semantic profiling, understanding their activities and links in the social ontology model [51]. It enhances recommendation systems' content and visual appearance by building user-oriented knowledge networks primarily based on user cognition. The field of user cognition has found substantial application in the music industry. A system having context as emotions is based on the fusion of user history and current emotions such as pleasure or displeasure [52]. It can be achieved by using a novel human-computer interaction in the form of wearable computing devices. It recognizes the emotions the user is feeling and sends signals to the system. Situation-aware music recommendation [53, 54] considers situational signals such as time of the day, ongoing activities and mood of the user, weather conditions, and the day of the week. For example, the song choices of a user would be different according to places like the gym and libraries. Furthermore, the cultural pattern analysis in music usage is studied and scrutinized under culture-aware music recommendations. It includes building models based on different cultural music. Their amalgamation into a recommendation system is crucial to upgrade the personalization and robustness of recommendations.

3.3 Summary

In the beginning, recommendation systems only used explicit ratings, collaborative and content-based approaches, and demographic information. Later, they became more intelligent and started using social data, moods, and emotions of users. Now, recommendation systems are moving toward extracting context-aware information from the Internet of Things, such as geographical and data like time and location. Incorporating various data types leads recommendation systems to apply a hybrid approach that gives better accuracy. Extracting and utilizing implicit user feedback is becoming a critical factor in improving the accuracy and precision of the recommendations. In addition, the user interface plays a crucial role in user interaction with the system. These factors help in providing better business insights.

4 Current Challenges and Trends in Music Recommendation

Listener emotions and song perception play a crucial role in music recommendation. Individual likes/dislikes, musical and cultural background, and mood contribute to the effectiveness of recommendation. Music consumption patterns have changed drastically in recent years, and online music consumption is becoming more popular among users. Frequent recommendation of popular songs is another challenge as it leads to popularity bias in the recommendation systems.

4.1 Music Consumption

The kind of music we listen to depends on our mood and emotions. The general mood and the emotional aspect of a user can be considered at a particular time of day. In psychological studies, user personality is considered the main reason for user behavior and variable preferences. The model named MUSIC [55] is one approach—Mellow, Unpretentious, Sophisticated, Intense, and Contemporary. The MUSIC preference attributes are calculated using attribute learning based on its acoustic content whenever a new song is encountered. Then from user ratings, the model gets to know the five factors of their personal preferences. Then taking the estimated factors of music and user, the model rates the songs and recommends them.

The arrival of streaming platforms is a significant change in the music industry. They generate the most considerable recorded music revenue [56]. Streaming services help users in music discovery with various tools. The tools are available for text searching for song playlist, artist, release-related metadata, grouping albums by themes or highlighting latest releases; song and album playlists created by various theme, genre, mood, region-wise popularity charts and different hyperlinks for artist and album name which can be clicked and divert the user to different parts of catalogue. In general, subscriptions provided by the music streaming companies contain ad-supported recommendations. The prior sequence of songs/ads needs matters when deciding the correct time for exploration or exploitation. For example, the AD-Song-AD sequence is likely to impact the user to change the station when an explored song is recommended after the sequence. The recent proposal by Lex et al. [57] uses the BLL equation from the cognitive architecture ACT-R approach for modelling popularity and temporal drift of music genre preferences. Furthermore, evaluate BLL on three groups of Last.fm users, separated based on their listening behavior to the mainstream LowMS, MedMS, and HighMS.

4.2 Popularity Bias

Implicit feedback to a recommendation system for a particular artist [58], can be improved by taking into consideration the number of days an artists song and the variety of songs heard by the user. The implicit matrix factorization can be trained with these signals to increase the artist's music consumption. [59] predicts the preference of a target user for a target artist and defines the preference by scaling the listening count in the range of [0, 1000]. It is done using four personalized recommendations factors as UserItemAvg, UserKNN, UserKNNAvg, and NMF. Further, it recommends the top 10 artists with the highest predicted preferences.

A similar study about music recommendation impacts the phenomenon of gender bias [60] by introducing bias in data, boosting the existing biases, and strengthening stereotypes. For the non popular artists [45], the factors that affect include (1) popularity bias when popular songs tend to get more and more attention and the rating system reinforces the popularity of the popular items and (2) superstar economy speculations: Music recommendation is biased on popular items because of their profits.

4.3 Trends in Music Recommendation

The future scope lies in advanced recommendation systems like context-awareness systems, group recommendation systems, systems based on social networks, and recommendation techniques based on computational intelligence, as displayed in Fig. 5. Computational intelligence-based recommendation techniques consist of approaches such as genetic algorithms, fuzzy logic, and Bayesian method. Social network-based recommendation systems are built upon the social engagement of one user with other users. Context-awareness recommendation systems use features such as geographical locations and times.

Future research directions for technology perspective are:

1. Full page optimization algorithm understands the user behavior. It thus personalizes how user experience is blended with a different component for the recommendation.
2. Personalizing how one recommends by thinking about pictures, depictions, metadata, associations with different factors, and so forth as components that can be customized.
3. Connecting indirect feedback or input to tons of contextual information.
4. Value-aware recommendations are where algorithms take the long haul value of the recommended item into consideration so that algorithm can move toward greater gains [61].

Another perspective, according to [62], is big data analytics. The users can gain cognizance from big data, retrieve fruitful information, and suggest user-

Fig. 5 Future directions of research

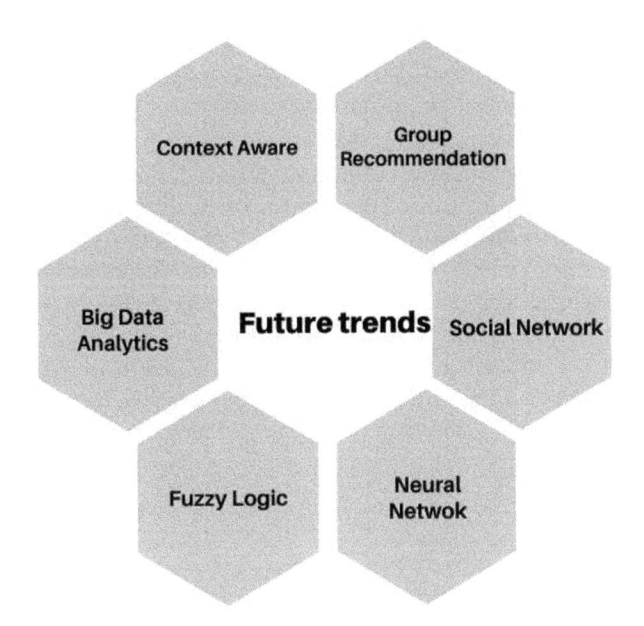

- Context Aware
- Group Recommendation
- Big Data Analytics
- Social Network
- Fuzzy Logic
- Neural Netwok

Future trends

centric recommendations. Cui et al. [63] suggests IoT service system involves large-volume, complex, high-dimensional, sparse, rich-in-content data which can be integrated with the novel intelligent optimization algorithm. As per [64, 65], future systems can gather data, analyze attributes attached with the music in order to extract significant features, and pre-process the metadata. First, a hidden Markov model is used to learn and then advance and estimate the parameters using the Baum-Welch algorithm. Next, the system generates random samples using the transition and emission probability distribution. Finally, evaluation is done by selecting the sample that has the highest likelihood. Several recommendation systems use single ratings in predictions. It is considered a limitation since the user's choice might consider several aspects, so the accuracy of the prediction can increase due to additional aspects. The future system can implement a deep learning-based collaborative filtering model with multi-criteria [66].

The design of computer technology should include the user's preference so that the interaction between the user and system is enhanced. A personalized and context-aware approach is likely to provide better recommendations. At present, the recommendation system evaluation focuses on evaluating the list of recommendations given by the system as a whole rather than an individual user. In addition, there is also a demand for recommendation systems for a group of users together. The latest trend in personalization is based on recognition of human activity. It intends to define and try new methods that automatically recognize human activities that exploit signals recorded by worn and environmental devices. Instead of segmenting users based on personalization rules, algorithms can be applied to deliver individual experiences. It is done by providing content or product recommendations. High customer demand and industrial competition have encouraged most businesses to look beyond mass production. As expected, affordable products are tailored according to their unique requirements, and thus, the demand has turned to mass personalization. So users want personalized appearances and functionalities along with it being affordable. In mass personalization, a significant challenge is designing the highest personalization at a low cost. A fuzzy-based system [67] provides intelligent and personalized recommendations for electronic products by considering the needs of the consumer in the form of features, their domain knowledge, and all temporary information provided by them. These factors compute the optimality and the quality of the product for the recommendation. Various domains of personalized recommendation systems face a variety of challenges. Some challenges faced by video streaming platforms such as YouTube mainly are the content uploaded by users have below par or non-existent metadata, tracking user engagement with the videos, and incorporating freshness in recommendations.

4.4 Summary

Specific challenges of the music recommendation systems are:

1. Catalogue aspects: New tracks are getting released constantly and are added to the catalogue. An important quality factor to consider for some musical genres might be the freshness of recommendations.
2. Preference information: Along with the challenge of correctly interpreting the vast amounts of implicit feedback, the additional problem is that the preferences can change over time.
3. Repeated recommendations: Repeatedly tuning to the same songs is, however, expected. Suppose such repeated use is to be supported. In that case, algorithmic approaches have to decide which tracks to recommend repeatedly and when to recommend these tracks.
4. Immediate consumption and feedback: The recommendations provided on a music streaming service are immediately consumed by the listeners. Many musical genres (like jazz, classical music, or pop) have specific audiences, and recommending trending popular items might easily lead to a bad user experience.
5. Context dependence and time variance: The ability to retrieve and include the contextual factors can be important for a recommender's quality perception and acceptance.
6. Purposes and taste of music listening: The recommended items, along with aligning to the current context, also need to satisfy the user's purpose.

The music users prefer and stream to is affected by their mood, social environment, or trends in the community. A significant number of recommendation systems depend on acquiring user evaluations to foresee unknown ratings. A hidden supposition in this methodology is that the user ratings majorly define user taste. Furthermore, user ratings are more predictable when items with the same ratings are gathered. Thus, extreme feedback is more reliable than mellow feedback.

5 Conclusion and Future Directions

This chapter provides an overview of different approaches used in the evolving recommendation systems. The different perspectives in technology, such as collaborative, content-based, and hybrid filtering models, help to identify the existing advantages and shortcomings. It is concluded that hybrid models are the best fit. In recommendation systems, user interface design is also very crucial. The user feedback and interactions with the interface designed considering implicit cognitive processes, helps boost business profits.

The hybrid models are likely to be used extensively soon, using benefits from individual models and overcoming the current limitations. Moreover, the hybrid model can be further enhanced by its amalgamation with future technology trends. New models are likely to emerge with changing needs, usage patterns, and improvements in the technologies. Specializing models for a customer segment with specific needs can lead to better business opportunities than a generalized model. The specialized, focused models can address the specific issues and need more effectively.

Future research must incorporate various advanced technologies such as big data analysis, IoT, machine learning, and artificial intelligence in recommendation systems, resulting in better personalization. Also, attempts should be made to counter the drawbacks of the current systems and improve their accuracy. As a result, new solutions will emerge in the coming years with a more personalized experience for the end users. In addition, it will be helpful to music and other industries such as e-business to improve the recommendation's effectiveness.

References

1. International Federation of the Phonographic Industry https://www.ifpi.org/ifpi-releases-music-listening-2019/ Last accessed 4 Nov. 2021
2. International Federation of the Phonographic Industry https://www.ifpi.org/ifpi-issues-annual-global-music-report-2021/ Last accessed 4 Nov. 2021
3. Kim, Namyun, Won-Young Chae, and Yoon-Joon Lee. Music recommendation with temporal dynamics in multiple types of user feedback. In *Proceedings of the 7th International Conference on Emerging Databases*, Springer, Singapore (2018)
4. Makarand Velankar, Amod Deshpande, and Parag Kulkarni. 3 application of machine learning in music analytics. In *Machine Learning Applications*, De Gruyter (2020)
5. Huihui Han, Xin Luo, Tao Yang, and Youqun Shi. Music recommendation based on feature similarity. In *2018 IEEE International Conference of Safety Produce Informatization (IICSPI)*, IEEE (2018)
6. Gang Li and Jingjing Zhang. Music personalized recommendation system based on improved knn algorithm. In *2018 IEEE 3rd Advanced Information Technology, Electronic and Automation Control Conference (IAEAC)*, IEEE (2018)
7. Rui Cheng and Boyang Tang. A music recommendation system based on acoustic features and user personalities. In *Pacific-Asia Conference on Knowledge Discovery and Data Mining*, Springer (2016)

8. Guoqiang Zhong, Haizhen Wang, and Wencong Jiao. Musiccnns: A new benchmark on content-based music recommendation. In *International Conference on Neural Information Processing*, Springer (2018)
9. Markus Schedl. Deep learning in music recommendation systems. *Frontiers in Applied Mathematics and Statistics* (2019)
10. Mohammad Soleymani, Anna Aljanaki, Frans Wiering, and Remco C Veltkamp. Content-based music recommendation using underlying music preference structure. In *2015 IEEE international conference on multimedia and expo (ICME)*, IEEE (2015)
11. Velankar Makarand and Kulkarni Parag. Unified algorithm for melodic music similarity and retrieval in query by humming. In *Intelligent computing and information and communication*, Springer (2018)
12. Velankar MR, Sahasrabuddhe HV, Kulkarni PA. Modeling melody similarity using music synthesis and perception. In*Procedia Computer Science* (2015)
13. Ahmet Elbir and Nizamettin Aydin. Music genre classification and music recommendation by using deep learning. *Electronics Letters* (2020)
14. Xinxi Wang and Ye Wang. Improving content-based and hybrid music recommendation using deep learning. In *Proceedings of the 22nd ACM international conference on Multimedia* (2014)
15. Parmar Darshna. Music recommendation based on content and collaborative approach & reducing cold start problem. In *2018 2nd International Conference on Inventive Systems and Control (ICISC)*, IEEE (2018)
16. Dip Paul and Subhradeep Kundu. A survey of music recommendation systems with a proposed music recommendation system. In *Emerging Technology in Modelling and Graphics*, Springer (2020)
17. Yading Song, Simon Dixon, and Marcus Pearce. A survey of music recommendation systems and future perspectives. In *9th International Symposium on Computer Music Modeling and Retrieval*, volume 4 (2012)
18. Greg Linden, Brent Smith, and Jeremy York. Amazon. com recommendations: Item-to-item collaborative filtering. *IEEE Internet computing* (2003)
19. Lalita Sharma and Anju Gera. A survey of recommendation system: Research challenges. *International Journal of Engineering Trends and Technology (IJETT)* (2013)
20. Diego Sánchez-Moreno, Ana B Gil González, M Dolores Muñoz Vicente, Vivian F López Batista, and María N Moreno García. A collaborative filtering method for music recommendation using playing coefficients for artists and users. *Expert Systems with Applications* (2016)
21. Sergio Oramas, Oriol Nieto, Mohamed Sordo, and Xavier Serra. A deep multimodal approach for cold-start music recommendation. In *Proceedings of the 2nd workshop on deep learning for recommender systems* (2017)
22. Juhan Nam, Keunwoo Choi, Jongpil Lee, Szu-Yu Chou, and Yi-Hsuan Yang. Deep learning for audio-based music classification and tagging: Teaching computers to distinguish rock from bach. *IEEE signal processing magazine* (2018)
23. Poonam B Thorat, RM Goudar, and Sunita Barve. Survey on collaborative filtering, content-based filtering and hybrid recommendation system. *International Journal of Computer Applications* (2015)
24. Gediminas Adomavicius and Alexander Tuzhilin. Toward the next generation of recommender systems: A survey of the state-of-the-art and possible extensions. *IEEE transactions on knowledge and data engineering* (2005)
25. Nirmal Jonnalagedda and Susan Gauch. Personalized news recommendation using twitter. In *2013 IEEE/WIC/ACM International Joint Conferences on Web Intelligence (WI) and Intelligent Agent Technologies (IAT)*, volume 3, IEEE (2013)
26. Carlos A Gomez-Uribe and Neil Hunt. The netflix recommender system: Algorithms, business value, and innovation. *ACM Transactions on Management Information Systems (TMIS)* (2015)
27. Yan Yan, Tianlong Liu, and Zhenyu Wang. A music recommendation algorithm based on hybrid collaborative filtering technique. In *Chinese National Conference on Social Media Processing*, Springer (2015)

28. Dhanashree A Phalke, Sunita Jahirabadkar, and Pune SPPU. A systematic review of near duplicate video retrieval techniques. *International Journal of Pure and Applied Mathematics* (2018)
29. Ayesha Butalia, AK Ramani, and Parag Kulkarni. Emotional recognition and towards context based decision. *International Journal of Computer Applications* (2010)
30. Shang H Hsu, Ming-Hui Wen, Hsin-Chieh Lin, Chun-Chia Lee, and Chia-Hoang Lee. Aimed-a personalized tv recommendation system. In *European conference on interactive television*, Springer (2007)
31. Dan Wu. Music personalized recommendation system based on hybrid filtration. In *2019 International Conference on Intelligent Transportation, Big Data & Smart City (ICITBS)*, IEEE (2019)
32. Ashu Abdul, Jenhui Chen, Hua-Yuan Liao, and Shun-Hao Chang. An emotion-aware personalized music recommendation system using a convolutional neural networks approach. *Applied Sciences* (2018)
33. Dan Wu. Music personalized recommendation system based on hybrid filtration. In *2019 International Conference on Intelligent Transportation, Big Data & Smart City (ICITBS)*, IEEE (2019)
34. Ferdos Fessahaye, Luis Perez, Tiffany Zhan, Raymond Zhang, Calais Fossier, Robyn Markarian, Carter Chiu, Justin Zhan, Laxmi Gewali, and Paul Oh. T-recsys: A novel music recommendation system using deep learning. In *2019 IEEE International Conference on Consumer Electronics (ICCE)*, IEEE (2019)
35. Javier Pérez-Marcos and Vivian López Batista. Recommender system based on collaborative filtering for spotify's users. In *International Conference on Practical Applications of Agents and Multi-Agent Systems*, Springer (2017)
36. Martijn Millecamp, Nyi Nyi Htun, Yucheng Jin, and Katrien Verbert. Controlling spotify recommendations: effects of personal characteristics on music recommender user interfaces. In *Proceedings of the 26th Conference on User Modeling, Adaptation and Personalization* (2018)
37. Vaibhav Khatavkar and Parag Kulkarni. Context vector machine for information retrieval. In *International Conference on Communication and Signal Processing* (2016)
38. Anagha Kulkarni, Vrinda Tokekar, and Parag Kulkarni. Discovering context of labeled text documents using context similarity coefficient. *Procedia computer science* (2015)
39. Vidhate D, Kulkarni P. Review on Context Based Cooperative Machine Learning with Dynamic Decision Making in Diagnostic Applications. In *International Conference on Computing* (2012)
40. Umberto Panniello, Alexander Tuzhilin, Michele Gorgoglione, Cosimo Palmisano, and Anto Pedone. Experimental comparison of pre-vs. post-filtering approaches in context-aware recommender systems. In *Proceedings of the third ACM conference on Recommender systems* (2009)
41. Khalid Haruna, Maizatul Akmar Ismail, Suhendroyono Suhendroyono, Damiasih Damiasih, Adi Cilik Pierewan, Haruna Chiroma, and Tutut Herawan. Context-aware recommender system: A review of recent developmental process and future research direction. *Applied Sciences* (2017)
42. Katrien Verbert, Nikos Manouselis, Xavier Ochoa, Martin Wolpers, Hendrik Drachsler, Ivana Bosnic, and Erik Duval. Context-aware recommender systems for learning: a survey and future challenges. *IEEE Transactions on Learning Technologies* (2012)
43. Gediminas Adomavicius and Alexander Tuzhilin. Context-aware recommender systems. In *Recommender systems handbook*, Springer (2011)
44. Xavier Amatriain and Justin Basilico. Recommender systems in industry: A netflix case study. In *Recommender systems handbook*, Springer (2015)
45. Christine Bauer, Marta Kholodylo, and Christine Strauss. Music recommender systems challenges and opportunities for non-superstar artists. In *Bled eConference* (2017)
46. Yueming Sun and Yi Zhang. Conversational recommender system. In *The 41st international acm sigir conference on research & development in information retrieval* (2018)

47. Kirsten Swearingen and Rashmi Sinha. Beyond algorithms: An hci perspective on recommender systems. In *ACM SIGIR 2001 workshop on recommender systems*,Citeseer (2001)
48. Joseph A Konstan and John Riedl. Recommender systems: from algorithms to user experience. *User modeling and user-adapted interaction* (2012)
49. Vincenzo Moscato, Antonio Picariello, and Giancarlo Sperli. An emotional recommender system for music. *IEEE Intelligent Systems* (2020)
50. Gawesh Jawaheer, Martin Szomszor, and Patty Kostkova. Comparison of implicit and explicit feedback from an online music recommendation service. In *proceedings of the 1st international workshop on information heterogeneity and fusion in recommender systems* (2010)
51. Francisco García-Sánchez, Ricardo Colomo-Palacios, and Rafael Valencia-García. A socialsemantic recommender system for advertisements. *Information Processing & Management* (2020)
52. Deger Ayata, Yusuf Yaslan, and Mustafa E Kamasak. Emotion based music recommendation system using wearable physiological sensors. *IEEE transactions on consumer electronics* (2018)
53. Markus Schedl, Hamed Zamani, Ching-Wei Chen, Yashar Deldjoo, and Mehdi Elahi. Current challenges and visions in music recommender systems research. *International Journal of Multimedia Information Retrieval* (2018)
54. Jiakun Fang, David Grunberg, Simon Lui, and Ye Wang. Development of a music recommendation system for motivating exercise. In *2017 International Conference on Orange Technologies (ICOT)*, IEEE (2017)
55. Himan Abdollahpouri and Steve Essinger. Towards effective exploration/exploitation in sequential music recommendation. *arXiv preprint arXiv:1812.03226* (2018)
56. Marcus O'Dair and Andrew Fry. Beyond the black box in music streaming: the impact of recommendation systems upon artists. *Popular Communication* (2020)
57. Elisabeth Lex, Dominik Kowald, and Markus Schedl. Modeling popularity and temporal drift of music genre preferences. *Transactions of the International Society for Music Information Retrieval* (2020)
58. Andrés Ferraro, Sergio Oramas, Massimo Quadrana, and Xavier Serra. Maximizing the engagement: exploring new signals of implicit feedback in music recommendations. In *Bogers T, Koolen M, Petersen C, Mobasher, Tuzhilin A, Sar Shalom O, Jannach D, Konstan JA, editors. Proceedings of the Workshops on Recommendation in Complex Scenarios and the Impact of Recommender Systems co-located with 14th ACM Conference on Recommender Systems (RecSys 2020); 25 Sep 2020; Brazil. Aachen: CEUR Workshop Proceedings; 2020.* CEUR Workshop Proceedings (2020)
59. Dominik Kowald, Markus Schedl, and Elisabeth Lex. The unfairness of popularity bias in music recommendation: a reproducibility study. *Advances in Information Retrieval* (2020)
60. Dougal Shakespeare, Lorenzo Porcaro, Emilia Gómez, and Carlos Castillo. Exploring artist gender bias in music recommendation. *arXiv preprint arXiv:2009.01715* (2020)
61. Xavier Amatriain and Justin Basilico. Past, present, and future of recommender systems: An industry perspective. In *Proceedings of the 10th ACM Conference on Recommender Systems* (2016)
62. Carson K Leung, Abhishek Kajal, Yeyoung Won, and Justin MC Choi. Big data analytics for personalized recommendation systems. In *2019 IEEE Intl Conf on Dependable, Autonomic and Secure Computing, Intl Conf on Pervasive Intelligence and Computing, Intl Conf on Cloud and Big Data Computing, Intl Conf on Cyber Science and Technology Congress (DASC/PiCom/CBDCom/CyberSciTech)*, IEEE (2019)
63. Zhihua Cui, Xianghua Xu, XUE Fei, Xingjuan Cai, Yang Cao, Wensheng Zhang, and Jinjun Chen. Personalized recommendation system based on collaborative filtering for iot scenarios. *IEEE Transactions on Services Computing* (2020)
64. Yuchen Dong, Xiaotong Guo, and Yuchen Gu. Music recommendation system based on fusion deep learning models. In *Journal of Physics: Conference Series*, IOP Publishing (2020)
65. Markus Schedl, Peter Knees, and Fabien Gouyon. New paths in music recommender systems research. In *Proceedings of the Eleventh ACM Conference on Recommender Systems* (2017)

66. Nour Nassar, Assef Jafar, and Yasser Rahhal. A novel deep multi-criteria collaborative filtering model for recommendation system. *Knowledge-Based Systems* (2020)
67. Yukun Cao and Yunfeng Li. An intelligent fuzzy-based recommendation system for consumer electronic products. *Expert Systems with Applications* (2007)

Music Recommender Systems: A Review Centered on Biases

Yesid Ospitia-Medina, Sandra Baldassarri, Cecilia Sanz, and José Ramón Beltrán

1 Introduction

The music market has undergone significant changes in recent years, mainly as a result of the digitization of sound and the emergence of new strategies adopted by the music distribution process, which has shown a transition from physical media to resources available through streaming services hosted in the cloud. The impact of this transition can be evidenced in the 2021 Global Music Report which summarizes statistical data from 2020: the global recorded music market grew by 7.4%, and global streaming revenues increased by 18.5% [22]. From the point of view of the challenges in the music market, this transition has generated new conditions in the business model, especially for artists and listeners, creating a much closer relationship for both, as well as new consumption possibilities [17, 40].

This evolution in the music market allows artist to skip many of the steps in music production and distribution processes, since nowadays musical productions can be made in a home studio and then distributed directly through the various digital platforms without any intermediaries such as distributors, aggregators, or retailers [17]. However, some musical pieces that are not produced through a

---

Y. Ospitia-Medina (✉)
Institución Universitaria Antonio José Camacho, Cali, Colombia

National University of La Plata, La Plata, Argentina
e-mail: yesid.ospitiam@info.unlp.edu.ar

S. Baldassarri · J. R. Beltrán
University of Zaragoza, Zaragoza, Spain
e-mail: sandra@unizar.es; jrbelbla@unizar.es

C. Sanz
National University of La Plata, La Plata, Argentina
e-mail: csanz@lidi.info.unlp.edu.ar

© The Author(s), under exclusive license to Springer Nature Switzerland AG 2023
A. Biswas et al. (eds.), *Advances in Speech and Music Technology*, Signals and Communication Technology, https://doi.org/10.1007/978-3-031-18444-4_4

professional recording studio under the supervision of experienced producers could reveal musical deficiencies, as well as low sound quality; consequently, the artist's chances of success in the music industry could be negatively impacted. In addition, although artists have several platforms with worldwide exposure on which to publish their songs, the fact that millions of artists have access to these platforms results in a high volume of musical content that grows constantly, generating a very high level of competition among artists to be known and heard by the public. For listeners, the current music market provides countless possibilities of consumption offered by huge music databases. This creates a major challenge in terms of how the user interacts with the platform, searches for songs, and successfully discovers songs that are to his or her liking [28].

Given the challenges explained above, music recommender systems (MRS) have emerged with the main objective of facilitating the user's navigation through large song repositories by providing suggestions for songs that are unknown to the user and that he/she will probably like [36]. Several strategies are currently available to achieve this objective, each one with its strengths and weaknesses. Some MRS implement only one strategy, while others implement various as a result of a hybrid design. The different strategies and the way they are used determine, to a certain extent, the success of the recommendation of musical pieces, so a thorough knowledge of their main characteristics and limits is very relevant. It is also important to note that, despite the significant development of MRS today, many artists, especially those who are not superstars, are dissatisfied with the actual chances of their songs being recommended in a way that can boost their commercial artistic career [4]. Similarly, in many cases, listeners express dissatisfaction with the MRS, mainly because the recommended songs do not appeal to them or because the MRS usually recommends the same group of songs, preventing the listener from discovering new musical content. This unsatisfactory treatment of data and users' preferences revealed by some recommender systems constitutes a novel research topic known as *biases*, which is of great interest to the current MRS research community.

In view of the above, the aim of this chapter is to analyze the current problems of bias that contribute to inappropriate recommendations in MRS, as well as to propose some guidelines to produce fair and unbiased systems. For this purpose, recent papers related to MRS have been selected, and for each of them, the recommendation strategies used, as well as the potential for generating biases, have been identified and analyzed. The analysis of biases is mainly focused on recommendation strategies involving a machine learning approach, such as content-based filtering, emotion-based filtering, and user-centric models, because the success of machine learning models is often highly dependent on the quantity, quality, and diversity of the data available in the datasets on which these models are trained, as well as on the strategies selected to cope with the cold-start problem. The case of non-superstar artists is one of the most interesting ones related to unfair treatment in MRS and is therefore reviewed in detail.

The main contributions of this chapter are listed below:

- An overview of the biases discussed in MRS literature, including preexisting, technical, and emergent biases.
- An analysis of MRS as a multi-objective problem with a multi-stakeholder perspective.
- A classification of related works by recommendation strategies and the kind of bias identified in each one.
- A review and analysis of the datasets available for the MIR scientific community.
- A very thorough discussion focused on understanding how biases impact MRS, taking into account the relationship among external factors, technical factors, classes of biases, and weaknesses on datasets.
- A proposal of a set of guidelines to handle biases in MRS.

This chapter is organized as follows. Section 2 presents the methodology followed in this research. Section 3 introduces music recommender systems, detailing the theoretical background of the most important recommendation strategies used nowadays, as well as providing an analysis of related works. Section 4 presents the theoretical background for biases and an analysis of related works. Section 5 analyzes some musical datasets available for the scientific community, considering a set of features that could be affected by technical biases. Section 6 proposes some guidelines to handle biases in MRS. Finally, Sect. 7 highlights the conclusions obtained from this chapter and some lines of future work.

2 Methodology

The problem studied in this chapter is related to the quality of the recommendations provided by MRS to users, which leaves a high degree of dissatisfaction among some artists who publish their songs and expect to be heard and listeners interested in accessing music content they like. To achieve a detailed analysis of this problem and propose some guidelines to mitigate its consequences, this chapter has considered five research questions which have been studied through three steps defined in the methodology: first, a literature review related to the questions; second, a discussion of the research problem based on the answers to each question; and third, the formulation of some guidelines for handling biases in MRS. So, the five research questions are:

1. What are the main characteristics of MRS?
2. How do MRS recommend new songs to users?
3. What are the principal biases in MRS?
4. What is the relationship between biases and recommendation strategies?
5. What is the relationship between biases and musical datasets?

For questions 1, 2, 3, and 4, the review process has considered papers published since 2018, although a few items outside this period have also been included because of their important contribution. The review process includes titles and abstracts over the whole literature explored, and the strings used to carry out the paper selections are **bias in computer science, music recommender systems, music recommender systems strategies**, and **bias in music recommender systems**.

For question 5, the review process has considered papers published since 2002, especially in ISMIR proceedings, taking into account the relevance of this community in providing novelty datasets. The strings used for the review process are **musical dataset, song dataset, sound dataset, emotion in music dataset**, and **audio dataset**. In this case, the review process also includes titles and abstracts over the whole literature explored.

In general, the literature review has allowed us to classify the articles into three categories: first, articles that explain the functionality, objectives, and strategies of MRS, which involves questions 1 and 2, in Sect. 3; second, articles that focus on explaining and discussing what kind of biases exist and how they affect MRS, which involves questions 3 and 4, in Sect. 4; and third, articles that present musical datasets, which involves question 5, in Sect. 5.

3 Music Recommender Systems

The main objective of an MRS is to suggest new songs to the user, and its effectiveness is measured according to the degree of acceptance (likes/dislikes) of the recommended song expressed by the user [44]. Although the majority of MRS literature is focused on the user's perspective, there are some works that highlight the importance of analyzing the point of view of the artists, which in general is determined by the real possibilities they have of promoting a commercial career [4]. Section 3.1 provides a theoretical background for MRS, and Sect. 3.2 presents a discussion of the state of the art after reviewing related works and classifying each one according to its implemented strategy.

3.1 Theoretical Background for MRS

In general, MRS combine the following elements: artists, listeners, items (songs), and recommendation strategies. Artists publish their songs on a music digital platform with the purpose of furthering their commercial careers. Listeners use the music digital platform to find specific songs that they want to listen to. They are also interested in discovering new songs that they might like. To accomplish this discovery process, MRS generate a match between a song and a specific listener through recommendation strategies [12]. The functionality and the possibilities of success in MRS are defined, in most cases, by the different recommendation strategies, which have been evolving in recent years [4, 23, 36, 50]. The strategies presented below do not define a taxonomy in the field of MRS. They are presented taking into account the different approaches found through a literature review process.

- **Collaborative filtering (CF):** CF generates automatic predictions according to a user's interests by collecting information about the preferences of a large set of users, in most cases from social networks [44].
- **Demographic filtering (DF):** DF is based on classifying the user profiles by criteria such as age, marital status, gender, etc. [50].
- **Content-based filtering (CBF):** CBF recommends songs based on their internal features, which can be low (signal-level sound features) or high level (musical features). There should be a relationship between the values of these features and the degree of acceptance by the user [36].
- **Hybrid filtering (HF):** HF works by combining the different types of filtering. In general, this approach provides better results compared to the implementation of a single type of filtering; however, it requires a detailed tuning and optimization process [12].
- **User context (UC):** UC includes any information that can be used to characterize the user's situation; this information can be obtained from different sources (personal information, information from sensors, information from the user's activity) [44].
- **Metadata (MD):** MD includes a group of data that describes the song [36]. This data can be classified by categories, for example, in [12], three groups of metadata are proposed: editorial, cultural, and acoustic.
- **Emotion-based filtering (EBF):** EBF considers the relationship between human emotions and the intrinsic features of music; based on this relationship, EBF aims to identify the values of features that generate an interest in the user and evoke a particular emotion [3, 23].
- **Personalized approach (PA):** PA consists of the design and implementation of user-centric models that allow the creation of highly effective recommendation experiences. Some authors recognize PA as a paradigm, so it could include strategies based on user interaction and context [12].
- **Playlist-based (PLB):** Playlists designed by the user include songs that are more relevant than those recommended by the system. From these lists, it is possible to study the features of the songs contained and then make recommendations based on similarity [19].
- **Popularity-based (PB):** This strategy generates a tendency to recommend songs that are commercially more famous or that have a highlighted presence in the music market through large investments in marketing strategies [14], creating a bias that affects non-superstar artists in the same market because their probability of being recommended is very low [5].
- **Similarity-based (SB):** Similarity-based systems calculate the degree of closeness between the features of one song and another; this degree of closeness is used to generate the recommendations [3].
- **Interaction-based (IB):** Interaction-based systems focus on analyzing the user's behavior in relation to the use of the system, considering aspects such as when the user generates a playback, which song is played and how many times, and the relationship between the song and particular days of the week, among others [26].

Generally, different types of logs are used to store the diverse events generated by the user in the system.

It should be emphasized that defining a taxonomy for recommendation strategies in the MRS field would be a challenge because in many of the reviewed papers, some strategies with the same goals are presented with different names. In addition, some authors present some higher-level strategies as a paradigm, for example, PA, but others simply refer to them as strategies.

3.2 Related Work on Recommendation Strategies

This section provides a review of a group of 18 MRS papers resulting from a search process in Google Scholar and a previous discussion with experts in music, affective computing, and music information retrieval to determine the search strings. For each paper, the recommendation strategies implemented or described are detailed, and the relation between each paper and the recommendation strategy implemented or analyzed is shown in Table 1.

Before analyzing in detail each work and its strategies, it is very important to highlight how the selection of the recommendation strategies varies from one work to another. This is a first finding that motivates the need to discuss and understand the convenience or suitability of each strategy in the field of MRS.

Among the different recommendation strategies, collaborative filtering (CF) is probably one of the most widely used, maybe due to its technical simplicity compared to other more sophisticated models. However, it is imperative to have a digital community and a representative flow of reliable information to deliver a minimally optimal performance. In contrast, metadata is one of the least used strategies in the reviewed works, and this occurs fundamentally when such metadata is not built automatically. There are some cases in which metadata is generated through a content-based filtering process in which, for example, sound features are automatically extracted from a song after which, using previously trained models, musical genre or evoked emotions, among others, are determined. Such cases really represent a hybrid filtering strategy and not a purely metadata strategy. Although CF are still discussed in recent papers [10, 15, 26, 44], all of them involve additional strategies so that hybrid filtering strategies are implemented.

It is very interesting that the most recent works [15, 36, 44, 51] explore strategies of personalized approaches, emotion-based filtering, content-based filtering, and user context. There is a tendency toward implementing a personalized approach to improve MRS, which is usually based on machine learning techniques since they allow the design of more dynamic strategies to generate recommendations through a learning process. This learning process is maybe the most important advantage of the machine learning approach in contrast with traditional systems that implement static rules, and it could be used to recognize emotions and make predictions based on the user's context. However, in spite of the contributions of the machine learning

Table 1 MRS literature review. *CF* collaborative filtering, *DF* demographic filtering, *CBF* content-based filtering, *HF* hybrid filtering, *UC* user context, *MD* metadata, *EBF* emotion-based filtering, *PA* personalized approach, *PLB* playlist-based, *PB* popularity-based, *SB* similarity-based, *IB* interaction-based

Year	Article	CF	DF	CBF	HF	UC	MD	EBF	PA	PLB	PB	SB	IB
2020	Shah et al. [44]	✓	–	–	–	✓	–	–	–	–	–	–	–
	Paul et al. [36]	–	–	✓	–	✓	✓	✓	✓	–	–	–	–
2019	Zheng et al. [51]	–	–	–	–	–	–	✓	✓	–	–	–	–
	Fessahaye et al. [15]	✓	–	✓	–	–	✓	–	–	✓	–	–	–
	Yucheng et al. [23]	–	–	–	–	✓	–	✓	–	–	–	–	–
	Bauer et al. [5]	–	✓	–	–	–	–	–	–	–	✓	–	–
	Andjelkovic et al. [3]	–	–	–	–	–	–	✓	–	–	–	✓	✓
	Chen et al. [10]	✓	–	–	–	–	–	–	–	–	–	–	–
	Ferraro et al. [14]	–	–	–	–	–	–	–	–	–	–	–	✓
	Katarya et al. [26]	✓	–	–	✓	✓	–	–	–	–	–	–	✓
2018	Garcia-Gathright et al. [19]	–	–	–	–	–	–	–	–	✓	–	–	✓
	Deshmukh et al. [12]	✓	–	✓	✓	✓	✓	✓	✓	–	–	–	–
	Schedl et al. [43]	–	✓	–	–	✓	–	✓	✓	✓	✓	–	✓
2017	Bauer et al. [4]	–	–	✓	–	–	✓	–	–	–	✓	–	–
2016	Cheng et al. [11]	–	–	–	✓	–	–	✓	–	–	–	–	–
	Vigliensoni et al. [50]	–	✓	–	–	✓	–	–	–	–	–	–	✓
	Katarya et al. [25]	–	–	–	–	✓	–	✓	–	–	–	–	–
2013	Bobadilla et al. [7]	✓	✓	✓	✓	–	–	–	–	–	–	–	–

approach to the most recent strategies, there is evidence of dissatisfaction from the point of view of listeners and artists [1], and the reason for this dissatisfaction is related to biases. This finding motivates an in-depth examination of the impact of biases in MRS described in the next section.

The importance of achieving an objective comparison between the different MRS has raised a great deal of interest in identifying the most appropriate evaluation process. However, the evaluation of music recommender systems is very difficult to define, since in most cases it will depend on the particular interests of the stakeholders involved in the music industry, the recommendation strategies implemented, and several other issues [43]. Moreover, considering the tendency of using recommendation strategies based on the machine learning field, many of the MRS metrics related to novelty, or serendipity of an item, are defined in terms of evaluation metrics commonly used in that field, such as accuracy, precision, recall, and root-mean-square error. In recent years, some novel metrics for the recommendation problem have emerged, and these so-called beyond-accuracy measures handle particularities of the MRS such as the utility, novelty, or serendipity of an item [24].

4 Biases in Music Recommender Systems

In its most general sense, the term bias refers to a subjective prejudice or inclination toward a particular person, thing, or idea. Bias is often unfair because, for example, it can benefit some stakeholders at the expense of others. This can even lead to discussions of morality [18]. This section about bias in MRS is divided into two subsections: Sect. 4.1 provides a theoretical background for biases, and Sect. 4.2 presents a discussion of the current state of bias in MRS.

4.1 Theoretical Background for Biases

A computer system with bias problems discriminates unfairly between some specific items by denying or decreasing the possibility of such items appearing in an interaction process between the end-user and a system [18]. In the case of MRS, a particular music recommender system that always recommends the most popular songs, and never or very rarely recommends songs produced by non-superstar artists, is a clear example of bias. It is also important to highlight the economic impact that biases can generate in a particular business model. In the case of the music industry, the problem is not only about the fame of the artist; the problem is also about the earnings that an artist can receive since the majority of music streaming services pay the artist according to the number of times his/her songs are played [4, 9]. According to Friedman and Nissenbaum, biases in computer science can be classified into three categories [18]: preexisting, technical, and emergent. These are briefly described below:

- **Preexisting biases:** Biases generated by social institutions, practices, and attitudes. This kind of bias is promoted by society, it has a direct relationship with culture, and it can be exercised explicitly or implicitly way by customers, system designers, and other stakeholders.
- **Technical biases:** Biases which have their roots in technical constraints or technical considerations. This kind of bias arises from technical limitations, which may be present in hardware, software, and peripherals. In the case of software, it is very important to analyze and deal with decontextualized algorithms, which promote unfair data processing.
- **Emergent biases:** Biases which can only be detected in a real context of use. This kind of bias appears sometimes after a design phase is completed, as a result of changing societal knowledge, population, or cultural values.

In most cases, computer systems, and science in general, try to help and improve different aspects of our life such as business models, entertainment services, health services, social policies, etc. Nevertheless, preexisting biases can produce a negative perception of computer systems from the end-user perspective because it is not clear to the end-user where and why he or she is subjected to unfair treatment. As regards technical biases, there are two relevant scenarios to analyze: first, the case in which computer systems promote biases due to several weaknesses in the design process of their algorithms and second, the case in which although technical staff have identified preexisting biases, they do not implement any technical improvement to mitigate them. Both scenarios have been considered in the works analyzed in the next section.

4.2 Related Work on Biases

This section analyzes and discusses nine papers selected for their novel contributions relevant to the discussion of biases in MRS.

The kind of bias and its relationship with each recommendation strategy addressed in each paper is shown in Table 2, revealing the following findings:

- In general, the recommendation strategies most affected by biases are collaborative (CF) and popularity-based filtering (PB).
- Seven out of nine works discuss preexisting biases, which suggests the importance of this kind of bias.
- Only one work discusses emergent bias, this being the least studied bias.

Table 2 MRS literature review focused on biases. *CF* collaborative filtering, *DF* demographic filtering, *CBF* content-based filtering, *MD* metadata, *PA* personalized approach, *PB* popularity-based, *SB* similarity-based

Year	Article	Bias	CF	DF	CBF	MD	PA	PB	SB
2020	Perera et al. [37]	Preexisting	–	–	–	–	–	✓	–
	Melchiorre et al. [30]	Preexisting	✓	✓	–	✓	–	–	–
		Emergent	✓	–	–	–	–	–	–
	Abdollahpouri et al. [1]	Preexisting	–	–	–	–	–	✓	–
		Technical	–	–	–	–	–	✓	–
	Sánchez-Moreno et al. [48]	Preexisting	✓	–	–	–	–	✓	–
		Technical	✓	–	–	–	–	✓	–
	Patil et al. [35]	Technical	–	–	✓	–	–	–	–
	Abdollahpouri et al. [2]	Preexisting	✓	–	–	–	–	✓	–
		Technical	✓	–	–	–	–	✓	–
	Shakespeare et al. [45]	Preexisting	✓	–	–	–	–	✓	–
		Technical	✓	–	–	–	–	✓	–
2019	Ferraro et al. [13]	Preexisting	–	–	–	–	✓	✓	–
		Technical	–	–	–	–	✓	✓	–
2018	Flexer et al. [16]	Technical	–	–	✓	–	–	–	✓

It is important to highlight some individual findings of each work included in Table 2, because they help to understand better how biases operate over recommendation strategies in some specific cases. A preexisting bias is identified in [37] related to the popularity-based (PB) strategy. Despite the rating of songs being influenced by marketing strategies, the rating data of the songs are used to deal with the cold-start[1] problem. In [30], both demographic filtering (DF) and collaborative filtering (CF) strategies with preexisting bias are found. In this case, the preexisting bias is generated due to a data mining process, in which the social network Twitter is the main source. In most cases, these data are incomplete and unreliable due to the inconsistency between the user's personality deduced from Twitter and the user's real behavior. There is a very close relationship between the behavior of society and Twitter, any change in society will also change the Twitter data, and any system that depends on this data will be affected in real time, which is the main feature of an emergent bias. In [1], a preexisting bias is highlighted in the popularity-based filtering (PB) strategy because the rating of songs is influenced by marketing strategies and, as a result, the number of times songs are played (play counters) by the listeners tends to rise. The recommender algorithms implemented by MRS use play counters as the main input. These algorithms do not implement any action to mitigate the popularity effect, so they also promote a technical bias. Exploiting social information from social networks is a key issue identified in [48] in the implementation of a collaborative filtering (CF) strategy based on a neighborhood similarity algorithm, which promotes a preexisting bias. The neighbors are found according to the similarity of user ratings considering only the same songs rated for both neighbors, usually the most well-known, and this reveals that the popularity-based (PB) strategy is also applied. Any other song that is not common between users is discarded by the neighborhood similarity algorithm, although there is a possibility that both users may like these songs. This discarding process generates a technical bias.

The analysis presented in [35] is focused on technical biases, especially on algorithms based on mathematical models such as singular value decomposition, Bayesian personalized ranking, autoencoders, and machine learning. These algorithms, typically implemented with content-based filtering (CBF), include a certain level of noise in their internal layers, which impacts their accuracy rates and, in most cases, negatively impacts user expectations. This paper does not analyze any preexisting bias or possible relations between preexisting bias and technical bias; this fact could be a weakness from the point of view of many stakeholders such as artists, listeners, software developers, and others, because there is not a more detailed vision of the problem that allows an understanding of the real impacts in the business model of the music industry.

According to [2], in recommender systems, a small number of items appear frequently in user's profiles, while, in contrast, a much larger number of less popular

---

[1] The cold-start problem occurs when it is not possible to make reliable recommendations due to an initial lack of ratings.

items appear very rarely. This bias has its roots in two different sources: data and algorithms. In the case of data, the rating process is based on the degree of fame of each artist (preexisting bias), which generates an imbalance property in the rating data. The algorithms are not designed to treat the imbalance property in the rating data, and they therefore over-recommend the popular items (technical bias) while at the same time reducing the chances of increasing the popularity of less popular items.

Gender discrimination, with roots in socio-cultural factors, is the main focus of the study of biases presented in [45]. There is a highly imbalanced distribution by gender according to the analysis of LFM-1b and LFM-360k datasets [8, 42], such that artists of the male gender constitute the majority (82%) of artists for whom gender can be identified.

Ferraro [13] explains how the cold-start problem in many cases is treated with strategies based on popularity ratings. This rating information depends fundamentally on data extracted from social networks, which amplify a preexisting bias. Also, Ferraro proposes to carry out a deeper analysis on the user perspective, implementing a user-centric evaluation that allows optimizing a multi-objective problem with a multi-stakeholder perspective, which will help to mitigate possible technical biases.

In [16], the ethical responsibility to produce fair and unbiased systems is proposed as a new challenge for the data mining community, highlighting the importance of reviewing and improving the conditions of datasets, as well as the design of algorithms in the machine learning field. In this paper, the most recommended songs are called hub songs, and the less recommended or never recommended songs are called anti-hubs. This classification of songs is a consequence of a weakness promoted by clustering strategies through a non-supervised machine learning approach, in which the most recommended songs are the nearest to a specific cluster center, while the less recommended songs are those farther away from the same center of the cluster.

5 Bias Analysis in Datasets

One of the most important causes of biases in MRS is related to datasets [2]. In view of this fact, this section presents a review of a group of nine datasets available for conducting research in MIR with the main purpose of understanding in-depth how biases are involved. The datasets have been analyzed taking into account the findings obtained in Sect. 4.2, which suggest that popularity is one of the most important causes of preexisting and technical biases in MRS, which not only affects listeners but also has a crucial impact on artists, especially in the case of non-superstars. This analysis of musical datasets is a novelty compared to other previous similar analyses for the following reasons: clips are evaluated to identify if they are complete songs with labeled musical structure, artists are included as relevant stakeholders, artists can emotionally label their songs, and non-superstar artists are identified. Moreover, emotional labeling of datasets is also included in this analysis, basically for two reasons: the relevance that emotion-based filtering showed in the most recent works studied in Sect. 3.2 and the strong relationship between emotion and music [46], which suggests the relevance of emotions in the design of novel musical descriptors that allow improving MRS performance.

The following is a detailed explanation of each criterion shown in Table 3:

- **Clips:** The number of audio files whose metadata has been included in the dataset.
- **Clip length:** The average duration of audio files.
- **Audio:** If the dataset is included the audio files or not.
- **Musical structure:** If the dataset includes or not metadata related to the complete identification of the musical structure of a song, the most typical structure being introduction, verse, chorus, and solo. The musical structure allows performing experiments based on the similarity of the parts of the structure, which is very useful considering that a song is an emotional experience that happens over time.
- **Affective model:** The kind of affective model used, which can be categorical or dimensional.
- **Non-superstar artist:** If the dataset includes songs produced by non-superstar artists or not.
- **Emotional labeling by artist:** If the artist has emotionally labeled their own songs or not.
- **Emotional labeling by listener:** If the listener has emotionally labeled songs or not.

A very important finding to highlight is that in seven of the nine reviewed datasets, GTZAN [49], Ballroom [21], MagnaTagATune [27], AudioSet [20], TUT Acoustic Scene [31], UrbanSound8k [41], and ESC-50 [38], the duration of the sound files varies between 1 and 30 seconds, limiting the possibilities for experimentation. In most cases, audio files with real songs are not available; instead, the audio files correspond to ambient sounds or perhaps small sound fragments with a little bit of musical content. In the case of the Million Song Dataset [6], the average length of the files is not very clear, and although it is a dataset of songs, these correspond to covers of famous songs and not original songs of non-superstar artists. In addition, the audio files are not available. The Mediaeval dataset [47] is very complete due to a large number of emotional annotations. However, there are no annotations for the different parts of the song structure (introduction, verse, chorus), and there is no data referring to a deeper analysis of the artist's perspective. The emotional recognition is performed over time by giving valence and arousal coordinates in a dimensional affective model every 500 milliseconds in [29]. Then, these coordinates are used in [34] to achieve an emotional classification in four quadrants, finding that the dataset is unbalanced with respect to the distribution of songs by quadrants. This represents a problem when implementing machine learning algorithms because these kinds of algorithms tend to recognize with more accuracy the majority class data, whereas they have a low accuracy rate with the minority classes. Another interesting point is that none of the datasets include songs

Table 3 Dataset review

Dataset	Year	Clips	Clip length	Audio	Musical structure	Affective model	Non-superstar artist	Emotional labeling by artist	Emotional labeling by listener
GTZAN [49]	2002	1000	30s	✓	✗	None	✗	✗	✗
Ballroom [21]	2006	698	≈30s	✓	✗	None	✗	✗	✗
MagnaTagATune [27]	2009	25,850	≈30s	✓	✗	None	✗	✗	✗
Million Song Dataset [6]	2011	1M	–	✗	✗	Categorical	✗	✗	✓
UrbanSound8k [41]	2014	8732	≤4s	✓	✗	None	✗	✗	✗
ESC-50 [38]	2015	2000	5s	✓	✗	None	✗	✗	✗
TUT Acoustic Scene [31]	2016	1560	30s	✓	✗	None	✗	✗	✗
Mediaeval [47]	2016	1744	≡45s	✓	✗	Dimensional	✗	✗	✓
		58	[46s, 627s]						
AudioSet [20]	2017	≈2.1M	10s	✗	✗	Categorical	✗	✗	✗

with a real artistic intention and with an interest on the part of the artist in joining the music industry, so there is no annotation of any kind by artists/composers that would allow a deeper analysis from a musical artist perspective.

Taking into account the above considerations for the case of musical recommender systems, the following particular limitations are identified in the existing datasets:

- In most cases, they include very short audio clips which are not really songs and consequently do not have a complete musical structure.
- There is no in-depth analysis from the point of view of the music involving the composer in order to understand his/her emotional intention and techniques when composing.
- The analyses are not focused on original songs by non-superstar artists, which generates a popularity bias and is nowadays one of the main objections of non-superstar artists regarding the performance of many music streaming services [4].
- There is no information available about the degree of balance of the data for the different classes defined through labeling processes.

6 Guidelines for Handling Biases in MRS

This section presents a set of guidelines for handling biases in MRS. These guidelines have been proposed taking into account the findings reported in the previous sections and could be considered for studying and mitigating the impact of biases in MER.

The guidelines detailed below have been discussed and formulated for each type of bias (preexisting, technical, and emergent).

For preexisting biases, it is important to highlight that in a traditional software development process, the functional requirements are based on a specific business model, in this case that of the music industry. The business model defines the business rules which take into account the interests and objectives of the stakeholders invited to participate in the interviews conducted by functional engineers [39]. In general, any final product will address the expectations of the stakeholders, so if some of them are not invited to participate in the product development process or are not considered relevant for the business model from the sponsors' point of view, they and their interests may not be taken into account, and this will generate a preexisting bias. According to the discussion and findings presented in the previous sections, artists who are not superstars should be considered as stakeholders for any project related to MRS. Unfortunately, the findings of this chapter suggest that there is no real interest in understanding their needs. Consequently, they are negatively affected by the popularity effect promoted by the current business models of the music industry.

In view of the above, the following guidelines are important for the handling of preexisting bias:

- Identify all the stakeholders involved in the business case and evaluate their interests.
- Analyze how the requirements and constraints defined by the business model impact each one of the stakeholders.
- Take into account the needs of all the stakeholders in the product development process.
- Maintain close communication with all stakeholders that allow making decisions focused on improving a fair treatment for all.

There are two ways of analyzing the roots of technical biases. On the one hand, there is the case where the technical bias is unavoidable because it is a consequence of a preexisting bias. The technical staff of the project follow orders and implement the business rules defined by stakeholders. On the other hand, there is the case where the conditions of the datasets, recommendation strategies, and algorithms present technical weaknesses. Considering the case of technical weaknesses, the following guidelines are important for the handling of technical bias:

- Understand in detail the business model and the data involved, as well as the point of view of each stakeholder.
- Apply a rigorous evaluation process to identify the most appropriate recommendation strategies taking into account the specific needs of the business model.
- Avoid the recommendation strategy based on popularity, especially if the business case includes non-superstar artists.
- Be careful with the information extracted from social networks to implement a recommendation strategy based on collaborative filtering. The quality of this information and the way it will be integrated into the MRS must be evaluated.
- Design metrics, or select some available metrics, to evaluate the quality of data according to the definitions of the business model and the recommendation strategies selected to implement in MRS.
- Design musical descriptors closer to the reality indicated by the artists, suggesting the importance of marking the different parts of the structure of the song and implementing recommendation strategies for each of these parts.
- Consider the emotion-based filtering strategy in MRS. To achieve better results, the music emotion recognition should be over time, taking into account that listening to music is a dynamic emotional experience given by the structure of the song [32].
- Be careful with unbalanced datasets, especially when they are involved in strategies based on the machine learning approach, because the learning process favors the majority class and consequently the predictions will show a big bias by classes. There are different ways to treat unbalanced data. One of them is implementing binary classifiers (one per class), although this does not produce a great improvement. Another way is implementing balancing strategies which could be over-sampling, under-sampling, and a combination of both [34].

- Analyze how to deal with the subjective information provided by human perception through labeling processes, especially in the case of emotional labeling, which directly impacts any future data analytic process for musical datasets [33].
- Consider always that music is an art and should be treated as such. Although music can be studied as a digital signal, there are some high-level features such as emotions and musical terms that should be understood in depth and included in any design process for recommendation strategies.

Emergent biases can only be detected in a real context of use, so a continuous monitoring process will be the key to manage them. The implementation of feedback strategies could be a good way to reveal new perceptions of unfair treatment from the point of view of each stakeholder identified in the past, as well as new stakeholders that might appear in a near future. This feedback information would serve as input for proposals for new changes in MRS to mitigate current emergent biases.

7 Conclusions and Future Work

This chapter has focused on identifying and analyzing the most typical biases in MRS. For that purpose, a literature review was carried out addressing different approaches to the study of recommendation strategies, biases, and audio datasets. The analysis identified the most recent recommendation strategies used in MRS, the kind of biases (preexisting, technical, emergent) present in each recommendation strategy, as well as the most common biases involved in musical datasets. The analysis also revealed important findings that help us to understand how and why biases are present in MRS, such as the fact that collaborative and popularity-based filtering are two of the strategies most affected by preexisting biases and that technical biases are more related to data conditions and algorithms. This detailed literature review and its subsequent analysis allowed us to propose a set of guidelines for handling biases in MRS, which will be useful for continuing improvements in the research field in MRS.

In future work, these guidelines will be considered in the design of a new recommender system for songs composed by non-superstar artists, providing a novel musical dataset for the scientific community, as well as a recommendation algorithm that mitigates the impact of biases. In addition, we will consider extending the state-of-the-art analysis with a focus on identifying and discussing the evaluation metrics available (or not available) in the articles selected for review, as well as conducting some novel experiments to analyze the magnitude of each type of bias for each dataset by applying each of the available recommendation strategies. An in-depth understanding of the relationships between preexisting biases and technical biases is considered very important and will therefore be the subject of future study.

Acknowledgments This research has been partially supported by the Spanish Ministry of Science, Innovation and Universities through project RTI2018-096986-B-C31 and the Government of Aragon through the AffectiveLab-T60-20R project. It has also been partially supported by the Computer Science School of the National University of La Plata (UNLP) through the Ph.D.

program in Computer Science and the Faculty of Engineering of the Antonio José Camacho University Institution.

References

1. Abdollahpouri, H., Burke, R., Mansoury, M.: Unfair Exposure of Artists in Music Recommendation (mar 2020), http://arxiv.org/abs/2003.11634
2. Abdollahpouri, H., Mansoury, M.: Multi-sided Exposure Bias in Recommendation. In: International Workshop on Industrial Recommendation Systems (IRS2020) in Conjunction with ACM KDD 2020 (jun 2020), http://arxiv.org/abs/2006.15772
3. Andjelkovic, I., Parra, D., O'Donovan, J.: Moodplay: Interactive music recommendation based on Artists' mood similarity. International Journal of Human-Computer Studies **121**, 142–159 (jan 2019). https://doi.org/10.1016/j.ijhcs.2018.04.004
4. Bauer, C., Kholodylo, M., Strauss, C.: Music Recommender Systems Challenges and Opportunities for Non-Superstar Artists. In: Digital Transformation – From Connecting Things to Transforming Our Lives. pp. 21–32. University of Maribor Press (jun 2017). https://doi.org/10.18690/978-961-286-043-1.3
5. Bauer, C., Schedl, M.: Global and country-specific mainstreaminess measures: Definitions, analysis, and usage for improving personalized music recommendation systems. PLOS ONE **14**(6), e0217389 (jun 2019). https://doi.org/10.1371/journal.pone.0217389
6. Bertin-Mahieux, T., Ellis, D.P., Whitman, B., Lamere, P.: The Million Song Dataset. In: Proceedings of the 12th International Conference on Music Information Retrieval (ISMIR 2011) (2011), http://millionsongdataset.com/pages/publications/
7. Bobadilla, J., Ortega, F., Hernando, A., Gutiérrez, A.: Recommender systems survey. Knowledge-Based Systems **46**, 109–132 (jul 2013). https://doi.org/10.1016/j.knosys.2013.03.012
8. Celma, Ò.: Music Recommendation and Discovery. The Long Tail, Long Fail, and Long Play in the Digital Music Space. Springer Berlin Heidelberg, Berlin, Heidelberg (2010). https://doi.org/10.1007/978-3-642-13287-2
9. Celma, O., Cano, P.: From Hits to Niches? Or How Popular Artists Can Bias Music Recommendation and Discovery. NETFLIX '08, Association for Computing Machinery, New York, NY, USA (2008). https://doi.org/10.1145/1722149.1722154
10. Chen, J., Ying, P., Zou, M.: Improving music recommendation by incorporating social influence. Multimedia Tools and Applications **78**(3), 2667–2687 (feb 2019). https://doi.org/10.1007/s11042-018-5745-7
11. Cheng, R., Tang, B.: A Music Recommendation System Based on Acoustic Features and User Personalities. Lecture Notes in Computer Science, vol. 9794, pp. 203–213. Springer International Publishing (2016). https://doi.org/10.1007/978-3-319-42996-0_17
12. Deshmukh, P., Kale, G.: A Survey of Music Recommendation System. In: International Journal of Scientific Research in Computer Science,, vol. 3, p. 27 (2018)
13. Ferraro, A.: Music cold-start and long-tail recommendation. In: Proceedings of the 13th ACM Conference on Recommender Systems. pp. 586–590. ACM, New York, NY, USA (sep 2019). https://doi.org/10.1145/3298689.3347052
14. Ferraro, A., Bogdanov, D., Choi, K., Serra, X.: Using offline metrics and user behavior analysis to combine multiple systems for music recommendation. In: Conference on Recommender Systems (RecSys) 2018, REVEAL Workshop (jan 2019)
15. Fessahaye, F., Perez, L., Zhan, T., Zhang, R., Fossier, C., Markarian, R., Chiu, C., Zhan, J., Gewali, L., Oh, P.: T-RECSYS: A Novel Music Recommendation System Using Deep Learning. In: 2019 IEEE International Conference on Consumer Electronics (ICCE). pp. 1–6. IEEE (jan 2019). https://doi.org/10.1109/ICCE.2019.8662028

16. Flexer, A., Dorfler, M., Schluter, J., Grill, T.: Hubness as a Case of Technical Algorithmic Bias in Music Recommendation. In: 2018 IEEE International Conference on Data Mining Workshops (ICDMW). vol. 2018-Novem, pp. 1062–1069. IEEE (nov 2018). https://doi.org/10.1109/ICDMW.2018.00154, https://ieeexplore.ieee.org/document/8637517/
17. Frejman, A.E., Johansson, D.: Emerging and Conflicting Business Models for Music Content in the Digital Environment. In: eChallenges e-2008. IOS Press, Stockholm (2008)
18. Friedman, B., Nissenbaum, H.: Bias in Computer Systems. ACM Transactions on Office Information Systems **14**(3), 330–347 (jul 1996). https://doi.org/10.1145/230538.230561
19. Garcia-Gathright, J., St. Thomas, B., Hosey, C., Nazari, Z., Diaz, F.: Understanding and Evaluating User Satisfaction with Music Discovery. In: The 41st International ACM SIGIR Conference on Research & Development in Information Retrieval. pp. 55–64. ACM, New York, NY, USA (jun 2018). https://doi.org/10.1145/3209978.3210049
20. Gemmeke, J.F., Ellis, D.P.W., Freedman, D., Jansen, A., Lawrence, W., Moore, R.C., Plakal, M., Ritter, M.: Audio Set: An ontology and human-labeled dataset for audio events. In: 2017 IEEE International Conference on Acoustics, Speech and Signal Processing (ICASSP). pp. 776–780 (2017). https://doi.org/10.1109/ICASSP.2017.7952261
21. Gouyon, F., Klapuri, A., Dixon, S., Alonso, M., Tzanetakis, G., Uhle, C., Cano, P.: An experimental comparison of audio tempo induction algorithms. IEEE Transactions on Audio, Speech, and Language Processing **14**(5), 1832–1844 (2006). https://doi.org/10.1109/TSA.2005.858509
22. IFPI: Global Music Report 2021. Tech. rep., IFPI, London (2021)
23. Jin, Y., Htun, N.N., Tintarev, N., Verbert, K.: ContextPlay. In: Proceedings of the 27th ACM Conference on User Modeling, Adaptation and Personalization. pp. 294–302. ACM (jun 2019). https://doi.org/10.1145/3320435.3320445
24. Kaminskas, M., Bridge, D.: Diversity, Serendipity, Novelty, and Coverage: A Survey and Empirical Analysis of Beyond-Accuracy Objectives in Recommender Systems **7**(1) (Dec 2016). https://doi.org/10.1145/2926720, https://doi.org/10.1145/2926720
25. Katarya, R., Verma, O.P.: Recent developments in affective recommender systems. Physica A: Statistical Mechanics and its Applications **461**, 182–190 (nov 2016). https://doi.org/10.1016/j.physa.2016.05.046
26. Katarya, R., Verma, O.P.: Efficient music recommender system using context graph and particle swarm. Multimedia Tools and Applications **77**(2), 2673–2687 (jan 2018). https://doi.org/10.1007/s11042-017-4447-x
27. Law, E., West, K., Mandel, M., Bay, M., Downie, J.S.: Evaluation of algorithms using games: The case of music tagging. In: In Proc. wISMIR 2009 (2009)
28. Lee, J.H., Downie, J.S.: Survey of Music Information Needs, Uses, and Seeking Behaviours: Preliminary Findings. In: ISMIR 2004, 5th International Conference on Music Information Retrieval, Barcelona, Spain, October 10-14, 2004, Proceedings. pp. 441–446 (2004), http://ismir2004.ismir.net/proceedings/p081-page-441-paper232.pdf
29. Medina Ospitia, Y., Beltrán, J.R., Sanz, C., Baldassarri, S.: Dimensional Emotion Prediction through Low-Level Musical Features. In: ACM (ed.) Audio Mostly (AM'19). p. 4. Nottingham (2019). https://doi.org/10.1145/3356590.3356626
30. Melchiorre, A.B., Zangerle, E., Schedl, M.: Personality Bias of Music Recommendation Algorithms. In: Fourteenth ACM Conference on Recommender Systems. pp. 533–538. ACM (sep 2020). https://doi.org/10.1145/3383313.3412223
31. Mesaros, A., Heittola, T., Virtanen, T.: TUT database for acoustic scene classification and sound event detection. In: 2016 24th European Signal Processing Conference (EUSIPCO). pp. 1128–1132 (2016). https://doi.org/10.1109/EUSIPCO.2016.7760424
32. Nielzen, S., Cesarec, Z.: Emotional Experience of Music as a Function of Musical Structure. Psychology of Music **10**(2), 7–17 (1982). https://doi.org/10.1177/0305735682102002
33. Ospitia-Medina, Y., Baldassarri, S., Sanz, C., Beltrán, J.R., Olivas, J.A.: Fuzzy Approach for Emotion Recognition in Music. In: 2020 IEEE Congreso Bienal de Argentina (ARGENCON). pp. 1–7 (2020). https://doi.org/10.1109/ARGENCON49523.2020.9505382

34. Ospitia-Medina, Y., Beltrán, J.R., Baldassarri, S.: Emotional classification of music using neural networks with the MediaEval dataset. Personal and Ubiquitous Computing (apr 2020). https://doi.org/10.1007/s00779-020-01393-4
35. Patil, M., Brid, S., Dhebar, S.: COMPARISON OF DIFFERENT MUSIC RECOMMENDATION SYSTEM ALGORITHMS. International Journal of Engineering Applied Sciences and Technology 5(6), 242–248 (oct 2020). https://doi.org/10.33564/IJEAST.2020.v05i06.036
36. Paul, D., Kundu, S.: A Survey of Music Recommendation Systems with a Proposed Music Recommendation System. Advances in Intelligent Systems and Computing, vol. 937, pp. 279–285. Springer Singapore, Singapore (2020). https://doi.org/10.1007/978-981-13-7403-6_26
37. Perera, D., Rajaratne, M., Arunathilake, S., Karunanayaka, K., Liyanage, B.: A Critical Analysis of Music Recommendation Systems and New Perspectives. In: Advances in Intelligent Systems and Computing, vol. 1152 AISC, pp. 82–87 (2020). https://doi.org/10.1007/978-3-030-44267-5_12
38. Piczak, K.J.: ESC: Dataset for Environmental Sound Classification. p. 1015–1018. MM '15, Association for Computing Machinery, New York, NY, USA (2015). https://doi.org/10.1145/2733373.2806390
39. PMI (ed.): A Guide to the Project Management Body of Knowledge (PMBOK Guide). Project Management Institute, Newtown Square, PA, 5 edn. (2013)
40. Rahimi, R.A., Park, K.H.: A Comparative Study of Internet Architecture and Applications of Online Music Streaming Services: The Impact on The Global Music Industry Growth. In: 2020 8th International Conference on Information and Communication Technology (ICoICT). pp. 1–6. IEEE (jun 2020). https://doi.org/10.1109/ICoICT49345.2020.9166225
41. Salamon, J., Jacoby, C., Bello, J.P.: A Dataset and Taxonomy for Urban Sound Research. In: Proceedings of the 22nd ACM International Conference on Multimedia. p. 1041–1044. MM '14, Association for Computing Machinery, New York, NY, USA (2014). https://doi.org/10.1145/2647868.2655045
42. Schedl, M.: The LFM-1b Dataset for Music Retrieval and Recommendation. In: Proceedings of the 2016 ACM on International Conference on Multimedia Retrieval. p. 103–110. ICMR '16, Association for Computing Machinery, New York, NY, USA (2016). https://doi.org/10.1145/2911996.2912004
43. Schedl, M., Zamani, H., Chen, C.W., Deldjoo, Y., Elahi, M.: Current Challenges and Visions in Music Recommender Systems Research. International Journal of Multimedia Information Retrieval 7(2), 95–116 (oct 2017). https://doi.org/10.1007/s13735-018-0154-2
44. Shah, F., Desai, M., Pati, S., Mistry, V.: Hybrid Music Recommendation System Based on Temporal Effects. In: Advances in Intelligent Systems and Computing, vol. 1034, pp. 569–577 (2020). https://doi.org/10.1007/978-981-15-1084-7_55
45. Shakespeare, D., Porcaro, L., Gómez, E., Castillo, C.: Exploring artist gender bias in music recommendation. In: 2nd Workshop on the Impact of Recommender Systems (ImpactRS), at the 14th ACM Conference on Recommender Systems (RecSys 2020). vol. 2697 (2020)
46. Sloboda, J.: The Musical Mind. Oxford University Press, New York, oxford psy edn. (apr 1986). https://doi.org/10.1093/acprof:oso/9780198521280.001.0001
47. Soleymani, M., Aljanaki, A., Yang, Y.H.: DEAM: MediaEval Database for Emotional Analysis in Music pp. 3–5 (2016), http://cvml.unige.ch/databases/DEAM/manual.pdf
48. Sánchez-Moreno, D., López Batista, V., Vicente, M.D.M., Sánchez Lázaro, L., Moreno-García, M.N.: Exploiting the User Social Context to Address Neighborhood Bias in Collaborative Filtering Music Recommender Systems. Information 11(9) (2020), https://www.mdpi.com/2078-2489/11/9/439
49. Tzanetakis, G., Cook, P.: Musical genre classification of audio signals. IEEE Transactions on Speech and Audio Processing 10(5), 293–302 (2002). https://doi.org/10.1109/TSA.2002.800560
50. Vigliensoni, G., Fujinaga, I.: Automatic Music Recommendation Systems: Do Demographic, Profiling, and Contextual Features Improve Their Performance?. In: Proceedings of the 17th International Society for Music Information Retrieval Conference. pp. 94–100. ISMIR, New York City, United States (Aug 2016). https://doi.org/10.5281/zenodo.1417073

51. Zheng, H.T., Chen, J.Y., Liang, N., Sangaiah, A., Jiang, Y., Zhao, C.Z.: A Deep Temporal Neural Music Recommendation Model Utilizing Music and User Metadata. Applied Sciences **9**(4), 703 (feb 2019). https://doi.org/10.3390/app9040703

Computational Approaches for Indian Classical Music: A Comprehensive Review

Yeshwant Singh and Anupam Biswas

1 Introduction

Over the past two decades, we have witnessed many technological advances such as improvements in data storage technology, the emergence of cloud storage, less noisy transmission, etc. Our way of creating, storing, and teaching art has also been affected by these changes. Nowadays, we have a tremendous amount of different types of data. It has led to the necessity of developing novel tools and techniques to analyze the data at scale.

Western music has been studied in research for quite a long time [50]. There are many approaches for both symbolic and content-based analysis. Each year, the number of contributions increases, as evident in Google Scholar search results shown in Fig. 1. A similar trend is observed for Indian music, where researchers in the past few years have also started focusing on and contributing approaches for tasks related to ICM. Still, the research area of computational techniques for the analysis of ICM is very young. There is a good scope of developing novel techniques and tools pertaining to ICM by leveraging the domain knowledge and learning-based methods. Over the last few years, several contributions in the low-level analysis of audio signals have started emerging, considering domain knowledge of ICM.

Extraction of helpful information, even low-level melodies like the predominant pitch, from recorded music performances is unreliable. As a result, computational methods mainly aim to retrieve the low-level melodic definitions in audio recordings and cannot extract a higher melodic representation. A low-level melody interpretation of audio data with the present state-of-the-art (SOTA) evaluation methods

---

Y. Singh (✉) · A. Biswas (✉)
Department of Computer Science and Engineering, National Institute of Technology Silchar, Silchar, India
e-mail: yeshwant_rs@cse.nits.ac.in; anupam@cse.nits.ac.in

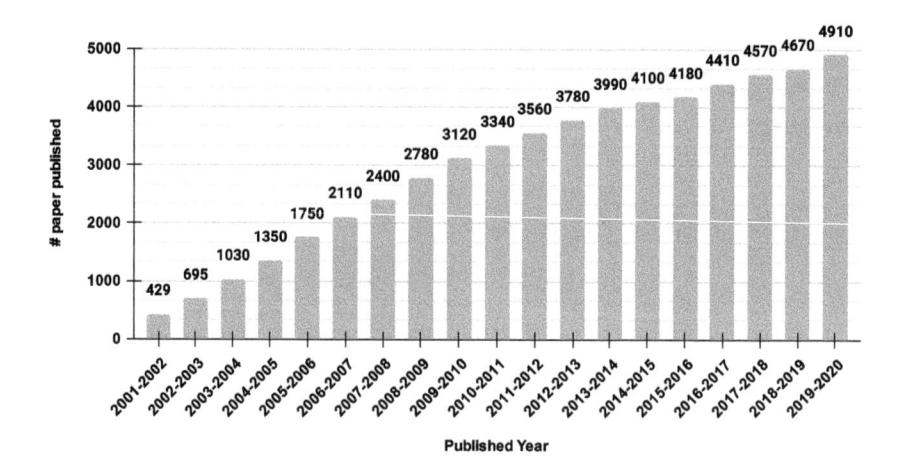

Fig. 1 Increasing trend in the research contribution in the field MIR. Source: http://scholar.google.com based on MIR keyword

for predominant pitch is made possible by some heterophonic features extraction of ICM data. The previous findings of the MIREX (international MIR evaluation campaign)[1] also suggest this by comparing the precision of different MIREX-2011 dataset algorithms obtained from INDIAN083, MIREX054, and MIREX09 0dB5. A feasible extraction of a fair predominant melody pitch in music recordings helps one concentrate on explaining the melodic aspects of musical performances to a higher degree [34]. In doing so, ICM provides an opportunity to extend the reach of melodic evaluation and interpretation by computational approaches beyond just pitch contours of music performances for explaining melodic aspects.

While obtaining the predominant pitch in ICM recordings is less challenging, getting a symbolic representation of its abstract form constitutes a difficult challenge. Transcription of melody is one way to get an abstract form by processing continuous pitch contour. It is still a complicated and unspecified task for ICM, mainly because of its meandering melodic characters. In addition, the process of discretization may lead to a loss of relevant information for defining and characterizing melodies and present challenges in its processing. Challenges arise from the absence of standard reference frequency in ICM for tuning instruments and vocals. The lead artist selects a comfortable Tonic frequency (Shadja) as the reference for all other accompanying artists and their instruments during a performance. Therefore, the Tonic differs between artists and may differ among artists' performances. These aspects make it difficult to process melodies explicitly through various musicians and performances.

---

[1] http://www.music-ir.org/mirex/wiki/MIREX_HOME.

Analysis of recurring melodic patterns has been influential in defining melodies and thus utilized in various algorithmic approaches for music information retrieval (MIR). Raag, the melodic framework of ICM, contains repeating melodic patterns. They serve as building blocks for melodies in the grammar of a Raag. ICM has numerous forms of melodic patterns with their well-defined functionality. Some patterns are melodic decorations, and some create opening stanzas of compositions that are musically related to Raags. Thus, ICM offers a fascinating chance to establish computer-based approaches for finding melodic patterns from audio collections and characterizing them. However, this task is daunting because of the changes in this artistic tradition.

We have considered the most fundamental task in ICM for our survey in this chapter. Tonic identification, melodic pattern/motif processing, and Raag identification/recognition are the essential tasks considered in this survey work. This survey aims to discover the challenge in the present techniques and motivate the upcoming researcher to focus on these research areas. The organization of this chapter is as follows: Sect. 1.1 introduces basic concept and terminology of ICM followed by some Indian regional music in Sect. 1.2. We present the literature survey in Sect. 2, which goes deep into three tasks of ICM as mentioned earlier. Tonic identification presented in Sect. 2.1 along with subdivision of methods by feature extraction, feature distribution, and Tonic selection. Then we present the works on melodic pattern processing in Sect. 2.2, breaking methods for pattern processing into melody segmentation, melody similarity, and melody representation. Finally, the works on Raag recognition are presented in Sect. 2.3), dividing each method for Raag recognition by their feature extraction and recognition method. Following the literature survey in Sect. 3, we present some of the datasets used in research for studying ICM. Evaluation matrices are another vital area of study which is essential to assess the performance of developed systems. Section 4 presents an overview of evaluation metrics used in music followed by open challenges and conclusion in Sects. 5 and 6, respectively.

1.1 Indian Classical Music

Indian classical music (ICM) refers to the classical form of music in India and neighboring countries. Majorly, there are two traditions in ICM. The tradition predominantly practiced in the northern part is called Hindustani classical music. In contrast, the tradition practiced in the southern part is known as Carnatic classical music. The Carnatic traditional style comprises short compositions, whereas the Hindustani style focuses on improvisation and exploration of Raags, making the performances quite long. However, there are more similarities than differences between the two traditions. Until the sixteenth century, these traditions were not separate. Both forms of ICM refer to singing the Swaras (called Sargam) instead of composition words with various ornamentations such as Meend, Gamak, Kan, and Khatka, as part of a Khayal style of music composition.

Raag and Taal are the two fundamental components of ICM. The Raag forms the fabric of a profoundly complicated melodic structure, built on a vast repertory of Swara and Shruti (notes and microtones). A Raag provides a palette for creating melodies from sounds, whereas the Taal provides an artistic structure for rhythmic improvisation utilizing time. The range between the Swaras in ICM is generally more significant than the Swaras themselves. Generally, ICM avoids Western classical notions and musical concepts like modulation, chords, counterpoint, and harmony.

As stated earlier, ICM is a highly developed and productive music tradition, with roots back in 1500 BC. It is a very well-explored music tradition with advanced and evolved music principles. The literature is full of theoretical text written for the musical principles of ICM. However, from a computational research standpoint, ICM is not thoroughly investigated. The current musicological work's developed music theories provide a rooted foundation for formulating MIR tasks.

Raag It is a key idea in ICM that is expressed in various ways. Despite being a notable and significant aspect of ICM, a definition of Raag, according to Walter Kaufmann [40], cannot be provided in one or two lines. Raag is a melodic structure that comprises Swara intonation, relative duration, and sequence of Swaras; in the same way, words can create expressions by creating an environment of expression. Specific regulations are mandatory in certain circumstances and optional in others. A Raag provides versatility, allowing the artist to use the primary expression or add ornamentation while still expressing the same core message but evoking a distinct emotional intensity.

A Raag is composed of Swaras on a scale (called thaat) organized into melodies with musical themes. The performance delivers a Rasa (mood, atmosphere, essence, and inner emotion) by following the distinct rules of a specific Raag. Each Raag has distinct musical characteristics like Vadi-Samvadi Swaras (the two most essential Swaras in a Raag), Aaroh-Avroh (ascending and descending sequence of Swaras), Chalan (melodic phrase), and many more. A Raag prescribes a particular sequence of the musician's progress (Pakad) from Swara to Swara. Thousands of Raags are theoretically feasible given five or more Swaras, but the ICM tradition has honed and generally relies on a few hundred in practice. The basic refined repertory of most performers consists of 40 to 50 Raags.

1.2 Forms of ICM

Regional music is mainly played and produced in specific regions and evolved through regional cultures and traditions. The audience of these regional forms of music is mostly its native people or neighboring regions. Regional music, therefore, encompasses regional folk music and traditional high culture or art music from a specific region. Regional music varies from prehistoric, sophisticated old, traditional folk to contemporary creations that strongly borrow ideas from ICM.

The context and intent are everything in regional music. Regional music is usually played at local events and brings together communities. Although it has a long history, the classical forms are highly disciplined, standardized, researched, and refined. Folk melodies and forms inspire ICM, but greater discipline and profound training are necessary for most repertories. Bollywood music has been inspired by ICM and has created numerous film songs based on various Raags and other popular modern styles of Indian music. It is intended mainly for amusement and entertainment, whereas many classical artists declare their music is for enlightenment.

2 Literature Survey

2.1 Tonic Identification

The primary step in analyzing ICM is to find out the Tonic of music performance. By looking at the tonal structure created by the Tonic, one may reasonably compare melodies between various artists and their recordings. This section examines the available Tonic identification approaches in ICM audio recordings. There were several efforts to identify a Tonic pitch automatically [4, 14, 31, 55, 66, 67]. These techniques differ primarily in the musical cues used for the Tonic identification and the style of music they focus on (Carnatic or Hindustani music, instrumental or vocal music). These techniques can be split into three basic processing blocks despite the differences. The first part is extracting tonal features, then performing feature distribution estimation, and, finally, selecting the Tonic. The single exclusion to this division of techniques is the method suggested by Sengupta et al. [67].

Tonic identification entails a tonal analysis of the music content, and the features derived by almost all techniques invariably pertain to pitch. Then in the second block, the feature distributions are evaluated by either estimating the Parzen window-based density or creating a variant of a histogram. The distribution of the feature is then utilized to detect the Tonic in the third block. The peaks in the distribution are usually the Raag Swaras or their harmonics; one of them correlates to the pitch of the Tonic. It is not always guaranteed that Tonic will be the highest peak of the distribution; thus, several approaches are used to select the peak related to the Tonic. A brief review of each of these processing block techniques is given in Table 1.

Feature Extraction Pitch-related information is extracted by all techniques from the music recordings for subsequent processing in the feature extraction. All techniques employ only one feature, the predominant pitch in the music, except Salamon et al. [66], Gulati et al. [31], and Karakurt et al. [39]. In order to utilize the tonal data of the Tonic enforcing drone instrument, Salamon et al. [66] utilize a multi-pitch salience feature. Furthermore, the multi-pitch salience feature and predominant melody are used by Gulati et al. [31].

Table 1 Survey of techniques on Tonic Identification in the context of ICM

Technique	Feature	Feature distribution	Tonic selection
Sengupta et al. [67]	Pitch [16]	NA	Error minimization
Ranjani et al. [55]	Pitch [5]	Parzen window PDE	GMM fitting
Gulati et al. [31]	Multi-pitch salience [65]	Multi-pitch histogram	Decision tree
	Predominant melody [64]	Pitch histogram	Decision tree
Bellur et al. [4]	Pitch [18]	Pitch histogram	Decision tree
Salamon et al. [66]	Multi-pitch salience [65]	Multi-pitch histogram	Decision tree
Bellur et al. [4]	Pitch [18]	GD histogram	Highest peak
Bellur et al. [4]	Pitch [18]	GD histogram	Template matching
Chordia and Senturk [14]	Pitch [18]	PCD variant	k-NN and Stat classifier
Karakurt et al. [39]	Predominant melody [64]	PD, PCD	k-NN
Manjabhat et al. [49]	Pitch [18]	PDF of pitch profile	Feedforward NN
Gaikwad et al. [25]	Pitch [5]	Pitch histogram	Highest peak
Pawar et al. [52]	Pitch [18]	Pitch histogram	Decision tree
Sinith et al. [71]	Pitch [9]	Harmonic ratio	Hardware implementation
Chapparband et al. [12]	Feature ensemble	NA	Neural network

Abbreviations: NA, not applicable; GD, group delay; PCD, pitch-class distribution; PD, pitch distribution; PDE, probability density estimate; Stat, statistical

We now present a picture of the algorithms for extracting f0 from music recordings using the several techniques discussed before. The pitch contours are obtained by Ranjani et al. [55] with Praat software14 [5]. The program uses the Boersma [6] method, which was mainly suggested for speech signals and utilized in the past for monophonic recordings. Bellur et al. [4] used the YIN pitch estimation algorithm devised by Cheveigné and Kawahara [18], which is based on the average magnitude difference function (AMDF). The automatic method for extracting the fundamental frequency f0 from a monophonic audio signal called the phase space analysis method (PSA) developed by Datta [16] is used by Sengupta et al. [67]. Similarly, Chordia and Senturk [14], Manjabhat et al. [49], Gaikwad et al. [25], and Pawar et al. [52] use various methods to extract pitch-related features.

One of the possible errors of the above estimating techniques for pitch estimation (strictly speaking f0) is that they are designed primarily for single-source monophonic sounds. It indicates that we might increase estimating errors by adding additional instruments to the mix. The predominant pitch estimation algorithm is one approach to overcome this difficulty. For the estimation of the predominant melody pitch sequence from the music recording, Gulati et al. [31] use melody's pitch information to determine the particular octave of the Tonic pitch. It is the same as in Salamon et al. [66], during the second step of their method (the first stage of the algorithm is used for identifying the Tonic pitch class).

Some approaches use a multi-pitch technique for Tonic identification. The approaches use pitch salience across time by calculating a multi-pitch time-frequency representation rather than extracting the predominant melodic part from the audio signal [65]. The reason for choosing multi-pitch analysis is twofold: firstly, the content under study is non-monophonic, and secondly, the drone instrument reinforces the Tonic constantly, which cannot be used by extracting one pitch value for each music recording.

Salamon et al. [66] and Gulati et al. [31] used the salience function that shows prominent tops in the pitch histogram, as the drone instrument is continuously reinforcing the Tonic in the performance of ICM and present in the music recordings. In Salamon et al. [66], the principal difference is to use the pitch histogram to recognize the Tonic. However, Gulati et al. [31] divide the work into two phases: firstly, with Salamon et al. [66] extension, the Tonic pitch class is identified, and then the correct Tonic octave is distinguished using predominant melody representation [31].

Feature Distribution Estimation A cumulative analysis is performed for the tonal feature by deriving different methods for Tonic identification. The pitch distribution function aggregates the pitch values from all analytical frames (a single value or multiple values per frame) and the occurrence rate (potentially weighted) of distinct pitch values in the audio extracts. The lone exception is Sengupta et al. [67], which computes the aggregate error function to pick the Tonic instead of examining the distribution of the features. Salamon et al. [66] and Gulati et al. [31] used the aggregates of the peaks' pitch value from the salience function for all frames into a histogram. The top ten peaks are used in each frame. The Tonic selection also considers the pitch content of the other accompanying instruments, notably Swaras played on the drone instrument. The selection of the frequency range of the salience peaks is limited to 100–370 Hz in the histogram (the standard frequency range is 100–260 Hz).

In some situations, the techniques mentioned above can utilize a peak of the fourth/fifth (Ma/Pa) Swaras above the Tonic to recognize the Tonic pitch. Therefore, the computation of histogram is above 260 Hz as the lead voice/instrument is, in many cases, significantly louder than the Tonic frequency reinforced by drone instruments. So while calculating the histogram, the weights of salience peaks are neglected, and only their counts are considered. As previously mentioned, the pitches created by the drone instrument (harmonics of the Tonic: Ma, Pa, or Ni) are evident by the histogram in the form of high peaks. The sound of the drone controls the accuracy of the results.

A histogram is built with a resolution of 1 Hz for 40,800 Hz frequency range and afterward post-processed with the group delay (GD) function by Bellur et al. [4]. The authors show that if the squared magnitude of the resonators is represented simultaneously by the pitch histogram, then GD functions can be used such that the peaks of the resultant histogram are better resolved. The delay function is also observed to accentuate the peaks with smaller bandwidth. It boosts the accuracy of the Tonic identifier, as the Shadja (Sa) and Pancham (Pa) are present without any pitch variations. The histograms processed are called GD histograms. The notion of segmented histograms is also proposed by Bellur et al. [4]. The authors suggest computing smaller parts of pitch contour and generating a GD histogram for each of these parts to utilize the omnipresence of Tonic. Since Tonic is present in every unit, the GD histograms improve the corresponding peak. The bins of individual histograms are then multiplied. It also helps reduce non-Tonic peaks that may not be available in all segments. The resultant histogram called the segmented GD histogram then selects the Tonic.

Parzen window estimator is utilized to calculate a pitch density function by Ranjani et al. [55] contrary to a histogram. This estimator is a non-parametric density estimator (or kernel density estimator). Kernel function regulates the regularity of the estimated density. It is commonly used to help in reducing the false discontinuities at the edges of the histogram bins and therefore help in the peak selection process. Furthermore, the separation of data into separate bins is not necessary. For the density of the retrieved pitch frequencies, the authors utilize Parzen window estimators with Gaussian kernels. Chordia and Senturk used various variants of pitch-class distribution over pitch tonal features. Similarly, Manjabhat et al. [49], Gaikwad et al. [25], and Pawar et al. [52] have used pitch tonal feature and analyzed the distribution using various distribution techniques like probability density function of pitch profiles or histogram of pitch classes. Karakurt et al. [39] have used predominant melody contours as a feature for Tonic identification and analyzed it over two distributions: pitch distribution and pitch-class distribution.

Tonic Selection Tonic selection considers the peaks extracted from the pitch distribution function that match the more frequent (or remarkable) audio signal pitches. Based on how the pitch is calculated, peaks will correspond to the Swaras of a given Raag or their harmonics. Thus, the selection of distribution peaks corresponding to the Tonic of the leading artist is reduced during the process. As previously stated, it is not necessarily the highest point in distribution that corresponds to the Tonic pitch.

Motivated by two musical indications in ICM, first, the relative Swara positions regarding the Tonic float nearby the mean ratio [44] and second, the Shadja and the Pancham are the immovable (Achal) Swaras, which signifies that they do not have any sharp or flat version [44], Ranjana et al. [55] approached the modeling of the pitch distribution using semi-continuous Gaussian mixtures [36]. Possible Tonic candidates are picked from the peaks of pitch density within an acceptable pitch field. The variation of the pitch distribution function around the peak is obtained by modeling each peak (Tonic candidate) using the Gaussian distribution. Tonic candidates have an output range of 100–250 Hz in the literature. If the editorial information of the audio sample is known prior, the size of the pitch depends further on the gender of the leading artist. 100–195 Hz is used for males and 135–250 Hz for female singers.

The identification technique used by Salamon et al. [66] is to find the peak of the multi-pitch histogram corresponding to the Tonic pitch class or Tonic pitch (in the former case). Since the Tonic is inherently related to all the prominent peaks, a collection of features are extracted to develop a classification model to identify the peak concerning the Tonic is used to compute the connections among the histogram peaks (height and distance). The authors derive distance and height features from the top ten peaks in the multi-pitch histogram. The authors show that the decision tree classifier C4.5 [54] provides the most excellent classification accuracy for the Tonic identification challenge.

In order to determine the histogram peak corresponding to the Tonic pitch class, Gulati et al. [31] utilize a similar classification-based technique. The proper Tonic octave is identified at the second step of processing and similarly classified. For each Tonic pitch candidate, 25 features are calculated (candidates with the same pitch class but different octaves). The features are the melody histogram values at a distance of 25 equally spaced places covering the Tonic pitch candidate. Whether a Tonic candidate is correct and on the proper octave, the categorization procedure represents a double problem. Using Weka's cluster software for data mining, Salamon et al. [66] trained the C4.5 decision tree for classification. Pawar et al. [52] also utilized a decision tree version and feature distribution for Tonic selection.

Error reduction approach is used by Sengupta et al. [67] to find a Tonic. It is a brute force technique in which many pitch values are examined as candidates for the Tonic pitch within a predefined frequency range. The cumulative deviation is calculated by employing three distinct tuning schemes supplied by a Tonic candidate between the pitch contour's steady-state regions and the pitch values from the nearest notes in these areas. The Tonic candidate is chosen as the Tonic for the musical extract, which results in the smallest divergence. Karakurt et al. [39] applied k-NN classifier for Tonic selection. Chordia and Senturk [13] also evaluated several Bayesian classifiers. The neural feedback network is employed by Manjabhat et al. [49] for Tonic selection from likely candidates.

The methods proposed by Bellur et al. [4] and Gaikwad et al. [25] are straightforward, and the pitch distribution is used for getting the highest peak for the Tonic selection. The bin value of the highest point of the segmented GD pitch histogram is chosen as the Tonic pitch for two out of three recommended models of their technique. The histogram's frequency range is 100–250 Hz. This range is further limited when the sex information of the leading artist is provided for an audio record. Bellur et al. [4] offer a template matching technique to find Tonic in addition to the simple highest peak strategy. It is similar to Ranjani et al. [55], who used GMM fitting, which makes use of minor variations in the pitch around the Shadja and the Pancham. The author's template for the tone candidate employs three octaves and considers Tonic pitch values and its fifth (Pa) in various octaves.

The literature proposes several approaches to Tonic identification, which vary significantly between each processing block. Many of these studies demonstrate successful results (above 90%). However, they cannot be directly compared as they are tested on diverse datasets with various measurement metrics and assessment configurations. None of these studies tried to compare the results with other research. If various approaches to musical material are to be fully understood, they have to be contrasted under one experimental setting and the same collection of music.

2.2 Melodic Pattern Processing

Processing melodic patterns in ICM audio data is another crucial task. Multiple tasks involving the calculation of melodic patterns/motifs, such as similarity matching of melodic patterns/motifs, recognition, and identification of new patterns, are meant by melodic pattern processing. Analysis of melodic patterns is a widely researched MIR and computer musicology research task. For ICM, however, this task has received attention just lately, despite the importance of melody motifs in the Raag grammar. Summarizing essential facts for comparing these techniques to current pattern processing techniques in ICM, it is observed that the following techniques deal with three closely related but separate pattern processing problems.

1. Pattern detection where the aim is to obtain other occurrences in audio test recordings when a melodic query pattern is supplied [22, 26, 28, 37, 61, 62].
2. Pattern distinction, where the goal is to find other cases from a pool of annotated melodic patterns in a query pattern [3, 56, 57].
3. Pattern discovery, where the aim is to find melodic patterns when there is a lack of ground truth melodic pattern annotations given for the collection of music recordings [21].

Pattern distinction may be seen as a sub-task to detect patterns in which the difference is the usage of melodic patterns per segment and the lack of unrelated melodic patterns in the search space. These two tasks need to be distinguished since they differ significantly in their difficulty. Furthermore, the techniques suggested that pattern distinction does not address computational complexity problems, frequently challenging pattern discovery and detection. Most of the current techniques are supervised approaches and concentrate on pattern detection or distinction. That can be ascribed particularly to the computational complexity of the problems involved in the process of pattern discovery.

By examining the current techniques for the three processing units, representing melody, segmenting melody, and computing similarities (or dissimilarities), it is observed that all techniques use a fine-grained continuous melody representation, with just a few exceptions [28, 61]. In the calculation of melodic similarities that are lost in the simple transcription of melodies, the transient melodic areas between the Swaras in the melody are discovered to be essential [17, 35]. By utilizing approaches such as symbolic approximation aggregate (SAX) [48] and behavioral symbol sequence (BSS) [77], Ross et al. [62] and Ganguli et al. [28] explore the abstraction melody's representation. Also, Ganguli et al. [28] propose a discrete melody representation that is based on a heuristic technique for quasi-melody transcription. These abstract representations have been shown to lower computing costs by an important factor. However, its precision is lower than a continuous representation of melody (taking best distance measure for both melody representation) [28, 62].

In addition, these discrete representations are assessed using a limited dataset that includes a particular singing style within the Hindustani ICM tradition.

Consequently, the usefulness of such abstract melody representation in Carnatic and Hindustani ICM traditions is dubious. Ishwar et al. [37] and Dutta and Murthy [21, 22] employ specific melodic characteristics of Carnatic music for abstract melodic representation. The authors only examine the stationary points of a continuous representation of a song (when the slope is zero). However, such a melody representation is too crude to compute a reliable melodic similarity and is thus used merely to reduce the search space and computation costs. A continuous melodic representation makes the final calculation. Overall, it is a stiff challenge for ICM to develop a melody depiction that can contain abstract melody properties that help calculate the melodic similarity. It is also noted that the most adaptable representation is the continuous melody representation, making minimum constraints on the melodic style. A brief review of each of the techniques for melody pattern processing is given in Table 2.

Melody Segmentation As previously stated, melody segmentation is a crucial part of pattern detection. In the symbolic representation of the music, many well-examined models for melody segmentation are available [8, 10, 60]. In contrast, segmentation models are not available in our awareness for ICM that work on audio directly. It leads to brute force segmentation or locally based distance alignment measures for the pattern detection in ICM, which do not need intentional melodic segmentation. For the audio recording, Ross and Rao [61] and Ross et al. [62] identify rhythmic or melodic sites to assess the position of prospective candidates to melodic patterns (Sama locations and Nyas Swara onsets). However, these techniques are unique to musical style, melodic design, and slow rhythm (Vilambit laya). For example, in a recording of Hindustani music, Sama placement can approximately identify the start of a Mukhda. However, the typical melodic phrases of Raags do not have any specific link to them.

Similarly, the method for the segmentation of Pa Nyas adopted by Ross and Rao [61] can only function with the melodic motif phrases that finish in Pa Swara and particularly in compositions with a slow tempo, which have a significant presence of the notion of Nyas Swara. Ganguli et al. [26] used a heuristic threshold-based Swara baseline for segmentation. Furthermore, it is a difficult task to recognize these markers in itself [33, 75]. Such techniques may thus not generalize and extend to various melodic patterns and vast collections of music. We note that the segmentation models for melodies in ICM lack phrase levels.

Melodic Similarity Another essential element in melodic pattern processing is the measurement of melodic similarity (or dissimilarity). Most techniques utilize a similarity measure based on dynamic programming. Ganguli et al. [28] use Smith-Waterman algorithm [72] to compute melodic similarity. Many researchers use various versions of dynamic programming-based dynamic time warping (DTW) algorithm for similarity measure [56, 57, 61, 62]. Ishwar et al. [37] and Dutta and Murthy [21, 22] employ similarity measure based on rough longest common subsequence (RLCS). The dynamic programming dominance for calculating similarity measures can be linked to the melodic patterns of ICM having significant numbers

Table 2 Summary of latest works on melodic pattern processing in the context of ICM

Technique	Task	Melody representation	Segmentation	Similarity measure	Speed-up	#Raags	#Rec	#Patt	#Occ
Ross and Rao [61]	Detection	Continuous	Pa Nyas	DTW	Seg.	1[a]	2	2	107
Ishwar et al. [3]	Distinction	Continuous	GT annotations	HMM	–	5[b]	NA	6	431
Ross et al. [62]	Detection	Continuous, SAX-12, 1200	Sama location	Euc. DTW	Seg.	3[a]	4	3	107
Ishwar et al. [37]	Detection	Stationary point, continuous	Brute force	RLCS	Two-stage method	2[b]	47	4	173
Roa et al. [56]	Distinction	Continuous	GT annotations	DTW	–	1[a]	8	3	268
Dutta and Murthy [21]	Discovery	Stationary point, continuous	Brute force	RLCS	–	5[b]	59	–	–
Dutta and Murthy [22]	Detection	Continuous	–	Modified RLCS	Two-stage method	1[b]	16	NA	59
Roa et al. [57]	Distinction	Continuous	GT annotations	DTW	–	1[a]	8	3	268
	Distinction	Continuous	GT annotations	HMM	–	5[b]	NA	10	652
Ganguli et al. [28]	Detection	BSS, transcription	–	Smith-Waterman	Dis.	34[a]	50	NA	1075
Ganguli et al. [26]	Detection	Continuous	Heuristic threshold	Smith-Waterman	Seg.	55[a]	75	NA	1754
Velankar et al. [78]	Distinction	Continuous	–	N-gram matching	–	6[a]	12	72	NA
Ganguli et al. [27]	Distinction	Continuous	Peaks of pitch histogram	Modified RLCS	Seg.	2[a]	12	NA	499
Alvarez et al. [1]	Distinction	Continuous	–	Residual LSTM	Frame Seg.	–	–	22,062	NA

[a] Hindustani music
[b] Carnatic music

Abbreviations: "–", not applicable; GT, ground truth; NA, not available; #Rec, number of recordings; #Patt, number of unique patterns; Seg, segmentation; Dis., discretization; Euc., Euclidean distance; #Occ, total number of annotated occurrences of the patterns of non-linear changes in timing, which can be further attributed to this musical tradition's improvisational nature. The computing sequence similarity (Euclidean distance) without time alignment is not significant for ICM [61].

However, there is no comprehensive comparison of the Euclidean distance with the dynamic programmatic similarity measurement for the exact music representation in the literature. Some of the current research also offers improved distance measurements. Dutta and Murthy [22] proposed to alter the intermediate stages involved in RLCS distance computation in order to make them more adaptable for melodic sequences. These changes have been claimed to enhance the system's accuracy while retaining the same reminder. However, only 59 pattern instances in 16 extracts from only 1 Raag are included in the study. Rao et al. [57] propose to learn an optimum form of the comprehensive restricted path in DTW-based distance measurement. However, as the authors describe, the learned global limitation has reduced the technique's performance. Furthermore, while limiting learning for a particular pattern category is done, this approach does not apply to unseen data, which is the case in discovering patterns.

Contrary to these time series techniques focused on matching, some approaches employ models matching statistical patterns. This task is considered comparable to the task of detecting the keyword in words by Ishwar et al. [37] and using hidden Markov models (HMMs) to conduct mainly the pattern classification. Promising outcomes are shown in the evaluations of the HMM system. However, it is relatively easy for the authors to distinguish patterns with no irrelevant pattern applicants in the search space. In addition, because the assessments have not considered a baseline system, it is left to a comparative evaluation of the HMM method to the other techniques.

Melody Representation We compare the available methods in terms of the other essential elements of melodic pattern processing. Computational complexity is a significant issue in identifying and discovering patterns for large datasets. Since, as mentioned above, similarities are mainly dependent on dynamic programming, computational complexity is also critical. The fact that these systems are not computationally impervious is owing to limited datasets utilized to test methods. The use of a narrow abstracted discrete melody representation improves the calculation efficiency of the technique by Ganguli et al. [28].

As previously noted, the performance of such a system is worse than that of a continuous representation of melody, and the scalability of such a method is doubtful to various musical materials. In addition, the performance of the melody transcription system is restricted to the correctness of such an approach. Another sort of optimization is to carry out the pattern identification task in two steps as recommended by Dutta and Murthy [22] and Ishwar et al. [37]. A rough melody representation can be utilized in the first step to determine areas in the audio recordings that most likely have the appropriate patterns. Such a crude depiction of melody dramatically decreases computation costs. After the search is separated, the second step is utilized to identify the pattern in a reliable way using a fine-grained continuous melody representation. Ishwar et al. [37] used a coarse melody representation, which uses the existence of Gamakas, that applies mainly to Carnatic music (and not Hindustani music). In addition, this approach does not theoretically calculate a lower limit, and the trimming is based on a threshold of empiric determination. This implies it is not ideal for applications like pattern detection, where it is a problem determining a musically appropriate threshold. In addition, a perfect (100%) recall cannot go to the first stage in that approach, and therefore, it might create a system bottleneck.

Other ideas to reduce computational complexity include top-down segments that use particular types of melodic and rhythmic features such as Sama and Nyas onsets [61, 62]. These events characterize the occurrences of specific sorts of musical phrases. However, as already mentioned, the method is adapted to specific melodic patterns and musical shapes in Hindustani music. In general, it is noted that there are no generalizable methods used in previous approaches to minimize the task's computing cost.

2.3 Raag Recognition

One of the essential tasks in the computational approaches for ICM is automatically recognizing the Raag of a given audio recording. The available computational techniques for automated Raag recognition in ICM audio datasets in this task depend on the test datasets' size (concerning the number of Raags). The set of Raags picked has a significant impact on the accuracy of a Raag recognition system. When the selected set of Raags in the dataset has a similar set of Swaras and are allied Raags, the task gets more complicated. Comparing absolute Raag recognition accuracies across research is not feasible since the techniques are tested on separate datasets (usually produced for a particular study and not publicly available). A brief review of the latest Raag recognition techniques is given in Table 3.

A look at some of the existing Raags recognition methods based on observing Swara set, Swara salience, Swara intonation, Aaroh-Avroh, and melodic phrases as melodic characteristics gives insights about each component of the techniques. Each technique can be analyzed in terms of standard processing blocks, including Tonic identification, feature extraction, the learning algorithm used to detect Raags, and other key dataset characteristics.

Feature Extraction It is observed that the set of Swaras is the most commonly used melodic attribute in a Raag. The Swara set is also one of the most basic features to extract (computationally). In both the implicit and explicit ways, many different approaches considered the Swara set as a feature for Raag recognition. In an audio recording, Ranjani et al. [55] and Chakraborty and De [11] explicitly extract the constituent set of Swaras. The Raag from the recordings is determined by comparing the estimated Swara set to the stored set for each Raag. The papers do not specify the specific technique to map the predicted Swara set to a unique Raag label. As one

Table 3 Summary of works on Raag recognition

Technique	Tonal feature	Tonic identification	Feature	Recognition method	#Raags	Dataset (Dur/Num)[a]	Audio type
Pandey et al. [51]	Pitch [5]	NA	Swara sequence	HMM and N-gram	2	–/31	MP
Chordia and Rae [13]	Pitch [76]	Manual	PCD, PCDD	SVM classifier	31	20/127	MP
Belle et al. [2]	Pitch [58]	Manual	PCD (parameterized)	k-NN classifier	4	0.6/10	PP
Shetty and Achary [70]	Pitch [73]	NA	#Swaras, Vakra Swaras	NN	20	–/90	MP
Sridhar and Geetha [74]	Pitch [47]	Singer identification	Swara set	String matching	3	–/30	PP
Koduri et al. [41]	Pitch [59]	Brute force	PCD	k-NN classifier	10	2.82/170	PP
Ranjani et al. [55]	Pitch [5]	GMM fitting	PDE	SC-GMM and set matching	7	–/48	PP
Chakraborty and De [11]	Pitch [68]	Error minimization	Swara set	Set matching		–/	
Koduri et al. [42]	Predominant pitch [66]	Multi-pitch-based	PCD variants	k-NN classifier	43	–/215	PP
Chordia and Senturk [14]	Pitch [9]	Brute force	PCD variants	k-NN and stats. classifiers	31	20/127	MP
Dighe et al. [19]	Chroma [46]	Brute force (Vadi based)	Chroma, timbre features	HMM	4	9.33/56	PP
Dighe et al. [20]	Chroma [46]	Brute force (Vadi based)	PCD variants	RF classifier	8	16.8/117	PP

(continued)

Table 3 (continued)

Technique	Tonal feature	Tonic identification	Feature	Recognition method	#Raags[b]	Dataset (Dur/Num)[a]	Audio type
Koduri et al. [43]	Predominant pitch	Multi-pitch-based	PCD (parameterized)	Different classifiers	45[b]	93/424	PP
Kumar et al. [45]	Predominant pitch	Brute force	PCD + N-gram	SVM classifier	10	2.82/170	PP
Dutta et al. [23][b]	Predominant pitch	Cepstrum-based	Pitch contours	LCSS with k-NN	30[d]	3/254	PP
Gulati et al. [32]	Predominant pitch	NA	TDMS	k-NN	40	124.5/480	PP
Manjabhat et al. [49]	PDF of pitch profile	NA	Pitch histogram	NN, GMM, decision tree	30[d]	–/538	PP
Chowdhuri [15]	Chroma	NA	Normalized chroma	Bi-LSTM	30	116.2/300	PP
John et al. [38]	Predominant pitch	Multi-pitch-based	Pitch contours	CNN	5	–/40	PP
Sinith et al. [71]	Pitch	NA	FSPD	HMM	17	11.33/340	MP
Gawande [29]	Spectral representation	NA	STFT	CNN + LSTM	30	116.2/300	PP
Roy et al. [63]	Stats. physics-based distribution	NA	Rank frequency distribution	N-τ plane	3	–	PP

[a] Only the largest dataset is mentioned in case of multiple datasets
[b] Technique performs Raag verification
[c] Authors conduct experiments on 3 Raags per experiment out of 45 Raags
[d] Authors only use 17 Raags in experiments

Abbreviations: NA, not applicable; "–", not available; Dur, duration of the dataset in hours; Num., number of recordings; PCD, pitch-class distribution; RF, random forest; SC-GMM, semi-continuous GMM; NN, neural network; MP, monophonic; PP, polyphonic

might expect, this is a naive method because numerous Raags share the same set of Swaras but differ in more complex melodic and temporal aspects.

One method to distinguish between Raags that share a standard set of Swaras is to examine the importance of the Swaras that make up the Raags in the analysis. The Swara set feature is implicitly included when computing Swara salience for all potential Swara frequencies. Therefore, this characteristic is virtually included in all of the methods. Chordia and Rae [13] offer a feature for identifying Raags that combines the salience of distinct Swaras in song. The authors use a 12-bin pitch-class distribution (PCD) calculated as a histogram of the pitch sequence to describe Swara saliences. This global feature has been demonstrated to work effectively on a large dataset and is resistant to pitch octave errors. For computing PCD, Chordia and Rae [13] implicitly consider the duration of the Swaras in a melody for estimating their salience. Koduri et al. [41] investigate two distinct ways to calculate PCD, each with a different interpretation of Swara salience. Chordia and Rae [13] use one of their recommended techniques to weigh the salience by the duration of the Swaras. The alternative method evaluates the salience of Swaras based on their frequency of recurrence, regardless of their length. According to observation, the earlier method yielded more accuracy.

The pitch distribution utilizing a finely seeded bin boundary expands the 12-bin PCD feature above. The intonation elements of the Swaras additionally include a high-resolution PCD with the Swara saliences. Such a fine-grained PCD is utilized in Kumar et al. [45], Belle et al. [2], Koduri et al. [42], and Chordia and Şentürk [14]. Through high-resolution PCD, these studies claim improved performance compared to a 12-bin PCD. The pitch distribution using the kernel density estimation approach (KDE), a version of the PCD feature, was utilized by Ranjani et al. [55], and it further increases the accuracy of Raag recognition. Chordia and Şentürk [14] and Ranjani et al. [55] refer to these variations as probability density estimates (PDE) and kernel density pitch distribution (KPD). Some of the techniques [32, 71] skip Tonic identification and develop Tonic invariant methods for feature extraction.

The above-stated high-resolution PCD feature inherently captures some features of the Swara intonation in a song. However, it is not easy to regulate the prominence of the particular intonation characteristics in the PCD feature space. In order to solve this issue, Koduri et al. [43] and Belle et al. [2] proposed to utilize a parametrized PCD version for each Swara in the melody. The following characteristics are extracted by Belle et al. [2] for each Swara: top position, mean position, variance, and overall probability. Similarly, Koduri et al.[43] extract six characteristics for each Swara: the peak location, peak amplitude, mean, variance, skewness, and kurtosis for Swara distribution. Manjabhat et al. [49] have used pitch histogram of probability density function's values for pitch profiles. Gulati et al. [32] developed a time delay melodic surface (TDMS) novel feature with capturing both tonal and temporal characteristics of melody contours. Similarly, Sinith et al. [71] have developed Fibonacci series-based pitch distribution for feature representation.

Carnatic music melodies feature Gamakas, where a pitch variation can reach up to 200 cents, even when presenting a single Swara. It is necessary to determine which Swara is played in a melody at a particular stage to capture the intonation features in such situations. Koduri et al.[43] also proposed an alternative technique to categorize the pitch contours depending on the melodic context for the calculation of a context-based Swara distribution. Subsequently, parametrization is done in this context-based Swara distribution to extract six characteristics. Koduri et al. [43] claim, in the Raag recognition challenge, features derived using context-based Swara distribution achieve higher performance. However, a disadvantage of the techniques mentioned is that the temporal aspects of the melody are not considered, which are essential for the characterization of Raag.

Recognition Method Many techniques exist that capture the temporal characteristics of melody statistically, mainly through modelling the Raags [13, 19, 45, 51, 74]. There is also the Aaroh-Avroh progression. Some of them calculate Swara sequence and apply HMM and n-gram approaches to represent time aspects [19, 45, 51, 71]. Some techniques calculate a Swara transition representation capturing temporal aspects, such as Chordia and Rae [13] using pitch-class dyad distribution (PCDD) with a SVM classifier. Shetty and Achary [70] use the Swara combination feature with neural network classifier for Raag recognition. Few of the techniques mentioned also include characteristic melodic phrases of Raags [51, 74]. For each Raag, they maintain a dictionary containing patterns of predefined melodic and detect events in a Swara sequence derived from the tests. However, the scalability of these techniques is unclear as the dataset of just two to three Raags has been analyzed.

The above approaches generally employ a discrete melody representation by simply measuring the predicted pitch contours at the level of Swara or utilizing more advanced melodic transcription techniques slightly [51]. Ranjani et al. [55] and Chakraborty and De [11] extract the constituent set of Swaras and use a set matching algorithm for recognition explicitly. As automated melodic transcription in ICM remains a challenge and a relatively unspecified process [79], this phase may bring mistakes that further propagate the final precise techniques and affect them. However, the impact of the mistakes in the steps of melody transcription on absolute correctness still has to be formally assessed quantitatively. Besides the problems associated with the transcript of melodies, the methods mentioned above are further restricted. There is a lack of features capturing the continual melodic transitions across the Swaras. Chalan of Raag describes the development of the melody from Swara to Swara (continual melodic changes) and is a distinctive feature. In the techniques that employ a continuous melody representation and use melody patterns to recognize Raags, these tiny seeds of the temporal elements of melody are examined. However, not many techniques follow this methodology due to the difficulty of identifying melodic patterns in continuous melody representation. The Raag is verified by Dutta et al. [23] utilizing the automatically found melodic patterns from certain parts (Pallavi lines) of the Carnatic music. Several methods use k-NN as an algorithm for classification [2, 14, 23, 32, 41]. A Raag verification system presupposes that a certain Raag is identified and examines whether the identified Raag is valid or not by the system. Thus, Raag testing may be seen as a subset of recognition of Raag with reduced task complexity.

All other techniques utilize pitch as a key to Raag recognition, except for the methods proposed by Dighe et al. [19, 20] and Chowdhuri [15]. There are several pitch estimate techniques in existing approaches for Raag recognition. Some algorithms [64] are mainly intended to operate with polyphonic audio material, although some are mostly suited for [5] monophonic speech signal. Therefore, a mistake in the system might result in an erroneous estimate of the pitch of the audio signals. However, as some of the approaches are assessed using monophonic content of audio, and the remaining approaches have polyphonic content, the results provided in the research are hard to conclude. A comparative assessment of a shared dataset will be necessary to verify this notion. As mentioned above, Dighe et al. [19, 20] techniques are the two exceptions for utilizing the pitch feature for Raag recognition and utilizing a 12-bin chroma feature along with HMM and RF classifier to complete the task. The chroma feature is commonly utilized for identifying key and mode tasks. All the audio signals' tonal components are included in the chroma calculation. For ICM, the sound of Tabla and Tanpura, which are typically in the backdrop, would mean strengthening the essential Swara Sa. Thus, the technique's performance can be reduced if the tonal components are disconnected from the underlying Raag. However, since none of the experiments are compared with a Raag recognition system which is pitch-based [19, 20], we cannot conclude without comparative evaluation.

An important step in recognizing Raag is to make the technique invariant to the Tonic pitch of an audio recording main artist. Many studies in recognition of Raag are either do a Tonic normalization by identifying their values manually or consider the performance that is set in a predetermined Tonic pitch [2, 13, 51, 70]. These approaches cannot be scaled to actual collections in real circumstances or another. According to the artists and their recordings, the Tonic pitch changes, making it cumbersome to extract information manually. Several techniques either use an external automated Tonic identification module [38, 42, 43] to overcome this restriction, or they explicitly identify a Tonic pitch before Raag is recognized [11, 55].

Another way of estimating the Tonic pitch and Raag of a recording together requires following a brute force technique [14, 41, 45], where distinct feature candidates that correspond to all potential Tonic values are examined (generally measured against the prominent Swara pitch values of the melody). The candidate with the best match is utilized to deduce both the label of Raag and Tonic. However, the knowledge of a trustworthy Tonic pitch in advance leads to considerably superior performance compared to a brute force joint estimate as demonstrated by Chordia and Şentürk [14]). It shows that an additional module that can consistently detect Tonic pitch may significantly enhance the Raag recognition performance. It might be beneficial for Tonic identification to utilize an external module as the corresponding acoustic characteristics might differ in assessing Tonic pitch and Raag. For example, the drone background Tanpura does not directly comply with the Raag recognition, but information may be used in recording [30] to accurately identify the pitch of the Tanpura. Some new methods take advanced architectured classifiers from deep learning and use them for recognition. Chowdhuri [15] used bi-directional long short-term memory (Bi-LSTM) classifier, while John et al. [38] used a convolution neural network-based classifier for recognition. Manjabhat et al. [49] used several classifiers over the pitch histogram for the recognition task.

A broad and sizeable music dataset generally lacking in the previous works on recognition of Raag is a crucial component of any data-driven research. For the recognition of Raag, several techniques are offered that employ several tonal characteristics and education methods. As observed, the number of Raags selected, the duration taken, and the numbers of the recordings of the audio content and its type vary considerably by existing techniques (monophonic or polyphonic). It is not easy to make definite conclusions with such heterogeneous datasets on the performance of the techniques in various research. In the same dataset and the same experimental configuration, even those datasets in the survey studies like in the Koduri et al. [41] did not conduct an extensive comparison. Therefore, it is possible to create various, large, and sharable datasets indicative of musical tradition. Another common factor in several current studies is the insufficient description of implementation details, in addition to the datasets. This issue gets much more challenging because no technique has provided code to the public to ensure that the research outcomes are reproducible.

3 Datasets

3.1 CompMusic Research Corpora

Dunya music corpus has been collected as a part of the CompMusic project by the Music Technology Group (MTG) [69]. The audio music files are collected from commercial quality audio compact discs (CDs) and stored in the mp3 format of 160 kbps. The metadata for each recording is stored at MusicBrainz [24]. This dataset contains approximately 400 CDs and 3500 tracks. The total time is roughly 800 hrs, which comprises Carnatic, Hindustani, and three other forms of music. The editorial metadata accompanying the datasets is carefully curated and verified labels. This dataset comprises approximately 600 male and female artists, with all the popular instruments. Parts of the corpus are open-sourced under a Creative Commons license. The recording has been done with the stereo channel with a sampling rate of 22.05 kHz with 16 bit PCM.

- **Carnatic Raag Dataset** The Carnatic test dataset is a part of the Dunya collection that focuses on the Carnatic style of ICM. The dataset contains 124.5 hrs of recording of 40 different Raags. Each Raag category has 12 music samples.
- **Hindustani Raag Dataset** Another test dataset from the collection is the Hindustani dataset. It focuses on the northern and middle parts of the ICM style. It has 116.2 hrs of recording of 30 Raags. Each Raag has ten recording samples.
- **Indian Music Tonic Dataset** The dataset comprises six small Tonic datasets and contains audio samples and annotations made manually of the leading artist's

Tonic pitch for each audio sample. Every sample accompanies the corresponding editorial metadata. A bulk of these datasets originate from the CompMusic corpora of Indian art music, which has an MBID for each recording. Other data may be accessed using the MBID with the Dunya API.[2]

- **Saraga: Research Datasets of Indian Art Music** This collection [7] is currently the most extensive annotated open data collection available for computational research on Indian art music. They comprise audio, editorial metadata, manual, and automatically extracted annotations for different aspects of melody, rhythm, and structure. The dataset can be accessed through the PyCompMusic API.[3]

4 Evaluation Metrics

Performance evaluation of the music-based systems is performed from two perspectives that are objective evaluation and subjective evaluation. Objective evaluation measures the performance of a system based on a chosen metric or set of metrics. At the same time, subjective evaluation is based on human subjectivity and preferences. This type of evaluation is very easy to be biased, and utmost caution is needed to perform them. They are rarely observed in available approaches for ICM.

4.1 Objective Evaluation

The following performance assessment measures are used in the literature to evaluate the effectiveness of the different approaches:

Let vector **g** and **t** be the one-dimensional estimated frequency sequence of melodic pitches and the sequence for the ground truth. Let **b** be a voicing indicator vector of which ith element is $b_i = 1$ when the frame is determined to be voiced (where the melody is present) with a matching ground truth **c**. The $\hat{b}_i = 1 - b_i$ represents un-voicing.

1. **Voice Recall (VR)**: The ratio of the frames is described as a voiced frame with the melodic frame of ground truth.

$$VR = \sum_i b_i c_i / \sum_i b_i$$

2. **Voicing False Alarm (VFA)**: The percentage of incorrectly computed frames as melodic frames with the non-melodic frames is called voicing false alarm.

---

[2] https://dunya.compmusic.upf.edu/.
[3] https://github.com/MTG/pycompmusic.

$$VFA = \sum_i b_i \hat{b}_i \Big/ \sum_i \hat{b}_i$$

3. **Raw Pitch Accuracy (RPA):**
   Part of the correct pitch of frames is correctly identified as melodic frames and pitch-correct but un-pitched with the frames that are melodic frames of the ground truth.

$$RPA = \Big\{\sum_i b_i \tau[\zeta(g_i) - \zeta(t_i)]\Big\} \Big/ \sum_i \hat{b}_i$$

where the τ describes the threshold feature:

$$\tau[a] = \begin{cases} 1 \text{ if } |a| \leq 50 \\ 0 \text{ if } |a| > 50 \end{cases}$$

Moreover, ζ maps a motivated frequency value axis where every semitone is divided into 100 cents. Frequency can be defined as a significant number of cents above the reference frequency g_{ref}.

$$\zeta(g) = \log_2 g / g_{ref}$$

4. **Overall Accuracy (OA):**
   Overall accuracy is nothing, except that melody and pitch are accurately marked with the frames. If L is the total frame number, the OA is

$$OA = \frac{1}{L} \sum_i c_i \tau[\zeta(g_i) - \zeta(t_i)] + \hat{b}_i \hat{c}_i$$

4.2 Subjective Evaluation

Users listen to and assess the whole impression based on their feeling of the existing music system, i.e., the subjectivity of the user toward music. Some subjective assessment metrics [53] are:

1. **Scoring**
   A click is deemed "hit" if it took place inside a specified time of the expert-defined boundary region (EDBR). We suppose that each "hit" is the response of the listener to the EDBR rather than an early or later EDBR getting a late response. In addition, if clicks fall shortly before or after the EDBR, they may be construed as anticipated or delayed reactions; they are not considered "hits." A "false alarm" was instead regarded as a click when it occurred in the "interstitial"

space, i.e., the off-boundary area between EDBRs. For hits, it was quantified and evaluated as a "promptness" of reaction concerning the start of the EDBR.

2. **Signal detection theory measures**
The hit and false alarm rates (HR, FAR) (as defined by their number of hits and false alarms, divided by the total number of "signal present" or "absent" trials (EDBR and non-EDBR)) are calculated to compare the ability of the listener to identify the diverse hierarchical grouping levels. Participant-specific computations for HR and FAR were performed independently for level 1 (section-level) and level 2 (phrase-level) EDBRs throughout the two hearings. Based on these rates, they also measured the general capacity of the participant to detect phrase boundaries throughout the Alap using a sensitivity index, described traditionally as

$$d' = z(\text{HR}) - z(\text{FAR})$$

3. **Promptness**
Each participant's assessment is performed based on how quickly they recognized the ending of the current phrase for every hit in each repeat. To do this, a "promptness" score is created for every particular hit that varied from 1 (if the click happened early at the EDBR, i.e., prompt answer) to 0 (if it occurred at the very end, a delayed response). They calculated either a $(a * x + b)$ linear score or a (c/x) reciprocal function with 0-1 values. These findings were comparable; therefore, they maintained the linear definition. For promptness measurement, just the initial click inside each EDBR was employed. The speed ratings were calibrated individually throughout the hits of EDBRs at each of the levels in order for the participants to determine if they replied more quickly to level 1 than level 2 or vice versa. For each participant in levels 1 and 2, the cumulative results for section and phrase awareness are reported.

4. **Repetition (listening) number**
All the analyses (average or total) of sample 1 and sample 2 were carried out because the discrepancies in the answers of the two listen in all crucial steps were modest. Due to its resilience to outliers, they have chosen the medium instead of the mean as a measure of central tendency (using means yielded similar results in this case) for the relationships of participant-wise performance (HR, FAR, d', and promptness) over hearings.

5 Open Challenges

Based on the literature survey of current approaches for computational analysis for ICM, we have found the following issues present in the research that are needed to be addressed:

- Most of the approaches for various tasks work on a few data samples that are not conclusive. Furthermore, those approaches that work on the sizable dataset are not publicly available most of the time. Sizable standard open-licensed datasets need to be prepared to increase models' quality and performances.
- The approaches focus heavily on extracting various features to solve a problem at hand. The extracted feature may or may not comprise the necessary latent patterns in its representation of the music samples. Low-level spectral representations of music need to be explored along with advances in deep learning for developing end-to-end automated systems.
- Current Swara (note) transcription systems are very raw and perform poorly over live concert recordings. The systems need to be improved by incorporating music segmentation and melodic pattern detection advances.
- None of the current approaches focus on performing multiple tasks concerning ICM simultaneously, and they only focus on a single task at hand. Advances in hardware and pattern learning have allowed the researcher to build systems that perform multiple operations simultaneously. So, multi-task learning needs to be explored in the ICM domain.

6 Conclusion

In this chapter, we performed a comprehensive survey of the current approaches for ICM computational analysis. All the approaches are analyzed from the perspective of three major tasks in ICM: Tonic identification, melodic pattern processing, and the most studied task, Raag recognition. In each task, approaches are broken down into logical processing blocks, and their techniques are elaborated for each block. In the end, we discussed the challenges still present in the research and future directions to improve the overall computational analysis field for ICM.

Acknowledgments This research was funded under grant number ECR/2018/000204 by the Science and Engineering Research Board (SERB), Department of Science and Technology (DST), of the Government of India.

References

1. Alvarez, A.A., Gómez, F.: Motivic pattern classification of music audio signals combining residual and lstm networks. International Journal of Interactive Multimedia & Artificial Intelligence **6**(6) (2021)
2. Belle, S., Joshi, R., Rao, P.: Raga identification by using swara intonation. Journal of ITC Sangeet Research Academy **23**(3) (2009)
3. Bellur, A., Ishwar, V., Murthy, H.A.: Motivic analysis and its relevance to raga identification in carnatic music. In: Serra X, Rao P, Murthy H, Bozkurt B, editors. Proceedings of the 2nd CompMusic Workshop; 2012 Jul 12–13; Istanbul, Turkey. Barcelona: Universitat Pompeu Fabra; 2012. p. 153–157. Universitat Pompeu Fabra (2012)

4. Bellur, A., Ishwar, V., Serra, X., Murthy, H.A.: A knowledge based signal processing approach to tonic identification in indian classical music. In: Serra X, Rao P, Murthy H, Bozkurt B, editors. Proceedings of the 2nd CompMusic Workshop; 2012 Jul 12–13; Istanbul, Turkey. Barcelona: Universitat Pompeu Fabra; 2012. p. 113–118. Universitat Pompeu Fabra (2012)
5. Boersma, P.: Praat, a system for doing phonetics by computer. Glot. Int. **5**(9), 341–345 (2001)
6. Boersma, P., et al.: Accurate short-term analysis of the fundamental frequency and the harmonics-to-noise ratio of a sampled sound. In: Proceedings of the institute of phonetic sciences. vol. 17, pp. 97–110. Citeseer (1993)
7. Bozkurt, B., Srinivasamurthy, A., Gulati, S., Serra, X.: Saraga: research datasets of indian art music (May 2018). https://doi.org/10.5281/zenodo.4301737
8. Bozkurt, B., Karaosmanoğlu, M.K., Karaçalı, B., Ünal, E.: Usul and makam driven automatic melodic segmentation for turkish music. Journal of New Music Research **43**(4), 375–389 (2014)
9. Camacho, A.: SWIPE: A sawtooth waveform inspired pitch estimator for speech and music. University of Florida Gainesville (2007)
10. Cambouropoulos, E.: Musical parallelism and melodic segmentation:: A computational approach. Music Perception **23**(3), 249–268 (2006)
11. Chakraborty, S., De, D.: Object oriented classification and pattern recognition of indian classical ragas. In: 2012 1st International Conference on Recent Advances in Information Technology (RAIT). pp. 505–510. IEEE (2012)
12. Chapparband, M., Kulkarni, M.G., Sameeksha, D., Krishna, A.V., Bhat, A.: Shruti detection using machine learning and sargam identification for instrumental audio. In: Advances in Speech and Music Technology, pp. 145–156. Springer (2021)
13. Chordia, P., Rae, A.: Raag recognition using pitch-class and pitch-class dyad distributions. In: ISMIR. pp. 431–436. Citeseer (2007)
14. Chordia, P., Şentürk, S.: Joint recognition of raag and tonic in north indian music. Computer Music Journal **37**(3), 82–98 (2013)
15. Chowdhuri, S.: Phononet: multi-stage deep neural networks for raga identification in hindustani classical music. In: Proceedings of the 2019 on international conference on multimedia retrieval. pp. 197–201 (2019)
16. Datta, A.: Generation of musical notations from song using state-phase for pitch detection algorithm. J Acoust Soc India XXIV (1996)
17. Datta, A., Sengupta, R., Dey, N., Nag, D.: A methodology for automatic extraction of meend from the performances in hindustani vocal music. Journal of ITC Sangeet Research Academy **21**, 24–31 (2007)
18. De Cheveigné, A., Kawahara, H.: Yin, a fundamental frequency estimator for speech and music. The Journal of the Acoustical Society of America **111**(4), 1917–1930 (2002)
19. Dighe, P., Agrawal, P., Karnick, H., Thota, S., Raj, B.: Scale independent raga identification using chromagram patterns and swara based features. In: 2013 IEEE International Conference on Multimedia and Expo Workshops (ICMEW). pp. 1–4. IEEE (2013)
20. Dighe, P., Karnick, H., Raj, B.: Swara histogram based structural analysis and identification of indian classical ragas. In: ISMIR. pp. 35–40 (2013)
21. Dutta, S., Murthy, H.A.: Discovering typical motifs of a raga from one-liners of songs in carnatic music. In: ISMIR. pp. 397–402 (2014)
22. Dutta, S., Murthy, H.A.: A modified rough longest common subsequence algorithm for motif spotting in an alapana of carnatic music. In: 2014 Twentieth National Conference on Communications (NCC). pp. 1–6. IEEE (2014)
23. Dutta, S., PV, K.S., Murthy, H.A.: Raga verification in carnatic music using longest common segment set. In: ISMIR. vol. 1, pp. 605–611. Malaga, Spain (2015)
24. Foundation, M.: Musicbrainz - the open music encyclopedia. https://musicbrainz.org/, (Accessed on 07/14/2021)
25. Gaikwad, C.J., Kodag, R.B., Patil, M.D.: Tonic note extraction in indian music using hps and pole focussing technique. In: 2019 10th International Conference on Computing, Communication and Networking Technologies (ICCCNT). pp. 1–5 (2019). https://doi.org/10.1109/ICCCNT45670.2019.8944486

26. Ganguli, K.K., Lele, A., Pinjani, S., Rao, P., Srinivasamurthy, A., Gulati, S.: Melodic shape stylization for robust and efficient motif detection in hindustani vocal music. In: 2017 Twenty-third National Conference on Communications (NCC). pp. 1–6 (2017). https://doi.org/10.1109/NCC.2017.8077055
27. Ganguli, K.K., Rao, P.: A study of variability in raga motifs in performance contexts. Journal of New Music Research **50**(1), 102–116 (2021)
28. Ganguli, K.K., Rastogi, A., Pandit, V., Kantan, P., Rao, P.: Efficient melodic query based audio search for hindustani vocal compositions. In: ISMIR. pp. 591–597 (2015)
29. Gawande, K.: Raga recognition in indian classical music using deep learning. In: Artificial Intelligence in Music, Sound, Art and Design: 10th International Conference, EvoMUSART 2021, Held as Part of EvoStar 2021, Virtual Event, April 7–9, 2021, Proceedings. vol. 12693, p. 248. Springer Nature (2020)
30. Gulati, S., Bellur, A., Salamon, J., HG, R., Ishwar, V., Murthy, H.A., Serra, X.: Automatic tonic identification in indian art music: approaches and evaluation. Journal of New Music Research **43**(1), 53–71 (2014)
31. Gulati, S., Salamon, J., Serra, X.: A two-stage approach for tonic identification in indian art music. In: Proceedings of the 2nd CompMusic Workshop; 2012 Jul 12–13; Istanbul, Turkey. Barcelona: Universitat Pompeu Fabra; 2012. p. 119–127. Universitat Pompeu Fabra (2012)
32. Gulati, S., Serrà Julià, J., Ganguli, K.K., Sentürk, S., Serra, X.: Time-delayed melody surfaces for rāga recognition. In: Devaney J, Mandel MI, Turnbull D, Tzanetakis G, editors. ISMIR 2016. Proceedings of the 17th International Society for Music Information Retrieval Conference; 2016 Aug 7–11; New York City (NY).[Canada]: ISMIR; 2016. p. 751–7. International Society for Music Information Retrieval (ISMIR) (2016)
33. Gulati, S., Serrà Julià, J., Ganguli, K.K., Serra, X.: Landmark detection in hindustani music melodies. In: Georgaki A, Kouroupetroglou G, eds. Proceedings of the 2014 International Computer Music Conference, ICMC/SMC; 2014 Sept 14–20; Athens, Greece.[Michigan]: Michigan Publishing; 2014. Michigan Publishing (2014)
34. Gulati, S., et al.: Computational approaches for melodic description in indian art music corpora. Ph.D. thesis, Universitat Pompeu Fabra (2017)
35. Gupta, C., Rao, P.: Objective assessment of ornamentation in indian classical singing. In: Speech, Sound and Music Processing: Embracing Research in India, pp. 1–25. Springer (2011)
36. Huang, X., Acero, A., Hon, H.W., Reddy, R.: Spoken language processing: A guide to theory, algorithm, and system development. Prentice hall PTR (2001)
37. Ishwar, V., Dutta, S., Bellur, A., Murthy, H.A.: Motif spotting in an alapana in carnatic music. In: ISMIR. pp. 499–504. Citeseer (2013)
38. John, S., Sinith, M., Sudheesh, R., Lalu, P.: Classification of indian classical carnatic music based on raga using deep learning. In: 2020 IEEE Recent Advances in Intelligent Computational Systems (RAICS). pp. 110–113. IEEE (2020)
39. Karakurt, A., Şentürk, S., Serra, X.: Morty: A toolbox for mode recognition and tonic identification. In: Proceedings of the 3rd International Workshop on Digital Libraries for Musicology. p. 9–16. DLfM 2016, Association for Computing Machinery, New York, NY, USA (2016). https://doi.org/10.1145/2970044.2970054
40. Kaufmann, W.: The ragas of north India. Indiana University Press (1968)
41. Koduri, G.K., Gulati, S., Rao, P.: A survey of raaga recognition techniques and improvements to the state-of-the-art. Sound and Music Computing **38**, 39–41 (2011)
42. Koduri, G.K., Gulati, S., Rao, P., Serra, X.: Rāga recognition based on pitch distribution methods. Journal of New Music Research **41**(4), 337–350 (2012)
43. Koduri, G.K., Ishwar, V., Serrà, J., Serra, X.: Intonation analysis of rāgas in carnatic music. Journal of New Music Research **43**(1), 72–93 (2014)
44. Krishnaswamy, A.: On the twelve basic intervals in south indian classical music. In: Audio Engineering Society Convention 115. Audio Engineering Society (2003)
45. Kumar, V., Pandya, H., Jawahar, C.: Identifying ragas in indian music. In: 2014 22nd International Conference on Pattern Recognition. pp. 767–772. IEEE (2014)

46. Lartillot, O., Toiviainen, P., Eerola, T.: A matlab toolbox for music information retrieval. In: Data analysis, machine learning and applications, pp. 261–268. Springer (2008)
47. Lee, K.: Automatic chord recognition from audio using enhanced pitch class profile. In: ICMC. Citeseer (2006)
48. Lin, J., Keogh, E., Lonardi, S., Chiu, B.: A symbolic representation of time series, with implications for streaming algorithms. In: Proceedings of the 8th ACM SIGMOD workshop on Research issues in data mining and knowledge discovery. pp. 2–11 (2003)
49. Manjabhat, S.S., Koolagudi, S.G., Rao, K.S., Ramteke, P.B.: Raga and Tonic Identification in Carnatic Music. Journal of New Music Research **46**(3), 229–245 (2017). https://doi.org/10.1080/09298215.2017.1330351
50. Murthy, Y.V.S., Koolagudi, S.G.: Content-based music information retrieval (cb-mir) and its applications toward the music industry: A review. ACM Comput. Surv. **51**(3) (Jun 2018). https://doi.org/10.1145/3177849
51. Pandey, G., Mishra, C., Ipe, P.: Tansen: A system for automatic raga identification. In: IICAI. pp. 1350–1363 (2003)
52. Pawar, M.Y., Mahajan, S.: Automatic tonic (shruti) identification system for indian classical music. In: Soft Computing and Signal Processing, pp. 733–742. Springer (2019)
53. Popescu, T., Widdess, R., Rohrmeier, M.: Western listeners detect boundary hierarchy in indian music: a segmentation study. Scientific reports **11**(1), 1–14 (2021)
54. Quinlan, J.R.: C4. 5: programs for machine learning. Elsevier (2014)
55. Ranjani, H., Arthi, S., Sreenivas, T.: Carnatic music analysis: Shadja, swara identification and raga verification in alapana using stochastic models. In: 2011 IEEE Workshop on Applications of Signal Processing to Audio and Acoustics (WASPAA). pp. 29–32. IEEE (2011)
56. Rao, P., Ross, J.C., Ganguli, K.K.: Distinguishing raga-specific intonation of phrases with audio analysis. ninad **26**, 64 (2013)
57. Rao, P., Ross, J.C., Ganguli, K.K., Pandit, V., Ishwar, V., Bellur, A., Murthy, H.A.: Classification of melodic motifs in raga music with time-series matching. Journal of New Music Research **43**(1), 115–131 (2014)
58. Rao, V., Rao, P.: Improving polyphonic melody extraction by dynamic programming based dual f0 tracking. In: Proc. of the 12th International Conference on Digital Audio Effects (DAFx), Como, Italy (2009)
59. Rao, V., Rao, P.: Vocal melody extraction in the presence of pitched accompaniment in polyphonic music. IEEE transactions on audio, speech, and language processing **18**(8), 2145–2154 (2010)
60. Rodríguez López, M., de Haas, B., Volk, A.: Comparing repetition-based melody segmentation models. In: Proceedings of the 9th Conference on Interdisciplinary Musicology (CIM14). pp. 143–148. SIMPK and ICCMR (2014)
61. Ross, J.C., Rao, P.: Detection of raga-characteristic phrases from hindustani classical music audio. In: Serra X, Rao P, Murthy H, Bozkurt B, editors. Proceedings of the 2nd CompMusic Workshop; 2012 Jul 12–13; Istanbul, Turkey. Barcelona: Universitat Pompeu Fabra; 2012. p. 133–138. Universitat Pompeu Fabra (2012)
62. Ross, J.C., Vinutha, T., Rao, P.: Detecting melodic motifs from audio for hindustani classical music. In: ISMIR. pp. 193–198 (2012)
63. Roy, S., Banerjee, A., Sanyal, S., Ghosh, D., Sengupta, R.: A study on raga characterization in indian classical music in the light of mb and be distribution. In: Journal of Physics: Conference Series. vol. 1896, p. 012007. IOP Publishing (2021)
64. Salamon, J., Gómez, E.: Melody extraction from polyphonic music signals using pitch contour characteristics. IEEE Transactions on Audio, Speech, and Language Processing **20**(6), 1759–1770 (2012)
65. Salamon, J., Gómez, E., Bonada, J.: Sinusoid extraction and salience function design for predominant melody estimation. In: Proc. 14th Int. Conf. on Digital Audio Effects (DAFX-11). pp. 73–80 (2011)

66. Salamon, J., Gulati, S., Serra, X.: A multipitch approach to tonic identification in indian classical music. In: Gouyon F, Herrera P, Martins LG, Müller M. ISMIR 2012: Proceedings of the 13th International Society for Music Information Retrieval Conference; 2012 Oct 8–12; Porto, Portugal. Porto: FEUP Ediçoes; 2012. International Society for Music Information Retrieval (ISMIR) (2012)
67. Sengupta, R., Dey, N., Nag, D., Datta, A., Mukerjee, A.: Automatic tonic (sa) detection algorithm in indian classical vocal music. In: National Symposium on Acoustics. pp. 1–5 (2005)
68. Sengupta, R.: Study on some aspects of the "singer's formant" in north indian classical singing. journal of Voice **4**(2), 129–134 (1990)
69. Serra, X.: Creating research corpora for the computational study of music: the case of the compmusic project. In: AES 53rd International Conference: Semantic Audio; 2014 Jan 27–29; London, UK. New York: Audio Engineering Society; 2014. Article number 1-1 [9 p.]. Audio Engineering Society (2014)
70. Shetty, S., Achary, K.: Raga mining of indian music by extracting arohana-avarohana pattern. International Journal of Recent Trends in Engineering **1**(1), 362 (2009)
71. Sinith, M., Murthy, K., Tripathi, S.: Raga recognition through tonic identification using flute acoustics. International Journal of Advanced Intelligence Paradigms **15**(3), 273–286 (2020)
72. Smith, T.F., Waterman, M.S., et al.: Identification of common molecular subsequences. Journal of molecular biology **147**(1), 195–197 (1981)
73. Sridhar, R., Geetha, T.: Swara indentification for south indian classical music. In: 9th International Conference on Information Technology (ICIT'06). pp. 143–144. IEEE (2006)
74. Sridhar, R., Geetha, T.: Raga identification of carnatic music for music information retrieval. International Journal of recent trends in Engineering **1**(1), 571 (2009)
75. Srinivasamurthy, A., Serra, X.: A supervised approach to hierarchical metrical cycle tracking from audio music recordings. In: 2014 IEEE International Conference on Acoustics, Speech and Signal Processing (ICASSP). pp. 5217–5221. IEEE (2014)
76. Sun, X.: A pitch determination algorithm based on subharmonic-to-harmonic ratio. In: Sixth International Conference on Spoken Language Processing (2000)
77. Tanaka, Y., Iwamoto, K., Uehara, K.: Discovery of time-series motif from multi-dimensional data based on mdl principle. Machine Learning **58**(2), 269–300 (2005)
78. Velankar, M., Deshpande, A., Kulkarni, P.: Melodic pattern recognition in indian classical music for raga identification. International Journal of Information Technology **13**(1), 251–258 (2021)
79. Widdess, R.: Involving the performers in transcription and analysis: a collaborative approach to dhrupad. Ethnomusicology **38**(1), 59–79 (1994)

Part II
Machine Learning

A Study on Effectiveness of Deep Neural Networks for Speech Signal Enhancement in Comparison with Wiener Filtering Technique

Vijay Kumar Padarti, Gnana Sai Polavarapu, Madhurima Madiraju,
V. V. Naga Sai Nuthalapati, Vinay Babu Thota,
and V. D. Subramanyam Veeravalli

1 Introduction

The five fundamental senses, i.e., hearing, sight, smell, taste, and touch, perceive the information from the environment, and the human brain processes this information to create a precise response. Sound acts as an information provider to these senses. The information that is transmitted has to be free of noises to get a better understanding of the external environment. Noise can be described as any unwanted information which hinders the ability of the human body to process the valuable sensory information. Hence, an uncorrupted sound becomes essential for proper interaction of humans with their external world. The primary focus is on speech signals which are information providers in various communication systems. During the transfer of signals, distortion by some unwanted signals causes loss of useful data and information stored in the signals. There are many real-world noise signals such as the noise of a mixer grinder, washing machine, and vehicles which have to be reduced to retrieve the wanted information. The frequency of speech signals ranges from 85 to 255 Hz. Typical male voice ranges in between 85 and 180 Hz, whereas the female voice ranges in between 165 and 255 Hz. Babies have even higher ranges of frequency reaching up to 1000Hz in a few cases [1].

Speech denoising refers to the removal of background content from speech signals. The goal of speech denoising is to produce noise-free speech signals from noisy recordings while improving the perceived quality of the speech component and increasing its intelligibility [2]. Speech denoising can be utilized in various applications where we experience the presence of background noise in communica-

---

V. K. Padarti (✉) · G. S. Polavarapu · M. Madiraju · V. V. Naga Sai Nuthalapati · V. B. Thota ·
V. D. Subramanyam Veeravalli
Department of Electronics and Communications Engineering, Velagapudi Ramakrishna
Siddhartha Engineering College, Vijayawada, India
e-mail: vijayakumar.padarti@vrsiddhartha.ac.in

© The Author(s), under exclusive license to Springer Nature Switzerland AG 2023
A. Biswas et al. (eds.), *Advances in Speech and Music Technology*, Signals and
Communication Technology, https://doi.org/10.1007/978-3-031-18444-4_6

tions, e.g., hearing aids, telecommunications, speech recognition applications, etc. [3].

A number of techniques have been proposed based on different assumptions on the signal and noise characteristics in the past, but in this chapter, we shall compare two main methods, Wiener filtering technique and neural network method. For neural network technique, we will consider two types of networks, fully connected network and convolutional neural network. We compute PSNR and SNR values for these three techniques to compare the denoised signal quality.

2 Background

2.1 Wiener Filtering Technique

One of the notable techniques of filtering that is widely used in signal enhancement methods is Wiener filtering. The key principle of Wiener filtering, essentially, is to take a noisy signal and acquire an estimate of clean signal from it. The approximate clean signal is acquired by reducing the mean square error (MSE) between the estimated signal and desired clean signal [4].

The transfer function of the Wiener system in frequency domain is

$$H(w) = \frac{P_s(w)}{P_s(w) + P_v(w)}. \quad (1)$$

where

$$P_s(w) = power\ spectral\ density\ of\ clean\ signal. \quad (2)$$

$$P_v(w) = power\ spectral\ density\ of\ noise\ signal. \quad (3)$$

Here, the signal s and noise v are considered to be uncorrelated and stationary.

The signal-to-noise ratio (SNR), which is used to detect the quality of a signal, is defined as

$$SNR = \frac{P_s(w)}{P_v(w)}. \quad (2)$$

Substituting SNR in the above transfer function, we obtain

$$H(w) = \left(1 + \frac{1}{\text{SNR}}\right)^{-1}. \quad (3)$$

One of the popular applications of the Wiener filtering technique is the Global Positioning System (GPS) and inertial navigation system. Wiener filter, which is also used in geodesy to denoise gravity records, is used in GPS to model only those time variabilities that are significant when adapted to noise level of data [5].

Signal coding applications is a field where Wiener filter is widely used in. In signal processing and broadly engineering applications too, Wiener filter is considered to be a great tool for speech applications due to its accurate estimation characteristic. This filter can further be adapted to serve different purposes like satellite telephone communication [6].

If we dive into the world of electronics and communication more, the Wiener filter has a range of applications in signal processing, image processing, digital communication, etc. like system identification, deconvolution, noise reduction, and signal detection [7].

Specifically in image processing, Wiener filter is a quite popular technique used for deblurring, attributed to its least-mean-squares technique. The blurriness in images that is caused as a result of motion or unfocused lens is removed using this filter. Additionally, since it returns mathematically and theoretically the best results, it also has applications in other engineering fields [8].

2.1.1 Algorithm

To denoise a speech signal using the Wiener filtering technique, we first fetch a clean audio signal file and a noise signal file from the audio datastore in MATLAB. We then extract a segment from the noisy signal and add it to the clean signal to make it a noisy speech signal which is given as input to the Wiener filter. The Wiener filter performs denoising of the speech signal, and then we visualize the output signal. Flowchart for the algorithm can be seen in Fig. 1.

In order to compute peak SNR and SNR values, the output and input signals are given to PSNR function which is a built-in function in MATLAB. The frequency response of the Wiener filter is such that, at frequencies where SNR is low, that is, noise power is high, the gain of the filter decreases, and the output is limited, causing

Fig. 1 Wiener method flowchart

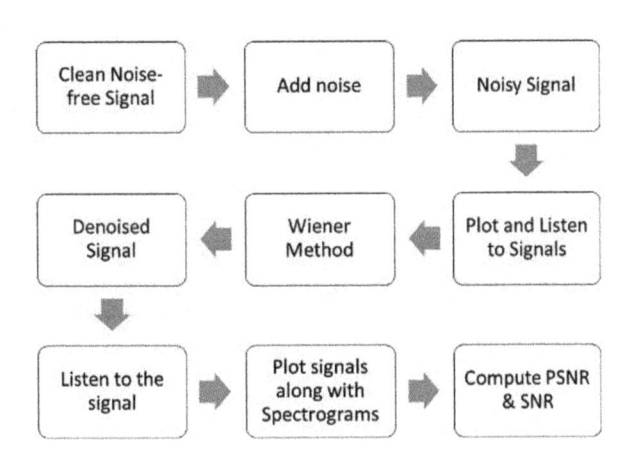

noise reduction. Correspondingly, for high SNR, that is, when signal power is high, the gain becomes nearly one (~1), and output sought is very close to input. Another drawback is that at all given frequencies, the Wiener filter requires a fixed frequency response. One more shortcoming in the Wiener filter is that before filtering, the power spectral density of both clean and noise signals has to be estimated. Noise amplification is also a problem [6, 9–11].

2.2 Deep Neural Networks

Deep learning is part of machine learning with an algorithm inspired by the structure and function of the brain, which is called an artificial neural network. Artificial neural networks are the statistical model inspired by the functioning of human brain cells called neurons. Deep learning is used in many fields such as computer vision, speech recognition, natural language processing, etc. [12].

A neural network mimics the human brain and consists of artificial neurons, also known as nodes. Group of nodes make a layer. There are three types of layers: the input layer, the hidden layer(s), and the output layer. There can be multiple hidden layers and it depends on the model. All the nodes are provided with information in the form of input. At each node, the inputs are multiplied with some random weights and are computed, and then a bias value is added to it. Finally, activation functions, such as rectified linear unit (ReLU) function, are applied to determine which neuron to eliminate.

While deep learning algorithms feature self-learning representations, they depend upon neural networks that mimic the way the brain processes the information. During the training process, algorithms use random unknown elements in the input to extract features, segregate objects, and find useful data patterns. Much like training machines for self-learning, this occurs at multiple levels, using the algorithms to build the models. Deep learning models utilize several algorithms. Although none of the networks is considered perfect, some algorithms are preferred to perform specific tasks. Some commonly used artificial neural networks are feedforward neural network, convolutional neural network, recurrent neural network, and autoencoders [13].

There are also some disadvantages of deep learning. Very large amount of time is required to execute a deep learning model. Depending upon the complexity, sometimes, it may take several days to execute one model. Also, for small datasets, the deep learning model is not suitable. There are various applications of deep learning such as computer vision, natural language processing and pattern recognition, image recognition and processing, machine translation, sentiment analysis, question answering system, object classification and detection, automatic handwriting generation, automatic text generation, etc.

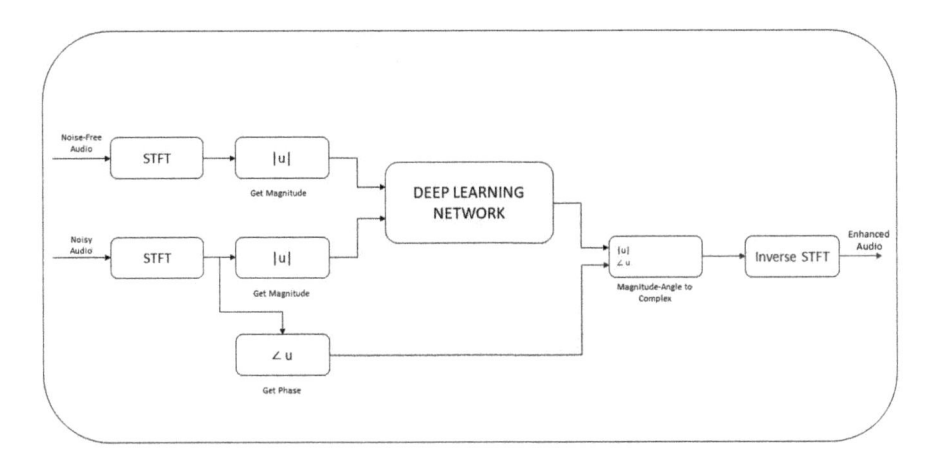

A Study on Effectiveness of Deep Neural Networks for Speech Signal Enhancement 125

Fig. 2 Neural network block diagram

2.2.1 Algorithm

We first fetch clean and noisy audio files from the audio datastore in MATLAB, and then we extract a segment from the noisy audio and add it to the clean audio signal. This will be the input given to the deep learning network. Neural network block diagram is shown in Fig. 2.

We utilize short-time Fourier transform (STFT) to transform the audio signals from time domain to frequency domain. The magnitudes are extracted and then fed to the neural network. Then the output signal which is the denoised and enhanced version of the input noisy signal is converted back into the time domain using the inverse STFT.

An exemplary speech signal is shown in Fig. 3. Clearly, it can be seen that the amplitudes vary significantly with time, i.e., there will be huge variations frequently in the signals like music and speech. This is the reason we utilize the short-time Fourier transform technique.

We have utilized two models of deep learning networks: fully connected and convolutional neural network. For any model, the network first needs to be trained so that it learns its function to segregate the noise segments from the audio segments. For training the model, we consider a sample signal and then set the required parameters such as learning rate, number of epochs, batch size, etc. Once the model completes its training, it has to be tested. For the testing phase, we feed the model with another set of samples which were not given in the training phase and observe the outputs. The flowchart for the artificial neural network is depicted in Fig. 4.

To compare the efficiency of the two models, we compute PSNR and SNR values using psnr function which is a built-in function in MATLAB. We also use another in-built function, sound(), to listen to the audio signals. Besides this, we also represent the signals with timing plots and spectrogram.

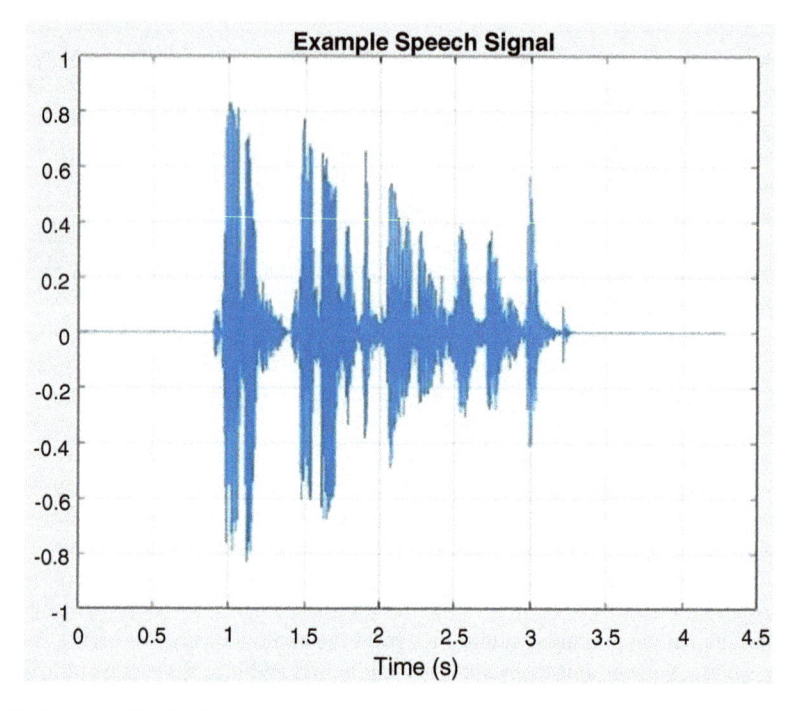

Fig. 3 Sample audio signal

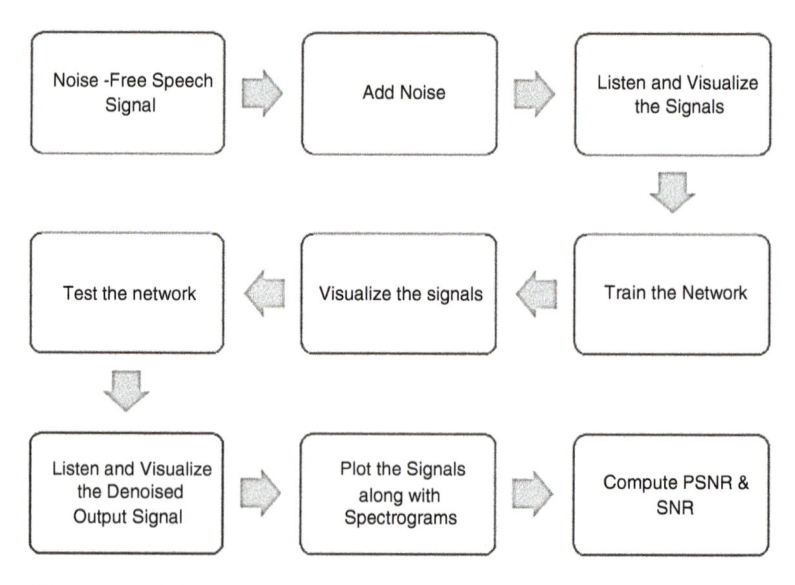

Fig. 4 Neural network flowchart

Fig. 5 Fully connected network

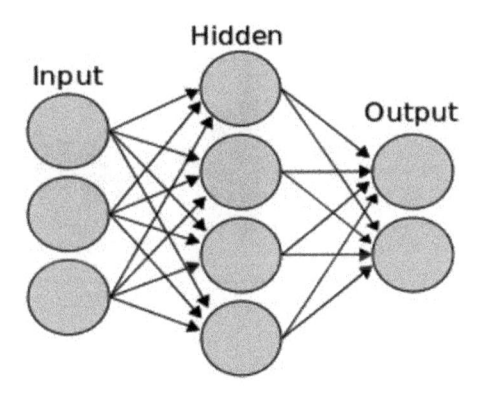

2.2.2 Fully Connected Network

A fully connected neural network consists of a series of fully connected layers that connect every neuron in one layer to every neuron in the next layer. For any network, there are three types of layers: input, hidden, and output layers. The information received from the input is given to the model, and then the model is trained using this data [14]. A fully connected network model is shown in Fig. 5.

We define the number of hidden layers in the model. For our model, we have 2 hidden layers with 1024 neurons each. The model is trained on the training dataset. Each of the hidden layers is followed by ReLU layers and batch normalization layers.

A clean audio file fetched from the audio datastore is corrupted with a noisy segment extracted from the noise signal. These signals are plotted in Fig. 6. Then these signals are passed to the network model, and the model is trained. The training process involves learning the model function by passing the model through the given dataset for 3 epochs (in order to avoid overfitting, we have limited to 3 epochs) with a batch size of 128 at an initial learn rate of 10^{-5}, and for every epoch, the learning rate decreases by a factor of 0.9.

2.2.3 Convolutional Neural Network (CNN)

Convolutional neural networks can be differentiated from other neural networks by their superior performance with image, speech, or audio signal inputs. They have three main types of layers: convolutional layer, pooling layer, and fully connected layer.

The first layer of a convolutional network is the convolutional layer. After the convolutional layer, we can have the additional convolutional layers or the pooling layers. The final layer is the fully connected layer. The CNN complexity increases with each layer, but the model gets more accurate outputs. Therefore, there should be some optimality.

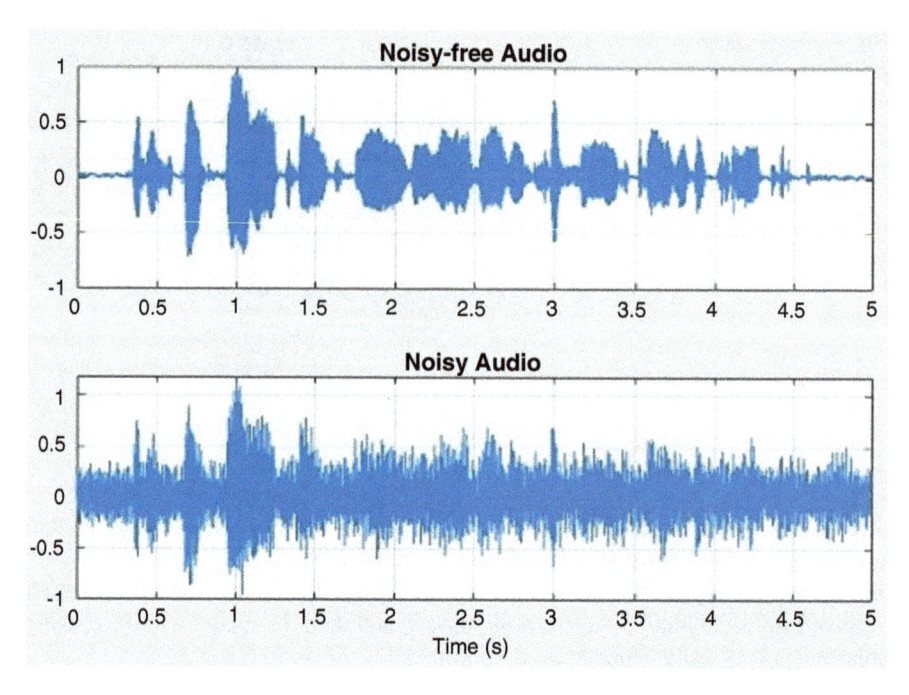

Fig. 6 Clean and noisy audio

The convolutional layer is the core building block of a CNN. It is where the major part of the computation occurs. It requires a few components: input data, a filter, and a feature map. Pooling layers, also known as downsampling, conduct dimensionality reduction, reducing the number of parameters in the input. It is similar to the convolutional layer in processing and filtering the input, but the difference is that this filter does not have any weights. Instead, the kernel applies an aggregation function to the values within the receptive field, populating the output array. A basic convolutional layer model is shown in Fig. 7.

There are two main types of pooling: max pooling which selects the maximum value and average pooling which computes the average value of data. A lot of information is lost in the pooling layer, but it also has a number of advantages to the CNN. They help to reduce complexity, improve efficiency, and limit risk of overfitting.

In the fully connected layer, each node in the output layer connects directly to a node in the previous layer. This layer performs the task of classification based on the features extracted through the previous layers and their different filters. While convolutional and pooling layers tend to use ReLU functions, FC layers usually use a SoftMax activation function to classify inputs appropriately, producing a probability from 0 to 1 [15].

Fig. 7 Convolutional layer

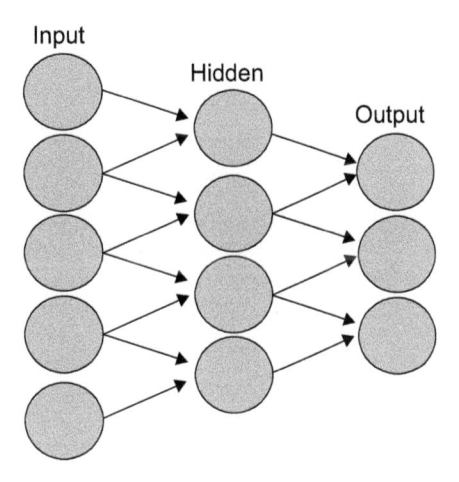

For our convolutional network model, we have defined 16 layers. Similar to the fully connected network, convolutional layers are followed by ReLU and batch normalization layers.

3 Results

3.1 Wiener Filtering Technique

The Wiener model is provided with a clean and a noisy signal, and it yielded the output as shown in Fig. 8. We can clearly see that a lot of high-frequency components in the original audio signal are lost when denoised. This leads to reduced quality. Also, the model resulted in a PSNR 20.0589 dB and SNR of 2.4825 dB which are considered to be low compared to standards. Results for different samples are shown in Table 1.

3.2 Fully Connected Network

3.2.1 Training Stage

In the training stage, our model is provided with training dataset and is made to learn its function. The training progress is visualized in Fig. 9, where we can see the internal approximation process.

After the training is complete, the time and frequency plots are visualized as in Fig. 10. We can see that the denoised version is almost approximately equal to the original clean signal.

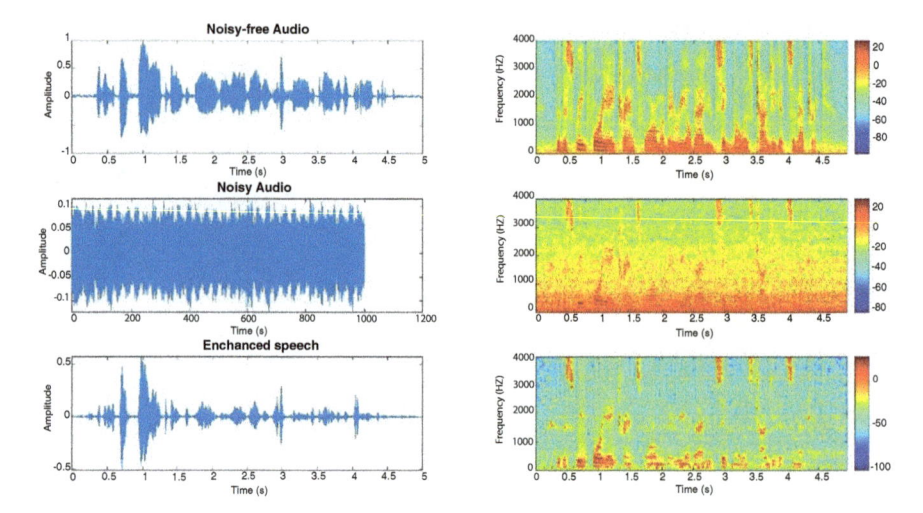

Fig. 8 Wiener method results

Table 1 Wiener method results

Sample	PSNR [dB]	SNR [dB]
1	20.0589	2.4825
2	20.7663	3.1900
3	20.4732	2.8969
4	20.8320	3.2557
5	21.2536	3.6772
6	20.6340	3.0577
7	20.2921	2.7157
8	20.6636	3.0872

3.2.2 Testing Stage

Once the model is trained, we test the model with a dataset which is not given in the training stage. When our fully connected model is tested, it resulted in a PSNR of 23.5416 dB and an SNR of 6.5651 dB, which are greater than those of the Wiener method, but yet they are lower than the acceptable standard values. The output time and spectrogram plots are visualized as in Fig. 11. The original and enhanced versions are nearly equal but not exactly equal, but when compared to the Wiener filtering technique, it can be seen that the fully connected network method yielded better results. Results for different samples are shown in Table 2.

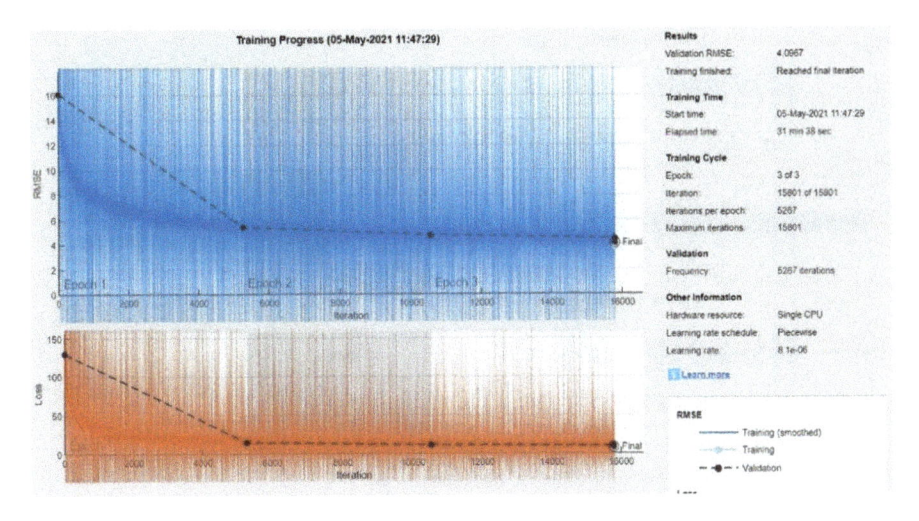

A Study on Effectiveness of Deep Neural Networks for Speech Signal Enhancement 131

Fig. 9 Fully connected network training progress

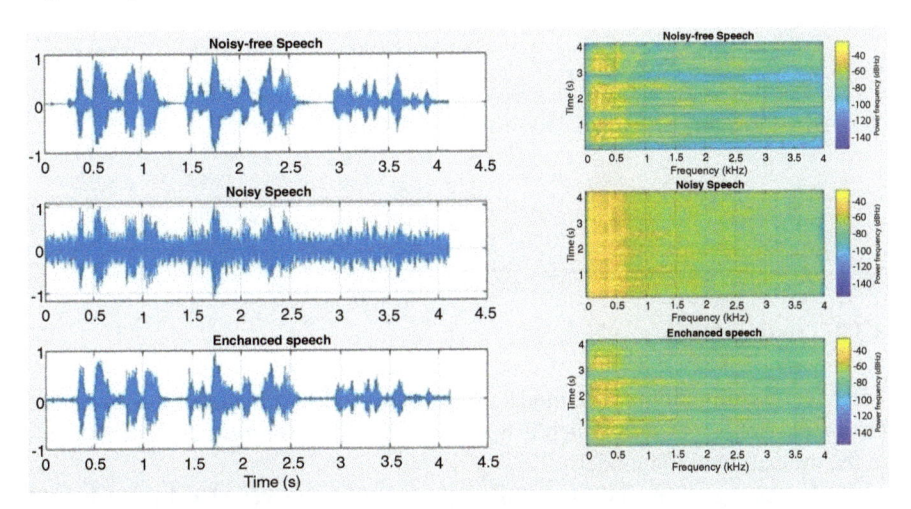

Fig. 10 Fully connected network result in the training stage

3.3 Convolutional Neural Network

3.3.1 Training Stage

Similar to that of the fully connected model, our convolutional model is first trained with a dataset. The training progress window is visualized in Fig. 12. It should be noted that the training process took longer time compared to fully connected model due to the number of layers and the complexity of the layers. Once training is completed, the time and frequency plots are visualized as in Fig. 13.

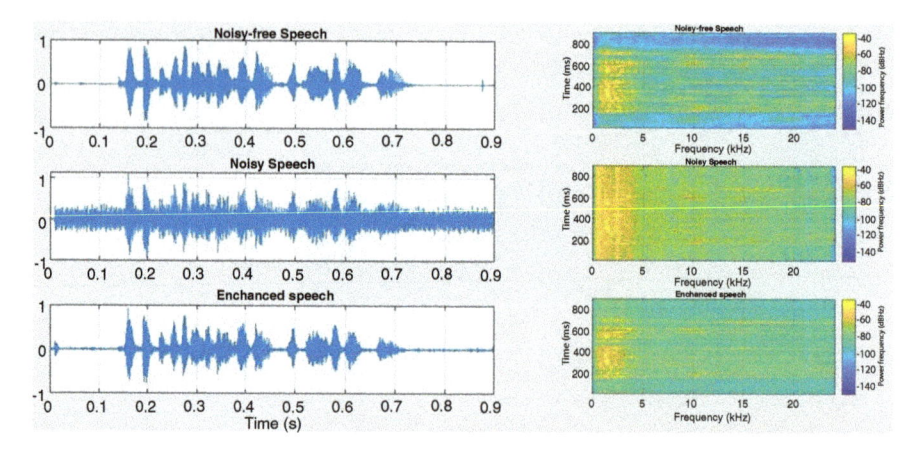

Fig. 11 Fully connected network result in the testing stage

Table 2 FC model results

Sample	PSNR [dB]	SNR [dB]
1	23.5416	6.5651
2	24.1829	7.2064
3	24.0188	7.0423
4	26.5571	7.7258
5	23.7959	6.8194
6	23.8697	6.8932
7	23.9390	6.9625
8	23.8521	6.8756

3.3.2 Testing Stage

After successfully training the model, we test the model with a testing dataset. The testing dataset is a dataset which is not provided in the training dataset. This helps in the evaluation of the model.

In the testing stage, the time and frequency plots are visualized as in Fig. 14 where the original clean audio signal, the corrupted noisy signal, and the enhanced speech signals are plotted. It can be clearly seen that our model effectively denoises the noisy signal and enhances its quality. When compared to the previous technique, i.e., Wiener filtering technique, where we have removed all the high-frequency components, we have removed only the noisy components and retained the original high-frequency components in the output signal. So, this model is effective compared to the Wiener method.

The convolutional model resulted in a SNR of 7.6137 dB and a PSNR of 26.4451 dB. When compared to Wiener and fully connected models, these values are higher. However, they are still lower than the acceptable standard values. Results for different samples are shown in Table 3.

A Study on Effectiveness of Deep Neural Networks for Speech Signal Enhancement 133

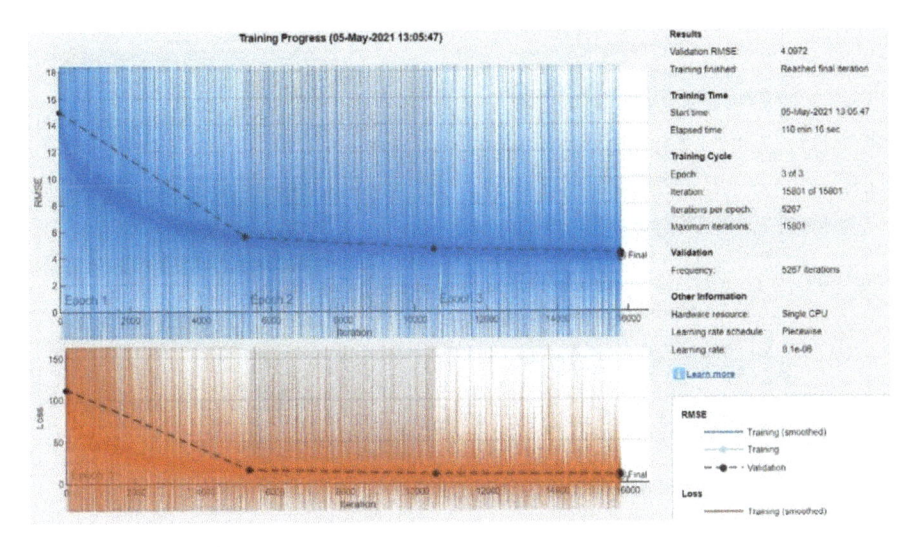

Fig. 12 Convolutional model training progress

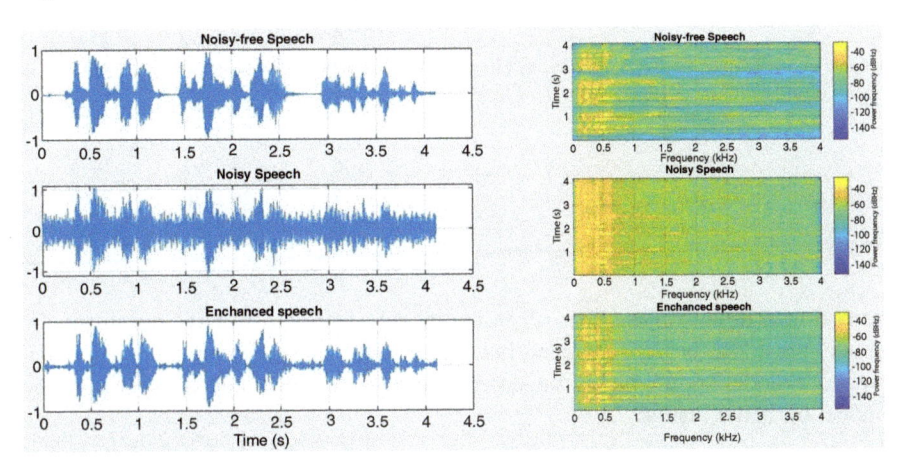

Fig. 13 Convolutional model result in the training stage

4 Conclusion

We have built three models to apply the Wiener filtering technique and neural networks for speech enhancement. The results from models for different samples are shown in Tables 1, 2, and 3. From the results obtained, we can clearly see that convolutional network performs better when compared to the other two models, but it requires very large amount of time for training and computation. We know that the resources are very limited and expensive to process such models. Besides this, requiring a very large computational time is a big disadvantage when it comes

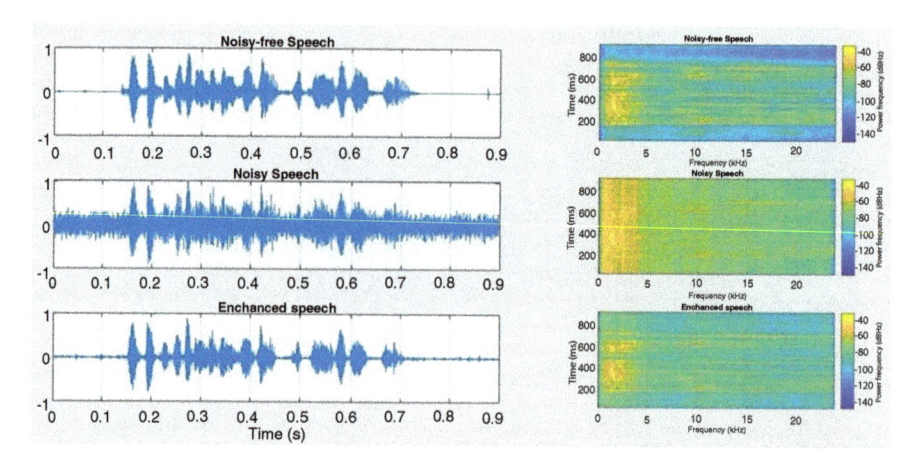

Fig. 14 Convolutional model result in the testing stage

Table 3 CNN model results

Sample	PSNR [dB]	SNR [dB]
1	26.4451	7.6137
2	25.9861	7.1547
3	26.1930	7.3617
4	26.3692	7.5378
5	26.3822	7.5508
6	25.9799	7.1485
7	26.2276	7.3962
8	26.5732	7.7461

to real-time applications. Also, when the model requires such huge resources, the model must also be very efficient, but the results obtained from the convolutional network model are not highly satisfactory. Therefore, we still need to optimize the model for better results.

References

1. G. K Rajini, V. Harikrishnan, Jasmin. M, S. Balaji, "A Research on Different Filtering Techniques and Neural Networks Methods for Denoising Speech Signals", IJITEE, ISSN: 2278-3075, Volume-8, Issue- 9S2, July 2019.
2. D. Liu, P. Smaragdis, M. Kim, "Experiments on Deep Learning for Speech Denoising", Interspeech, 2014.
3. F. G. Germain, Q. Chen, and V. Koltun, "Speech Denoising with Deep Feature Losses", arXiv:1806.10522v2 [eess.AS], 14 Sep 2018.
4. M. A. Abd El-Fattah, M. I. Dessouky, S. M. Diab and F. E. Abd El-samie, "Speech enhancement using an adaptive wiener filtering approach", Progress in Electromagnetics Research M, Vol. 4, 167–184, 2008.

5. A. Klos, M. S. Bos, R. M. S. Fernandes, "Noise-Dependent Adaption of the Wiener Filter for the GPS Position Time Series." Math Geosci 51, 53–73 (2019). Available at https://doi.org/10.1007/s11004-018-9760-z.
6. S. China Venkateswarlu, K. Satya Prasad, A. Subbarami Reddy, "Improve Speech Enhancement Using Weiner Filtering", Global Journal of Computer Science and Technology Volume 11 Issue 7 Version 1.0 May 2011.
7. This is from an online internet source which is available at https://en.wikipedia.org/wiki/Wiener_filter#Applications, as on 28th May 2021.
8. This is from an online internet source which is available at https://www.clear.rice.edu/elec431/projects95/lords/wiener.html, as on 28th May 2021.
9. Chavan, Karishma and Gawande, Ujwalla. (2015), "Speech recognition in noisy environment, issues and challenges: A review", 1-5.10.1109/ICSNS.2015.7292420.
10. Article from an online internet source available here, as on 28th May 2021.
11. A. Dubey and M. Galley, "Experiments with Speech Enhancement Techniques", available here, as on 28th May 2021.
12. This is from an online internet source which is available here, as on 28th May 2021.
13. This is from an online internet source which is available at https://www.simplilearn.com/tutorials/deep-learning-tutorial/deep-learning-algorithm, as on 28th May 2021.
14. This is from an online internet source which is available here, as on 28th May 2021.
15. This is from an online internet source which is available at https://www.ibm.com/cloud/learn/convolutional-neural-networks, as on 28th May 2021.

Video Soundtrack Evaluation with Machine Learning: Data Availability, Feature Extraction, and Classification

Georgios Touros and Theodoros Giannakopoulos

1 Introduction

In recent years, there has been a multitude of attempts to create models that touch upon domains of artistic creativity. Such a domain is that of creating, or choosing, music that accompanies visual content, i.e., video soundtracks. This is a type of artistic task that is usually taken up by dedicated professionals. Especially in the film industry, a typical film would have a composer and a music supervisor working separately to create and select the musical content that best accentuates the themes, energy, and emotion of a particular scene. Their works are interconnected, but a close collaboration between them is not always the case.

In this chapter, we focus on methods that combine music and video, in order to approach the task of choosing satisfactory accompaniment music for video content. To that end, we extract a selection of handcrafted features from three different content modalities: audio, video, and symbolic representations of music. In our research, we could not find other works that combined all three of these. Our experimentation suggests that there is potential in this approach, as it combines high-level with low-level information: features from sound timbre and audio spectra are combined with features from song structure and composition; color histograms are combined with object and optical flow features. Given the complexity of the task, and the lack of available open data, our main focus was creating a large enough data-set that contains data from all the necessary modalities and forms (songs in raw audio form, transcriptions of these songs in MIDI form, and video excerpts containing these songs). Our contribution is not only to gather and clean the data but also to develop an open-source scalable process to create and manage such data-sets, which could potentially be given to the research community for further expansion.

G. Touros (✉) · T. Giannakopoulos (✉)
National Center for Scientific Research "Demokritos", Athens, Greece

© The Author(s), under exclusive license to Springer Nature Switzerland AG 2023
A. Biswas et al. (eds.), *Advances in Speech and Music Technology*, Signals and Communication Technology, https://doi.org/10.1007/978-3-031-18444-4_7

We also proceed to extract features from this data-set and examine their suitability using a rudimentary classifier, which matches the different modalities with each other. We experiment with tuning the model, but the lack of large-scale data in our proof of concept prevents us from reaching high levels of accuracy.

Our work could potentially help in many use cases in the field. An improved version of our classifier could be used to create a platform for video soundtrack selection. The end-users could be video editors that need temporary music while they wait for the music supervisor to provide them with a fitting, licensed piece of music or music supervisors who want to validate their choice of music or let the algorithm help them decide between some options. Such a system could also serve as a platform for artists that want to expose their music to such an audience (video editors or film producers), by providing an evaluation of the goodness of fit of their music, with regard to the visual content. Further to this, such a model could be used in order to create a soundtrack recommendation system from a database of existing tracks, which in turn could be used either as a plug-in on video editing software or as a stand-alone application.

Furthermore, there is much literature around video processing, audio processing, and classification. To our knowledge, there are not as many papers when it comes to video soundtrack generation or classification. A common thread in similar research is music retrieval for user-generated videos (UGV). In [1], a system for creating automatic generation of soundtracks for outdoor videos of users is proposed. The system is built on contextual data based on the geo-location in which the video was shot. This contextual data contains geographic tags, and mood tags, collected from OpenStreetMap and Foursquare. In [2], a system for recommending soundtracks for user outdoor videos is proposed, based on geographic, visual, and audio features. The visual features are based on color only and are combined with tags of the mood of the specific area. These are then combined with music, combining the user's previous listening history with mood tags and audio features of the track. A similar approach is followed in [3]. In [4], the authors propose a process that recommends the soundtrack and edits the video simultaneously. Their approach uses a multi-task deep neural network to predict the characteristics of an ideal song for the video and then retrieves the closest match from a database. The track is then aligned to the video using a dynamic time warping algorithm and concatenates the video given a cost function, trained on an annotated corpus.

Another research thread tries to create music recommendation systems that could be used as plug-ins in video editing software or in order to create music videos. In [5], a method and a system are proposed for the recommendation of soundtracks by video editing software. The proposed system is based on emotional and contextual tags given by the end-user and then retrieves and combines relevant loops of preexisting content. In [6], a music video generation system is developed, which utilizes the emotional temporal phase sequence of the multimedia content to connect music and video. It is trained on annotated data that is mapped to an arousal-valence scale that is tracking the shifts in emotion along the content of the medium. The multimedia are then matched according to the time series of the emotional shifts, using string matching techniques. In [7], a model that synchronizes the climax of a video clip and a music clip is proposed. The model is trained on annotated data for the audience perception of climax in music and video and applies dynamic programming to synchronize the climax in both modalities. In [8], a deep learning approach is introduced for cross-modal content retrieval. The features are extracted from audio and video, and a system for cross-modal queries is developed, relying on content rather than tags or metadata.

The rest of this chapter is organized as follows: In Sect. 2, we describe in detail our attempt to address the data availability issue. We propose a method of acquiring, cleaning, and managing the data of all three relevant modalities, and we report on the specifics and challenges of implementing this method. In Sect. 3, we discuss issues of representation and encoding of data for music and video, and we describe in depth the feature extraction process based on various alternative representations and modalities. In Sect. 4, we experiment with the creation of a classifier that could discriminate between real and fake examples of soundtracks, given data and features that were collected in the previous chapters. We present the results of our experiments, and we highlight some promising directions. In Sect. 5, we discuss future work, focusing on how our data collection pipeline could be improved for building a more robust model. In Sect. 6, we sum up and conclude our work.

2 Data Collection Pipeline

The main challenge we faced was data availability. We could not find any open data-set that includes all three necessary modalities or that was appropriately scalable. We therefore proceeded to create our own data collection pipeline, which can be found in our GitHub repository,[1] along with a detailed description of each module. All code is written in Python 3, Bash, and SQL, run and tested in Ubuntu Linux 20.04. There are some dependencies with third-party software beyond those that were mentioned in the requirements document in the repo. These are FFmpeg, MySQL, CUDA Toolkit (optional), MuseScore 3, FluidSynth, QjackCtl, Qsynth, and LilyPond.

2.1 Collection, Cleanups, and Storage

The MIDI data was downloaded from composing.ai[9], containing 124,470 files. The main problem of this data-set is its lack of structure. The files come from different data sources, each with their own naming conventions and directory structures, making it hard to determine which songs are included in the collection. We used regular expressions to keep valid MIDI files and perform a cleanup of the

---

[1] https://github.com/GeorgeTouros/video-soundtrack-evaluation.

names. As ground truth for name legitimacy, we chose the popular music streaming platform Spotify. The platform provides a web-based API, which we access using the relevant library spotipy[10]. We used that to get a proper name of each MIDI filename, as well as relevant artist and URL information. The match rate is currently at 61%, resulting in a MIDI data-set of 43,567 MIDI files. The collection consists mostly of pop and rock songs, which is a big limitation for the task at hand. While a lot of films use these as background music for film montages, productions usually avoid them, given that obtaining usage rights is expensive. This limits the pool of films and scenes we can draw from. Each film usually also has its own unique score, which in most cases carries the bulk of the scenes and the emotional core of the film.

The audio data is based on a personal collection of MP3 files. This provides some challenges, as will be explained later. The collection has an initial catalog of 55,800 files from different genres and of varying audio quality. They are arranged in directories according to genre, country, and artist. As was the case with the MIDI files, the files do not follow strict naming conventions for the song titles, and the sampling rates and bit rates vary.

Video data is based on a personal film collection. The collection includes 106 films and 37 episodes from television series. The files are in various types: avi, mp4, and mkv. As was the case with the other two modalities, the naming conventions and image quality of the raw data are inconsistent.

Matching MIDI to Audio Data In order to match the audio and MIDI files, we utilized a rule-based method to connect our MIDI with our audio data, via our external knowledge base, Spotify. The matching resulted in **3109 audio-MIDI matches**. While the number of results seems adequate, there are some concerns. The most prominent for the feature extraction process was that MIDI data is often corrupt or otherwise inappropriate for further processing.

An important performance indicator is the recall of the Spotify search API. This depends on the level of cleanliness of the original file names, as well as how well our custom regex-based cleaner works. This is evident by the fact that in the audio files (which had much cleaner initial file names), recall is considerably higher than in the (much "dirtier") MIDI files. More specifically, in the audio files, recall reaches 83.3%, whereas in the MIDI catalog, it's only 65.2%. This means that we lose a lot of MIDI data when we apply the matching process.

On the other hand, the precision of the Spotify search API depends on the level of accuracy of the original files, i.e., whether the file name reflects the actual content of the file. Due to the size of the data-set, we weren't able to precisely compute the precision of the method. Nevertheless, while performing manual curation of the matched files, in order to create fake examples for our classifier, we came across plenty of files that were wrongly matched. This has a few implications. Firstly, there were plenty of audio files that weren't the original versions of the songs (most often being covers in other genres or live versions). Furthermore, there were cases where the MIDI and audio files had the same song title, but were referring to different songs with the same name. Since the Spotify search API yields the most popular result, one of them (or both) was falsely identified. Finally, a lot of jazz songs have titles that closely resemble other popular songs; therefore, there were cases where the audio was falsely identified as a more popular song by the Spotify API.

2.2 Finding Music Within Videos

The ultimate goal in our data collection pipeline is finding combinations of videos and songs, in both audio and symbolic formats. We decided to follow an approach that would allow expanding to videos that aren't necessarily listed in external film knowledge bases; we created a script which, given a database of song fingerprints (implemented in a Python library called pyDejavu [11]), breaks the video in chunks and compares the audio against the said fingerprints, using the same parameters as the ones that were used to build the fingerprint database. If a match is found with the same song in more than three consecutive video chunks, the clip is stored in our catalog. Our experimental setup used the following settings: *video chunk size,* 5 s; *audio sample rate,* 16 KHz; and *audio channels,* 1.

In order to increase the recall of the method, we apply the following rule: For every three video chunks, if the first and third are matched with the same song, then the match is imposed on the middle one, too. Given our objective, we also impose a minimum size of three chunks to each extracted video clip, in order to maintain a balance between keeping irrelevant minuscule clips and missing the opportunity to get larger clips by combining chunks together.

Ideally, we would also want to maximize precision. To that end, we use the song fingerprint offsets per video chunk, which are calculated by pyDejavu. These offsets demonstrate which part of the song corresponds to the matched video chunk. We then calculate the mode (most frequent value) of these offsets for each clip (which is a collection of at least three chunks) and flag those clips that have no mode. In a perfect match scenario, the mode of these offsets would be equal to the chunk length. Overall, the method yields **68 video clips**.

On Fingerprinting As explained in [12], an audio fingerprint is a compact content-based signature that summarizes an audio recording. Audio fingerprinting technologies extract acoustic relevant characteristics of a piece of audio content and store them in a database. When presented with an unidentified piece of audio content, characteristics of that piece are calculated and matched against those stored in the database. Using fingerprints and matching algorithms, distorted versions of a single recording can be identified as the same music title.

The implementation of audio fingerprints in pyDejavu uses the spectrogram as the basis for creating the fingerprint. As described by the creator of the library [13], the algorithm finds peaks in the spectrogram, which are defined as time-frequency pairs that correspond to an amplitude value which is the greatest in a local "neighborhood" around it. Other such pairs around the peak are lower in amplitude and thus less likely to survive noise. To find the peaks, pyDejavu is implementing a combination of a high-pass filter and local maxima structs from the Python library SciPy. The spectrogram peak frequencies along with the time difference between them are then passed through a hash function (SHA-1), representing a unique fingerprint for this song. In order to save space, the SHA-1 hash is cut down to half its size (just the first 20 characters) and then converted to binary, reducing the fingerprint's size from 320 bits down to 80 bits. After the database is filled with the fingerprints of the available songs, a new audio can be matched using the same hashing method.

An important factor in the success of the matching process is hash alignment. When doing the original fingerprinting of a sample, the *absolute* offset, with regard to the beginning of the song, is stored. When the captured sound that is to be compared with the database is fingerprinted, the offset is *relative* to the start of the sample playback. If we make the assumption that the playback speed and sample rates are identical between the songs in our database and the input, then it follows that the relative offset should be the same distance apart. Under this assumption, for each match, the difference between the offsets is calculated as

$$D = O - S, \qquad (1)$$

where O is the offset from the original track and S is the sample offset from the recording. This always yields a positive integer since the database track will always be at least the length of the sample. All of the true matches will have this same difference. The system then looks over all of the matches and predicts the song ID that has the largest count of a particular difference.

We used the following settings when importing pyDejavu: the sampling rate is 44,100 samples per second; the FFT window size is 4096 samples; the FFT overlap ratio was 0.5; the fan value[2] was set to 15; the minimum amplitude of peaks was 10; and the minimum number of cells around an amplitude peak was 10. As these settings are not exposed in the library's API, we had to apply the changes locally. Prior to initializing the fingerprint database, we applied pre-processing in the original data, in order to bring the audio files in the same sample rate and number of channels as the videos.

Challenges and Considerations While the proof of concept for utilizing pyDejavu was initially promising, we came across some issues that should be taken into consideration in the future. The most evident was execution time. Filling in the database takes a lot of time, especially when using settings that favor high accuracy. We have identified the main bottleneck is the data input in the MySQL database. Using the settings described above, we calculate an average of 104,882 fingerprints per song, and the total running time to parse 3109 songs was approximately 210 h, even though we had 3 instances of the script working in parallel. We determined that it takes almost 45 min to calculate fingerprints for a

---

[2] Degree to which a fingerprint can be paired with its neighbors.

batch of 200 songs, but it takes 12 h to perform the database insert for each batch. This translates to 3 min per song insert, which makes sense, if each song insert translates to 100 thousand row inserts. Each experiment would take approximately a whole week to run, making it very hard to experiment on the whole pipeline end to end.

Another concern that arises is the alignment of samples between the audio signal that is being searched and the fingerprint database. The assumption is made that the playback speed and sample rates are identical between the songs in our database and the input; however, this is not always the case. While running the pipeline, we encountered cases where the song that was used in the video would not be matched, even if it existed within the database. The algorithm did surprisingly well in very noisy scenes, whereas it would miss quite obvious and prominent clips. We realized that the song in the film was in fact played at a different speed to the original, which also resulted in a slightly different pitch. This goes against the assumption that was made above, and it leads to a failure to match songs in a lot of films.

An additional issue is that due to the nature of the data-set, it is often the case that music is used in an inconsequential way. Such would be the cases when the song is so far in the background of the audio mix or is so generic that virtually any song could be used in its place. Furthermore, it is possible that the song is only used in the end credits of a film or TV episode. In these cases, when taken out of context, the visual content wouldn't really carry enough information to be easily classifiable. Due to the data-set size, we decided to keep such clips. Furthermore, we needed to intervene manually in two situations: once after the audio-video matching process is completed, in order to ensure a correct match, and once before extracting the features, in order to create the examples for the negative class of the classifier. We chose to handcraft fake examples for the negative class to ensure that the resulting data would be balanced and that the visual content would indeed not match the song.

Finally, reflection on our design choices reveals a few sources of bias. As the raw content is based on our privately owned collection, it is limited by our own taste. This could be problematic when trying to train a general-purpose classifier, introducing an element of cultural bias. The data-set is comprised almost exclusively of Hollywood films and series, and the songs that were available in both formats were mostly Western pop and rock ranging from the 1960s to the 2000s. Additionally, the notion of a song being a good fit for a video or not is inherently subjective. In our work, we have implicitly assumed that the choices of the original music directors were good fits, even though some other music director could have chosen something else. The same is true for the manufactured "bad" examples which are only bad because we judged them as such.

3 Data Representation and Feature Extraction

This section focuses on the issue of data representation, i.e., how musical and visual content is represented. We briefly discuss the issues that need attention when choosing representation, encoding, and feature extraction strategies for each of the relevant modalities. Especially for the sound domain, given that our task concerns music and not just audio signals, we have the choice of representing it either as raw audio or as some higher-level symbolic representation. The techniques for processing and transforming each one of these are quite different, with the former leaning more toward the realm of signal processing and the latter being closer to the domain of knowledge representation. For an in-depth explanation of the different uses of both types of features in deep architectures, we recommend the work of [14].

This choice is particularly important as each type of representation reveals different aspects of the content. Audio features based on signal processing might reveal more about timbre and texture of a piece (such as the energy of the signal or frequencies present), while symbolic representations allow us to extract elements that belong to musicology, such as the flow of harmony, cadence, and melodic structures. Furthermore, if the resulting classifier was to be used as a method of evaluation, or an objective function for an architecture that generates music for videos, it should be able to handle both types of representations. Therefore, including the musical content in both modalities is very important.

3.1 Audio Representations and Feature Extraction

We will first examine raw audio representations. For an in-depth analysis of how physical audio signals are captured, represented, and transformed within the digital domain, we recommend the works of [15] and [16]. The collection of handcrafted features are extracted using the modules provided in the pyAudioAnalysis library in Python [18]. The features are based on various representations of the audio signal, from both the time and spectral domain and Mel-frequency cepstrum coefficients (MFCCs) and chroma vectors. Also a there is a beat detection feature based on time distances between successive local maxima of the waveform. An in-depth explanation is given in [19].

In order to extract features from audio files, we apply mid-term windowing. The audio signal is initially split into mid-term segments (windows), typically 1 to 10 seconds long (depending on the application). Subsequently, a short-term processing stage is carried out. In this process, the audio signal is broken into short-term windows (a.k.a. frames) which can be overlapping. The analysis is then done on a frame-by-frame basis, which is a practical way of dealing with the non-stationarity of the audio signal, which usually has abrupt changes over time. The product of this step is a sequence of features, which is then used for computing feature statistics, e.g., the mean zero crossing rate of the window. In the end, each mid-term segment is represented by a set of statistics. Our implicit assumption in this process is that the mid-term windows exhibit uniform behavior with respect to audio type; therefore, extracting these statistics is reasonable. After the short-term and mid-term steps produce a vector of feature statistics per segment (e.g., 2 s long), these statistics are then long-term averaged, in order to provide a single vector representation of the whole signal. Through this process, temporal evolution details are sacrificed in order to obtain the most notable features of the music signal. Based on this method, we extract a selection of 134 handcrafted features. We decided to follow the recommendations of [17] and adopt a 2 s mid-term window with a 50% overlap, combined with a short-term window of 0.05 s and 50% overlap. After calculating the mid-term features, we store long-term averaging statistics for each feature, which include the mean, standard deviation, and average delta between frames.

3.2 Symbolic Representations of Music and Feature Extraction

We will now shift our focus to symbolic representations of music. These types of representations are not concerned with the audio as a signal, but instead are focused on higher-level information concerning musical concepts (such as notes, chords, etc.) which we will briefly describe in the following section. Our overview of the symbolic representations draws mainly from the work of [14]. Before we move to the feature extraction process, we provide some brief definitions of basic aspects of music, in order to establish a common terminology.

Notes are symbolic units of musical notation, which is defined by pitch, duration, and dynamics. *Rests* are representations of intervals of silence in a musical score. They behave much like notes; only they do not have pitch or dynamics components. *Interval* between two notes is their relative distance in terms of pitch, as quantized in semitones. For example, the major third interval means that two notes are four semitones apart. Intervals are the building blocks to create *chords*, which are sets of at least three notes. Conceptually, they are built by combining intervals (e.g., the major chord is a combination of a major triad and a perfect fifth). In this context, there are two possible representations, either *implicit* and *extensional*, enumerating the exact notes composing it, or *explicit* and *intensional*, by using a chord symbol that combines the pitch class of the root note and the type of the chord (e.g., major, minor, diminished, etc.). In our process, we use both in order to extract statistics about each song.

Rhythm is an indispensable aspect of music and its execution, though it is quite hard to define in pragmatic terms. The basic unit of rhythm, or musical pulsation, is the *beat*. Groups of beats are called *measures*. The number of beats per measure and the duration between successive beats constitute the rhythmic signature of the measure, a.k.a. *time signature* or *meter*. Usually, it is expressed as the fraction of the number of beats within the measure to the beat duration. Some frequent meters are 2/4, 3/4, and 4/4.

Using the MIDI protocol, up to 16 channels of information can be carried through a single link, between multiple instruments and computers. The two most important messages for our concern (i.e., the expression and storing of music) are *Note on* which indicates that a note is played and *Note off* which indicates that a note ends. The note events are embedded into track chunks, which are data structures that contain a delta-time value specifying the timing information and the event itself. The delta-time value is usually a relative metrical time, which is the number of ticks from the beginning. The number of ticks per quarter note is defined in a reference in the file header.

As explained in [14], while transcribing music to a symbolic representation, it is important to make some considerations on the aspect of time. The most fundamental is temporal scope, i.e., the way data will be interpreted by the model with respect to time: *global*, wherein the temporal scope is the whole musical piece; *time step*, wherein the temporal scope is a local time slice of the musical piece, corresponding to a specific temporal moment; and *note step*, wherein there is no fixed time step and the granularity of the model is the note. When either a global or time step temporal scope is used, the granularity of the time step must be defined. Usually, the time step is set to a relative duration, which typically is the smallest duration of a note in the corpus. This ensures that all notes will be represented at their proper duration, with a whole number of time steps. On the other hand, this also increases the number of time steps that need to be processed, regardless of the duration of actual notes. As we identify later, the feature extraction library that we used has some difficulty parsing the time step granularity.

Symbolic Feature Extraction and Concerns In our feature extraction process, we use the `music21` Python library [20], as found in [21], to parse and translate MIDI files to MusicXML. This parsing process often results in errors, mostly in regard to temporal synchronization and part instrumentation.[3] It is often the case that the MIDI files found freely on the Web contain corrupt sections or do not have the proper numbering system for instrument selection. Given that we are using `music21` in conjunction with `MuseScore 3` [22] for MusicXML compiling, we realized that in order to resolve this error, we had to make considerable changes to the way that the library is handling multi-track MIDI files. At the time of writing this, the issue had not been resolved; therefore, the feature extraction process is designed in order to work around it. We have decided to do our best to minimize these shortcomings from our end, without altering any code from the library itself, and let the models decide which features carried meaningful information. We enhanced the process by doing some pre-processing using the `mido` [23] library. We identified that the main library had an issue with percussion instruments, interpreting them as regular tuned instruments, and their messages were decoded as regular pitches, essentially destroying all the melodic interval features. We removed them from each MIDI file by parsing the track names with `mido`. Another issue with the library is that it is not optimized for speed, making the extraction process take a lot of time. We have calculated that it took around 6 min per MIDI file in our machine, which made the feature extraction process quite expensive in terms of computational time.

---

[3] We would like to thank Cuthbert Labs for collaborating on navigating these issues.

The `music21` library has two main groups of feature extractors. One is based on `jSymbolic` [24], a Java-based program for computational musicology. The other group is native feature extractors, developed in Python for `music21`. The categories of features of `jSymbolic` are described in the following paragraphs, drawing from the original publication in [25]. The native group of feature extractors contains a small collection of handcrafted features that mostly fall into the same categories as `jSymbolic`. Some of them, like *Quality Feature*, improve upon `jSymbolic`'s implementation of the same concept. In these cases, we accept the native class by default and skip the `jSymbolic` version of the feature.

Pitch statistics are features that describe the amount of co-occurrence between various pitches, in terms of both absolute pitches and pitch classes. These are used to also flag the amount of tonality in a piece (how much it follows rules of tonal music), the range of pitches in the piece, as well as pitch variety. *Melodies and horizontal intervals* examine the kinds of melodic intervals that are present, as well as the amount of melodic variation of the piece. Furthermore, the melodic contours are measured, as well as common phrases. *Chords and vertical intervals* concentrate on vertical intervals, the types of chords they represent, and the presence and velocity of harmonic movement.

Rhythm features are calculated based on the time intervals between note attacks and the duration of individual notes, in order to identify rhythmic patterns and variations. It contains features about rhythm measure, both on a musical piece level and on a bar-by-bar level, for more complex pieces. *Instrumentation* features are, in theory, relevant to the types of instruments that are present and the emphasis that is given to each instrument. In our implementation, we found these features to be weaker, as they are dependent on accurately parsing track metadata from the MIDI files, which we have already identified as problematic. We have therefore chosen to exclude them from the feature extraction process.

Texture features compute the level of polyphony within the piece. They also measure the amount of interaction between those voices, identifying voice independence or equality. *Miscellaneous features* are based on the metadata of the musical piece, such as *composer popularity* and *language feature*, as well as a few features specific to classical music, such as *Landini cadence*.

3.3 Video Representation and Feature Extraction

The third and final modality of interest for the task at hand is video. Getting into details of video processing and defining the relevant nomenclature is beyond the scope of this chapter, but we recommend [26] and [27] for an in-depth explanation. Based on the aspects described in the following paragraphs, we have used the modules found in the `multimodal_movie_analysis` [28] library in Python, which in turn is based on `OpenCV` and `PyTorch`, to extract a selection of features. The library is an extension of the work that was done in [29] and [30].

Color Features For the purposes of data acquisition and feature extraction, it suffices to understand the two common ways of representing color (color models): *The RGB color model* is based on a Cartesian coordinate system, whose axes represent the three primary colors of light (red, green, and blue), normalized to the [0,1] range. *The HSV color model* is based on the human perception of color. Its foundation is the psychophysics of color and is obtained by looking at the RGB color cube along its main diagonal. The advantage of this way of representing color is that it is closer to the human perception. We extract statistics for each color histogram from both color models.

Flow Features and Shot Detection The pattern of apparent motion of the objects in an image between two consecutive frames is named optical flow. This perception of motion can stem either from the movement of the object itself within the 2D frame or from the movement of the camera. To extract the features, the implementation of OpenCV that was used is based on the Lucas-Kanade method of solving the above equation. The method, proposed in [31], is using a 3x3 area around the point and makes the assumption that neighboring pixels will have the same motion. Further to these features, we also utilize a threshold on frame transformations, such as grayscale versions of the frames, optical flow, and color value changes, to detect large-scale changes between frames. The changes are interpreted as changes of the shot and kept as *shot detection* features.

Face Detection A pre-trained machine learning model called *Haar cascade classifier* is used to detect faces. It is based on a method proposed in [32] and extracts Haar features by subtracting the sum of pixels for areas of the image. The classification algorithm is based on Adaptive Boosting (*AdaBoost*), which calculates the weighted sum of many weak classifiers. The "cascading" part of the model's name is owed to the fact that features are extracted in groups and the classification is applied in sequence, only if the first classifier fails.

Single Shot MultiBox Detector In order to detect objects within frames, we made use of a pre-trained deep learning model. Based on the method described in [33], a pre-trained convolutional neural network is used to detect objects of 92 classes. The implementation that we used is derived from a ResNet-50 model (instead of the VGG applied in the paper), which is trained on the COCO data-set, a popular data-set for object detection, containing photos of multiple objects from multiple angles. The code for applying the model, as well as an explanation on the specifics of the model, is found in PyTorch Hub [34].

4 Building a Classifier

4.1 Exploring the Data-Set

Dimensions The data-set is comprised of 136 data points in total. These points are split in two classes "match" and "no-match," depending on whether they are real examples of video soundtracks that come from the process described in Sect. 2.2 or hand-picked mismatched examples. The goal of the classifier is to discriminate the matching soundtracks from the mismatched, constructed examples. Overall, the data-set is balanced, as there are 68 matches and 68 fake examples. We notice that, while the unweighted mean in the data-set is exactly 50%, if we group the clips by originating film, the baseline is slightly increased. The median number of clips per film is 3, and the mean rate of positive clips per film is 52%. Therefore, the method of cross-validation by leaving one film out each time seems to make sense, and our baseline threshold against pure chance (within the sample of data that is available) is 52%. The feature extraction process described in the previous section results in 455 attributes per data point. More specifically, there are 138 features from the audio content, 73 features from the symbolic content (MIDI), and 244 features from the visual content.

Attribute Completeness and Scaling In order to be able to apply further statistical tests, we first test the completeness of the features. It seems that there is no serious issue with data completeness, as there are no columns with more than 10% empty values and only three features that had 5% of empty values, which originate from the symbolic modality. The completeness issue is mild enough that we will only use a simple imputation technique, using the mean value of the column. We wouldn't expect to greatly affect the classification results by using more complex imputation techniques. There are ten features whose values are all zero. They come mostly from the set of object detection features; therefore, we expect that to depend on the diversity of the visual content (e.g., no sport scenes included, no outdoor backgrounds detected, etc.). As expected, features concerning MIDI file metadata are included, which is probably connected to the issues we identified with how `music21` parses them. We decided to remove all ten features from further tests.

The maximum value in the data-set is 8105.14, while the minimum value is -28.58. The mean value in the data-set is 10.0. This shows us that there are big outliers and potential scaling issues in our data-set. We address this issue by applying simple scaling methods, either MinMax scaling (in most cases) or standardization (when applying PCA for dimensionality reduction).

Attribute Relevance We ran a Spearman correlation test, a non-parametric test for statistical dependence between two variables, between all the features and the class. We sort that list based on the absolute value of the correlation and collect the ten features which rank the highest, displayed in Table 1. It is evident that there is no strong monotonic correlation between a feature and the class. It is worth noting

Table 1 Features with the highest absolute value of Spearman's correlation coefficient with the target variable

Feature name	Absolute Spearman correlation
chroma_7_std	0.26
delta chroma_7_std	0.24
RelativeStrengthOfTopPitchClassesFeature	0.24
chroma_7_mean	0.23
delta energy_entropy_mean	0.22
delta chroma_std_std	0.20
delta chroma_10_std	0.19
VariabilityOfNoteDurationFeature	0.18
RelativeStrengthOfTopPitchesFeature	0.18
mfcc_3_mean	0.18

that although the number of visual features is larger than the sum of the other two modalities combined, none of them belong in the top 5% of features with the highest monotonic correlation with the target variable. While this is not a definitive measure of feature importance, this observation could perhaps lead to further future improvements in the data extraction process.

We also performed a mutual information test between each (scaled) feature and the target variable, using a non-parametric method based on entropy estimation from k-nearest neighbors distances. This test yielded 162 features with non-zero mutual information with the target. Out of these, only nine are from the visual modality, and they all come from the object detection group. This is a bit worrying, as these features are very specific to the videos that were used, and we would expect them to be volatile to the types of video clips that were extracted during our collection process. If in the future a larger and more diverse data-set is collected, we would expect the lower-level visual features (such as color histograms) or flow features to also become more important.

Train-Test Split Methodology Given the length of our data-set, we would need some form of cross-validation in order to get the best possible results from this small set of data. To avoid the danger of overfitting, we created splits that always include all clips that come from the same film in either the training set or the test set. Ideally, we would hold out videos from a few movies, in order to create a final held-out test set. As the video clips originate from only 33 films, there were only 33 possible folds that we could create; therefore, holding some of these films for a final test set was not possible. Knowing that this is not the best solution, we decided to create the folds and store the predicted class, the probability of the prediction, and the real value. We then computed the metrics of interest using all the validation set predictions. Scaling, imputation, and decomposition techniques were fitted to the training set of each fold. An issue that we identified with this method is that whenever the held-out film has a lot of relevant clips, the results are significantly affected. We would expect this volatility to decrease as the data-set size increases.

Table 2 Baseline results

		DT	LR	kNN	SVM	RandF	BTr	AdaBoost	XGBoost	GradB	ExTr	NB
Accuracy		0.5	0.43	0.53	0.41	0.46	0.42	0.56	0.49	0.53	0.51	**0.57**
Precision	0	0.5	0.43	0.53	0.43	0.47	0.43	0.56	0.49	0.52	0.51	**0.65**
	1	0.5	0.43	0.53	0.38	0.46	0.41	**0.56**	0.48	0.54	0.53	0.55
Recall	0	0.6	0.43	0.54	0.56	0.53	0.49	0.56	0.60	0.66	**0.72**	0.32
	1	0.4	0.44	0.51	0.26	0.40	0.35	0.56	0.37	0.40	0.31	**0.82**
f1 measure	0	0.55	0.43	0.54	0.49	0.50	0.46	0.56	0.54	0.58	**0.60**	0.43
	1	0.44	0.44	0.52	0.31	0.43	0.38	0.56	0.42	0.46	0.39	**0.66**

Bold values represents the highest result for each metric (rows)
Note: rows are metric and columns are algorithms

In order to perform hyper-parameter tuning, we took the most promising settings and compared them against each other, using the Leave-One-Film-Out method on the whole data.

Models Tested We tested a few classification algorithms, from various model families. Given the length of the data-set, neural networks were deemed inappropriate. Before any optimization, we tested the algorithms in their default settings. The algorithms used are decision tree *(DT)*, logistic regression *(LR)*, k-nearest neighbors *(kNN)*, support vector machines *(SVM)*, random forest *(RandF)*, bagged trees *(BTr)*, AdaBoost *(AdaB)*, XGBoost *(XGB)*, Naïve Bayes *(NB)*, Extra Trees *(ExTrees)*, and gradient boosting ensemble *(GradBoost)*. The Naïve Bayes classifier is the winner with 57% mean average and 66% f1 measure. The results are summed up in Table 2.

Model Selection We performed various experiments in order to create candidate pipelines. These include a scaling stage, a dimensionality reduction stage, and an estimator stage. Dimensionality reduction was crucial, given the high number of features available. Apart from PCA, we tried excluding features based on mutual information, as well as using a Pearson correlation test for cross-correlation between features. We present here the three candidate pipelines, which belong to different families of classifiers. The parameters checked as well as the results are demonstrated in Table 3. The winning pipeline is standard scaling, PCA, and Extra Trees, with an accuracy of 60% and macro-average f1 measure of 58%. The complete classification report is given in Table 4.

Results Discussion We attempted to approach our methodology in a way that would reduce bias while taking advantage of the limited data-set as much as possible. This latter constraint stopped us from using a held-out set of films in order to estimate how well the models generalize. Given that we only had samples from 33 films, removing any of them from the set would have a big impact in the learning process. The limitations of the data-set also stopped us from testing deep architectures.

Table 3 Estimator pipeline selection

Pipeline	Parameter	Values	Accuracy	Macro f1
Standard scaling, PCA, Extra Trees	Number of estimators	50	0.49	0.48
		100	0.54	0.54
		150	**0.60**	**0.58**
		200	0.58	0.56
MinMax scaling, mutual information dimensionality reduction, Naïve Bayes	Number of features kept	50	0.43	0.42
		100	0.48	0.45
		150	0.54	0.49
		200	0.51	0.47
MinMax scaling, Pearson correlation dimensionality reduction, kNN	Number of neighbors	1	0.53	0.53
		2	0.51	0.49
		3	0.53	0.53
		4	0.48	0.46

Bold values represents the highest result for each metric (columns)
Note: rows are alternative data pipelines and columns are metric

Table 4 Classification report for winning estimator

	0	1	Macro Avg
Precision	0.64	0.57	0.61
Recall	0.43	0.76	0.60
F1 measure	0.51	0.65	0.58
Accuracy			0.60

Furthermore, the mediocre results of algorithms, such as SVM, that usually perform well indicate that probably more data is needed for the classifier to learn the relationships between so many variables. This makes sense, given the original dimensions that we are working with. It is also worth noting that the mutual information and Spearman correlation tests performed in Sect. 4.1 scored the visual features quite low. Perhaps this is an indication that the model doesn't have enough data yet, in order to determine the association between visual features and music. To determine if this is a valid assumption, we would need to perform some tests by isolating pairs of modalities, but time would not allow for this to happen.

Despite those challenges, a fine-tuned pipeline including a scaling, a dimensionality reduction, and a classification step achieved results that were better than chance. We believe that this is a promising indication that more data would allow a future researcher to build a much more robust model. This being a proof of concept, we consider the results adequate enough to conclude our own research. While, overall, the model would not be suitable for cases where high precision matters, at this stage, it could serve as an initial filtering classifier for a soundtrack retrieval system.

5 Extensions and Future Work

Expanding the Data Collection Pipeline Despite our best efforts, the issues that we identified in Sects. 2.1 and 2.2 made it hard to gather a large enough data-set in terms of data points. One solution could be crowdsourcing. The way that the pipeline is built allows for bootstrapping more data from the community, without sharing any of the raw underlying content. The robustness of the process in different computer setups needs further testing.

Scaling is hindered by efficiency and speed issues. The most time-consuming element during our own experiments was the bottleneck of inserting hundreds of thousands of lines per song in the fingerprint table. Optimizing the database for inserts would probably solve this issue. Further to that, song detection in videos also took a lot of time. The process can be modified to run in a distributed way, which would probably be an easy win to reduce the time between experiments. Another time-consuming task was feature extraction, especially for the symbolic features. We have identified a few problems with the underlying library that we used, but it should not be too hard to modify our extractor class so that it runs more than one feature extraction processes at the same time.

Finally, adding additional modalities for the participating content such as text, in the form of lyrics or subtitles, could be interesting. Research on whether there are semantic or topic-related similarities between the lyrics of a song and the lines of dialogue that are uttered during the scene seems to have potential. Another interesting direction would be to add metadata about the scene using the results of a classifier that labels scene content as features, as done in [35].

Data Quality Another issue that we identified was that often the actual content of a file was not the same as what the filename would suggest. This problem is very important, as slight errors in labelling can lead to errors in matching and essentially create severe issues in later stages of the pipeline. We address data corruption for each file type in the following paragraphs.

Audio Files suffered from the existence of live performances or covers of songs in the data-set. Even though we tried to impose some filtering in the initial parsing and cataloging of the data, when we were reviewing the final catalog, we came across plenty of mislabeled files. One solution could be to filter out whole subsets of songs, based on metadata filtering either from the audio files themselves or from external metadata sources such as Spotify.

MIDI Files were a particularly problematic data source in this pipeline. The files that are freely available on the Web often suffer from erroneous choice of instruments, incorrect notes, or overall file corruption. Furthermore, the naming conventions are inconsistent, and the existence of metadata, like artist name or song name, within the file, is very rare. One approach could be adding the "Lakh MIDI Dataset" by [36]. Furthermore, one could try parsing the files with a library like mido to identify the instrument names and the playback duration of each file and storing those in the appropriate table. This would allow running some diagnostics on the health of the file and throwing away corrupt files at an early stage. Improving the text cleaning rules that are applied on the names of the files before performing a query in an external knowledge base, or trying different knowledge bases to retrieve a proper ID for the file, could also help.

Video Files were mostly consistent, in terms of quality. The biggest improvement would be to use a more sophisticated cleanup method to get better file names and compare that to an external knowledge base in order to get more metadata or lists of soundtracks.

Reducing Bias The lack of diversity in our data manifests itself in a number of ways. A crucial one is better handling of UTF-8 characters in our script, so that audio and MIDI files that have non-ASCII characters in their title can be fed to the knowledge base query in a robust way. Further to that, we could use an external knowledge base that has song names in their original language. Furthermore, crowdsourcing the data would mean that people of more diverse aesthetic and cultural background bring their own taste in film and music into the raw data-set. Finally, dropping the manual creation of fake examples in favor of some automated or semi-automated process, could reduce bias. Some kind of bootstrapping method could be applied, but the specifics of such a method could be the object of another study.

Song Detection in Videos Increasing the precision and recall of our method could be achieved using an external knowledge base in order to get the names of the songs that are present in each film. This could serve as ground truth for a process that accepts or rejects what the song detection script yields as result. Furthermore, the common practice in films of altering the playback speed (and therefore the pitch) of a song is detrimental to the success of the fingerprint algorithm which is based on the spectrogram of the sound and therefore the relationship between frequency and time. When these are altered, the peaks of the spectrogram are in different frequencies and at different timestamps. Solving this problem could be the object of a future study, as it would probably mean that other, more robust algorithms, compared to the implementation of `pyDejavu`, should be tested. Other proposed algorithms exist, such as [37] which uses auto-correlation, [38] that uses wavelets, and more recently [39].

Algorithm Improvements Given the dimensionality of the data-set, we would argue that the acquisition of more and better data is much more important than building more sophisticated algorithms. Nevertheless, our experiments led to three classification settings which could, potentially, be combined into a voting ensemble. Moreover, acquiring more data might make deep architectures a suitable setting for experimentation, due to the large number of features extracted. Finally, one of our biggest concerns around the classifier's performance is what exactly it is learning. The mutual information tests that we applied in Sect. 4 show that the visual content features have significantly lower mutual information with the target variable than the features of the other modalities. We have hypothesized that this is due to the lack of diversity in the visual content and the small size of the data-set. If more data were collected, this hypothesis should be put to the test in future work.

6 Conclusion

In this chapter, we got involved with the task of video soundtrack evaluation. We proposed an end-to-end data collection and feature extraction pipeline, which takes raw video, audio, and MIDI files as its input, matches them using text processing and audio fingerprinting, and creates a multimodal feature library. To our knowledge, this is the first attempt to combine these three modalities.

We managed to create a small proof of concept using some limited data and built a classifier with the task of discriminating between real and fake soundtracks. The results of this classifier were adequate, as the resulting accuracy is better than random choice, though we believe that the performance could be improved by assembling a larger data-set. Overall, the model would not be suitable for cases where high precision matters, but at this stage could serve as an initial filtering classifier for a soundtrack retrieval system.

Finally, we explored some solutions to the various challenges and problems of both the data collection process and of the resulting classifier and presented some interesting directions for future research.

References

1. Yu, Y., Shen, Z. & Zimmermann, R. Automatic Music Soundtrack Generation for Outdoor Videos From Contextual Sensor Information. Proceedings Of The 20th ACM International Conference On Multimedia. pp. 1377–1378 (2012)
2. Shah, R., Yu, Y. & Zimmermann, R. Advisor: Personalized Video Soundtrack Recommendation by Late Fusion With Heuristic Rankings. Proceedings Of The 22nd ACM International Conference On Multimedia. pp. 607–616 (2014)
3. Shah, R., Yu, Y. & Zimmermann, R. User Preference-aware Music Video Generation Based on Modeling Scene Moods. Proceedings Of The 5th ACM Multimedia Systems Conference. pp. 156–159 (2014)
4. Lin, J., Wei, W., Yang, J., Wang, H. & Liao, H. Automatic Music Video Generation Based on Simultaneous Soundtrack Recommendation and Video Editing. Proceedings Of The 25th ACM International Conference On Multimedia. pp. 519–527 (2017)
5. Herberger, T. & Tost, T. System and Method of Automatically Creating an Emotional Controlled Soundtrack, Google Patents, US Patent 7,754,959 (2010)
6. Lin, J., Wei, W. & Wang, H. EMV-matchmaker: Emotional Temporal Course Modeling and Matching for Automatic Music Video Generation. Proceedings Of The 23rd ACM International Conference On Multimedia. pp. 899–902 (2015)
7. Sato, H., Hirai, T., Nakano, T., Goto, M. & Morishima, S. A Soundtrack Generation System to Synchronize the Climax of a Video Clip With Music. 2016 IEEE International Conference On Multimedia And Expo. pp. 1–6 (2016)

8. Hong, S., Im, W. & Yang, H. Cbvmr: Content-based Video-music Retrieval Using Soft Intra-modal Structure Constraint. Proceedings Of The 2018 ACM On International Conference On Multimedia Retrieval. pp. 353–361 (2018)
9. https://composing.ai/dataset (accessed 2020-05-02)
10. Spotify-Python API connector https://spotipy.readthedocs.io/en/2.16.1/
11. pyDejaVu Audio Fingerprint Library https://pypi.org/project/PyDejavu/
12. Cano, P., Batlle, E., Kalker, T. & Haitsma, J. A Review of Audio Fingerprinting. Journal Of VLSI Signal Processing Systems For Signal, Image And Video Technology. 41, 271–284 (2005)
13. Drevo, W. Audio Fingerprinting With Python And Numpy. (2013), https://willdrevo.com/fingerprinting-and-audio-recognition-with-python/ (accessed 2020-12-05)
14. Briot, J., Hadjeres, G. & Pachet, F. Deep Learning Techniques for Music Generation–A Survey. ArXiv Preprint ArXiv:1709.01620. (2017)
15. Kim, H., Moreau, N. & Sikora, T. MPEG-7 Audio and Beyond: Audio Content Indexing and Retrieval, John Wiley & Sons (2006)
16. Giannakopoulos, T. & Pikrakis, A. Introduction to Audio Analysis: a MATLAB®Approach, Academic Press (2014)
17. Karim, D., Adnen, C. & Salah, H. An Optimization of Audio Classification and Segmentation Using GASOM Algorithm. International Journal of Advanced Computer Science and Applications. 9, 143–157 (2018)
18. The pyAudioAnalysis library https://github.com/tyiannak/pyAudioAnalysis
19. Giannakopoulos, T. pyaudioanalysis: An Open-source Python Library for Audio Signal Analysis. PloS One. 10, e0144610 (2015)
20. The Music21 Library http://web.mit.edu/music21/ (accessed 2020-12-14)
21. Cuthbert, M., Ariza, C. & Friedland, L. Feature Extraction and Machine Learning on Symbolic Music Using the music21 Toolkit.. Ismir. pp. 387–392 (2011)
22. The MuseScore website https://musescore.org/en (accessed 2020-06-01)
23. Mido Python https://mido.readthedocs.io/en/latest/ (accessed 2020-09-02)
24. The jSymbolic program http://jmir.sourceforge.net/ (accessed 2020-09-02)
25. McKay, C., Tenaglia, T. & Fujinaga, I. jSymbolic2: Extracting Features From Symbolic Music Representations. Late-Breaking Demo Session Of The 17th International Society For Music Information Retrieval Conference. (2016)
26. Marques, O. Practical Image and Video Processing Using MATLAB. John Wiley& Sons (2011)
27. Bovik, A. Handbook of Image and Video Processing. Academic press (2010)
28. Multimodal Movie Analysis Python Library https://github.com/tyiannak/multimodal_movie_analysis (accessed 2020-12-20)
29. Bougiatiotis, K. & Giannakopoulos, T. Multimodal Content Representation and Similarity Ranking of Movies. ArXiv Preprint ArXiv:1702.04815. (2017)
30. Psallidas, T., Koromilas, P., Giannakopoulos, T. & Spyrou, E. Multimodal Summarization of User-Generated Videos. Applied Sciences. 11, 5260 (2021)
31. Lucas, B., Kanade, T. & Others An Iterative Image Registration Technique With an Application to Stereo Vision. Proceedings Of Imaging Understanding Workshop. pp. 121–130 (1981)
32. Viola, P. & Jones, M. Rapid Object Detection Using a Boosted Cascade of Simple Features. Proceedings Of The 2001 IEEE Computer Society Conference On Computer Vision And Pattern Recognition. CVPR 2001. 1 pp. I-I (2001)
33. Liu, W., Anguelov, D., Erhan, D., Szegedy, C., Reed, S., Fu, C. & Berg, A. SSD: Single Shot Multibox Detector. European Conference On Computer Vision. pp. 21–37 (2016)
34. PyTorch Hub https://pytorch.org/hub/nvidia_deeplearningexamples_ssd/ (accessed 2020-12-01)
35. Giannakopoulos, T., Makris, A., Kosmopoulos, D., Perantonis, S. & Theodoridis, S. Audio-visual Fusion for Detecting Violent Scenes in Videos. Hellenic Conference On Artificial Intelligence. pp. 91–100 (2010)
36. Raffel, C. Learning-based Methods for Comparing Sequences, With Applications to Audio-to-MIDI Alignment and Matching. Columbia University (2016)

37. Haitsma, J. & Kalker, T. Speed-change Resistant Audio Fingerprinting Using Auto-correlation. 2003 IEEE International Conference On Acoustics, Speech, And Signal Processing, 2003. Proceedings.(ICASSP'03). 4 pp. IV-728 (2003)
38. Baluja, S. & Covell, M. Waveprint: Efficient Wavelet-based Audio Fingerprinting. Pattern Recognition. 41, 3467–3480 (2008)
39. Yao, S., Niu, B. & Liu, J. Enhancing Sampling and Counting Method for Audio Retrieval with Time-Stretch Resistance. 2018 IEEE Fourth International Conference On Multimedia Big Data (BigMM). pp. 1–5 (2018)

Deep Learning Approach to Joint Identification of Instrument Pitch and Raga for Indian Classical Music

Ashwini Bhat, Karrthik Gopi Krishnan, Vishal Mahesh, and Vijaya Krishna Ananthapadmanabha

1 Introduction

Indian classical music consists of two basic genres, namely, Hindustani and Carnatic, and is known to have unique raga structures. As a result of this, the standard computer-based analysis used in Western music has to be modified to incorporate the salient features of these genres.

Though the two genres developed along different lines, the Indian classical instruments such as bansuri and sitar (Hindustani) are known to share a lot of similarities with flute and veena (Carnatic), respectively [1]. The instruments have evolved over the years to suit contemporary needs as well as the growing demand for their revival in response to the market demands of generation Z. Despite this recent development, instruments with similar characteristics still pose a challenge and make identification difficult. The pitch which is also known as "shruthi" is one of the most complex concepts of Indian music and is often considered the backbone of Indian music [2].

In the case of Indian classical music, the rhythmic instruments such as mridangam, tabla, ghatam (earthen pot), etc. are also tonic in nature. Thus, melodic analysis and rhythmic analysis are heavily dependent on the pitch, which in turn is difficult to identify via conventional music information retrieval (MIR).

Raga is a sequence of notes known as "swaras" which produce a melody. Each raga consists of a characteristic sequence of notes unique to it which helps express various moods. Conventional techniques for raga identification involve extracting the sequence of swaras and comparing it with pre-existing templates for the raga. This is a tedious process as it involves pitch detection before note extraction.

A. Bhat (✉) · K. G. Krishnan · V. Mahesh · V. K. Ananthapadmanabha
Department of Electronics and Communication, PES University, Bengaluru, Karnataka, India
e-mail: ashwinib@pes.edu; vijayakrishna@pes.edu

© The Author(s), under exclusive license to Springer Nature Switzerland AG 2023
A. Biswas et al. (eds.), *Advances in Speech and Music Technology*, Signals and Communication Technology, https://doi.org/10.1007/978-3-031-18444-4_8

2 Related Work

Identification of pitch, especially for Indian classical music, is a herculean task. Tonic identification is hence pivotal for swara and raga identification. Ragas are unique features that help identify songs and are broadly classified as Hindustani raga and Carnatic raga. The basic component of raga can be broken down into scales with a given set of swaras/notes. The pattern of recurring swaras is used to identify the raga.

The classical signal processing method involves spectrogram analysis followed by frequency selective extraction using fast Fourier transform (FFT) [3]. Sinusoidal extraction is applied to extract the spectral peaks which represent the pitches of the voice. The salience function is employed to construct a pitch histogram. Using the tonic selection technique, the pitch is identified from the most frequently occurring frequencies in the histogram.

FFT is used to identify the frequency content among the noise in the audio tracks prior to peak identification. Following this, the dominant frequency is identified as the pitch of the audio track [4]. Another unique method in pitch extraction is the extraction of tonal features such as the fundamental frequency. Pitch distribution is performed using probability density functions or pitch histograms, and the following pitch is cited [4]. After pitch identification, swaras are identified by finding the onset frequencies and comparing them to the pitch frequencies of the song.

Several classical approaches are explored to pitch identification [5–10], which is very tedious and requires very careful analysis. The pitch and the swara must also be determined in order to identify the raga. Classical approaches put a huge emphasis on the preprocessing and quality of audio samples used [11, 12]. It is difficult to extend the work for the increased dataset. Traditional methods are therefore too restrictive and complex.

We also explore various machine learning methods which have been implemented for the above complex tasks. There is a direct connection between the tasks and musical features. Bhat et al. [13] provide a comparative study comparing the performance of different deep learning models such as ANN, CNN, Bi-LSTMs, and XGboost for raga identification. XGboost was found to give the highest performance. The chapter also gives a visual representation of feature importance for the XGboost model.

3 Preprocessing

Data is at the heart of neural networks. For optimal results, it is essential to have an ensemble of different genres to make our models as diverse as possible. Before feature extraction, audio clips are segmented into 30-second-long samples. Segmentation was performed with PyAudio library. Audio clips were obtained from the Dunya dataset and consisted of Hindustani and Carnatic instrumental audio tracks.

Table 1 Music features which are extracted in preprocessing. NSS refers to non-source separated and SS to source separated

Feature	Formulae	Range NSS	Range SS
Zero crossing rate	$zrc = \dfrac{1}{L-1} \sum_{L-1}^{t-1} 1\mathbb{R}_{<0}(s_t s_{t-1})$	[0.0315–0.2351]	[0.0236–0.1937]
Chroma stft	$X(m,k) = \sum_{n=0}^{N-1} z(n+mH)u(n)exp\left(-\dfrac{2\Pi kn}{N}\right)$	[0.1359–0.4947]	[0.412–0.5032]
RMSE	$RMSE = \sqrt{\dfrac{\sum_{n}^{i=1}(P_i - O_i)^2}{n}}\, S$	[0.00024–0.4453]	[0.0001–0.452]
Spectral centroid	$centroid = \dfrac{\sum_{n=0}^{N=1} f(n)k(n)}{\sum_{n=0}^{N=1} k(n)}$	[842.0582–2852.8195]	[830.2313–2421.1343]
Spectral bandwidth	$BW = \left(\sum_{k} S(k)(f(k)-f_c)^p\right)^{1/p}$	[945.3606–3223.2640]	[463.8354–2390.8166]
Roll-off	$rolloff = 0.8 \sum_{n}^{i=1} S_t(n))$	[1373.3750–7280.0413]	[568.3632–5481.1585]
Mel-frequency spectral coefficients (1–20)	$mfcc = \sum_{n=0}^{N} X[h\tau + n]W[n]e^{-j\frac{2\pi kn}{N}}$	[−718.5999–117.53577]	[−855.6945–218.9529]

3.1 Feature Extraction

For the purpose of dataset creation, 26 standard audio features (shown in Table 1) were extracted via the Librosa library [14]. The class labels corresponding to the tasks are encoded as integers. Audio songs are visually represented via spectrograms, and songs are linked to tasks through features. Pitch classes of chroma shift resemble the swara notes of Indian classical music.

3.2 Top Feature Identification

XGboost is a decision tree-based gradient boosting machine learning algorithm. These models automatically provide an estimate of feature importance [13]. This was used to identify the top ten features. These features which play the most significant role in the identification process are grouped to form an optimized comma-separated values (CSV) file (shown in Table 2).

Table 2 Top 10 features for instrument and pitch based on results from XGboost

Sl. no.	Top 10 features Feature for pitch and instrument	Feature for raga
1	Chroma stft	Chroma stft
2	RMSE	Spectral centroid
3	Spectral centroid	Zero crossing rate
4	Zero crossing rate	Mfcc0
5	Mfcc1	Mfcc3
6	Mfcc4	Mfcc6
7	Mfcc8	Mfcc7
8	Mfcc10	Mfcc9
9	Mfcc13	Mfcc12
10	Mfcc14	Mfcc13

3.3 Feature Extraction

For the purpose of dataset creation, 26 standard audio features (shown in Table 1) were extracted via the Librosa library [14]. The class labels corresponding to the tasks are encoded as integers. Audio songs are visually represented via spectrograms, and songs are linked to tasks through features. Pitch classes of chroma shift resemble the swara notes of Indian classical music.

3.4 Top Feature Identification

XGboost is a decision tree-based gradient boosting machine learning algorithm. These models automatically provide an estimate of feature importance [13]. This was used to identify the top ten features. These features which play the most significant role in the identification process are grouped to form an optimized comma-separated values (CSV) file (shown in Table 2).

3.5 Source Separation

- Melodic source separation refers to the process where the instrumental pitch component of the audio tracks is carefully separated from the clippings [15]. This gives us a purer version of the songs whose melodic features are more distinct. This technique was applied to all the audio samples, and the corresponding features were extracted to create a new set of CSV file.

4 Methodology

The instrument, pitch, and raga classification is performed using four different models, namely, recurrent neural network (RNN), one-dimensional convolution neural network (CNN), and XGboost.

All the models were trained for three different datasets, the original dataset with 26 features, the top 10 optimized features, and a source-separated dataset. The performance of the models is then evaluated for all three tasks (instrument, pitch, and raga identification) for all three datasets.

4.1 Convolution Neural Network (CNN)

A conventional neural network is a deep learning algorithm that classically takes images as input and assigns weights and bias to different objects in the image for differentiation, classification, etc. They are commonly used in image processing and computer vision applications. The same training idea can however be extended to CSV datasets for classification.

A one-dimensional CNN was employed for the same. Kernel size is set to 3. The model hence iterates over three features at an instance till all features are covered (for each epoch). This allows the model to learn from the input features.

4.1.1 Model Details

ReLu activation function is used for the input and hidden layers and SoftMax activation for the output layer.

ADAM optimizer has been used with pre-set values for beta1, beta2, and epsilon and a learning rate of 0.0001. Instrument models were trained for 50 epochs, pitch for 60 epochs, and raga for a hundred epochs. Model parameters were changed several times to maximize accuracy and minimize losses.

The sparse categorical cross-entropy loss function has been utilized, as the output is in the form of a singular integer (prediction gives a single output) and not in the form of arrays.

Categorical cross-entropy is commonly used for multi-class classifications. The labels are mutually exclusive (each data point belongs to only one class). Hence, a sparse categorical cross-entropy loss function has been utilized as it saves memory and improves speed. This loss function compares the predicted and true label classes to calculate the loss.

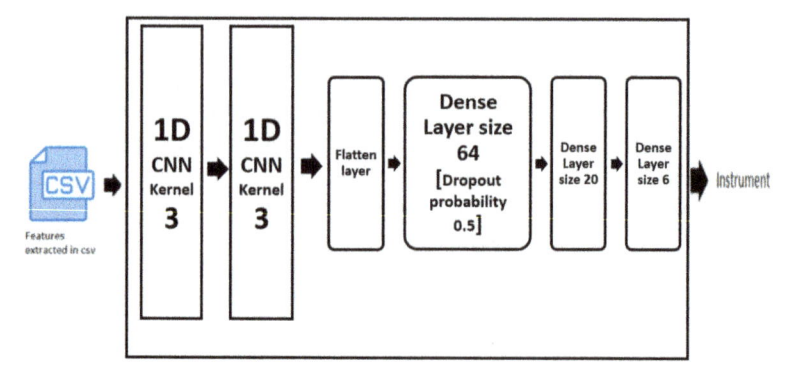

Fig. 1 CNN model flow for instrument identification

4.1.2 Customization of Models

Instrument Identification

The model consists of an input layer, an output layer, and six hidden layers. There are 2 Conv-1D layers with filter sizes of 32 and 64, 3 flattening layers, 0.5 probability dropout layers, and dense layers with 64 and 20 neurons, respectively. The output layer consists of six neurons corresponding to the six instruments (Fig. 1).

Pitch Detection

The model consists of an input layer, an output layer, and seven hidden layers. Hidden layers consist of 3 Conv-1D layers with filter sizes of 32, 64, and 128, followed by flattening layers, dropout layers with 0.5 probability, and dense layers containing 64 neurons and 20 neurons, respectively. In the output layer, there are 12 neurons corresponding to the 12 classes of pitch.

Raga Recognition

The model consists of an input layer, an output layer, and eight hidden layers. The hidden layer consists of 3 1D convolution layers having filter sizes 32, 64, and 128, respectively, and a kernel size of 3. These layers are followed by flattening layers, dropout layers (probability of 0.5), and dense layers containing 64 neurons, 32 neurons, and 20 neurons, respectively. The output layer is composed of 15 neurons corresponding to 15 raga classes.

4.2 Recurrent Neural Networks (RNN)

A recurrent neural network or RNN for short is a feedforward neural network that has internal memory. RNN uses sequential or time series data as input. It is mainly used for applications such as natural language processing, speech recognition, language modeling and generation of images from text, etc. RNNs have very high processing capabilities due to the presence of internal memory. This allows them to remember the previous state and draw relations between the different data points and share parameters such as bias and weights across the different layers.

LSTM (long short-term memory) forms the basic building block of RNN. They are used in many configurations, one to many, many to one, and many to many to name a few. In our models, we have utilized the many-to-many configurations.

4.2.1 Model Details

Tanh activation function has been used for the LSTM layers and SoftMax activation for the dense layers. Tanh or the hyperbolic tangent function is a non-linear function with a range from -1 to 1. It is similar to a sigmoid in shape but has a larger range.

ADAM **optimizer** has been used with pre-set values for beta1, beta2, and epsilon and a learning rate of 0.0001. Instrument models were trained for 100 epochs, pitch for 150 epochs, and raga for 200 epochs. Model parameters were changed several times to maximize accuracy and minimize losses.

Categorical cross-entropy is commonly used for multi-class classifications. The labels are mutually exclusive (each data point belongs to only one class). Hence, a sparse categorical cross-entropy loss function has been utilized as it saves memory and improves speed. This loss function compares the predicted and true label classes to calculate the loss.

4.2.2 Customization of Models

Instrument Identification

The model consists of an input layer, an output layer, and three hidden layers. Unlike the previous models, the input was directly fed to the RNN network. Hence, the input layer also functions as the first RNN layer with an output shape of tensor (26,560). There are 300, 100, and 50 LSTM units in the following hidden layers. The output layer is a dense layer consisting of six neurons corresponding to the six instruments (Fig. 2).

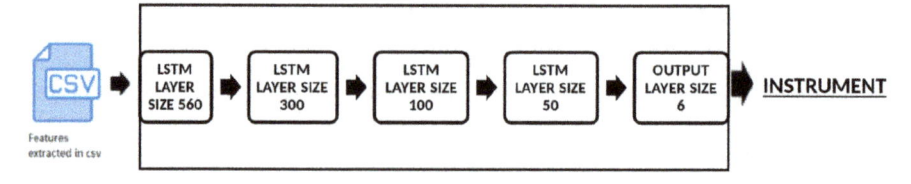

Fig. 2 RNN model flow for instrument identification

Pitch Detection

The model consists of an input layer, an output layer, and seven hidden layers. Hidden layers consist of 3 Conv-1D layers with filter sizes of 32, 64, and 128, followed by flattening layers, dropout layers with 0.5 probability, and dense layers containing 64 neurons and 20 neurons, respectively. In the output layer, there are 12 neurons corresponding to the 12 classes of pitch.

Raga Recognition

The model consists of an input layer, an output layer, and eight hidden layers. The hidden layer consists of 3 1D convolution layers having filter sizes 32, 64, and 128, respectively, and a kernel size of 3. These layers are followed by flattening layers, dropout layers (probability of 0.5), and dense layers containing 64 neurons, 32 neurons, and 20 neurons, respectively. The output layer is composed of 15 neurons corresponding to 15 raga classes.

4.3 Extreme Gradient Boosting (XGboost)

XGboost is a decision tree algorithm based on gradient boosted decision trees. Due to its excellent performance in analyzing small structured data, it is widely used in machine learning applications. In contrast, neural networks perform better than XGboost when it comes to the analysis of unstructured data and the visualization of the data.

A significant benefit of XGboost is that it is exceptionally fast and consumes less computational resources. It also has several optimizations and enhancement algorithms. As a result, the data can be visualized and analyzed more easily. There are several hyperparameters that play a role in the functioning of the model. Tuning of these hyperparameters depends on the task at hand and the size of the dataset. It is necessary to choose a range of values before discretizing them in order to achieve optimal results when tuning these hyperparameters. The tuned hyperparameters can be found in Table 3.

Table 3 XGboost model parameters

Parameter	Optimized feature's value
Eta	0.04
Max depth	3
Min child weight	1
Gamma	0.0
Colsample_bytree	0.7

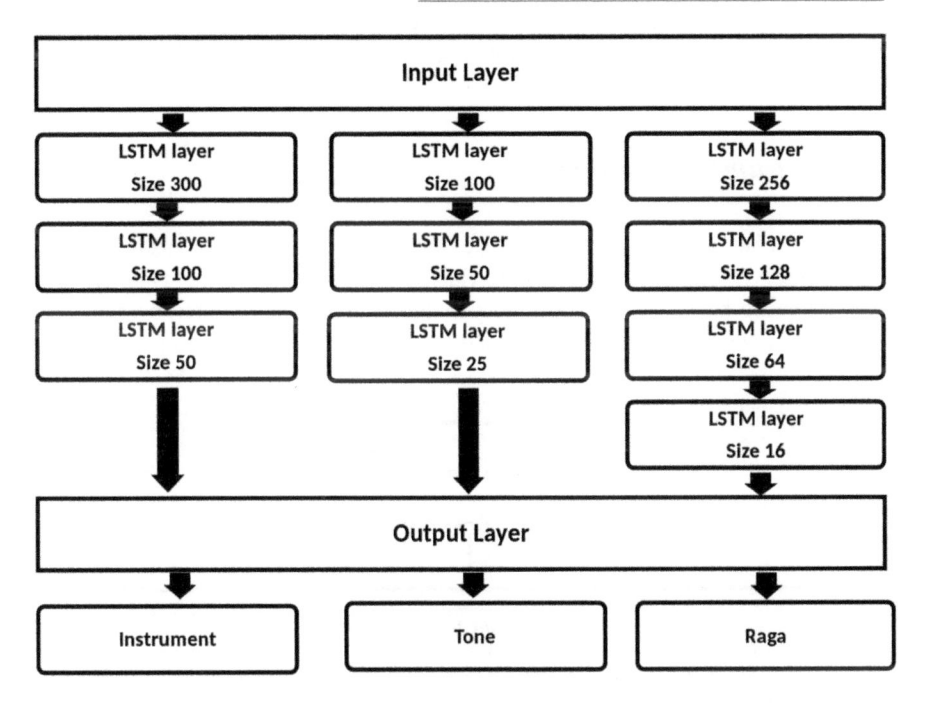

Fig. 3 Flow diagram for combined model

4.4 Combined Model

A combined neural network model was created for the three tasks. In the combined model, one input is taken into account, and three different predictions are given corresponding to instrument, pitch, and raga. The model flow is shown in Fig. 3.

4.4.1 Model Details

Similar to RNN, tanh activation function is being utilized in the LSTM layers and SoftMax for dense output layers. ADAM optimizer is used with a learning rate of 0.0001.

5 Results

For each task, three datasets containing an equal number of samples were created. There are 6 instruments, 12 pitch classes, and 16 raga classes considered. In order to extract features, audio samples were obtained from the Dunya dataset [11] and segmented into 30-second samples.

For the combined model, 864 train samples and 216 test samples were used. Figure 4 provides a detailed description of the training and test samples used.

5.1 Instrument Identification

Table 4 lays out the mapping of the above integers and the results are as in Fig. 5.

It is observed that there is a slight overlap for violin (5) and veena (4). Since they are both string-based instruments, their sound qualities and features overlap, and hence, RNN, XGboost, and the combined model identify the instrument accurately

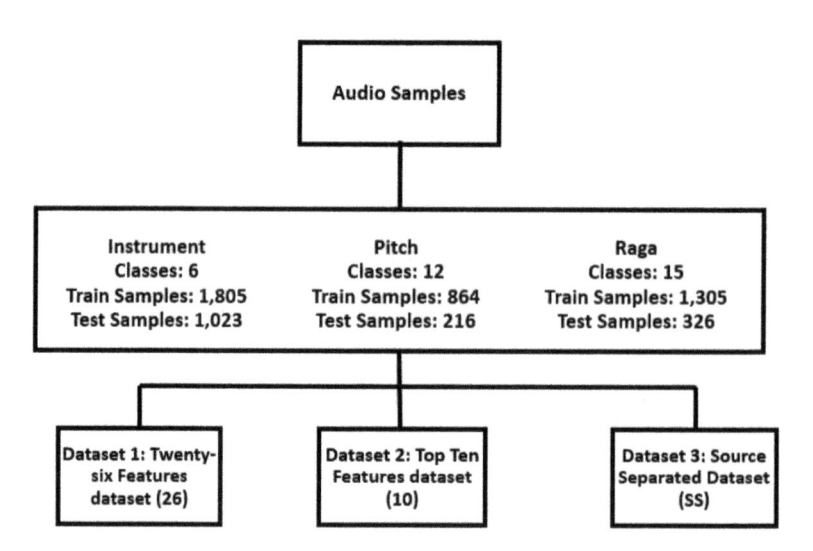

Fig. 4 Datasets for the three tasks

Table 4 Instrument identification model output labels

Output number	Instrument
0	Flute
1	Mandolin
2	Nadaswaram
3	Saxophone
4	Veena
5	Violin

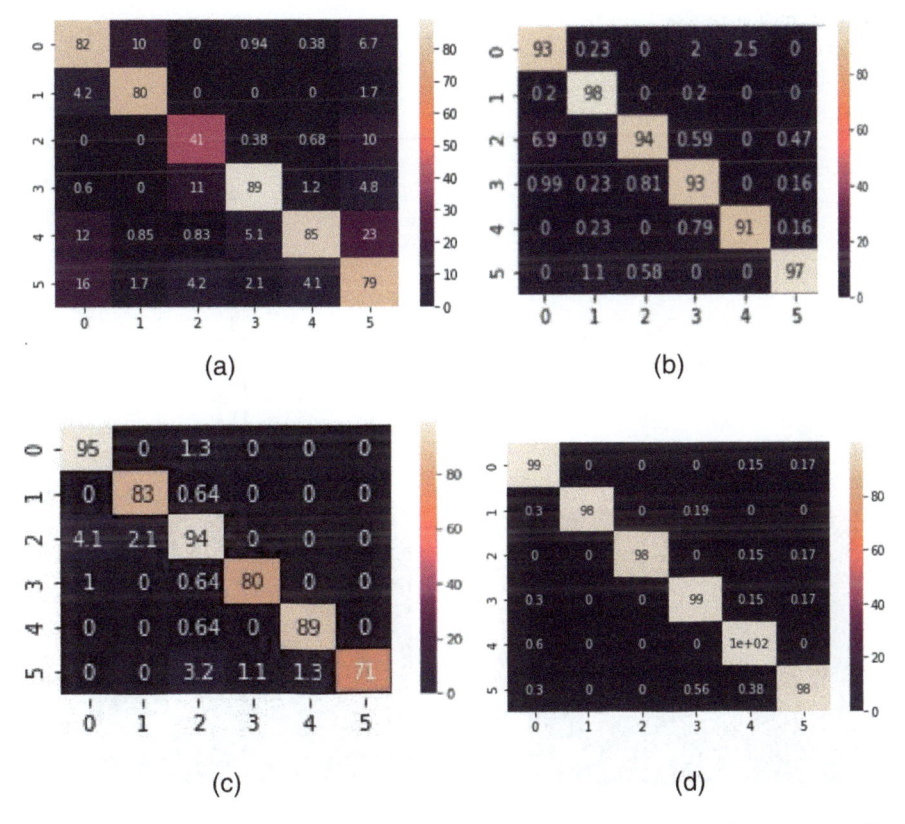

Fig. 5 Confusion matrix for instrument identification: (**a**) CNN, (**b**) RNN, (**c**) XGboost, and (**d**) combined model

without much overlap between the different classes. CNN however performs poorly for instrument recognition.

5.2 Pitch Detection

Table 5 lays out the mapping of the above integers and the results are as in Fig. 6. In Indian classical music, it is particularly difficult to identify pitch, especially for D and D#. Due to their similar frequency and sound, they are difficult to identify even by professionals. A clear indication of this behavior can be observed in the overlap observed between classes 5 and 6 (corresponding to D and D#). RNN and the combined model however are able to identify them more accurately compared to CNN and XGboost. The models are able to distinctly identify the other classes without much overlap (Fig. 6).

Table 5 Pitch detection model output labels

Output number	Pitch
0	A
1	A#
2	B
3	C
4	C#
5	D
6	D#
7	E
8	F
9	F#
10	G
11	G#

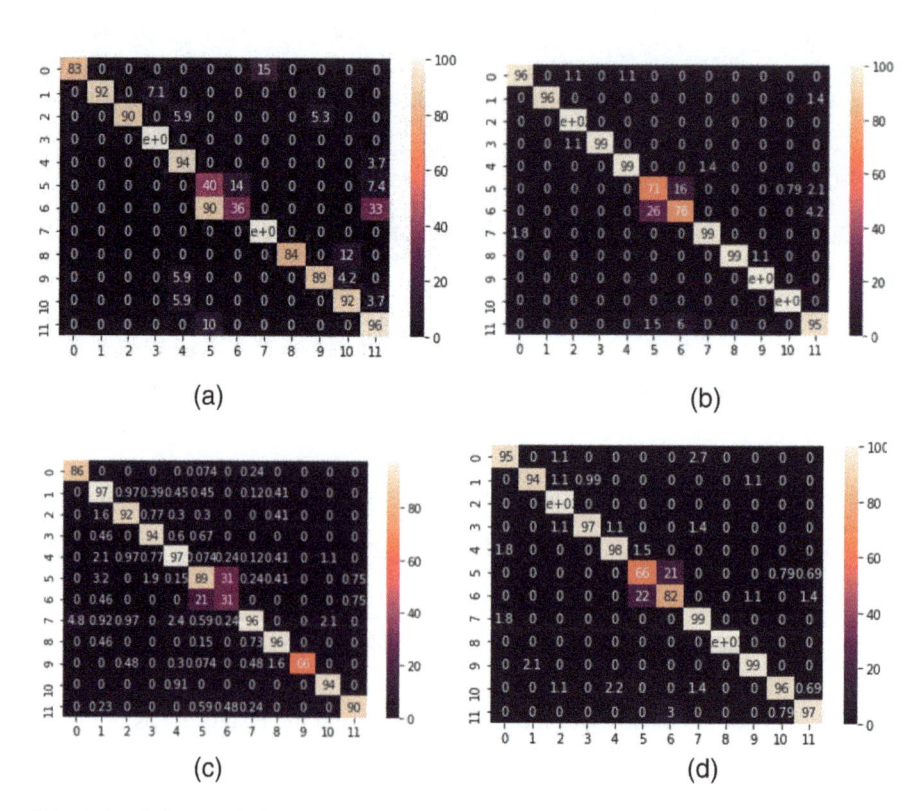

Fig. 6 Confusion matrix for pitch detection: (**a**) CNN, (**b**) RNN, (**c**) XGboost, and (**d**) combined model

5.3 Raga Detection

Table 6 lays out the mapping of the above integers and the results are as in Fig. 7.

The models are able to distinctly identify all raga classes. In comparison with the other models, CNN performs very poorly. Other models, however, do not exhibit the same behavior. There is no doubt that raga detection is the most complex task among the three considered. There are similarities between the characteristics of a specific instrument and those of pitch. A raga may, however, be performed on more than one instrument and have a different pitch, making its identification more challenging.

The combined model was tested with recorded songs of artists such as N. Ramani Fig. 8a and Lalgudi Vijayalakshmi Fig. 8b.

Loss and accuracy of the three different models for the three different tasks for the three datasets have been tabulated in Table 7.

By evaluating accuracy and loss, we can evaluate the performance of the models. Models with higher accuracy and a lower loss typically perform better. We observe that the RNN model performs well across all datasets, while CNN's performance is unsatisfactory, particularly when it comes to identifying ragas.

The above table shows that RNN exhibits the highest accuracy and low losses for all three datasets.

In all models, performance drops significantly for datasets containing the top ten features. This poor performance is observed as the remaining features also play a significant role in detection in addition to the top ten identified features. Despite the additional operation of source separation, source-separated datasets display slightly poor performance.

Table 6 Raga recognition model output labels

Number	Raga
0	Bahudari
1	Bauli
2	Behag
3	Brindavana Saranga
4	Gambira Nata
5	Hamsadhvani
6	Kalyani
7	Kapi
8	Madymavati
9	Nata Kurinji
10	Purvikalyani
11	Riti Gaula
12	Saramathi
13	Sindhu Bhairavi
14	Yamuna Kalyani

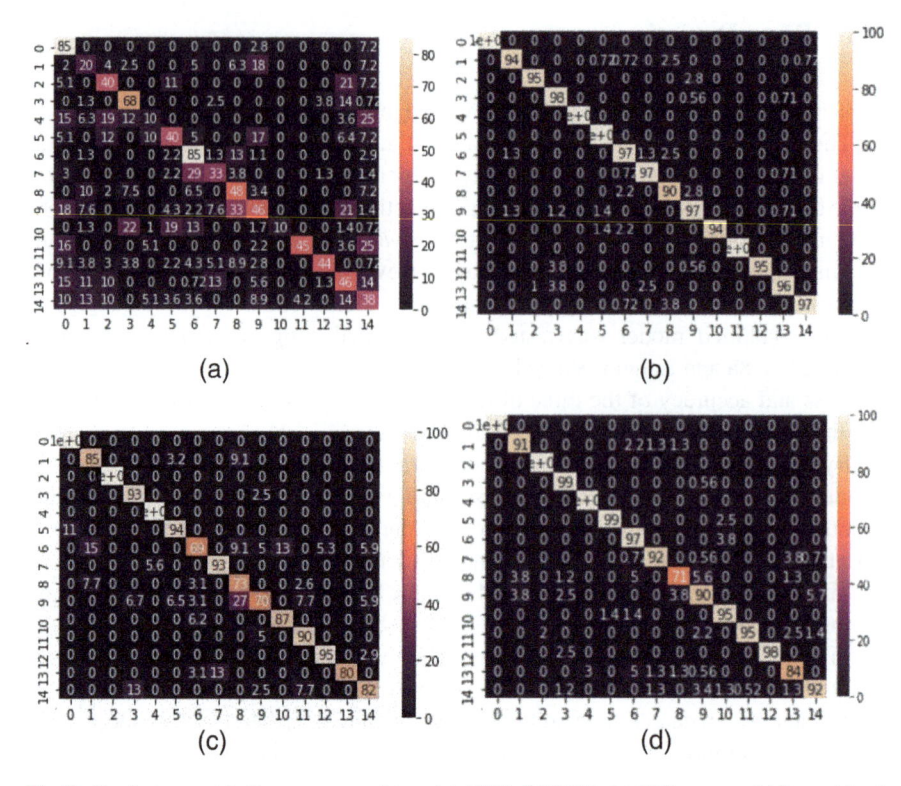

Fig. 7 Confusion matrix for raga recognition: (a) CNN, (b) RNN, (c) XGboost, and (d) combined model

```
print(f'INSTRUMENT IS {instrument}, BEING PLAYED ON| PITCH {pitch} ON RAGA {raga}')
INSTRUMENT IS Flute, BEING PLAYED ON PITCH D# ON RAGA Kalyani
```
(a)

```
print(f'INSTRUMENT IS {instrument}, BEING PLAYED ON PITCH {pitch} ON RAGA {raga}')
INSTRUMENT IS violin, BEING PLAYED ON PITCH C# ON RAGA brindavana saranga
```
(b)

Fig. 8 Output of the combined model

6 Conclusion

A tedious task of music analysis is the identification of pitch, raga, and instruments in Indian classical music. This chapter has proposed three different models to achieve the desired results, viz., the identification of the music attributes. We compare CNN, RNN, and XGboost models based on their performance when random music is to be preprocessed as an input. Further, with the help of XGboost

Table 7 Model performance comparison with 26 features, top 10 features, and source-separated features

Model	Dataset	Instrument Accuracy	Loss	Tone Accuracy	Loss	Raga Accuracy	Loss
CNN	26	87.8%	0.113	76%	0.453	70%	1.540
	10	68%	0.382	65%	1.334	52%	2.502
	SS	74%	0.421	72%	0.64	70%	1.75
RNN	26	98.56%	0.011	99%	0.010	98.1%	0.009
	10	85%	0.289	83%	0.210	82%	0.320
	SS	96%	0.0251	95%	0.012	94%	0.056
XGboost	26	92%	0.760	90.54%	0.84	87%	0.96
	10	79%	1.558	74%	1.378	70%	1.960
	SS	89%	0.996	86%	0.910	78%	1.2

feature importance, the ten best features are identified, and models are trained for the same. Source separation is also performed, and the model performance is compared.

RNN model clearly outperforms all models with high accuracy over varied test cases. All three tasks were simultaneously performed by the combined model. This model had an accuracy of 97.2Out of the three datasets, datasets containing only ten features were seen to have compromising accuracy values. In contrast, the original dataset containing all 26 features had a higher level of accuracy than the source-separated dataset. According to this theory, the tonic nature of the rhythmic section of Indian classical music also contributes to the melodic analysis. Thus, the cumbersome task of source separation required by most of the signal processing techniques can be totally avoided in the process of melodic analysis. Further, the sequential stages such as tonic detection, swara generation, and mapping of phrases involved in recognizing a particular raga can be performed parallelly which reduces the time complexity.

When considering studio-recorded audio tracks, models trained with source-separated datasets might perform well, while in real-life concert tracks, their performance might be questionable. This aspect can be explored in the future.

References

1. Ashwini, Krishna, A.V., Mahesh, V. and Karrthik, G.K., 2020. Tone detection for Indian classical polyphonic instrumental audio using DNN model. International Journal of Forensic Engineering, 4(4), pp. 310–322.
2. Pawar, M.Y. and Mahajan, S., 2019. Automatic Tonic (Shruti) Identification System for Indian Classical Music. In Soft Computing and Signal Processing (pp. 733–742). Springer, Singapore.
3. Salamon, Justin & Gulati, Sankalp & Serra, Xavier. (2012). A MultiPitch Approach to Tonic Identification in Indian Classical Music. Proceedings - 13th International Society for Music Information Retrieval Conference (ISMIR 2012).

4. Gulati, Sankalp & Bellur, Ashwin & Salamon, Justin & H.G., Ranjani & Ishwar, Vignesh & Murthy, Hema & Serra, Xavier. (2014). Automatic Tonic Identification in Indian Art Music: Approaches and Evaluation. Journal of New Music Research. 43. 10.1080/09298215.2013.875042.
5. S. Samsekai Manjabhat, Shashidhar G. Koolagudi, K. S. Rao & Pravin Bhaskar Ramteke (2017) Raga and Tonic Identification in Carnatic Music, Journal of New Music Research, 46:3, 229–245, DOI: 10.1080/09298215.2017.1330351
6. D. Poojary, A. Pinto, A. Zahied, Anusha, and A. Shetty, "Automatic Tonic Identification in Indian Art Music", IJRESM, vol. 3, no. 7, pp. 210–212, Jul. 2020.
7. Chordia, Parag, and Sertan Sentürk. "Joint Recognition of Raag and Tonic in North Indian Music." <i>Computer Music Journal</i> 37, no. 3 (2013): 82–98. Accessed August 19, 2021. http://www.jstor.org/stable/24265515.
8. Madhusdhan, Sathwik Tejaswi and G. Chowdhary. "Tonic Independent Raag Classification in Indian Classical Music." (2018).
9. S. Gaikwad, A. V. Chitre and Y. H. Dandawate, "Classification of Indian Classical Instruments Using Spectral and Principal Component Analysis Based Cepstrum Features," 2014 International Conference on Electronic Systems, Signal Processing and Computing Technologies, 2014, pp. 276–279, doi: 10.1109/ICESC.2014.52.
10. Raga and Tonic Identification in Carnatic Music https://idr.nitk.ac.in/jspui/handle/123456789/12733
11. Gulati, S., Serrà, J., Ganguli, K. K., Şentürk, S., & Serra, X. (2016). Time-delayed melody surfaces for Raga recognition. In Proceedings of the 17th International Society for Music Information Retrieval Conference (ISMIR), pp. 751–757. New York, USA.
12. Alekh, Sanchit. (2017). Automatic Raga Recognition in Hindustani Classical Music.
13. Bhat, A., Krishna, A.V. and Acharya, S., 2020. Analytical Comparison of Classification Models for Raga Identification in Carnatic Classical Instrumental Polyphonic Audio. SN Computer Science, 1(6), pp. 1–9.
14. McFee, B. et al., 2015. librosa: Audio and music signal analysis in python. In Proceedings of the 14th python in science conference.
15. Ashwini, B., Ramesh, N., Naik, S.M. and Krishna, A.V., 2017, November. Lead source separation for Indian instrumental audio. In TENCON 2017–2017 IEEE Region 10 Conference (pp. 1121–1126). IEEE.

Comparison of Convolutional Neural Networks and K-Nearest Neighbors for Music Instrument Recognition

S. Dhivya and Prabu Mohandas

1 Introduction

Music is one of the most popular forms of art that is practiced and listened to by billions of people all over the world. Music can improve mood, can decrease pain and anxiety, and can benefit our physical and mental health in numerous ways. Musical instrument recognition is the task of instrument identification by virtue of its audio [2]. Automatic recognition of musical instruments forms the basis of more complex tasks like melody extraction, music information retrieval, recognizing the dominant instruments from polyphonic audio [1], and so on. The task of efficient automatic music classification is of vital importance and forms the basis for various advanced applications of AI in the musical domain like music genre classification, automatic music transcription, and recommender systems.

Music instrument recognition enhances the performance of other MIR tasks. It helps to find the type of musical instrument used which would significantly improve the performance of other MIR tasks like automatic music transcription, music genre identification, and source separation. It will be very helpful for the people who are working on music data, and also for the present-day music companies, it can assist them on music recommendations for their users. Music information retrieval (MIR) is about retrieving information from music. MIR systems add significant value to existing music libraries and make them more easily accessible. They help in automatic music classification, indexing, searching, and organization [3].

S. Dhivya (✉)
Department of Computer Science and Engineering, Easwari Engineering College, Chennai, India

P. Mohandas
Intelligent Computing Lab, Department of Computer Science and Engineering, National Institute of Technology, Calicut, India
e-mail: prabum@nitc.ac.in

Machine learning helps systems to learn from data, identify patterns, and make decisions with minimal human interaction. Machine learning for audio signal processing has attracted a large amount of attention recently for its uses in speech recognition. Machine learning techniques provide numerous ways to perform music categorization as per need. Music instrument classification can be done easily on monophonic sounds than polyphonic ones, where multiple instruments played together. Classifying instances into three or more classes is called multi-class classification. A multi-class classifier is implemented which takes an audio stream as the input and outputs the class of the musical instrument present in the stream. Most work is done on monophonic music which is less challenging. The timbre of the instruments is studied which in turn gives patterns for classification. The methods for music instrument recognition can be classified as traditional machine learning techniques and deep learning techniques. Deep learning techniques for music instrument recognition have been evolving rapidly in the last decade.

The primary goal is to classify six instruments from the given music data. A convolutional neural network (CNN) and a κ-nearest neighbor (KNN) classifier are implemented to perform the classification. In the convolutional neural network classifier, the input audio stream is pre-processed to extract the Mel spectrogram. The features for the Mel spectrogram are used to perform the classification. The input of the model is the Mel spectrogram, and the output is an index corresponding to the predicted class. For the κ-nearest neighbor classifier, the MFCC feature vectors are calculated, the number of neighbors is set, and the classification process is done.

1.1 Motivation

Music instrument recognition can help in finding what kinds of instruments are present in a music clip and can distinguish the instruments with one another. The motivation behind this work is to come up with a system that can help musicians extract a particular instrument sound. It can help people who are working on music in music data transcription and identification. It can help present-day music companies with recommendations for their users. It allows us to perform various music information retrieval tasks like pitch, timbre separation, genre classification, automatic music transcription, and source signal separation. It assists people involved in musicology, psycho acoustics, signal processing, and optical music recognition.

1.2 Objective

- In the development of a model to train different audio files, the model should classify what instruments are used in the audio.
- A method to label unlabelled audio files to avoid manual annotation.
- Training of CNN and KNN models to perform music instrument recognition.
- Performance analysis of both the models to get better understanding

1.3 Organization

The proposed work analyzes the performance of CNN and k-nearest neighbor classifier. The entire chapter is organized as follows. Section 2 reviews the most popular existing works. The proposed methodology is explained in detail in Sect. 3. The experimental setup of the work is explained in Sect. 4. Section 5 discusses the results of the experiments. Finally, the chapter concludes with Sect. 6.

2 Literature Review

In musical instrument recognition using CNN and SVM [4], the classification task was performed on the IRMAS dataset [5]. The IRMAS dataset consists of musical audio excerpts with annotations of predominant instruments present in the file. The music and instrument classification using deep learning techniques [6] implemented a multi-class classifier that identifies instruments in music streams. They use Google's AudioSet which provides human-labelled data. It has a set of 10-second clips from YouTube, labelled with the audio instruments and any other sound label it contains. The musical instrument classification using neural networks [3] implemented an automatic classification of musical instrument sounds with a dataset of 4548 files from 19 instruments of MIS database—the University of Iowa Musical Instrument Samples [7]. In deep convolutional neural networks for predominant instrument recognition in polyphonic music [8], the music instrument classification in polyphonic music is accomplished. They also used the IRMAS dataset. An artificial neural network is implemented for classification in [9]. They use the full London Philharmonic Orchestra dataset which contains 20 classes of instruments belonging to the 4 families—woodwinds, brass, percussion, and strings. Kratimenos et al. [16] explored a variety of data augmentation techniques focusing on different sonic aspects, such as overlaying audio segments of the same genre, as well as pitch- and tempo-based synchronization. Eronen et al. [17] set up a system for pitch-independent musical instrument recognition. A wide set of features covering both spectral and temporal properties of sounds was investigated, and their extraction algorithms were designed. Patil S.R. [18] has described a system for musical instrument recognition in monophonic audio signals where the single sound source is active at a time using a Gaussian mixture model (GMM). Ghosh et al. [19] proposed a decision tree-based model for the automatic recognition of musical instruments.

Singh et al. [4] used a combination of convolutional neural network and support vector machine. The SVM uses MFCC for feature extraction. The audio excerpts used for training will be pre-processed into images (visual representation of frequencies in sound). The results obtained from both the CNN and SVM are added to get the weighted average, which gave better performance in terms of instrument identification. Lara Haidar-Ahmad [6] implemented a model which consists of a CNN which takes input as an audio stream that is pre-processed to extract the Mel spectrogram and outputs the class of pre-selected instruments. They focus on 3 instruments and classify audio streams into 1 of 4 classes: "Piano," "Drums," "Flute," or "Other"; around 8000 samples were trained. Lara Haider-Ahmed [6] obtained a precision of 70%, a recall of 65%, and a F1 score of 64%. In [3], probabilistic neural networks were used for classification for its flexibility and the straightforward design. The dataset used consists of 4548 tunes from 19 instruments of the MIS database. Probabilistic neural networks were used as classifiers. Mel-frequency cepstral coefficients (MFCCs) were used as features. Multi-level quantization was applied to the features before doing the classification. The accuracy of 92% was achieved. Kratimenos et al. [16] utilized convolutional neural networks for the classification task, comparing shallow to deep network architectures and an ensemble of VGG-like classifiers, achieving slightly above 80% in terms of label ranking average precision (LRAP) in the IRMAS test set. Eronen et al. [17] validated the usefulness of the features test data that consisted of 1498 samples covering the full pitch ranges of 30 orchestral instruments from the string, brass, and woodwind families, played with different techniques. The correct instrument family was recognized with 94% accuracy and in 80% of cases for individual instruments. Patil S.R. [18] obtained an accuracy of 93.18% for a combination of MFCC as a feature and GMM as a classifier. Ghosh et al. [19] obtained an accuracy of 84.02% by decision tree (DT) for a set of nine instruments belonging to different families. The accuracy for predicting the correct string instrument family is 96.07% and for wind instrument the overall prediction accuracy is 90.78%.

Han et al. [8] use a convolution neural network for the predominant instrument recognition. The model is trained on the single labelled predominant instrument. They used dataset of 10k audio files. It consisted of 11 instruments. Convolutional neural networks were found to be more robust than conventional methods and thus obtained an F1 measure of 0.602 for micro achieving 19.6% performance improvement compared with other algorithms. Mahanta et al. [9] achieved an accuracy of 97% on the full dataset containing all 20 classes of different musical instruments. Table 1 shows the comparison of performance for different models implemented for music instrument recognition.

2.1 Performance Issues

Mahanta et al. [9] proposed a deep artificial neural network model that efficiently distinguishes and recognizes 20 different classes of musical instruments, even across instruments belonging to the same family. The model trains on the full London Philharmonic Orchestra dataset which contains 20 classes of instruments belonging to the 4 families, viz., woodwinds, brass, percussion, and strings. They use only the Mel-frequency cepstral coefficients (MFCCs) of the audio data.

Table 1 Comparison of models

Authors and year	Model	Objective	Dataset	Accuracy/ F1 score
Hing, Dominick Sovana, and Connor Settle[10] 2020	CNN	A multi-class instrument classifier using CNN	6705 training samples and 1400 test samples from IRMAS [5]	70.3%
Yun, Mingqing, and Jing Bi[11] 2018	LSTM	A music instrument classifier using RNN with log Mel spectrogram	A dataset of 14 instruments with 200 training samples	81%
J. Liu and L. Xie[12] 2010	SVM	SVM-based classifier of musical instruments using MFCC features	2177 clips of 13 Chinese instruments and 13 Western instruments	95.44%
S. Prabavathy et al. [13] 2020	KNN	Proposed a KNN model for music instrument classification	1284 samples were used from 16 musical instruments	98.22%
Anhari, Amir Kenarsari[14] 2020	RNN	Multi-instrument classifier using an attention-based bi-directional LSTM	20k audio clips from the OpenMic dataset	F1 score of 0.83
Kingkor Mahanta et al. [9], 2021	ANN	An ANN model was trained to perform classification on 20 classes of musical instruments	13,679 examples divided among 20 classes of musical instruments	99.7%

The dataset was divided into training and validation or testing sets in the ratio 8:2 using stratified splitting, such that the number of examples from each of the 20 classes split proportionally into 2 sets. The training and test sets contained 10,943 training examples and 2736 test examples, respectively, after the split. MFCC features are extracted, from the constant length examples and feeding them into an ANN model to make predictions. The model uses an ANN architecture with 1690 input neurons which are connected to the first dense hidden layer having 512 neurons followed by ReLU activation function. The second and third hidden layers contain 1024 and 512 neurons, respectively, both followed by the ReLU activation function. A dropout layer with a rate of 0.3 is then added to induce regularization and avoid overfitting. After the dropout layer, the values pass through 2 more hidden layers containing 128 and 64 neurons, respectively, and a dropout layer with a 0.2 rate. The final output layer has 20 neurons for each class. They use the rectified linear unit (ReLU) activation function for all the hidden layers. It simply activates the neurons containing a positive value after the aforementioned computations.

$$y = max(0, x) \qquad (1)$$

The softmax function is used in the output layer. It provides the confidence scores of each class using

$$\Sigma(z_i) = e^{z_i} / \Sigma_{j=1}^{K} e^{z_j} \qquad (2)$$

The scores add up to 1. The class having the highest confidence score is the model's predicted class for a particular set of input features. The model achieved an accuracy of 97%. During model training, the training accuracy peaked 0.9913 and validation accuracy 0.9726. The dataset is quite imbalanced as most instruments belong to a particular family, so data augmentation measures may be adopted to deal with the imbalance problem. Different learning rates and optimizers can be tried to produce different results. Expanding the target space by supporting the recognition of even more instruments including the piano or the ukulele would be a notable improvement.

2.2 Problem Statement

- MFCCs and Mel spectrograms provide excellent visual perceptions of sound; thus, CNNs may prove to be quite efficient than ANN.
- The dataset may be imbalanced, and most of the instruments belonged to one particular class of the family.
- Other optimizers and activation functions can give better results.
- Lots of variables in the pre-processing stage can be tweaked to provide better results

3 Proposed Methodology

The identification of instruments present in an audio track plays a vital role in music information retrieval as it provides information about the composition of music. Music instrument recognition in polyphonic music is a challenging task. The proposed work employs a CNN and k-nearest neighbor classifier to identify the musical instrument present in a monophonic audio file. This section gives a detailed description of the proposed methodology.

3.1 Proposed Block Diagram

Figures 1 and 2 depict the block diagram of the proposed work. The input audio file is loaded to the processing module, and the output is the class of the musical instrument it belongs to. In the CNN model, the audio file is converted to a Mel spectrogram, and the extracted features are sent to the CNN training module. Inside the training module, the features go through the convolutional layer gets convoluted, then through the dropout layer and then the ReLU activation function. In the KNN

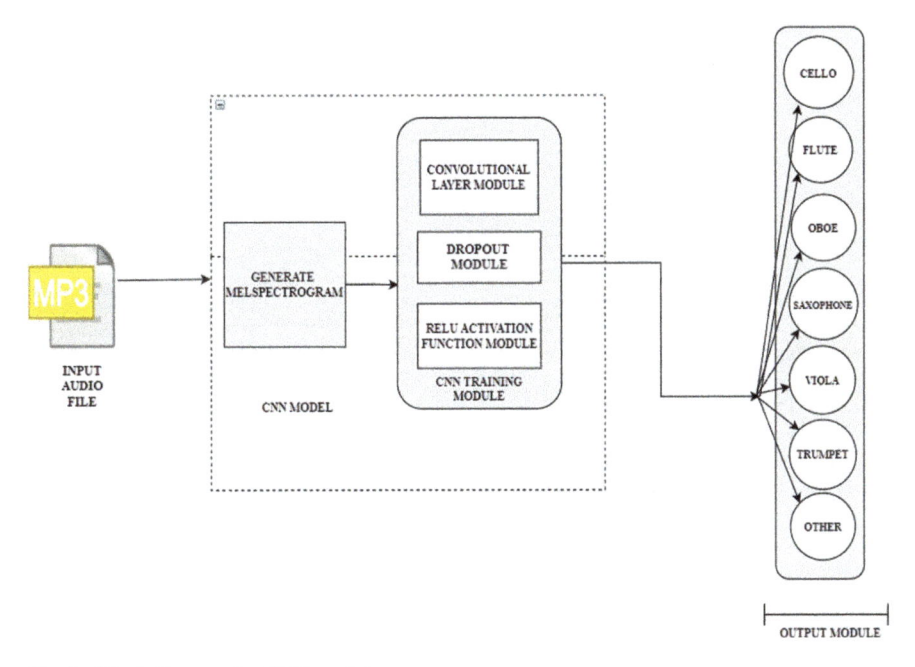

Fig. 1 Block diagram for CNN model

model, the audio file is resampled, and the audio features are calculated. The audio files are normalized, and the MFCC feature vectors are calculated using the Librosa module and are inputted into the KNN classifier for classification.

3.2 CNN-Based Approach

Figure 3 depicts the CNN model architecture. It consists of three convolutional layers followed by a pooling layer, an activation function, and a fully connected layer. The CNN model takes an image as the input. The audio files undergo some transformations so that they can be inputted as an image in the CNN model. In deep learning, a convolutional neural network (CNN/ConvNet) is a class of deep neural networks, which is most commonly used on images. It is composed of many layers of neurons. The first layer extracts basic features such as horizontal or diagonal edges which are passed on to the next layer. The next layer then detects more complex features like corners or combinational edges. It identifies even more complex features as we move deep into the network. CNN is patterned to process multidimensional array data in which the convolutional layer takes a stack of feature maps, like the pixels of those color channels, and convolves each feature map with a set of learnable filters to obtain a new stack of output feature maps as input. Based on the activation map of the final convolution layer, the classification layer outputs a

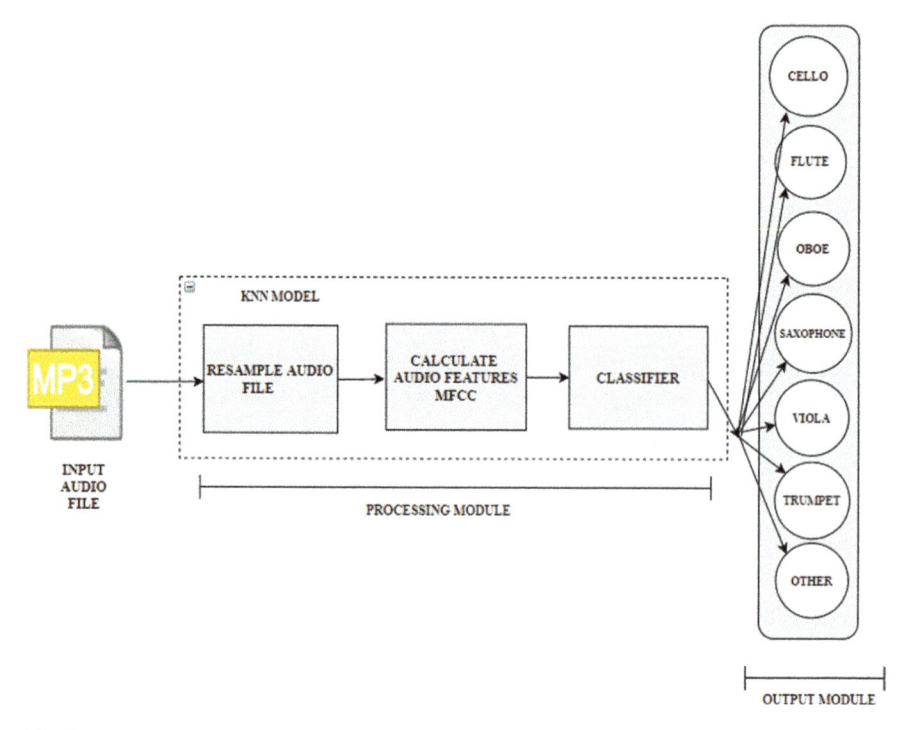

Fig. 2 Block diagram for KNN model

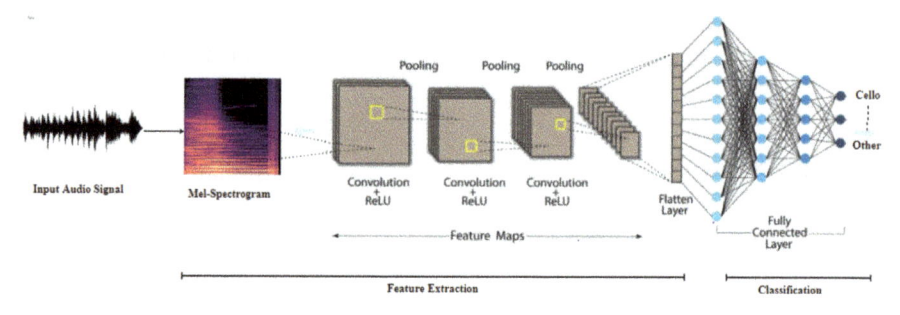

Fig. 3 CNN model architecture

set of numerical values between 0 and 1 that predicts which class the image belongs to. Figure 4 represents the two-dimensional representation of an audio file.

The Mel scale is the logarithmic transformation of the frequency of a given signal. It is difficult for humans to differentiate higher frequencies than lower frequencies. Even if the distances between the differences of the two sounds are the same, the human perception of the difference is not the same. Hence, the Mel scale is fundamental in machine learning applications of audio.

Transformation from Hertz scale to Mel scale:

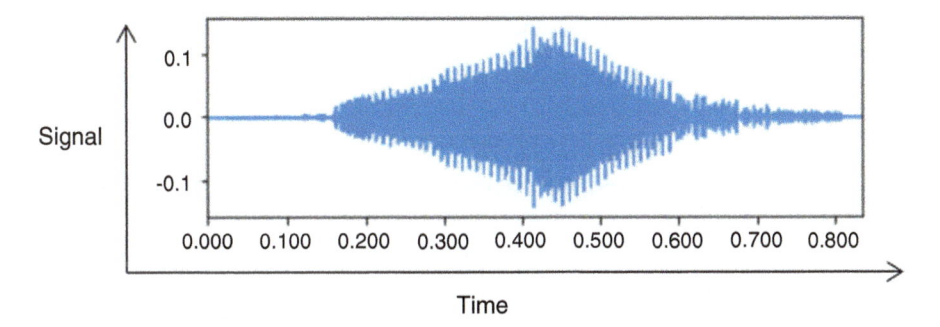

Fig. 4 Two-dimensional representation of an audio file

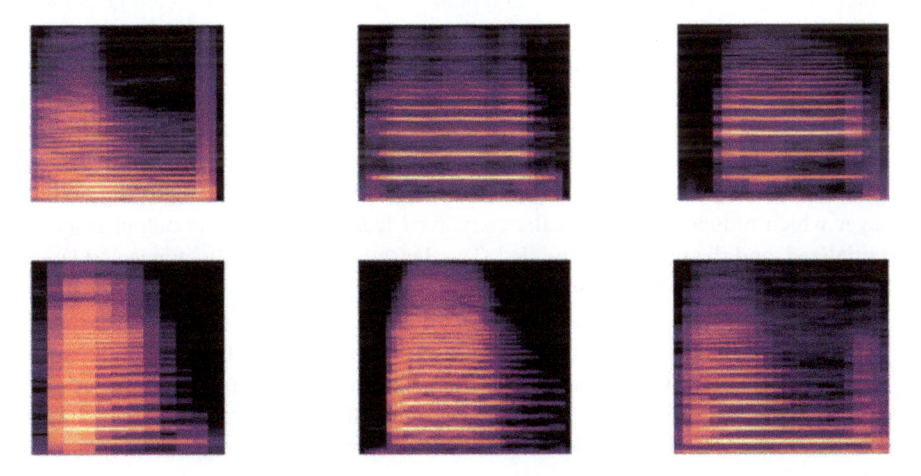

Fig. 5 Mel spectrograms of each musical instrument—cello, flute, oboe, saxophone, trumpet, and viola

$$m = 1127 * log(1 + f/700) \tag{3}$$

Equation 3 is a formula to transform Hertz scale to Mel scale from O'Shaughnessy's book. The Mel-frequency cepstral coefficients (MFCCs) of a signal are used to describe the overall shape of a spectral envelope.

Mel spectrogram is a spectrogram that is converted to a Mel scale. A spectrogram is a visualization of the frequency spectrum of a signal, where the frequency spectrum of a signal is the frequency range that is contained by the signal. Each audio file in the dataset is converted into a spectrogram to perform the classification. Figure 5 depicts the Mel spectrogram generated for each class of musical instrument present in the dataset.

In a CNN, the input of a shape (number of inputs) × (input height) × (input width) × (input channels) becomes a feature map of shape (number of inputs) × (feature map height) × (feature map width) × (feature map channels), after passing through a convolutional layer.

Convolutional layer generally has the following attributes:

- The number of filters the convolutional layers will learn from.
- The dimensions of the kernel and the size of the input
- The activation function to be applied after performing convolution

The model uses 32 filters, has a kernel size of 3*3, and uses ReLU activation function.

The pooling layer is responsible for reducing the spatial size of the convolved feature. The pooling layer resizes the input spatially, using the MAX operation. The MaxPool operation downsamples the input along its dimensions by taking the maximum value over an input window which is defined by the pool size for each channel of the input. The model uses a pool size of 3*3. Fully connected layers are responsible for connecting all neurons in one layer to neurons of another layer.

After uploading and reprocessing all the audio files, the labels of each sample are appended. The dataset is split into training, testing, and validation sets. The input convolutional layer followed by the second and the third convolutional layers is initialized. After the image is passed into the input convolutional layer, it gets convoluted to a different size. The feature map passes through the pooling layer which reduces the size of the convolved feature. Finally, the output layer is initialized, and the model is compiled. The dense layer or the fully connected layer connects every other neuron and all the extracted feature maps together. The model is trained for the given number of epochs.

The input is the mp3 audio file, and the output is the class of the musical instrument in the monophonic audio file. The musical instrument classes are initialized. All the audio files are loaded and the labels of each file are initialized. The dataset is split into training set, validation set, and test set. The CNN model is initialized and compiled. The model is trained for the specified number of epochs.

3.3 KNN-Based Approach

K-nearest neighbors (KNN) is one of the simplest algorithms used for both classification and regression problems. Classification is done by a majority vote to its neighbors. Figure 6 depicts the KNN model architecture. The feature extraction process is done from the input audio file, and the features are sent to the KNN classifier. All the audio files are normalized, and the MFCC features are calculated for each audio clip using the Librosa module. In the KNN classifier, the value of κ is initialized to the selected number of neighbors. The distance between the feature vectors of each pair of the audio clip is calculated and sorted. For κ entries from the sorted data, the mode of κ labels will be returned for classification problems. All the audio samples are loaded and pre-processed, and their respective labels are appended. The value of κ is initialized, and the Euclidean distances between the κ number of nearest neighbors are calculated. The distances of the inputs are sorted. For the κ-nearest neighbors, simple majority is applied. The process is first

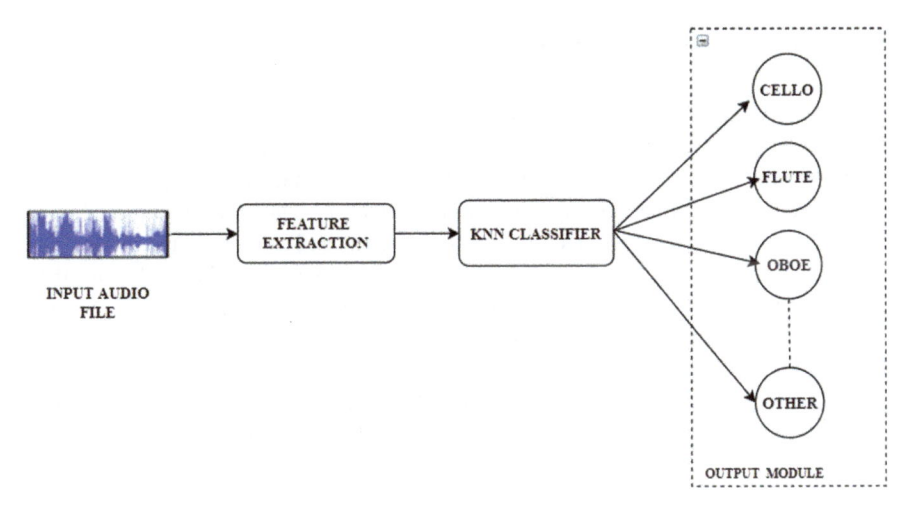

Fig. 6 KNN model architecture

performed for $\kappa = 1$; after finding the best value of κ from the error vs κ value graph, the process is repeated for that value of κ.

The input is the mp3 audio file, and the output is the class of the musical instrument in the monophonic audio file. The musical instrument classes are initialized. All the audio files are loaded, and the labels of each file are initialized. All the labels are encoded to numerical values to normalize the labels. The dataset is split into training set and test set. The KNN model is initialized and compiled. The best value of κ is found out for the model based on the error vs κ value graph, and the model is compiled again for that value of κ.

4 Experimental Setup and Analysis

4.1 Dataset and Annotations

The dataset consists of musical instrument samples from the Philharmonic website [15]. It is a balanced dataset and it consists of six different classes. The dataset consists of 600 files. The classes are "cello," "flute," "oboe," "saxophone," "trumpet," and "viola." Each class consists of 100 recordings of each instrument. All audio files are in .mp3 format. The size of the dataset is 8.16 MB. Dataset is divided into testing and training set. We pre-process the data before using it. To process, we use a sample rate of 44,100 Hz, an fft size of 2048, and a hop length of 512. The dataset includes musical audio excerpts with annotations of the musical instrument present.

4.2 Model Training and Testing

The given dataset [15] is split into training, validation, and testing set. The train set has 60% of the data, and the test set and validation set have 20% each. The model is trained on the training set. The CNN model requires an image as the input. The audio files have to be visualized using some transformations. The audio is pre-processed to extract the Mel spectrograms. Mel spectrogram is used as input of the model. The Mel spectrograms of all the audio files are stored separately. These files are then trained in the CNN model. Three convolutional layers which consist of 32, 64, and 128 filters, respectively, are used to produce feature maps. ReLU activation function was implemented after each convolutional layer. Three max pooling layers are used to reduce dimensionality without padding. A dropout rate of 0.25 was applied to reduce overfitting. ADAM optimizer with a 0.0001 learning rate was used, and the CNN model was trained up to 30 epochs. Categorical cross-entropy was used as a loss function to optimize results. After the model is trained, the accuracy and loss of the model are calculated. For the KNN algorithm, the dataset is loaded, and the features and feature vectors are calculated. The features are scaled using StandardScaler function. The data is then split into train (75%) and test data (25%). The KNN classifier is first trained for $\kappa = 1$, and the performance is evaluated. The best value of κ is calculated from the error rate vs κ value plot, and the model is trained for that value of κ. The performance is analyzed, and the confusion matrix is plotted.

5 Results and Discussion

5.1 Performance Evaluation of CNN Model

The proposed work is evaluated using different parameters. The training data of the Philharmonic dataset has 360 audio samples, and the validation and testing data have 120 each. We calculate the loss, accuracy, val_loss, and val_accuracy for the CNN model as shown in Table 2. The loss function keeps decreasing with every epoch, and the accuracy keeps increasing. The training set gave an accuracy of 97% at the end of the 30th epoch. From the plot of accuracy in Fig. 7a, it can be seen that the model has not over-learned the training dataset, showing comparable skill on both the training and validation datasets. From the plot of loss in Fig. 7b, it can be seen that the model has comparable performance on both training and validation datasets.

The test dataset gave an accuracy of 99.1% and a loss value of 0.24.

Table 2 Performance evaluation of CNN model

Epoch	Loss	Accuracy	val_loss	val_accuracy
1	2.2948	0.1787	1.7985	0.1417
2	1.7934	0.1892	1.7900	0.1500
3	1.7808	0.2036	1.7166	0.3000
4	1.7390	0.2792	1.5918	0.4917
5	1.5956	0.3752	1.3686	0.5333
6	1.3288	0.4498	1.1500	0.5750
7	1.2465	0.4992	1.0673	0.6333
8	1.0706	0.6139	0.8537	0.7083
9	0.7975	0.7186	0.6244	0.7583
10	0.5712	0.7933	0.3536	0.8750
11	0.4287	0.8451	0.3099	0.9000
12	0.3388	0.8715	0.3042	0.9000
13	0.3661	0.8762	0.1645	0.9500
14	0.2531	0.9180	0.1773	0.9500
15	0.1930	0.9466	0.2428	0.9083
16	0.2002	0.9299	0.0947	0.9583
17	0.2264	0.9185	0.0600	0.9917
18	0.1820	0.9325	0.1194	0.9500
19	0.1623	0.9268	0.0621	0.9667
20	0.1998	0.9333	0.0518	0.9750
21	0.2379	0.9225	0.1336	0.9500
22	0.2067	0.9251	0.0831	0.9750
23	0.1224	0.9632	0.0312	0.9833
24	0.1788	0.9502	0.1110	0.9500
25	0.1674	0.9580	0.0418	0.9917
26	0.0936	0.9716	0.0357	0.9917
27	0.1142	0.9529	0.0956	0.9750
28	0.1053	0.9649	0.0589	0.9750
29	0.0628	0.9862	0.0687	0.9833
30	0.0637	0.9763	0.1009	0.9667

5.2 *Performance Evaluation of KNN Model*

The KNN model is evaluated using different metrics. Precision, F1 score, recall, accuracy, and support are calculated. Table 3 shows the classification report for $\kappa = 1$. The error vs κ value plot is plotted to find the best value of κ so that the model is not overfitted.

From the plot in Fig. 8, it can be seen that the least stable error rate occurs around $\kappa = 7$; hence, $\kappa = 7$ gives the best model. The classification report for $\kappa = 7$ is shown in Table 4. Table 5 shows the comparison of F1 score, accuracy, recall, precision, and the number of wrong predictions for 150 samples.

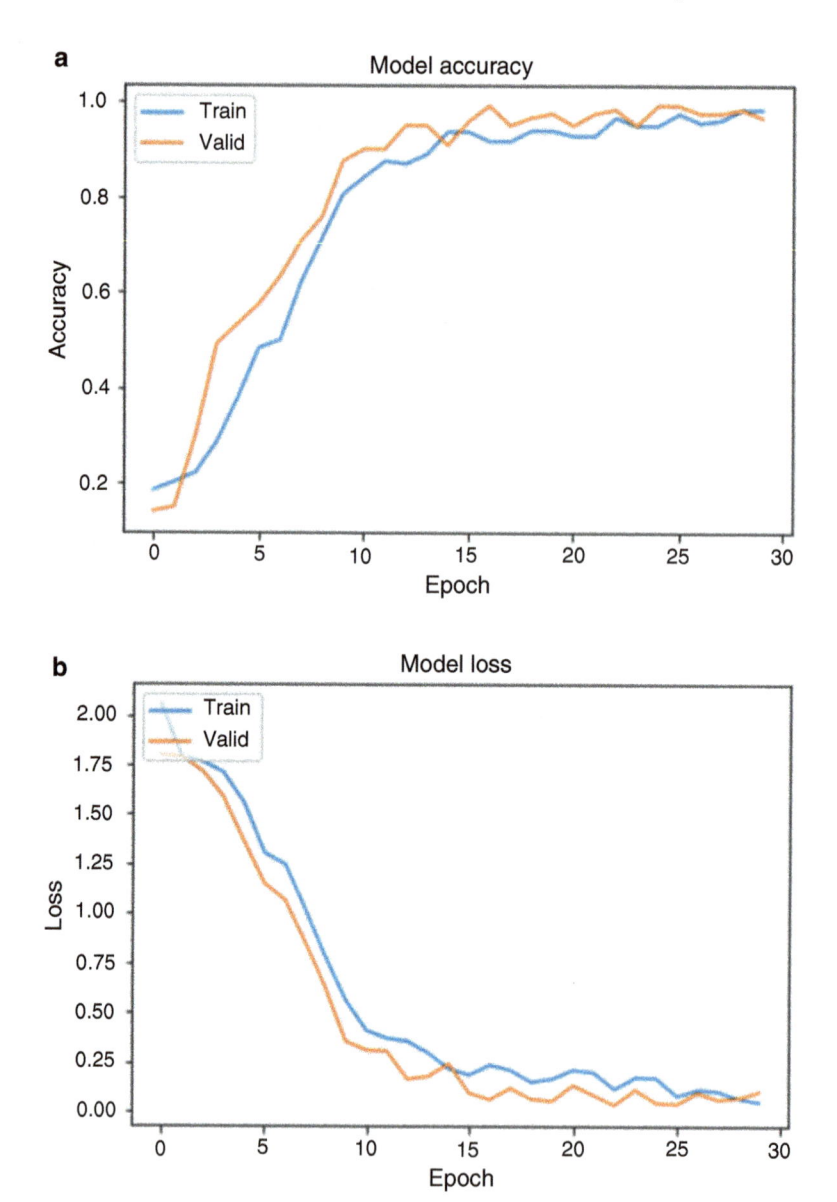

Fig. 7 Performance plots for CNN. (**a**)Model accuracy plot. (**b**) Model loss plot

A confusion matrix is a table that is often used to describe the performance of a classification model (or "classifier") on a set of test data for which the true values are known. Figures 9 and 10 show the confusion matrix for $\kappa = 1$ and $\kappa = 7$, respectively. Table 6 shows the comparison of accuracy for the CNN and KNN models for the 150 test audio samples.

Table 3 Classification report for $\kappa = 1$

Index	Precision	Recall	F1 score	Support
0	1.00	0.96	0.98	25
1	1.00	1.00	1.00	25
2	1.00	0.96	0.98	25
3	1.00	1.00	1.00	25
4	0.96	1.00	0.98	25
5	0.96	1.00	0.98	25

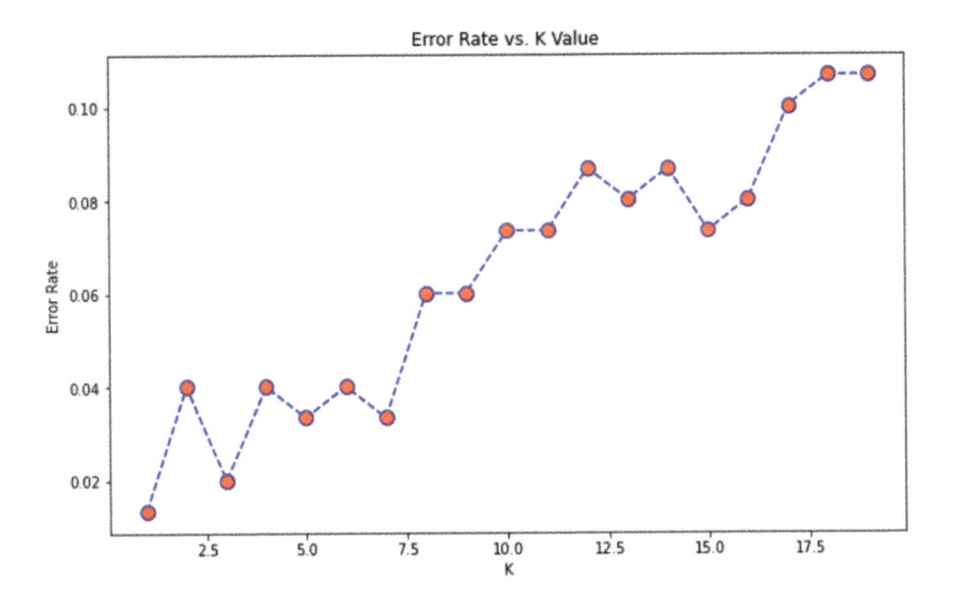

Fig. 8 Error vs κ value plot for KNN

Table 4 Classification report for $\kappa = 7$

Index	Precision	Recall	F1 score	Support
0	0.96	1.00	0.98	25
1	0.96	0.96	0.96	25
2	1.00	0.96	0.98	25
3	1.00	0.96	0.98	25
4	0.93	1.00	0.96	25
5	0.96	0.92	0.94	25

Table 5 Comparison of results for $\kappa = 1$ and $\kappa = 7$

K value	Accuracy	Recall	Precision	F1 score	No. of samples	Wrong predictions
1	0.99	0.99	0.99	0.99	150	2
7	0.97	0.97	0.97	0.97	150	5

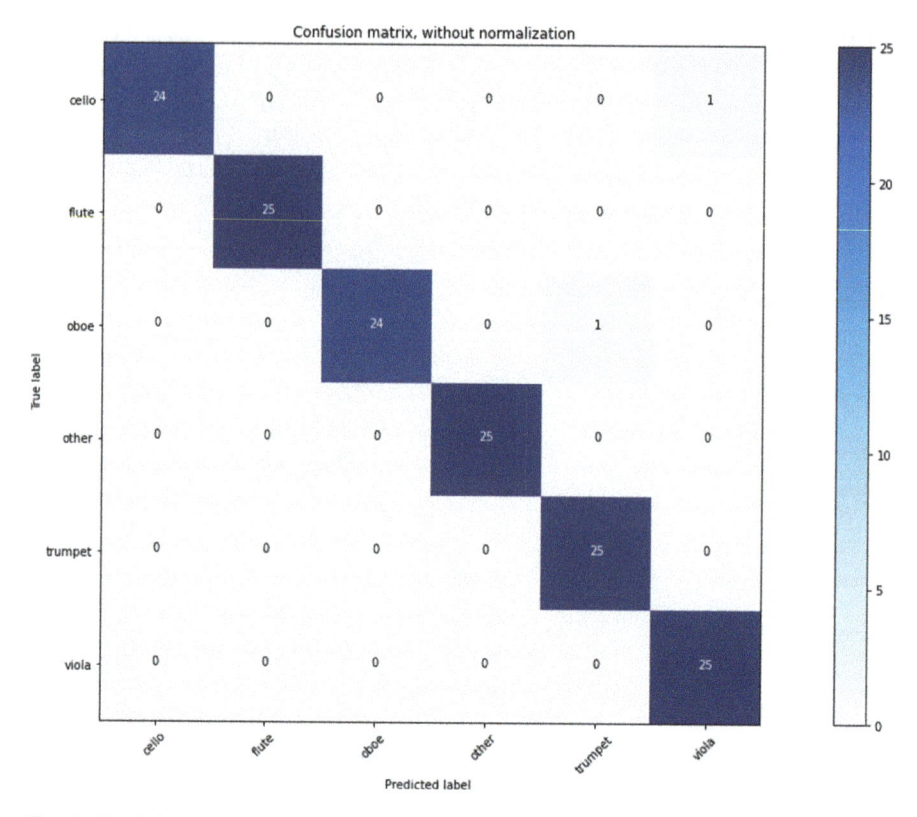

Fig. 9 Confusion matrix for $\kappa = 1$

Table 6 Comparison of results

Model	Accuracy	Number of samples
KNN ($\kappa=1$)	0.99	150
KNN($\kappa = 7$)	0.97	150
CNN	0.9917	120

The CNN algorithm gave an accuracy of 99.17% on 120 test samples, while the KNN algorithm ($\kappa = 7$) gave an accuracy of 97% on 150 test samples. Both the algorithms performed well for the unknown test samples.

6 Conclusion

After all the explanatory analysis of the result given, it is clear that both the models provided a satisfactory result. Both classification models performed with high accuracy. The performance of both the models has been analyzed carefully. The Mel spectrogram representation of music provided sufficient features and information for

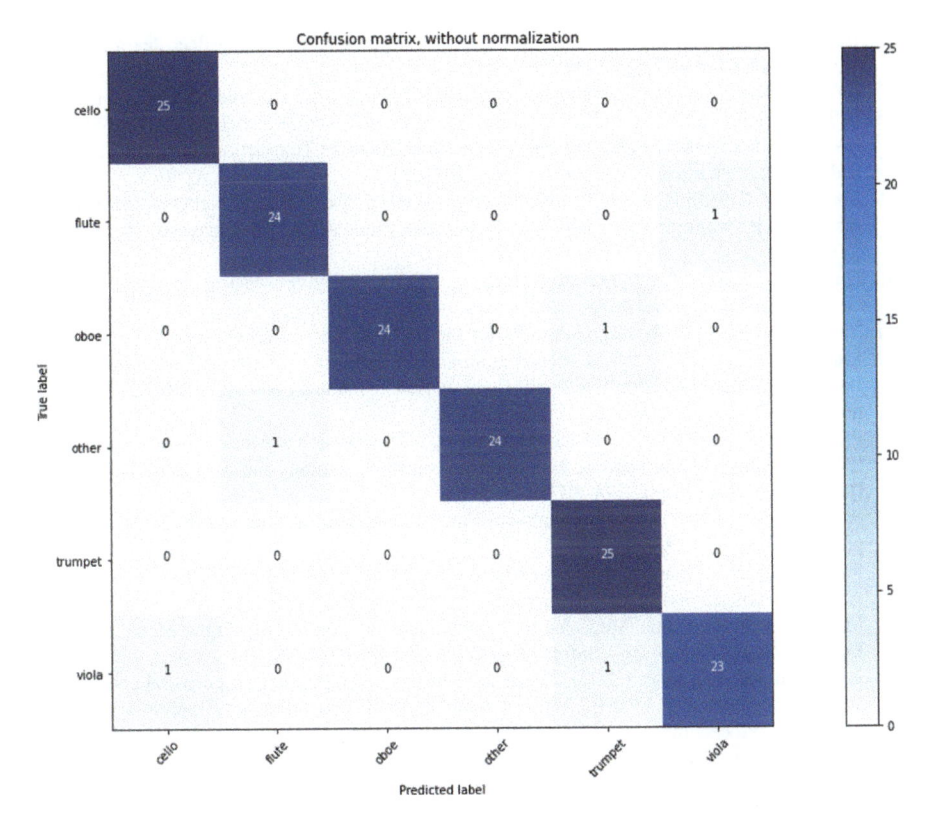

Fig. 10 Confusion matrix for $\kappa = 7$

the convolutional neural network model to accurately differentiate between musical instruments with very different timbres. After the 30 epochs, the research found the excellent result with 99.17% accuracy for the 120 samples used in the CNN model. The KNN model showed 97% accuracy for $\kappa = 7$, for the 150 test samples.

References

1. Essid, S., Richard, G., & David, B,(2005), "Instrument recognition in polyphonic music based on automatic taxonomies," in IEEE Transactions on Audio, Speech, and Language Processing, vol. 14, no. 1, pp. 68–80.
2. Pham, J., Woodford, T. and Lam, J.(2009), Classification of Musical Instruments by Sound.
3. Mustafa Sarimollaoglu, Coskun Bayrak, (2006)," Musical Instrument Classification Using Neural Networks,", Proceedings of the 5th WSEAS International Conference on Signal Processing, Istanbul, Turkey, pp. 151–154.
4. Prabhjyot Singh, Dnyaneshwar Bachhav, Omkar Joshi, Nita Patil,(2019),"Musical Instrument Recognition using CNN and SVM," in International Research Journal of Engineering and Technology (IRJET),vol. 06, no,3, pp. 1487–1491.
5. IRMAS Dataset, https://www.upf.edu/web/mtg/irmas. Last accessed 12 Sept 2021

6. Haidar-Ahmad, Lara.(2019),"Music and instrument classification using deep learning technics", Recall, 67(37.00), pp.80–00.
7. IOWA Dataset, http://theremin.music.uiowa.edu/MIS.html. Last accessed 12 Sept 2021
8. Y. Han, J. Kim and K. Lee,(2017), "Deep Convolutional Neural Networks for Predominant Instrument Recognition in Polyphonic Music," in IEEE/ACM Transactions on Audio, Speech, and Language Processing, vol. 25, no. 1, pp. 208–221.
9. Mahanta, S.K., Khilji, A.F.U.R. and Pakray, P.,(2021)," Deep Neural Network for Musical Instrument Recognition Using MFCCs," in Journal Computación y Sistemas, vol 25, no 2, pp. 351–360.
10. Hing, D.S. and Settle, C.J.(2020), Detecting and Classifying Musical Instruments with Convolutional Neural Networks.
11. Yun, M. and Bi, J.(2018), Deep Learning for Musical Instrument Recognition.
12. Liu, J., Xie, L., (2010),"SVM -based automatic classification of musical instruments," in International Conference on Intelligent Computation Technology and Automation, vol. 3, pp. 669–673.
13. Prabavathy, S., Rathikarani, V.,Dhanalakshmi, P., "Classification of Musical Instruments using SVM and KNN," in International Journal of Innovative Technology and Exploring Engineering (IJITEE), vol. 9, no. 7,pp.2278–3075.
14. Anhari, A.K.,(2020)," Learning multi-instrument classification with partial labels,", arXiv preprint arXiv:2001.08864.
15. Philharmoic Dataset, https://philharmonia.co.uk/. Last accessed 12 Sept 2021
16. A. Kratimenos, K. Avramidis, C. Garoufis, A. Zlatintsi, P. Maragos, (2020) ,"Augmentation Methods on Monophonic Audio for Instrument Classification in Polyphonic Music," in 28th European Signal Processing Conference (EUSIPCO), pp. 156–160.
17. A. Eronen and A. Klapuri,(2000), "Musical instrument recognition using cepstral coefficients and temporal features," in IEEE International Conference on Acoustics, Speech, and Signal Processing. Proceedings, vol.2, pp. II753-II756.
18. Patil S.R., Machale S.J.,(2020),"Indian Musical Instrument Recognition Using Gaussian Mixture Model," in Techno-societal 2018, Springer, Cham, (pp. 51–57).
19. A. Ghosh, A. Pal, D. Sil, S. Palit,(2018), "Music Instrument Identification Based on a 2-D Representation," in International Conference on Electrical, Electronics, Communication, Computer, and Optimization Techniques (ICEECCOT), pp. 509–513.

Emotion Recognition in Music Using Deep Neural Networks

Angelos Geroulanos and Theodoros Giannakopoulos

1 Introduction

Nowadays, accessing music content online is extremely easy as the volume of this content is growing exponentially and is readily available to everyone. Various digital music streaming services (e.g., Spotify, Apple Music, etc.) are available on our computers, mobile phones, smart speakers, recommender systems, etc. However, as much as the streaming of music is continuous, it often does not match the emotional state of the listener at the moment. In such cases, playing a suggested list based on a musical genre is not the best choice, as the variations in the evoked emotions received by the listener can be large, even in tracks of the same genre or even by the same artist. The main problem, before we even get to talking about machine listening algorithms, is that emotions have great difficulty in their classification due to the lack of universal definitions. So this "emotional confusion" [1–4] that is prevalent in various scientific fields is expected to be transferred to machine learning. For example, extracting handcrafted features from audio signals and driving them into a classical classifier, e.g., SVM, can give good results in the general classification of musical genres (e.g., jazz, classical music, rock, etc.). It is far from certain that it will be equally successful if asked to match the corresponding happy, sad, tender, angry, etc. parts of the same samples, since the properly defined or, otherwise, commonly accepted ground truth is missing. Concretizing the above general concern about emotion recognition, in this study, we created deep learning models, tested techniques, and compared them by having as ground truth a small set of just 360 movie music samples. The unique feature of the ground truth is that it is fully labeled and classified into musical emotions, as identified by experts in the

---

A. Geroulanos (✉) · T. Giannakopoulos
Institute of Informatics and Telecommunications, NCSR Demokritos, Athens, Greece
e-mail: tyianak@iit.demokritos.gr

© The Author(s), under exclusive license to Springer Nature Switzerland AG 2023
A. Biswas et al. (eds.), *Advances in Speech and Music Technology*, Signals and Communication Technology, https://doi.org/10.1007/978-3-031-18444-4_10

field of music in the context of psychological-musicological study experiments by Eerola and Vuoskoski [5].

1.1 Related Works

1.1.1 Emotion Recognition, Using the 360-set

The study [6] used as ground truth the second smallest set of the Eerola and Vuoskoski study which contains only 110 samples and is part of the 360-set. By analyzing the audio signals, they extracted features and used SVM to classify them. From the results, among other things, they confirmed the findings of the psychological experiments, such as the high overlap between the anger-fear emotions. In the study [7] once again with the 360-set as a reference, they tried to extract a model from a neural network in the style of the VGG architecture which can musically justify its predictions by the so-called mid-level perceptual features. These features such as rhythmic complexity or harmonic character (major-happy, minor-sad) have musical meaning and can be detected by listeners without musical knowledge. The study [8] is the last one we identified to make use of the 360-set. It presents a new method for music emotion recognition that involves chroma spectrograms with VGG16 and AlexNet architectures.

1.1.2 Emotion Recognition

In the textbook *Music Emotion Recognition* [9], among others, a new method is mentioned which focuses on personalized music emotion recognition. This means, instead of modeling any objectivity, it asks the user to annotate (via a user interface) part of his music collection so that the algorithm (regression) can be trained on his collection. In the experiment of the paper [10], volunteers were subjected to an electroencephalogram while listening to 16 songs selected from a specific song database. Once they were finished, they re-listened to the songs and labeled them based on the dimensional model of emotions. The data along with that of the electroencephalogram were fed into a convolutional neural network (CNN) whose ability to recognize musical emotion (valence and Arousal) was measured by comparing it to a traditional machine learning classifier, an SVM, which was outperformed by the CNN. In the paper [11], two problems were investigated simultaneously: the similarity of some musical compositions to a specific one and the classification of the musical emotion. Emotion recognition albeit a multi-class problem, here, was deconstructed into a multi-binary classification problem that was addressed by an SVM classifier trained from handcrafted features previously extracted from the audio samples. In the paper [12], a CNN was used for music emotion recognition using as input the spectrograms of the previously extracted music fragments to be classified with quite good results in the two sets tested. In the paper [13] on a set of 1000 songs labeled with the attributes of valence and arousal, classifications were performed with a CNN that had as input the spectrograms of the songs.

1.1.3 Transfer Learning

In [14], a CNN model pre-trained on ImageNet [15] is used in order to test it on music classification. Although ImageNet is an "unfamiliar" set in terms of song spectrograms, the study showed that transfer learning worked in this case. In the paper [16], a pre-trained ConvNet feature (as it was named) was proposed as a transfer learning method. It is a combined feature vector from the activations of the feature maps of the multiple levels of a trained convolutional and is used as a pre-trained model for classification tests (speech-music separation, music emotion prediction, music genre separation) or regression tests on other similar sets. The study [17] analyzed pre-trained models used for Devanagari handwritten alphabets recognition using learning transfer from CNNs, namely, AlexNet, DenseNet, VGG, and Inception, where they were used as feature extractors. The most efficient (in terms of accuracy) appeared to be Inception v3 with AlexNet trained faster and with 1% lower accuracy. In the paper [18], transfer learning was used to classify emotion through photographs of people extracted from movies. They used deep convolutional network models pre-trained in ImageNet using the fine-tuning technique. In study [19] to identify objects passing through X-ray baggage screening, a CNN pre-trained on a large and general image set was used to compensate for the lack of data (images through X-rays). The result of the transfer learning was to identify a pistol in baggage with 98.92% accuracy.

1.1.4 Data Augmentation (via GANs)

The study [20] addressed the problem of missing data, in a speech emotion recognition problem, using generative adversarial networks (GANs). Specifically, they improved a conditional GAN architecture to produce spectrograms for the minority class. Results on two large datasets showed an improvement of 5–10% when this method was used. The paper [21] suggests using CycleGAN as an image generator of the sample-deficient class to test (multi-class) emotion classification with sufficient sample presence. A CNN was used as the classifier. A 5–10% increase in classification performance was confirmed when using augmentation with CycleGAN. In [22], a mix-up data augmentation technique is used to augment GAN in both representation learning and synthetic feature vector creation. The proposed framework can learn compressed emotional representations as well as generate synthetic examples that aid in within-corpus and cross-corpus evaluation. This could improve the training size of speech emotion recognition for performance improvement.

1.2 Chapter Contribution

In this chapter, we use deep learning techniques to identify musical emotion and to classify accordingly the musical fragments. We emphasize the data availability issue, and we adopt two datasets: the first contains 17,000 pop rock song excerpts which are divided into feature classes as classified by Spotify's API ("big-set"), and the second set is the ground truth of our work as it is annotated by music experts in the context of a large psychological-musicological research by Eerola and Vuoskoski. For classifying the audio samples into musical emotional classes, we extracted handcrafted features from the audio signals and classified them with traditional classifiers such as SVM, k-NN, random forest, and Extra Trees. We then converted all samples of both sets to Mel scale spectrograms to make inputs to the deep convolutional networks of the tests. Six architectures (AlexNet, VGG16bn, Inception v3, DenseNet121, SqueezeNet1.0, ResNeXt101-32x8d) with an equal number of ImageNet pre-trained models were used in the first round of experiments for classifications via two transfer learning scenarios: in the first scenario, we "freeze" the weights of the network layers except for the classifier layer, and in the second scenario, we fine-tune the network and update the weights of all network layers. This was followed by a second round of experiments where we pre-trained our models on the "big-set" and perform iterative classification tests with the same architectures. Finally, in the third experiment, we create artificial data to augment the "360-set" using generative adversarial networks, specifically StyleGAN2-ADA. From the resulting new set, we re-train models and test them on a series of emotion classifications. We evaluate the performance of all the above rounds of experiments with the macro-avg f1-score obtained by the classifications on the test set of the reference set.

This chapter is organized as follows: In Sect. 2, we describe the features used in both traditional machine learning and deep learning methods used in this chapter to recognize musical emotions. The adopted deep learning architectures for training the emotional classifiers and for augmenting artificial samples are presented in Sect. 3. The experimental setup, the datasets used for training and validation, and the experimental results are presented in Sect. 4, while conclusions are provided in Sect. 5. Note that all results presented in this chapter are also available in the open-source form in this repository: https://github.com/ageroul/music_emotion_recognition_cnn.

2 Feature Extraction

Feature extraction is essential in music information retrieval applications since its goal is to extract features that must be informative concerning the task under study. Features are extracted through audio signal processing, and they result in a feature representation that is then used by the core machine learning algorithms (in our case, music emotion classifiers). In general, there are two categories of feature extraction techniques used in music emotion recognition:

(a) Handcrafted features focus on achieving a certain discrimination ability with regard to the particular classification task, so we could say that usually, they carry some higher level of information. They are used in the context of the "traditional" machine learning pipeline where the feature representations have the role of carrying some "prior" knowledge (provided during the feature engineering process).
(b) Low-level representations that are directly used by the machine learning algorithms. These representations (e.g., the raw audio signal or the spectrogram) are used as input in deep neural networks which act as "learnable" feature extractors.

2.1 Handcrafted Features

In order to extract audio features, the sound signal must first be converted from continuous (analog) to discrete (digital). Audio features are then extracted, and they can be classified into three main categories [23] as follows:

Low level: Low-level features can be extracted either directly from the sound signal or through its transformation, e.g., through Fourier transform. They are not particularly meaningful at the listener level, but their extraction is easy and widely used.

Medium level: They consist of features whose representation has more musical meaning than those of the low level. Such representations relate to the melodic, harmonic, and timbral aspects of music.

High level: This category refers to musical features that cannot be directly produced by the audio-musical signal. It is more concerned with symbolic representations of music, such as a music score that describes all the parts of musical composition in notes. Another such representation is the one generated by the MIDI (Musical Instrument Digital Interface) protocol.

In this chapter, we focus on extracting features directly from the audio signal itself, so we will focus on the low-level audio features case. The extraction process starts with the so-called short-term windowing of the original signal. The length of each such frame ranges from 10 to 100 ms depending on the application and signal type. Frames may also overlap sometimes. In our experiments, we use a non-overlapping framing with equal frame length and frame step values of 50 ms. Then, and for each such frame, we extract a set of features, which derives from either time, frequency, or cepstral domain [24]:

– *Time Domain*: When the features are derived directly from the signal, we say that they are related to the time domain like the ZCR (zero-crossing rate), energy, and energy entropy. The ease of signal analysis they offer is important, but very

often, it is required to combine them with more sophisticated features from the frequency domain.
- *Frequency Domain*: When we perform a discrete Fourier transform (DFT) on an audio signal, we project it into the Fourier spectral (or frequency) domain, and so we get a representation of the frequency content of the sound. Some features of this type are the spectral centroid, spread, entropy, and spectral rolloff.
- *Cepstral Domain*: The so-called cepstral features, for example, the group of Mel-frequency cepstral coefficients (MFCCs) and the chroma vector, are derived from the cepstrum, an anagram of the spectrum and defined as the inverted Fourier metric of the logarithm of the spectrum.

The results presented in this chapter use the low-level features extracted by *pyAudioAnalysis* [25] which is an open-source Python library for audio signal analysis.

2.2 Mel Spectrograms

The core methodology presented in this chapter uses CNNs to classify music signals to emotional classes. CNNs can function on feature matrices from the handcrafted features described above. However, it is more common in the field of audio and speech analytics to use spectrograms or Mel spectrograms as input images in CNNs when classifying sounds [26]. Spectrograms are simply time-frequency representations of the audio signal, extracted using short-term Fourier transform calculation. Mel scale is a logarithmic conversion of the signal frequency whose representation is closer to the human way of perceiving sound than the linear one. Thus, the Mel spectrogram represents the frequency domain of a signal as a function of time except that the frequencies are expressed in the Mel scale, instead of the linear one of the "simple" spectrogram. In the results presented in this chapter, converting the sound samples of the sets into Mel spectrograms was done using Librosa [27], a Python package for audio and music signal analysis.

3 Music Emotion Recognition Using Deep Learning

The practical difference between machine learning and deep learning in audio classification lies in the fact that machine learning algorithms use some handcrafted features which are sometimes quite sensitive when deployed. For example, one could extract MFCCs from the audio signal [28], as such, assuming they provide sufficient information for a particular classification and train a classifier (e.g., SVM) to label a dataset based on them alone. On the contrary, the basic idea of deep learning is that it hierarchically learns all features through multiple layers of a neural network trained directly from the data. Through the nonlinearity of the activation functions of the multi-layer neural network, the connections are fully "trainable" which is not the case in a single-layer network.

3.1 CNN Architectures Used for Music Emotion Recognition

This study investigates, through deep learning techniques, the ability of well-known CNN architectures (AlexNet, VGG16bn, Inception v3, ResNeXt101-32x8d, SqueezeNet1.0, DenseNet121) to recognize music emotion when the set has a small amount of data, with sets of different distributions and sometimes unbalanced. The techniques used are transfer learning and data augmentation via generative adversarial networks (GANs).

AlexNet In 2012, AlexNet [29] won the ILSVRC (ImageNet Large Scale Visual Recognition Challenge) image classification competition. Its architecture is made up of eight layers, the first five of which are convolutional and the last three of which are fully connected, and the last is a softmax activation function. The kernel sizes of the convolutional layers are 11×11, 5×5, and 3×3, with max pooling layers for image downsampling and dropout layers to avoid overfitting. Some features that had not previously been used in CNN, such as in LeNet, were used in AlexNet. Thus, AlexNet employs the ReLU activation function for the nonlinear part (nonlinearities), as opposed to the tanh and sigmoid activation functions used in more traditional neural networks. ReLU was used in the network's hidden layers, which resulted in faster training because its derivative is not as small as tanh and sigmoid, and thus the updated weights (during backpropagation) do not disappear. The dropout technique was used in the network's two fully connected layers for the first time on a large scale in AlexNet. The neurons in the layer that have been dropped out do not participate in training during either forward or backward propagation. As a result, each neuron is forced to learn more useful features in relation to a large number of other random groups of neurons.

VGG16 (Batch Normalization) The Visual Geometry Group at Oxford University created VGG [30] in 2014 (where it placed second at the ILSVRC the same year). Its basic components are identical to those of LeNet and AlexNet, with the exception that VGG is a deeper network with additional convolutional, pooling, and fully connected layers. VGG appears to be superior to AlexNet since it has replaced the big filters (11x11 and 5x5 in the first two convolutional layers) with several 3x3 dropout filters. The VGGs are a family of similar architectures, with the numbers next to VGG representing the number of layers (VGG11, VGG13, VGG16, and VGG19). We employed VGG16 in its batch norm version in this work's experiments.

Batch Normalization is the process that reduces the internal covariance shift. The change in the distributions of the internal nodes in a deep network during training is referred to as the internal covariate shift. Eliminating it will result in speedier training. This is accomplished by normalizing the means and variances of the layer inputs. It also lessens the derivatives' dependence on the magnitude of the parameters or their initial values. This enables us to work at larger training rates without concerns of divergence. Furthermore, batch norm smoothes the model, decreasing the need for dropout.

Inception v3 The Inception network architecture was first introduced by Szegedy et al. with GoogLeNet [46] (Inception v1), and other versions followed in the following years. In this work, we used the Inception v3 version [31] (Szegedy et al.). Inception v3 through the segmentation of the large convolutions of the previous versions tries to reduce the computational cost without affecting the generalization of the model. Thus, an $n \times n$ convolution is factorized into a combination of $1 \times n$ and $n \times 1$ convolutions (for example 1×7 and 7×1 filters replacing a 7×7) and 1×1 convolutions are used before them as bottlenecks. The bottlenecks through the split-transform-merge procedure drive the input to three to four different feature maps of smaller or equal dimensions to the input and then drive them through 3×3 or 5×5 convolutions to smaller three-dimensional maps. Although the network eventually has a depth of 48 layers, the computational cost is only 2.5 higher than Inception v1 and much more efficient than VGG.

ResNeXt101-32x8d ResNeXt [32] was presented in 2017 at the CVPR (Computer Vision and Pattern Recognition) Conference by Saining Xie et al. and is a state-of-the-art model in the field of ImageNet classification. It performs better than ResNet as it is an upgraded version of ResNet and it combines the convolutional layer stacks of VGG and the split-transform-merge idea used by inception models. In ResNeXt, there are four levels, and each level has a few residual blocks. Each such residual block has increased width where more filters create multiple parallel paths and the set of these paths was called cardinality. Essentially, we take a residual block with narrowing and make it less deep but wider. While ResNeXt has the same number of parameters as a ResNet counterpart, the features extracted by ResNeXt have better performance in the ImageNet classification task than those of ResNet meaning that they have a much stronger capability. In addition, ResNeXt101-32x8d (i.e., 101 layers; cardinality, 32; bottleneck width, 8), which we also use in the experiments in this work, achieves top performance in ImageNet without fine-tuning and without extra training data.

SqueezeNet 1.0 The SqueezeNet architecture [33] was presented in the 2016 paper by F. Iandola et al. who proposed a small network that achieved the performance of AlexNet on ImageNet but with 50x fewer parameters. Also, with compression techniques of the model, they managed to compress it to less than 0.5 MB which is 510x smaller than AlexNet. The main building block of SqueezeNet is the so-called Fire module. The Fire module consists of a squeeze convolutional layer that has only 1×1 filters that are driven by an extended layer that has a mix of 1×1 and 3×3 filters. The Fire module has three hyperparameters where their adjustment helps the squeeze layer to reduce the number of inputs to the 3×3 filters based on three strategies. The architecture of SqueezeNet starts with a standalone convolutional layer, is followed by eight Fire modules, and ends with a final convolutional. The number of filters is increased gradually per Fire module and from the beginning to the end.

DenseNet 121 The DenseNet architecture [34] was presented in 2018 by G. Huang et al., and it is based on the assumption that CNNs can be even deeper, more accurate, and more efficient at training if they have shorter connections between nearby input and output layers. Thus, in the DenseNet architecture, each layer is connected to every other layer. For every L levels, there are $L(L+1)/2$ direct connections. At each layer, the features of all the previous layers become its input, and its features become input to the subsequent layers. To make this feature merging work, the corresponding feature maps must also be of the same dimensions or undergo downsampling to achieve it. Thus, the creators of DenseNet created and placed so-called dense blocks within which the size of the feature map remains the same. Each dense block consists of a series of convolutional 1×1 and 3×3 layers, the conv blocks. The convolution and pooling are done on different layers located between the dense blocks called transition layers and include a batch norm, a 1×1 conv, and a 2×2 average pooling layer. The DenseNet model we used in the experiments of this work is DenseNet121.

3.2 Transfer Learning

To properly train a deep neural network, we need large amounts of data, much larger than that needed by machine learning algorithms. However, very often, neither the retrieval of massive data sets is feasible, nor is the required computing power available. The basic idea behind the transfer learning technique [35–37] is: "Store knowledge gained by solving one problem and use it to solve another similar problem in a shorter time and at a lower cost." More specifically, let us assume that for an image classification problem, it is common practice to use a pre-trained model derived from training a deep CNN on a very large dataset (e.g., ImageNet of 1.2 million images, 1000 classes). The pre-trained model can then be used in other classification projects adapted appropriately for specific projects. What is transferred through the pre-trained model are the weights of the original network after it has been trained.

3.2.1 Transfer Learning Scenarios

The following transfer learning strategies are possible and tested in this chapter:

- Use source model as *feature extractor*: the last layer (the classifier layer at most cases) is removed from the pre-trained (source) model, adapted to the needs of the current classification task (target), and the weights of the remaining layers are "frozen." For example, the ImageNet classes are 1000, while the ongoing task requires 5 classes, so we correct accordingly. This technique is most useful when the datasets (pre-training and training/testing sets) are similar. One can even use a machine learning classifier, e.g., SVM, instead of the last layer of the model.
- *Fine-tuning*: in this scenario, the classifier layer is corrected as well, but the weights of the previous layers are also fine-tuned during backpropagation. Either all or some of the layers can be updated, and the rest can be left "frozen" (with the pre-training weights). This works as the first few layers of a CNN identify the general features of an image (e.g., image edge detection) that are most likely to be common even in completely unrelated sets of images.
- *Classifier*: without any additional training or changes to any of its layers, the model is used as a classifier. This scenario assumes enough similarity of the sets of both the pre-training and the test on the new classification.

3.2.2 Choosing a Scenario

Deciding which of the above scenarios to adopt on a case-by-case basis depends mainly on two factors: the size of the set and its similarity to the pre-training set. In general, we follow the following guidelines:

- *The new set is small and similar to the original*: If fine-tuning occurs, there is a risk of overfitting, so the feature extractor scenario is preferred.
- *the new set is large and similar to the original*: having many samples almost eliminates the risk of overfitting so the scenario of fine-tuning all (or most) of the layers of the model prevails.
- *the new set is small and very different from the original*: from one side, we have a small set so overfitting is quite possible in the case of fine-tuning the whole network, and on the other side, due to the diversity of the sets, freezing the last layers is not a good idea as they contain weights adapted to the details of the original set. A halfway solution is to train the classifier (maybe an SVM) with the weights of only some initial layers of the model.
- *The new set is large and very different from the original set*: the scenario of fine-tuning the whole network over the initial weights of the pre-trained one is preferred. If the set is too large, then the weights can be trained from the beginning with random initial values. A valuable point to note concerning the choice of learning rate when fine-tuning weights is choosing a value small enough to avoid disturbing the already well-initialized weights from pre-training.

3.3 Data Augmentation Using GANs

The dataset used as "golden" ground truth for evaluating the proposed music emotion recognition classifiers, named (360-set), when tested in the five-class classification task (discrete emotions), and after its segmentation into train, validation, and test sets, left very few samples for training. The number of samples left is very few even for transfer learning and fine-tuning; therefore, a major effort of the work presented in this chapter focuses on the data augmentation technique to enlarge the train-validation sets. In particular, we adopted the StyleGAN2-ADA (SG2A) which we trained with the set of train+val samples of each class (i.e., anger, fear, happy, sad, tender) of the 360-set. SG2A was trained using the transfer learning method from a pre-trained NVIDIA model.

StyleGAN2-ADA (2020) [38] is the latest generative adversarial network (GAN) created by Teo Karras and the NVIDIA research team after Progressive GAN [39] (2017), StyleGAN (2018), and StyleGAN2(2019). StyleGAN [40] is an extension of the basic GAN architecture but can split over the individual features of the generated image. They are the evolution of Progressive GANs that were already capable of synthesizing high-resolution images with the gradual development of the discriminator and the generator networks during the training process. SG2A achieves the production of very-high-resolution images from short training sets without overfitting problems through a new method called adaptive discriminator augmentation (ADA).

Figure 1 shows the concept of the ADA method [38] where the augmentation processes are shown in blue, the network being trained in green, the loss functions are in orange, and the effect of probability (p) is on the transformations on the right.

As can be seen, all images received by the discriminator (D) are augmented with a default probability (p) for each augmentation which occurs randomly, and the performance of the discriminator is estimated from these images. On the other hand, the generator (G) is trained to produce only clean images as long as the probability (p) remains below the threshold of safety. The authors' experiments on StyleGAN2-ADA showed that leakage in G starts to occur when p is very close to one with the safety threshold being when $p < 0.8$. That is, the higher the probability, the more augmentations and the more variant set we get. However, manually adjusting the hyperparameter p is a difficult process so the creators set out to automate the ideal value of the hyperparameter. This process of controlling the augmentation intensity was called ADA and is an adaptive method based on correcting the value of a heuristic r_t.

As we can see in Fig. 2, the more there is sufficient overlap of the distributions (real—generated) of the images in the output of the discriminator (D), the better it will be and the better the predictions will be. The heuristic based on this overlap (easily measured throughout the training) computes the percentage of real images for which $D(x) > 0$. If the average is very high (meaning that the distributions no longer overlap), then the value of p is adjusted for more image augmentation; if it is too low, p is adjusted for less augmentation.

Fig. 1 ADA flowchart and the augmentation probability (p)

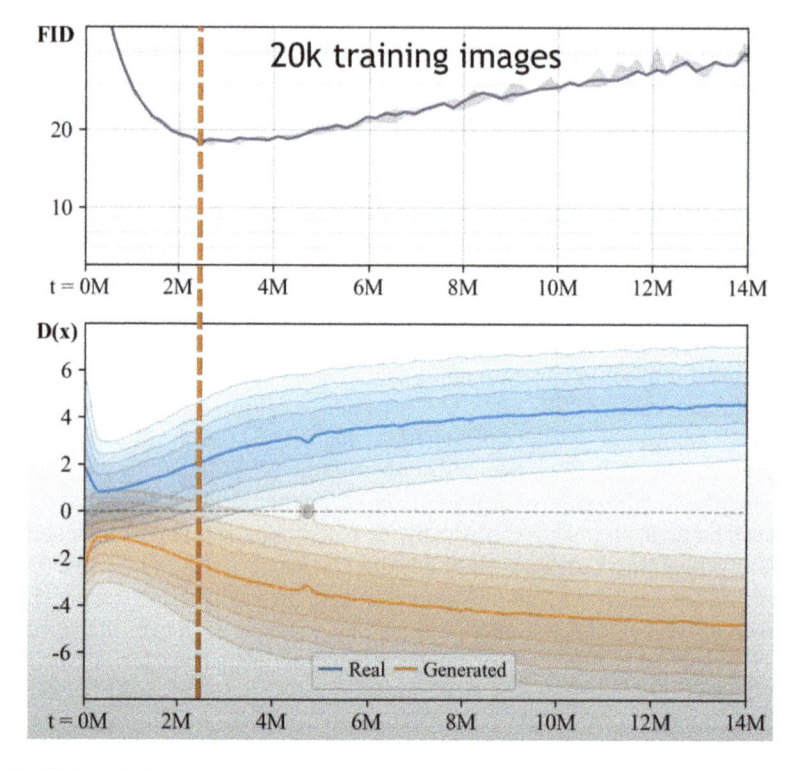

Fig. 2 ADA method

4 Experiments

For evaluating all classification, transfer, and augmentation techniques described above, we have used two datasets, namely, "big-set" and "360-set" (see the next subsection). The first was used to generate pre-trained models and the second as a golden ground truth for testing and tuning the pre-trained models.

The data generated by StyleGAN2-ADA were used as inputs for the 360-set emotion classification tests (anger, fear, happy, sad, tender) as the initial number of samples was too small for deep learning and benchmark tests were performed. All models used (ResNeXt101-32x8d, AlexNet, VGG16bn, SqueezeNet1.0, DenseNet121, Inception v3) are pre-trained on the ImageNet dataset and are derived from the torchvision.models [41] machine learning library model package of PyTorch. In summary, the classification tests for the 360-set are valence, energy, tension, and emotion (anger, fear, happy, sad, tender) in six different architectures and two scenarios, i.e., $4 \times 6 \times 2 = 48$ tests. Similarly, the classification tests for the big-set are valence and energy in six different architectures and two scenarios, i.e., $2 \times 6 \times 2 = 24$ tests.

The abovementioned series of tests were then repeated; only this time, the models were pre-trained on the big-set spectrograms. Finally, a comparison of their results was made. The code of the experiments is written in Python and was executed in the Google Colaboratory [42] platform environment in the form of a Jupyter Notebook as the use of the GPUs offered by Colab, albeit with some limitations, was considered necessary for training the models and generating artificial samples with StyleGAN2-ADA.

4.1 Dataset Origin

For the experiments of this study, two sets of musical data were used, which for convenience we will call big-set and 360-set.

4.1.1 Big-Set

Big-set includes 17,000 audio clips from songs in rock and pop styles, each lasting 10 seconds, from a random part of each song. The format of the files is wav, mono, with a sample rate of 8 kHz. The entire big-set belongs to a private collection (not publicly available) with their metadata coming from Spotify API, which allows users to explore and extract music features from music databases by choosing from the platform's millions of tracks. The feature categories are as follows and are specific to each musical excerpt: acousticness (whether the track is acoustic), danceability (whether it is danceable), energy (intensity of energy activity), instrumentalness (whether it contains vocals), liveness (whether there is an audience in the recording), loudness (the average volume in dB), speechiness (speech detection), valence (musical vigor), and tempo (rhythmic character estimation). Of these attributes, useful to us were energy and valence. In the Spotify API, energy and valence are broken down as follows: Energy: Energy is measured on a scale of 0.0 to 1.0 and represents a measure of the perceived intensity and activity of the music. Energetic pieces are mostly perceived as fast, loud, and noisy. For example, a death metal piece has high energy, but a Bach prelude has very low energy. Perceived features that characterize energy include dynamic range, volume, timbre, the rate of musical events per second, and overall entropy. Valence: Valence is measured on a scale of 0.0–1.0 and expresses the positivity derived from the musical piece. Segments with high valence sound more positive (joy, euphoria), while pieces with low valence sound more negative (sadness, anger, melancholy). All music excerpts in the big-set are fully tagged with the aforementioned Spotify API features.

4.1.2 360-Set

The 360 set consists of music clips in mp3 format, stereo, sampled at 44.1 KHz, lasting between 15 and 30 seconds and are exclusively instrumental i.e. they do not contain singing or recitation of lyrics. The 360-set excerpts are film soundtracks from 110 films selected from and used by Eerola and Vuoskoski [5] on their study on the discrete and dimensional model of musical emotion classification. The original name of the set is "Stimulus Set 1" and it is freely accessible. The selection of excerpts was made by experts (professional musicians, professors, and students of music university faculties) who classified-ranked them in five categories of the discrete model (happy, sad, tender, fear, angry) and three categories of the dimensional model (valence, energy, tension) of emotion classification in music. The 360-set was chosen as the ground truth for this research because it is fully annotated by experts in the music field, which is rare for publicly available sets.

4.2 Dataset Pre-processing

Sound samples must be converted into Mel spectrograms in order to be used as inputs for CNNs. In addition, the following steps were taken in order to ensure the uniformity of the experiments: (a) "big-set" remained unchanged (i.e., wav, mono, 8 kHz, 10 sec) (b) "360-set" had a reduced sampling rate, the sample duration, and the audio was changed to mono (mp3, mono, 8 kHz, 10sec). In Eerola and Vuoskoski's study, "Stimulus Set 1" (360-set) was the predecessor of their main experiment which resulted in an even smaller set of 110 tracks called "Stimulus Set 2" from which they removed one class, the so-called surprise that was present in "Stimulus Set 1." This happened as the experts failed to identify it as a standalone emotion which was shown by the very low scores they gave to the extracts that (supposedly) should have been dominated by surprise over the other emotions. The ratings and the ranking of the extracts are published in the file "mean ratings set1." Thus, the 30 "surprise" parts were classified by us in the other emotion classes according to the highest expert rating they obtained in the auditory test. Another issue that had to be addressed was the reduction of the duration of the excerpts so that they all had a duration of, as mentioned above, 10 seconds. So, we kept the first 10 seconds of each track.

4.2.1 Splitting the Datasets into Training, Validation, and Test

Having converted all sound samples into Mel spectrograms, the big-set and 360-set were separated into training, validation, and test sets as shown in Tables 1 and 2: *training set* which is used to train the model, *validation set* which is used to estimate the generalization of the model where, according to the losses, fine-tuning of the model is performed, and *test set* which is used after the training and contains only samples unknown to the model.

4.3 Hyperparameter Selection and Settings

The most important hyperparameter is the learning rate of the optimization function: a small learning rate means more reliable training, but convergence is achieved after a long time. On the other hand, a high rate can mean fast training but usually fails as the changes in weights are so large that the algorithm eventually is not able to discover a good minimum of the cost function. Finding the ideal training rate is a difficult task that usually requires many trials, and there are some techniques (such as annealing and scheduling) that help in this direction, of course. In this work, two of such techniques were used since the number of tests, limited computational resources, and lack of time did not allow a thorough and separate tuning of each architecture tested (stochastic gradient descent (SGD) with momentum = 0.9 was used as the optimization function in all tests, which has generally performed well on the networks we selected):

Table 1 Splitting the 360-set into train, val, and test sets

Train, validation and test sets of 360-set

Valence	Train	Val	Test	Tension	Train	Val	Test	Energy	Train	Val	Test	Emotions	Train	Val	Test
Positive	62	26	50	High	64	28	50	High	61	27	50	Anger	6	5	40
Neutral	47	21	50	Medium	33	15	50	Medium	48	20	50	Fear	32	21	40
Negative	38	16	50	Low	49	21	50	Low	37	17	50	Happy	15	10	40
												Sad	21	14	40
												Tender	22	14	40

Table 2 Splitting the big-set into train, val, and test sets

Train, validation, and test sets of big-set

Valence	Train	Val	Test	Energy	Train	Val	Test
Positive	3752	938	513	High	5739	1435	780
Neutral	4762	1191	661	Medium	4416	1104	628
Negative	3882	971	541	Low	2192	548	308

- *PyTorch learning rate finder.* This is a PyTorch implementation [43] of the "learning range test" as described in Leslie N. Smith [44]. During this test, the learning rate increases linearly or exponentially between two thresholds. The lower bound lets the network start convergence, and eventually as the rate increases, at some point, it will become too high, and then the network will diverge. Training with cyclic training rates instead of fixed values achieves improved accuracy without the need for tuning and often with fewer iterations.
- *ReduceLROnPlateau.* The torch.optim is a PyTorch package that includes various optimization algorithms. One of them is the ReduceLROnPlateau [45] which reduces the training rate when a metric stops improving. The algorithm reads a selected metric, and if no progress is made for a certain number of epochs, then the training rate is reduced.

4.4 Experiment Summary and Aggregated Results

We will summarize the process of the three experiments involving the techniques of transfer learning and the generation of artificial samples to understand the elements of Table 3.

Experiment X This is used as baseline experimentation to evaluate the ability of handcrafted features and traditional ML algorithms to recognize music emotions. It was entirely executed, as mentioned earlier, using the pyAudioAnalysis library [25] which has, among others, several machine learning classifiers (SVM, SVM_RBF,

Table 3 The best results (*macro-avg f1-score*) of the experiments are shown; the colors correspond to the different test sets

Experiment	Type	Model	Pretrained on	Tested on	macro avg f1-score
ENERGY					
A	DL	VGG16bn(whole)	ImageNet	Big-set	0.7
A	DL	VGG16bn(whole)	ImageNet	360-set	0.62
B	DL	VGG16bn(whole)	Big-set	360-set	0.65
X	ML	SVM		360-set (train&val)	0.52
VALENCE					
A	DL	VGG16bn(whole)	ImageNet	Big-set	0.64
A	DL	ResNeXt-101_32x8d (whole)	ImageNet	360-set	0.63
B	DL	AlexNet (whole)	Big-set	360-set	0.62
X	ML	SVM		360-set (train&val)	0.56
TENSION					
A	DL	ResNeXt-101_32x8d (whole)	ImageNet	test-set 360-set	0.65
X	ML	Random Forest		360-set (train&val)	0.53
EMOTIONS					
A	DL	Densenet121(whole)	ImageNet	test-set 360-set	0.55
C	DL	SqueezeNet1.0 (whole)	StyleGAN2-ADA	test-set 360-set	0.58
X	ML	SVM		360-set (train&val)	0.4

k-NN, Extra Trees, random forest) where we carried out ground truth classifications of the 360-set, for comparison reasons.

Experiment A We performed four feature classification tasks belonging to the two main types of music emotion classification (discrete and dimensional). The discrete model is a five-class classification task of emotions (anger, fear, happy, sad, tender), and the dimensional model is a three-class classification task of three classes: energy (high, medium, low), valence (positive, neutral, negative), and tension (high, medium, low). The six used architectures (ResNeXt101-32x8d, AlexNet, VGG16bn, SqueezeNet1.0, DenseNet121, Inception v3) and their pre-trained models were taken from the torchvision library. The classifications were performed in both sets with two variations of each model: either by updating the weights of all its layers (whole) or by "freezing" everything except the classifier layer (freeze). Algorithms were used to find an optimal learning rate and to optimize the initially selected one during training. Max number of epochs was 20 in each training.

Experiment B Same scenario as Experiment A, but this time we pre-trained the models on the big-set instead of ImageNet and performed two classifications on the energy and valence features since they are common to both sets. Table 3 shows those that gave the best result in the macro-avg f1-score.

Experiment C The five-class feature emotions had the characteristic of having the smallest number of samples per class. Thus, we used the very recently published StyleGAN2-ADA to create artificial samples (Mel scale spectrograms) to classify and compare them with the best model from Experiment A. Transfer learning was also used in StyleGAN2-ADA and specifically pre-trained model on the FFHQ image set. Table 3 shows their best performance on each feature.

F1-score was chosen because it is a metric that gives more emphasis to minor classes and, due to the calculation of the harmonic mean, it rewards models that have similar values in the precision and recall or sensitivity metrics. This property is especially useful in the emotions feature, which was divided into five classes of only a few dozen samples each. Note that in all DL experiments (A, B, C), we used transfer learning methods and discovered that the most prevalent models were the ones we called "whole," i.e., those that were fine-tuning the weights of all layers on top of the already existing pre-trained network-derived weights. More specifically, with this transfer learning method, the parameters of a pre-trained model are adjusted to the data of the new set which may have come from a different distribution. This is the case, e.g., in Experiment A when VGG is trained on the big-set containing only spectrogram images, while the model was pre-trained on the ImageNet of 1000 classes and millions of images but not including spectrograms. Of course, at the classifier layer, we performed changes to adjust it to the three classes of energy and valence. The following conclusions are directly drawn from the results shown in Table 3:

1. *Energy*: The pre-trained (on ImageNet) model VGG16bn when trained on the 360-set and then tested on the test set of the 360-set gave an f1-score of 0.62 while when pre-trained on the big-set, trained on the 360-set, and tested on the test-set of the 360-set gave an f1 of 0.65. That is, the transfer learning worked more efficiently in the second case where the pre-training has been done on a much smaller number of samples but with higher relativity compared to the test set. In addition, the execution time of the training was significantly faster.
2. *Valence*: ResNeXt101-32x8d pre-trained on ImageNet, when trained on 360-set and tested on the 360-set test set, gave an f1-score of 0.63. With the same training and test, AlexNet pre-trained on the big-set gave an f1-score of 0.62, i.e., lower by only 0.01. We can see that the architecture of just 8 layers performed almost similarly to a 101-layer network with much more modern architecture. The Colab training time for AlexNet was 1/3 that of ResNeXt.
3. *Tension*: The tension feature was only present in the 360-set; therefore, no benchmarking was performed with another deep learning (DL) model other than the pre-trained (in ImageNet) ResNeXt101-32x8d which we trained and tested on the 360-set. However, based on the f1-score of 0.65 that we obtained from the model, we can relatively safely assume that if the tension feature was present in the big-set, it would give a similar result as it gave in valence.
4. *Emotions*: A first observation is that the emotions is the only feature of our tests that has five classes and the fewest samples per class as it is only found in the 360-set (of 360 samples). From Experiment A, the model with the highest f1-score (0.55) was DenseNet121 in its "whole" form. Motivated by the lack of samples commented above, we proceeded to perform Experiment C, i.e., the generation of artificial samples (spectrograms) using NVIDIA's state-of-the-art GAN StyleGAN2-ADA. The SG2A model was also pre-trained on the FFHQ image set. So, from the 160 total samples of the train+validation part of the 360-set, after several days of training, 5000 samples (1000/class) were produced. The newly generated set was then trained and tested on classification projects as in Experiment A. SqueezeNet1.0 in its "whole" version was selected as the best model (pre-trained on ImageNet) where it yielded an f1-score of 0.58. That is, its performance was slightly better (by 0.03) than DenseNet trained with the real samples. This difference may not seem big enough to be worth that much training time (which is shorter if run on a modern GPU or array). However, if we consider that the anger class, for example, fed SG2A with only 11 samples while the smallest image set on which SG2A was tested by its creators contained 1336 images (MetFaces set), we can see the potential if applied to sets whose samples are really rare to find. We believe that if we subjected the model (SqueezeNet) to further tuning of its hyperparameters, the results would be even better. This was not possible as the experiments were numerous and their purpose was comparative.

5 Conclusion

In this study, we investigated deep learning techniques to achieve music emotion classification. We adopted 2 music sets, one with 17,000 samples annotated "in-the-wild" (big-set) and a "golden ground truth" dataset of 360 samples annotated by expert participants in a musicological-psychological experiment (360-set), which we converted into Mel scale spectrograms.

We first used ImageNet pre-trained models of deep convolutional neural networks ResNeXt101-32x8d, AlexNet, VGG16bn, SqueezeNet1.0, DenseNet121, and Inception v3 and showed that although pre-trained on an unfamiliar image set from the requested image set, they can perform well in classification if the weights of all their layers are updated, not just those of the classifier. Also, by doing our pre-training from the samples of big-set, we created models and showed that they yielded better (for energy $+3\%$) or slightly worse (valence -1%) results in classifications to those of the original ones, with the difference that the performance was achieved either with light-weight architectures, e.g., AlexNet instead of ResNeXt, or in a shorter training time of the same neural.

Finally, in the five-class classification of the 360-set where the samples for training were dramatically reduced, we augmented them by generating new artificial spectrograms using a pre-trained model of StyleGAN2-ADA. With the artificially augmented samples, we trained all CNNs of the experiments and extracted the corresponding models. In the classification which followed using the new models, we kept the best one, and the results were superior (emotions $+3\%$) to those of the original model, showing that in the similar case of classification of sparse data, which can be converted to image, transfer learning together with the generation of artificial samples gives satisfactory results. In a future extension of this work, we would like to attempt a reduction of the emotion classification problem to a multi-class and multi-label problem. Also, another possibility would be to try to incorporate sequential deep learning techniques.

References

1. Kleinginna, P.R., Kleinginna, A.M.: A categorized list of emotion definitions, with suggestions for a consensual definition. Motiv Emot. 5, 345–379 (1981)
2. Vuoskoski, J.K., Eerola, T.: The role of mood and personality in the perception of emotions represented by music. Cortex. 47, 1099–1106 (2011)
3. Dufour, I., Tzanetakis, G.: Using Circular Models to Improve Music Emotion Recognition. IEEE Trans. Affective Comput. 1–1 (2018)
4. Juslin, P.N.: Musical Emotions Explained: Unlocking the Secrets of Musical Affect. Oxford University Press (2019)
5. Eerola, T., Vuoskoski, J.K.: A comparison of the discrete and dimensional models of emotion in music. Psychology of Music. 39, 18–49 (2011)
6. Petri, T.: Exploring relationships between audio features and emotion in music. Front. Hum. Neurosci. 3, (2009)

7. Chowdhury, S., Vall, A., Haunschmid, V., Widmer, G.: Towards Explainable Music Emotion Recognition: The Route via Mid-level Features. arXiv:1907.03572 [cs, stat]. (2019)
8. Bilal Er, M., Aydilek, I.B.: Music Emotion Recognition by Using Chroma Spectrogram and Deep Visual Features: IJCIS. 12, 1622 (2019)
9. Yang, Y.-H., Chen, H.H.: Music emotion recognition. CRC; Taylor & Francis [distributor], Boca Raton, Fla.: London (2011)
10. Keelawat, P., Thammasan, N., Numao, M., Kijsirikul, B.: Spatiotemporal Emotion Recognition using Deep CNN Based on EEG during Music Listening (2019)
11. Tao Li, Ogihara, M.: Content-based music similarity search and emotion detection. In: 2004 IEEE International Conference on Acoustics, Speech, and Signal Processing. p. V-705–8. IEEE, Montreal, Que., Canada (2004)
12. Liu, X., Chen, Q., Wu, X., Liu, Y., Liu, Y.: CNN based music emotion classification (2017)
13. Liu, T., Han, L., Ma, L., Guo, D.: Audio-based deep music emotion recognition. Presented at the 6TH INTERNATIONAL CONFERENCE ON COMPUTER-AIDED DESIGN, MANUFACTURING, MODELING AND SIMULATION (CDMMS 2018), Busan, South Korea (2018)
14. Palanisamy, K., Singhania, D., Yao, A.: Rethinking CNN Models for Audio Classification (2020)
15. Deng, J., Dong, W., Socher, R., Li, L.-J., Kai Li, Li Fei-Fei: ImageNet: A large-scale hierarchical image database. In: 2009 IEEE Conference on Computer Vision and Pattern Recognition. pp. 248–255. IEEE, Miami, FL (2009)
16. Choi, K., Fazekas, G., Sandler, M., Cho, K.: Transfer learning for music classification and regression tasks (2017)
17. Aneja, N., Aneja, S.: Transfer Learning using CNN for Handwritten Devanagari Character Recognition. 2019 1st International Conference on Advances in Information Technology (ICAIT). 293–296 (2019)
18. Ng, H.-W., Nguyen, V.D., Vonikakis, V., Winkler, S.: Deep Learning for Emotion Recognition on Small Datasets using Transfer Learning. In: Proceedings of the 2015 ACM on International Conference on Multimodal Interaction. pp. 443–449. ACM, Seattle Washington USA (2015)
19. Akcay, S., Kundegorski, M.E., Devereux, M., Breckon, T.P.: Transfer learning using convolutional neural networks for object classification within X-ray baggage security imagery. In: 2016 IEEE International Conference on Image Processing (ICIP). pp. 1057–1061. IEEE, Phoenix, AZ, USA (2016)
20. Chatziagapi, A., Paraskevopoulos, G., Sgouropoulos, D., Pantazopoulos, G., Nikandrou, M., Giannakopoulos, T., Katsamanis, A., Potamianos, A., Narayanan, S.: Data Augmentation Using GANs for Speech Emotion Recognition. In: Interspeech 2019. pp. 171–175. ISCA (2019)
21. Zhu, X., Liu, Y., Qin, Z., Li, J.: Data Augmentation in Emotion Classification Using Generative Adversarial Networks (2017)
22. Latif, S., Asim, M., Rana, R., Khalifa, S., Jurdak, R., Schuller, B.W.: Augmenting Generative Adversarial Networks for Speech Emotion Recognition. In: Interspeech 2020. pp. 521–525. ISCA (2020)
23. Raś, Z.W., Wieczorkowska, A.A. eds: Advances in Music Information Retrieval. Springer Berlin Heidelberg, Berlin, Heidelberg (2010)
24. Giannakopoulos, T., Pikrakis, A.: Introduction to audio analysis: a MATLAB approach. Academic Press is an imprint of Elsevier, Kidlington, Oxford (2014)
25. Giannakopoulos, T.: pyAudioAnalysis: An Open-Source Python Library for Audio Signal Analysis. PLoS ONE. 10, e0144610 (2015)
26. Dalmazzo, D., & Ramirez, R.: Mel-spectrogram Analysis to Identify Patterns in Musical Gestures: a Deep Learning Approach. MML 2020, 1
27. McFee, B., Raffel, C., Liang, D., Ellis, D., McVicar, M., Battenberg, E., Nieto, O.: librosa: Audio and Music Signal Analysis in Python. Presented at the Python in Science Conference, Austin, Texas (2015)

28. Choi, K., Fazekas, G., Cho, K., Sandler, M.: A Tutorial on Deep Learning for Music Information Retrieval (2018)
29. Krizhevsky, A., Sutskever, I., Hinton, G.E.: ImageNet classification with deep convolutional neural networks. Commun. ACM. 60, 84–90 (2017)
30. Simonyan, K., Zisserman, A.: Very Deep Convolutional Networks for Large-Scale Image Recognition (2015)
31. Szegedy, C., Vanhoucke, V., Ioffe, S., Shlens, J., Wojna, Z.: Rethinking the Inception Architecture for Computer Vision. In: 2016 IEEE Conference on Computer Vision and Pattern Recognition (CVPR). pp. 2818–2826. IEEE, Las Vegas, NV, USA (2016)
32. Xie, S., Girshick, R., Dollar, P., Tu, Z., He, K.: Aggregated Residual Transformations for Deep Neural Networks. In: 2017 IEEE Conference on Computer Vision and Pattern Recognition (CVPR). pp. 5987–5995. IEEE, Honolulu, HI (2017)
33. Iandola, F.N., Han, S., Moskewicz, M.W., Ashraf, K., Dally, W.J., Keutzer, K.: SqueezeNet: AlexNet-level accuracy with 50x fewer parameters and <0.5MB model size (2016)
34. Huang, G., Liu, Z., Van Der Maaten, L., & Weinberger, K. Q.: Densely connected convolutional networks. In Proceedings of the IEEE conference on computer vision and pattern recognition (pp. 4700–4708) (2017)
35. Weiss, K., Khoshgoftaar, T.M., Wang, D.: A survey of transfer learning. J Big Data. 3, 9 (2016)
36. Tan, C., Sun, F., Kong, T., Zhang, W., Yang, C., & Liu, C. (2018, October).: A survey on deep transfer learning. In International conference on artificial neural networks (pp. 270–279). Springer, Cham.
37. Gao, Y., Mosalam, K.M.: Deep Transfer Learning for Image-Based Structural Damage Recognition: Deep transfer learning for image-based structural damage recognition. Computer-Aided Civil and Infrastructure Engineering. 33, 748–768 (2018)
38. Karras, T., Aittala, M., Hellsten, J., Laine, S., Lehtinen, J., Aila, T.: Training Generative Adversarial Networks with Limited Data (2020)
39. Karras, T., Aila, T., Laine, S., Lehtinen, J.: Progressive Growing of GANs for Improved Quality, Stability, and Variation (2018)
40. Karras, T., Laine, S., Aila, T.: A Style-Based Generator Architecture for Generative Adversarial Networks (2019)
41. torchvision.models - Torchvision 0.10.0 documentation, https://pytorch.org/vision/stable/models.html
42. Google Colaboratory, https://colab.research.google.com/notebooks/intro.ipynb?utm_source=scs-index
43. GitHub - davidtvs/pytorch-lr-finder: A learning rate range test implementation in PyTorch, https://github.com/davidtvs/pytorch-lr-finder
44. Smith, L.N.: Cyclical Learning Rates for Training Neural Networks. In: 2017 IEEE Winter Conference on Applications of Computer Vision (WACV). pp. 464–472. IEEE, Santa Rosa, CA, USA (2017)
45. ReduceLROnPlateau — PyTorch master documentation, https://pytorch.org/docs/master/generated/torch.optim.lr_scheduler.ReduceLROnPlateau.html
46. Szegedy, C., Liu, W., Jia, Y., Sermanet, P., Reed, S., Anguelov, D., Erhan, D., Vanhoucke, V., Rabinovich, A.: Going deeper with convolutions (2014)

Part III
Perception, Health and Emotion

Music to Ears in Hearing Impaired: Signal Processing Advancements in Hearing Amplification Devices

Kavassery Venkateswaran Nisha, Neelamegarajan Devi, and Sampath Sridhar

1 Introduction

The human auditory system is engineered to process complex acoustic signals such as speech and music. But when the hair cells in the cochlea (sensory end organ of hearing) are damaged, the coding of sound signals gets adversely affected resulting in perceptual deficits. The perceptual deficits in hearing due to hair cell loss (a condition called sensorineural hearing loss—SNHL) are often compensated using hearing aids. Hearing aids are auditory prostheses, which aid in speech and music perception in individuals with SNHL. Both speech and music are complex acoustic signals [1]. Music is a composition of sounds in time possessing melody, harmony, rhythm, and timbre [2]. Music is characterized by its physical and psychological characteristics. Frequency, intensity, duration, growth, decay, and vibrato contribute to its physical characteristics. The psychological elements such as pitch, timbre, melody, and harmony contribute to the tonality and quality of music. Tempo and rhythm are the temporal aspects of the music attributed to its psychological characteristics [3]. Apart from acoustical and psychological aspects of music perception, music also impacts emotional state, and listening to music is a motivating activity. Music influences mood and behaviors. Music effectively stimulates the auditory cortex and other areas related to attention, semantic processing, memory, motor functions, and emotional processing. Music was found to have positive effects on human cortical plasticity, moods, and quality of life. This chapter briefly highlights that perception of music is different from speech and elaborates the impact of hearing loss on music perception and then discusses the drawbacks of digital amplification devices in providing quality percept of music, by accounting for the engineering

K. V. Nisha (✉) · N. Devi · S. Sridhar
Department of Audiology, All India Institute of Speech and Hearing, Mysuru, India
e-mail: nishakv@aiishmysore.in; deviaiish@aiishmysore.in

© The Author(s), under exclusive license to Springer Nature Switzerland AG 2023
A. Biswas et al. (eds.), *Advances in Speech and Music Technology*, Signals and Communication Technology, https://doi.org/10.1007/978-3-031-18444-4_11

limitations about digital signal processing in such devices. The chapter also throws light on cochlear implants, a popularly successful rehabilitation option for the hearing impaired, which provides limited benefits in music perception with certain recent advancements. The sections of this chapter have been organized similarly so that it spotlights the perception of music in the hearing impaired and then how it is intervened with hearing aids and cochlear implants.

Hearing aids, being more promising management for SNHL, are engineered to improve the intelligibility of speech. However, the same engineering strategies in hearing aids are also supposed to help SNHL enjoy music. This places the hearing impaired at a great disadvantage as there are several differences in the acoustic features of speech and music. The fundamental frequency of music helps to perceive the pitch of a melody component in music. However, in speech, it contributes to prosody perception. In nontonal languages (including Indian languages like Hindi, Kannada, and Tamil), the contribution of fundamental frequency in speech perception is minimal [4]. Spectral regularities in music are more than that of speech. The fundamental frequency does not vary within a note in music but changes rapidly within an articulation in speech [5]. Temporal regularities of a complex sound help in rhythm perception. Rhythm is well characterized and precise in musical notes, but not in speech [6]. Speech is often from a single source, usually from the speaker. In contrast, music often involves multiple sources depending on the instruments used and type of music [7]. So, source segregation and scene analysis carried out by the human auditory system for music is much more complex than speech. The dynamic range of intensity for music falls around 100 dB, and for speech, it's between 30 and 35 dB. The dynamic range of frequency for music perception is large (in the order of 4/5 Hz to 31 kHz) compared to speech (which falls between 0.1 and 16 kHz) [4]. Music has a slower modulation rate of 12 Hz than speech, which has a rate of 45 Hz [8, 9]. The crest factor, which is defined as the difference in dB between the peak in a spectrum and the root mean square (RMS) value, is 12 dB in speech and 18 to 20 dB in music [10]. These differences in spectral and temporal characteristics of music from speech challenge the efficiency of digital hearing aids in processing music.

2 Music Perception in Hearing Impaired

The human cochlea and auditory neural structures (starting from auditory nerve till auditory cortex in the brain) are dynamic in perceiving speech and music with its highly variable psychoacoustic properties. Frequency, intensity, and duration, the essential components of sounds, are encoded at different levels of the auditory system. Every acoustic signal is represented along the length of the cochlea, with different frequencies of the signal encoded at different locations of the cochlea. This specific frequency coding for perceiving pitch is known as the tonotopic organization, shown in Fig. 1 [11]. These representations are then converted into a pattern of neural signals in auditory neurons, where the frequency and duration of

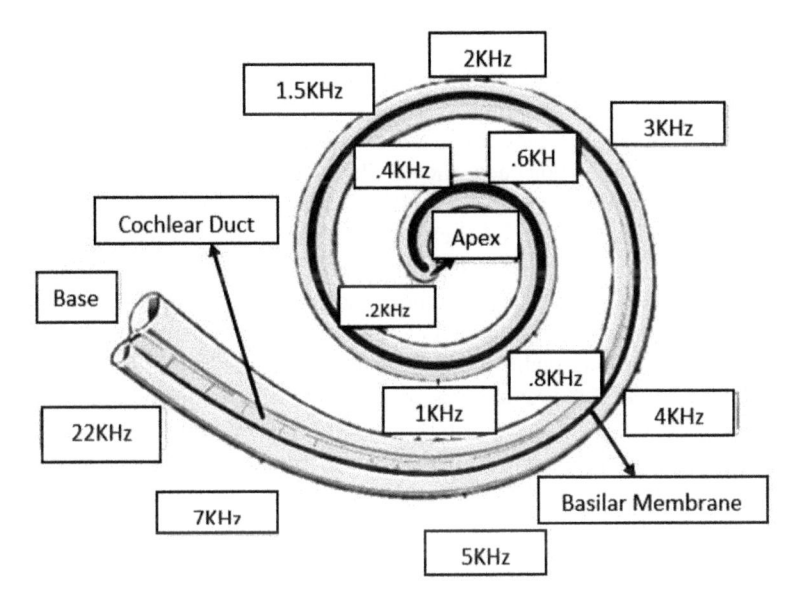

Fig. 1 Tonotopic organization of cochlea

the signal are coded in the form of the signal transmission rate. Finally, perceptual representations of pitch and timbre are extracted and interpreted from these central auditory system patterns. Hearing loss caused by damage to the cochlea (sensory or cochlear hearing loss) and auditory nerve (neural hearing loss), commonly known as SNHL, can result in deficits in pitch coding and poor perception of rhythm and timber leading to poor music perception. SNHL results in poor audibility, reduced frequency resolution, affected temporal resolution, impaired pitch perception, and impaired nonlinear effects [12]. Lack of audibility caused by SNHL loss affects music perception by impairing the ability of the listener to hear softer notes and melodies those well beyond the audible range of the listener. This is known as reduced dynamic range. On the other hand, reduced frequency resolution caused by the broadening of auditory filters in the cochlea results in impaired pitch perception. The frequency resolution of the auditory filters decides the resolution of the fundamental frequency, which in turn denotes the pitch of a complex tone such as music [13]. In addition, SNHL has a more detrimental effect on melody perception due to poorer frequency resolution and pitch discrimination caused by broader cochlear filters. Pitch discrimination is based on pitch contours (succession of tones), which along with musical intervals, form the bases for melody and harmony perception in music. Sequential musical interval contributes to melody, and simultaneous interval contributes to harmony. Pitch contours may be falling or rising, but their recognition depends on the perceptual distance between the pitches for which pitch discrimination is essential. Individuals with SNHL tend to have pitch distortions when listening to the melody, whether or not the timbre changes [14]. Meter, in music, is perceived based on the rhythmic patterns of strong and weak beats. These patterns are characterized by the number of beats in a group (two betas as in duple or three beats in triple). To recognize meter, level, pitch, and duration discrimination is vital, which is affected in individuals with SNHL[14]. A survey by Madsen and Moore [21] reported that hearing aid users complained about distortion, acoustic feedback, inappropriate gain, unbalanced weightage for frequency, and reduced tonality while listening to music. This results from the impaired perception of pitch, timbre, rhythm, and other psychoacoustic characteristics of music with hearing aids. Reduced frequency resolution affects pitch perception and other aspects of music, such as timbre. Timbre is often defined as "the attribute of auditory sensation in terms of which listener can judge those two sounds having the same loudness and pitch are dissimilar" [12]. Timbre in music helps to differentiate if the same notes are played in different instruments based on the quality of the notes. Timbre depends on the spectral shape of the music, which is represented as resolved frequencies in the cochlea. So, reduced frequency resolution alters the perception of the long-term spectral shape of music, thus impairing the timbre perception [16]. Timbre discrimination is poorer in individuals with sloping SNHL when compared to individuals with flat SNHL because the just noticeable difference for frequency in individuals with sloping hearing loss is high than in individuals with flat hearing loss [17]. Perception of sharpness is a timbre attribute for sensitivity to energy in high frequency harmonics of music. It is compromised in individuals with SNHL resulting from poor frequency resolution, especially at a higher frequency region, which is essential to perceive higher harmonics [18].

3 Music Perception with Hearing Aids

Hearing aids are electroacoustic devices that amplify sounds by converting acoustic signals into electrical signals and thus altering them according to the needs of individuals. It applies complex digital signal processing techniques to achieve it. After which, the signal will be converted back to its acoustic form and delivered to the individual's ear. Permanent hearing loss such as SNHL is not treatable but can be managed with conventional hearing aids. Hearing aids amplify sound and compensate for the loss of audibility in such individuals [19]. Most modern hearing aids contain single or multiple microphones to receive external sound inputs, a preamplifier with a low-pass filter for each microphone, an analog to digital signal converter for each microphone, a digital signal processor, a receiver, and a battery housed in a case. There are various styles of hearing aids, the one worn behind the ear to the one that confines to the ear canal, as shown in Fig. 2. Smaller the hearing aid, limited is the output level of the aid. But recently, even smaller hearing aids have got advanced signal processing circuits [19]. Fitting the hearing aid closed (using an ear mold or dome) or open also alters its output characteristics [20]. But recently, the open fitting has been outdated by in the ear hearing aids such as completely in the canal (CIC) and invisible in the canal (IIC). Hearing aids are primarily designed for speech perception. But many hearing aid users are professional musicians, whose

BEHIND-THE-EAR

RECEIVER-IN-THE-CANAL

IN-THE-EAR

COMPLETELY-IN-THE-EAR

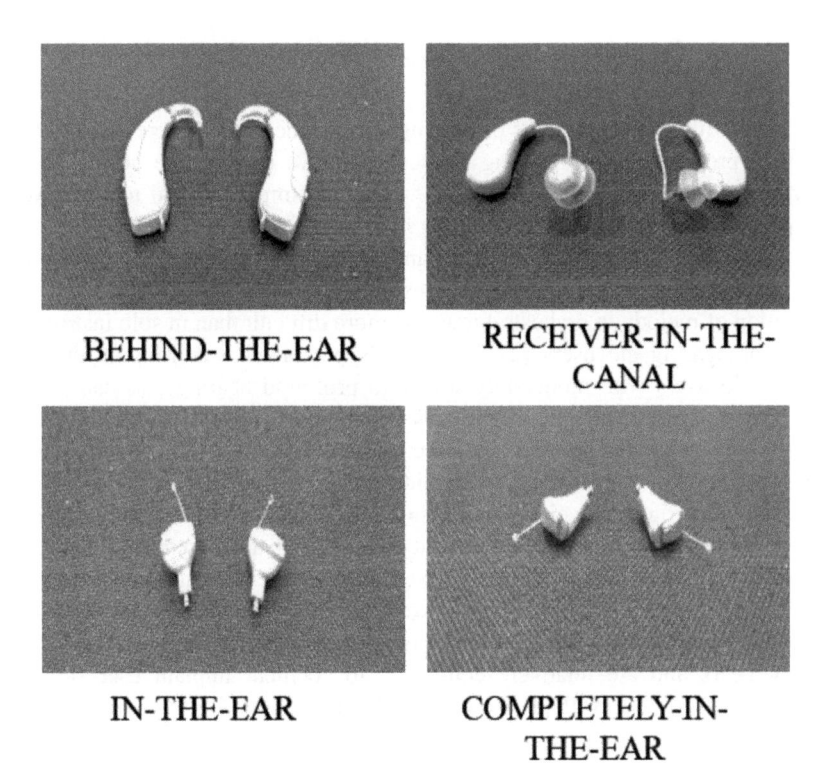

Fig. 2 Styles of hearing aids

livelihood depends on the hearing aid's ability to process music. Also, nonmusicians need to listen to the various non-speech sounds around them in day-to-day life. As described in the previous section, music is very much different from speech. Hearing aids should provide undistorted, high-fidelity, and noise-free music perception for hearing aid users. Considering the characteristics of microphones, amplifiers, signal processors, and receivers used in current hearing aids, the capability of hearing aids is compromised when it is being presented with the music signal [21]. About 30% of the participants in the study on hearing aid users reported that hearing aids hinder their music perception and half of them reported that music was either too loud or too soft [22]. Listening to live music is much more difficult than recorded music for hearing aid users [21]. The benefits of hearing aid in music appreciation also depend on the degree of hearing loss. The higher the degree of hearing loss, the poorer the music enjoyment and the music was less melodic [23]. Musicians exhibit superior perceptual abilities for music than nonmusicians. A musician who is a hearing aid user also experiences challenges while participating in music activities. Though the music perception is not completely restored with hearing aid, some musicians, especially instrumentalists, tend to use a hearing aid to hear the conductor's instruction, which is mostly speech [24]

3.1 Music Perception Difficulties in Hearing Aid Users

Individual with varying degrees of hearing loss (moderate to severe or moderately severe to profound) using hearing aids can perceive rhythm like normal hearing individuals when the stimuli sequence is made of simple musical tone. However, they perform poorer than normal hearing individuals when the stimuli are complex with a sequence of multiple pitch tones and musical instruments [25]. One sixth of individuals using hearing aid face troubles in the melody perception in music [22]. Perception of melody in orchestral music is more difficult than in solo instrumental music for hearing aid users [21]. Melody is perceived with rhythm and pitch. Hearing aid users with moderately severe to profound hearing loss can perceive pitch differences if the vowel sung differs in f_0 by one octave but cannot appreciate the difference when the f_0 differs by three fourth of an octave [26]. Identification of pitch was found to be better in hearing aids than in cochlear implants [27]. Timbre helps to deferentially perceive musical instruments playing the same notes in the same pitch. Closed-set musical instrument identification and ensemble identification in hearing aid users with moderately severe to profound hearing loss were poorer than normal hearing individuals, also worse for multiple instruments than single instrument test. Hearing aid users have poorer timbre perception than their normal hearing peers and are relatively equivalent to cochlear implant users [25, 27]. Hearing aid users have difficulty perceiving rhythm in complex musical notes, pitch differences if they differ by half of an octave or less, and timbre in simple and complex ensembles.

3.2 Signal Processing Approaches for Music Perception in Hearing Aid

In the modern digital hearing aids with multiple memories, individuals can choose to have a separate program for music perception. Such programs can have different signal processing algorithms, set particularly for music processing. This might help with differences in the way hearing aids handle speech and music. But, only 40% of hearing aid users prefer a separate music program. Firstly, such programs do not yield benefits in loudness, clarity, tonality, and less distortion [21]. Different signal processing approaches have different effects on music understanding.

3.2.1 Front-End Processing to Increase Input Dynamic Range

The unpleasant experiences of hearing aid users, such as reduced tonality and distortion with music, result from limitations of hearing aid components such as a microphone, preamplifier, analog to digital (A/D) converters, digital signal processors, and receivers. The front-end processing of the signal in a hearing aid occurs at the level of microphone and A/D converter. Due to several engineering factors, the dynamic input range of this front-end circuit is compromised to a level of about 96 dB SPL, which is sufficient for speech but not for music perception (as the dynamic range of music is up to 110 dB SPL) [28]. The dynamic range in hearing aids is usually regulated using the peak clipping strategy (a strategy to limit the input when the level of incoming sounds exceeds the saturation sound pressure level of the hearing aid). This strategy which occurs at the front-end processing cannot be rectified or improved with any software intervention in later stages of hearing aid signal processing. The decreased input dynamic range can be managed technologically by the use of a range of microphones like those using electret material [29].

Another way of reducing the peak clipping caused by the A/D converter is by reducing its load. It can be achieved by introducing a microphone that is less sensitive to a low frequency signal. A microphone with a 6 dB low frequency roll-off would serve this purpose. This approach aims to reduce the low frequency energy load in the A/D converter. In such a case, the unamplified low frequency sound can be heard through an open fit. But this method might cause increased gain in low frequency as the compression system tries to compensate for the reduction in the low frequency energy caused by the front-end processing. This in turn increases the hearing aid's noise floor. A low level expansion with a higher threshold point for the low frequency band alone can be used to counteract it [30]. Also, if compression and expansion of the music signal in the front-end processing, especially on either end of the A/D converter, are feasible, this issue can be avoided completely. For this purpose, input limiting, automatic gain control can be used in the front end to avoid peak clipping of the music signal in the A/D converter.

3.2.2 Digital Signal Processing to Improve Power Consumption

After the front-end processing, the pre-processed signal will be ready to undergo acoustic modifications based on the program set for the user. It is carried out by a digital signal processor (DSP), which is the seat for major signal modifications. DSP in hearing aids runs in megahertz, whereas personal computers run in gigahertz. The lower processing limits the number of DSP computations that a hearing aid can perform on a speech sample. Reduced processing speed results in fewer algorithms to improve the signal, which may not be sufficient for amplifying the required quality of the music signal [31]. Power consumption is the main contributing factor for this reduced processing speed. In modern digital hearing aids, class D amplifiers are used for low-power consumption. They are better than class A and class B amplifiers used in analog hearing aids, but they may cause output circuit noise, which the DSP cannot suppress. This might cause distortion.

3.2.3 Receiver Characteristics to Reduce Distortion

Fitting a hearing aid suitable for hearing loss without much distortion is very much crucial for music perception. A vast majority of the input circuit noise is caused when the D/A converter in the receiver tries to demodulate the analog signal of the DSP to fit its range [31]. Ideally, a receiver is a transducer with linear and distortionless output up to the maximum level measured in the real ear. Still, sometimes it may face soft saturation approximately around 5 dB below the saturation SPL with a pure tone. The bandwidth of the receiver is also important. Higher bandwidth helps in better music perception as it accommodates higher harmonics without causing distortion [29]. Recent advancement in the hearing aid uses infrared signal-based technology to overcome the receiver saturation and increase the input dynamic range. This hearing device, called a tympanic contact activator (TCA), is a custom mold light-based contact device placed in the ear canal. It receives the amplified signal from behind the ear processor as an infrared signal (see Fig. 3) and converts the infrared signal into vibratory patterns in the tympanic membrane. As the infrared signals directly translate into vibrations, this device is an ideal solution for receiver saturation in hearing aids. It was reported to be safe and can stay up to 122 days in the patient's ear without causing inflammation or

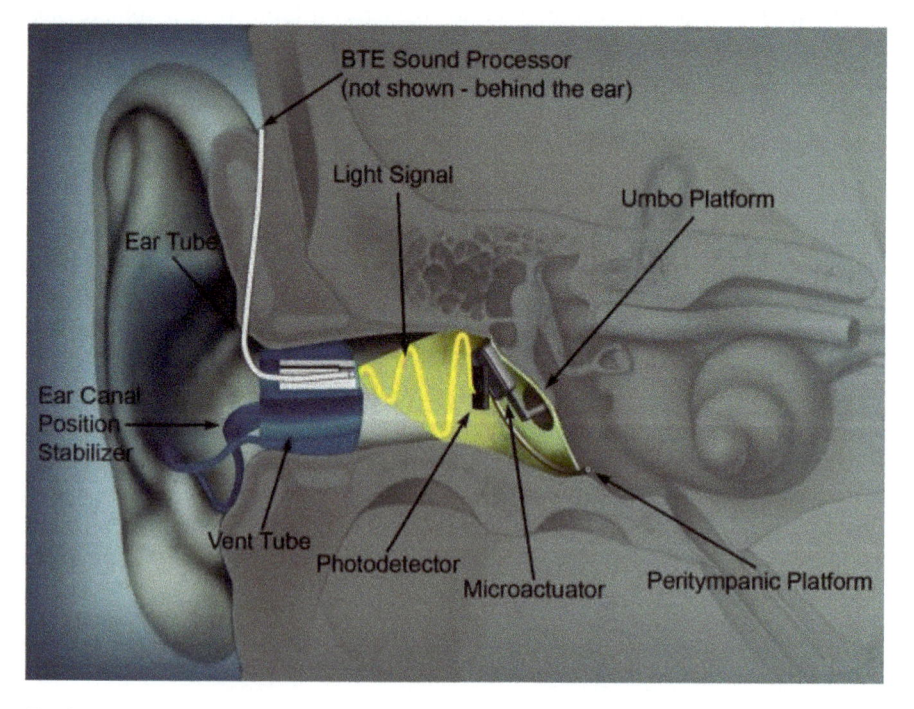

Fig. 3 Tympanic contact actuator (TCA), a custom mold light-based contact hearing device. Source: (Source: Reproduced with permission; Fay et al. [32])

infection. It was measured that TCA is capable of providing a maximum output of 110 dB SPL within the frequency range 0.25 kHz and 10 kHz. The average maximum gain (without causing feedback) produced was 40 dB with a functional gain through 10 kHz in extended bandwidth [32]. The increased dynamic range can have advantages in music perception (see 3.2. front-end processing).

3.3 Parameters of Digital Signal Processing in Hearing Aids for Music Perception

3.3.1 Bandwidth

Digital hearing aids can process signals up to 6 kHz, which is sufficient for good speech perception. But music perception requires extended bandwidth beyond 6 kHz to amplify music without distortion in higher harmonics, which is essential for sound quality and timbre. Simulated studies have shown that extended bandwidth up to 9 kHz is preferred by individuals with normal hearing and individuals with mild to moderate degree of hearing loss when listening to music processed with wide dynamic range compression (WDRC). But individuals with high frequency hearing loss and mild to severe hearing loss do not prefer extended bandwidth [33]. High frequency cutoff also depends on the prescriptive formula. While the Cambridge Method for Loudness Equalization 2 (CAM2) prescribes gain consistently up to 10 kHz, the National Acoustic Laboratories—NonLinear (NAL NL 1) and Desired Sensation Level Version 5 (DSL V.5) prescribes gain up to 6 kHz, with additional choices in programs with a high frequency cutoff at 8, 10, and 12 kHz [33–35]. Individuals with high frequency hearing loss prefer CAM 2 prescriptive formula with reduced high frequency gain rather than increased high frequency gain [34].

3.3.2 Compression

Time constants (attack and release time) of a compression circuit in a DSP may alter the timbre perception. Linear amplification produces the best sound quality because it does not cause temporal distortion. Slow-acting automatic gain control (AGC) will not distort the short-term changes in the spectral pattern of music because gains across frequencies change slowly [35]. On the other hand, WDRC can cause temporal distortion based on whether it is a fast-acting or slow-acting compression circuit [36]. Hearing aid users prefer linear amplification and slow-acting WDRC for classical music, whereas fast-acting WDRC is preferred for rock music. Schematic representation of WDRC is depicted in Fig. 4. More channels with different compression thresholds and ratios (CR) may cause temporal smearing of music stimuli. When listening to classical music, there is no effect of the channel, whereas for listening to rock music three channels were preferred over 18 channels [37]. Adaptive dynamic range optimization (ADRO) usually has 32 or 64 channels,

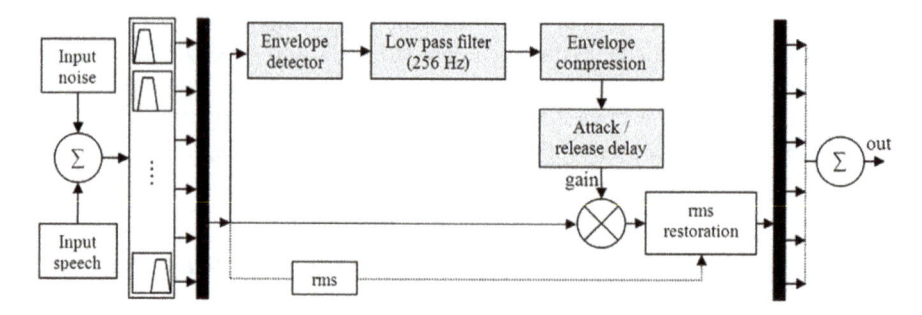

Fig. 4 Schematic representation of the wide dynamic range compression algorithm. The gray blocks are bypassed if the CR is set to 1. (Source: Reproduced with permission; Rhebergen, Maalderink & Dreschler [38])

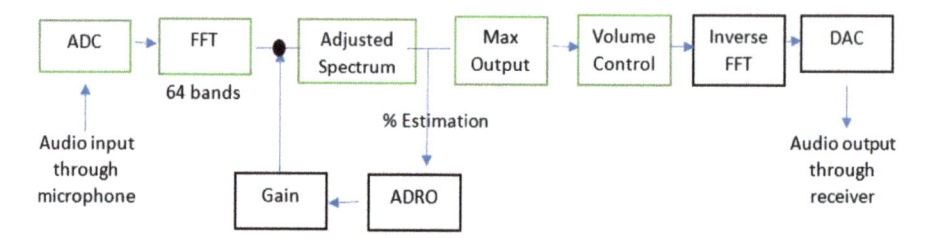

Fig. 5 Sound processing stages for a typical 64-channel implementation of ADRO in a hearing aid. ADC-analog to digital converter; DAC-digital to analog converter; FFT-fast Fourier transform

with slow-acting compression. A schematic representation of how ADRO works is depicted in Fig. 5. It helps to perceive soft sounds as soft and reduces low levels of noise. Because of the slower time constants, it minimizes the temporal distortion. Unlike WDRC, it does not compress the short-time dynamics of the music. Music quality ratings were higher for ADRO when compared to WDRC. ADRO aid users commented that music sounded sharp, brighter, and detailed, while WDRC aid users commented it was warm and had more bass.

3.3.3 Circuit Delays

Signal delays in hearing aids are also a concern when studying aided speech perception. Groth and Søndergaard studied the effect of circuit delays on music perception [39]. They reported that a delay of 2, 4, and 10 msec in an open-fit hearing aid does not cause perceived differences in music by the listeners [39].

3.3.4 Frequency Lowering

A frequency lowering algorithm is used when the high frequency hearing loss (specifically steeply sloping) prevents hearing aid users from perceiving high frequency information of the stimuli. This algorithm aids in music perception as it helps to preserve the high frequency harmonics, which are important for timbre and pitch perception. Among various frequency lowering algorithms, linear frequency transposition (LFT) and nonlinear frequency compression (NLFC) are studied using music stimuli. The difference between frequency transposition and frequency compression is illustrated in Fig. 6. Children with severe to profound high frequency hearing loss prefer to use a program LFT rather than a program with no LFT when listening to music stimuli. Adults with a moderately severe degree of high frequency hearing loss couldn't appreciate the difference with LFT when listening to music stimuli [40, 41]. An individual with a moderate to severe degree of hearing impairment, when provided with NLFC, could appreciate more musical instruments with very varying timbres. Still, with NLFC identification of simultaneous instruments, melodies and songs were likely to be performed [25]. These frequency lowering algorithms are ineffective in improving music perception because LFT confuses with unresolved higher harmonics transposed to lower frequency region and NLFC results in inharmonicity resulting in the poor perception of timbre. Harmonic frequency lowering (HFL), a frequency lowering algorithm

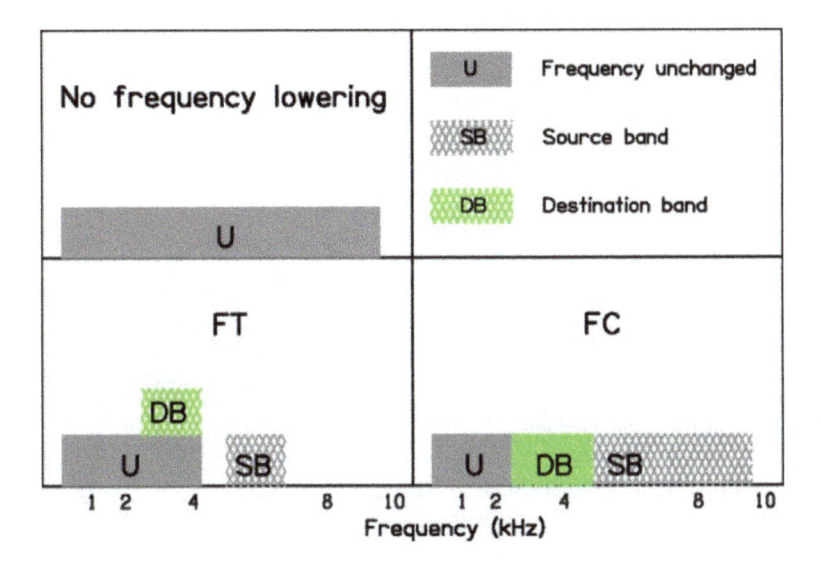

Fig. 6 Schematic representation of frequency transposition (FT, bottom left) and frequency compression (FC, bottom right). For FT, the source band (SB) and the destination band (DB) have the same width. For FC, the destination band is narrower than the source band. FT=frequency transposition; FC=frequency compression. (Source: reproduced with permission; SalorioCorbetto, Baer & Moore [43])

developed for music, combines frequency transposition and frequency compression. It applies a cutoff frequency to separate the high frequency components from the low frequency components. Following this, the high frequency components are linearly compressed by a factor of 2, weighed and mixed with the original signal, maintaining the harmonic ratio. HFL was found to give better music details than NLFC. Also, HFL did not have any detrimental effect on the music processing and provided audibility of higher harmonics which compression algorithm was used [42]

3.3.5 Feedback Canceller

Acoustic feedback hinders music listening in more than one third of individuals using hearing aid [21]. This is because of intense higher harmonics in the music signal. These intense higher harmonics get misclassified as feedback by the acoustic feedback suppression algorithms in hearing aids. They may reduce the gain in higher harmonics of the music (see Fig. 7). Acoustic feedback cancellation algorithms may cancel some of the harmonics in music and add some extraneous spectral components. Hearing aids are equipped with a special feature named "music mode" to overcome these feedback issues where the feedback algorithm adapts to the feedback path [44]. However, the music mode was proven ineffective as it adapts to the feedback path slowly, and it may be also compromised when the feedback path changes. Even with a constant feedback path, hearing aids, when presented with opera music, cause feedback [45, 46]. Other promising technology involves

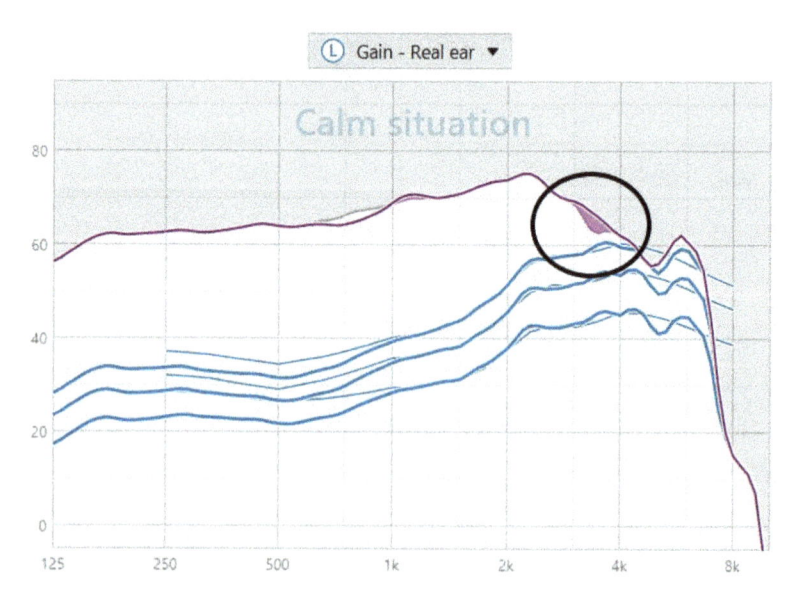

Fig. 7 Picture describing reduction in high frequency harmonics (circled region) by acoustic feedback cancellation

decorrelating the input and output music signal by introducing a frequency or phase shift, which in turn reduces adaptation of feedback canceller. However, this technology potentially affects the pitch and timbre of the music [44, 45]

3.3.6 Directional Microphone

Directional microphones help to improve the signal to noise ratio when listening to speech in noisy situations. But music can be treated as noise by such algorithms. Omni-directional and fixed directionality were beneficial for music perception. In contrast, adaptive directionality hauls music since it considers music as noise [28].

3.3.7 Noise Reduction Algorithms

The noise reduction algorithm works on modulation depth and rates of the incoming signal. It identifies signals with high modulation depth and speech like modulation rate as the wanted signal. In contrast, signals like music, usually with less modulation depth and other rates, are treated unwanted. These algorithms suppress music from instruments like guitar, saxophone, and piano [46]. Also, an algorithm based on modulation depth could cause temporal smearing in the signal's envelope, affecting source segregation, pitch, and timbre perception in music [29, 47]. Digital noise reduction (DNR) algorithms also reduce low frequency gains when it detects nonspeech signal like music, causing impaired pitch perception for hearing aid users. When music perception was assessed in hearing aid users using the Korean version of the clinical assessment of music perception test in DNR on and DNR off conditions, pitch discrimination at 226 Hz was poorer in DNR on condition than in DNR off condition. So, it was concluded in the study that music perception or sound quality did not improve with the activation of DNR in hearing aids [48]

3.3.8 Environmental Classifier

Environmental classifier (EC) algorithms are built to automatically change the hearing aid's DSP characteristics based on the signal input from the environment and the hearing aid user experience. Modern hearing aid classifies the acoustic environment whether it has only speech, speech in the presence of background noise, only noise, and music. For this purpose, the EC algorithm will be trained with a vast catalogue of stimuli by the manufacturer. 80–90% of the time, it classifies music correctly. Classical music is classified better than pop/rock music. But sometimes, certain types of music, like pop music, are misclassified as noise or speech in noise. Similarly, reverberant or amplitude compressed speech are misclassified as music [49]. In a situation where speech and music occur together, automatic environmental classifiers may not work properly. In such situations, whether music is considered as noise or a wanted signal depends on individual preferences. So, in such a situation, manually switching to music or speech in a noise program is recommended than using an automatic environmental classifier. Also, when these algorithms are trained individually, it works more efficiently than when trained with a database of stimuli. This is because when an individual listens to a particular genre of music and the algorithm is trained with that particular music, it easily classifies other stimuli as noise. Also, among the various manufacturers, the technicality behind their algorithm and choice of DSP feature engaged when a particular scene is being classified varies [50]. Hence, it becomes difficult in practice to use a manufacturer trained algorithm.

4 Music Perception by Individuals with Cochlear Implants

According to Graeme Clark [51], "the multiple channel cochlear implant (bionic ear) is a device that restores useful hearing in severe to profoundly deaf people when the organ of hearing situated in the inner ear has not developed or is destroyed by disease or injury. It bypasses the inner ear and provides information to the hearing centres through direct stimulation of the hearing nerve." Every cochlear implant comprises an external sound processor and an internal implant. The external sound processor is similar to a hearing aid in front-end sound processing. It picks the sound through a microphone, processes with a digital signal processing circuit, converts the output of DSP into electromagnetic signals, and delivers it to the internal implant through a transmitting coil, which will be placed over the scalp. The internal implant unit receives those electromagnetic signals through a receiving coil and codes the signals into electrical stimulation, which will be delivered to the auditory nerve through an electrode inserted in the cochlea (Fig. 8). When the microphone picks up a sound, it is converted into an electrical signal and converted into power spectrum by the DSP. The signal's power spectrum is calculated with Fourier transformation and bank of band-pass filters. The power of the signal at every band in every millisecond is analyzed. The number of bands usually corresponds to the number of intra-cochlear electrodes in the implant. Each intra-cochlear electrode represents a stimulating channel, and hence the number of channels in a cochlear implant corresponds to the number of intra-cochlear electrodes. The signal level at each frequency band is converted to an appropriate level of electrical current to be delivered to the respective electrode. The current is delivered in the form of brief, temporally nonoverlapping electrical pulses. The scheme in which these pulses are delivered depends on the coding strategy. Coding strategies help to represent the incoming sound in an intra-cochlear electrode array. The rate at which these pulses were delivered used to be constant earlier in generic cochlear implant coding strategies. A more recent strategy like continuous interleaved sampling (CIS) stimulates all available intra-cochlear electrodes together. In contrast, an advanced strategy like advanced combination encoder (ACE) stimulates a particular number of electrodes depending on the number of spectral maxima occurring in the signal [52].

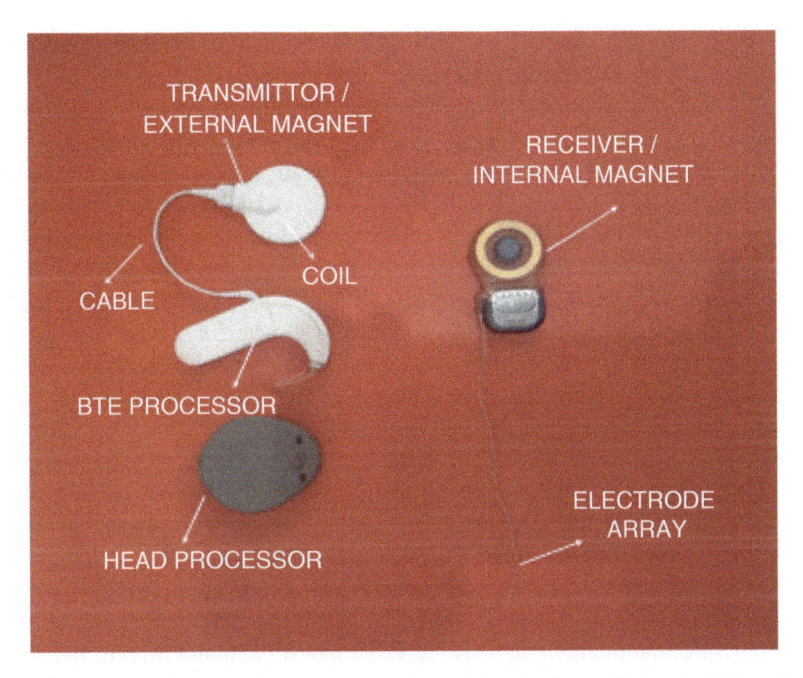

Fig. 8 Picture representing various component of a cochlear implant with a behind the ear (BTE) processor

Unlike CIS and ACE, which use an endless number of channels, HiRes120 is a coding strategy based on the virtual channel technique [53]. This technique uses current steering to create stimulations through virtual channels between two fixed electrodes. It helps to resolve frequencies between two fixed channels in a more precise way. The current steering technique used in HiRes120 is the Two-Electrode Current Steering Strategy (TECSS), which controls only two adjacent electrodes. Four-Electrode Current Steering Strategy (FECSS) is a step ahead by controlling four electrodes to narrow down the stimulating region and increase precision [54]. A flexible hybrid stimulating strategy has been proposed by Choi et al. combining both TECSS and FECSS techniques [55].

The coding strategy and the engineering aspects of cochlear implants have improved over decades, facilitating higher fidelity and improved contact between the electrode and soft tissues in the cochlea. The rate of stimulation or the rate at which the electrical pulse train is delivered is also exceeded with advancement in the power backup of the implant. An increase in the number of channels has also been a great advantage in cochlear implant benefits. Though these changes were intended for improving speech perception, recent research has turned its concern on perceived sound quality. Cochlear implant users can discriminate rhythm patterns with 93% accuracy, which is equivalent to hearing aid users who score 94% in the same task. Similarly, the rhythm perception score was close to 95% when tested with a hearing aid before and tested with cochlear implants after implantation [23, 26].

The cochlear implant users scores around 75% in melody recognition when the melody has rhythm, lyrics, and harmony cues. The score declined to 34% when the lyrics were removed from the melody [56]. When adult cochlear implant users were asked to discriminate musical sounds in three different conditions, single (solo) instruments, solo instruments with background accompaniment, and musical ensembles, the scores obtained were 61%, 45%, and 43% correct, respectively. Cochlear implant users find it easier to process simple acoustic properties of sound from a single instrument rather than a complex musical piece [26].

Envelope-based sound coding strategy such as CIS transforms only envelope information in the signal into electrical stimulation, wherein precise fine structure information is missed. The envelope is important for speech perception, whereas fine structure is important for pitch and timbre in music [57]. In a study using multidimensional scaling for the perception of timbre space in music, cochlear implant users gave more weightage to the spectral envelope, which is similar to a normal hearing individual perceiving vocoded music. However, when a normal hearing individuals perceive unprocessed natural music, they do not give more weightage to the spectral envelope. This difference has been attributed to the lack of precise spectral information by the temporal fine structure cues [58]. Fine structure processing (FSP) is an advanced coding strategy that encodes signal's detailed frequency information, helping to perceive temporal pitch and timbre. Individuals using FSP were able to perceive rhythm and discriminate instruments better than individuals using CIS [59]. Magnusson [60] reported that music perception with FSP and high-definition CIS (HDCIS, a strategy advanced than CIS, reported to provide better spectral information than CIS) were not statistically significant. Still, four out of twenty subjects preferred FSP for music experience [60]. When cochlear implant users rate music quality in pleasantness, naturalness, richness, fullness, sharpness, and roughness, FSP helps them perceive music as exactly as they want it to sound HDCIS. But acclimatization also plays an important role [61]. Emotion is an aspect of music, important for the perception of its meaning. It requires various acoustic features of music. An auditory prosthesis like cochlear implants should convey these cues for the perception of emotion in music. The cochlear implant users could detect different levels of emotional arousal conveyed by music, especially with a music program in which the currents levels (determined by the electrical pulse magnitude, sometimes even by the pulse width) are manipulated to increase the dynamic range of low frequency channels.

Bimodal listening in cochlear implant users refers to an individual with a cochlear implant fitted with an acoustic hearing aid for low frequency hearing. Lack of low frequency pitch and temporal fine structure sensitivity is attributed to poorer music perception in cochlear implant users. A web-based music reengineering study where cochlear implant users themselves adjusted treble frequencies, bass frequencies, percussion emphasis, and reverberation then rated on enjoyment. They preferred heightened bass, reverberation, and treble across musical genres [62]. Post-lingual deaf adults with bimodal cochlear implant users have been shown to perceive music significantly better than individuals using cochlear implants alone [63]. Melody identification was significantly better in bimodal cochlear implant users than in individuals using only cochlear implants [64].

Tactile aids are prosthetic devices that transmit information in signals through touch sensations. Though it cannot transmit complex information, it successfully conveys low frequency information such as the pitch of a signal such as music. After the emergence of the multichannel cochlear implant, the use of tactile devices in auditory rehabilitation ceased. But recently, few experiments demonstrated that new-generation tactile aids can augment music perception in cochlear implants. It improves pitch perception and thus provides better music quality [65].

5 Conclusions

Hearing aids fitted with a larger input dynamic range, lower input noise floor, minimal peak clipping or limiting, extended bandwidth at low frequency end, and prescription formulas specific for music will help in providing better music perception for hearing aid users. Auditory prosthesis like cochlear implants with better spectral resolution capacity such as the one capable of encoding fine structure information in music may improve music experience in cochlear implantees. Bimodal hearing with a cochlear implant and hearing aid has shown little hope as a promising future direction for improving music appreciation. Hearing aid technology, best suitable for music perception in individuals with a higher degree of hearing loss coupled with a cochlear implant, might provide encouraging outcomes.

References

1. Seinfeld, S., Figueroa, H., OrtizGil, J., SanchezVives, M. V: Effects of music learning and piano practice on cognitive function, mood and quality of life in older adults. Frontiers in Psychology. 4, 810 (2013). https://doi.org/10.3389/fpsyg.2013.00810.
2. American Heritage Dictionary Entry: Music, https://ahdictionary.com/word/search.html?q=Music, last accessed 2021/07/06.
3. Olson, H.: Music, physics and engineering. Dover Publications, New York (1967).
4. Russo, F.A.: Perceptual considerations in designing hearing aids for music. Hearing Review. 74–78 (2006).
5. Wolfe, J.: Speech and music, acoustics and coding, and what music might be "for." Proceedings of the 7th International Conference on Music Perception and Recognition, 2002. 10–13 (2002).
6. Chasin, M., Russo, F.A.: Hearing Aids and Music. Trends in Amplification. 8, 35–47 (2004). https://doi.org/10.1177/108471380400800202.
7. Chasin, B.M.: Can your hearing aid handle loud music? A quick test will tell you. Hearing Journal. 59, 22–24 (2006). https://doi.org/10.1097/01.HJ.0000286305.14247.7f.
8. Ding, N., Patel, A., Chen, L., Butler, H., Luo, C., Poeppel, D.: Temporal Modulations Reveal Distinct Rhythmic Properties of Speech and Music. bioRxiv. 059683 (2016). https://doi.org/10.1101/059683.
9. Ding, N., Patel, A.D., Chen, L., Butler, H., Luo, C., Poeppel, D.: Temporal modulations in speech and music, (2017). https://doi.org/10.1016/j.neubiorev.2017.02.011.

10. Chasin, M., Hockley, N.S.: Some characteristics of amplified music through hearing aids, (2014). https://doi.org/10.1016/j.heares.2013.07.003.
11. Moore, B.C.J.: Cochlear Hearing Loss: Physiological, Psychological and Technical Issues: Second Edition. John Wiley and Sons (2007). https://doi.org/10.1002/9780470987889.
12. Moore, B.C.J.: Cochlear Hearing Loss: Physiological, Psychological and Technical Issues: Second Edition. (2008). https://doi.org/10.1002/9780470987889.
13. Nelson, D.A.: High level psychophysical tuning curves: Forward masking in normal hearing and hearing-impaired listeners. Journal of Speech and Hearing Research. 34, 1233–1249 (1991). https://doi.org/10.1044/jshr.3406.1233.
14. Kirchberger, M.J., Russo, F.A.: Development of the adaptive music perception test. Ear and Hearing. 36, 217–228 (2015). https://doi.org/10.1097/AUD.0000000000000112.
15. Madsen, S.M.K., Moore, B.C.J.: Music and hearing AIDS. Trends in Hearing. 18, (2014). https://doi.org/10.1177/2331216514558271.
16. Van Summers, V., Leek, M.R. The internal representation of spectral contrast in hearing impaired listeners. Journal of the Acoustical Society of America. 95, 3518–3528 (1994). https://doi.org/10.1121/1.409969.
17. Emiroglu, S., Kollmeier, B.: Timbre discrimination in normal hearing and hearing impaired listeners under different noise conditions. Brain Research. 1220, 199–207 (2008). https://doi.org/10.1016/j.brainres.2007.08.067.
18. Fitz, K., McKinney, M.: Music through hearing aids: Perception and modeling. In: Proceedings of Meetings on Acoustics. p. 1951 (2010). https://doi.org/10.1121/1.3436580.
19. Moore, B.C.J., Popelka, G.R.: Introduction to Hearing Aids. In: Popelka, G.R., Moore, B.C.J., Fay, R.R., and Popper, A.N. (eds.) Hearing aids. pp. 1–19. Springer International Publishing (2016). https://doi.org/10.1007/978-3-319-33036-51.
20. Stone, M.A., Paul, A.M., Axon, P., Moore, B.C.J.: A technique for estimating the occlusion effect for frequencies below 125 Hz. Ear and Hearing. 35, 49–55 (2014). https://doi.org/10.1097/AUD.0b013e31829f2672.
21. Madsen, S.M.K., Moore, B.C.J.: Music and hearing AIDS. Trends in Hearing. 18, (2014). https://doi.org/10.1177/2331216514558271.
22. Leek, M.R., Molis, M.R., Kubli, L.R., Tufts, J.B.: Enjoyment of music by elderly hearing impaired listeners. Journal of the American Academy of Audiology. 19, 519–526 (2008). https://doi.org/10.3766/jaaa.19.6.7.
23. Looi, V., Rutledge, K., Prvan, T.: Music Appreciation of Adult Hearing Aid Users and the Impact of Different Levels of Hearing Loss. Ear and Hearing. 40, 529–544 (2019). https://doi.org/10.1097/AUD.0000000000000632.
24. Vaisberg, J.M., Martindale, A.T., Folkeard, P., Benedict, C.: A qualitative study of the effects of hearing loss and hearing aid use on music perception in performing musicians. Journal of the American Academy of Audiology. 30, 856–870 (2019). https://doi.org/10.3766/jaaa.17019.
25. Uys, M., van Dijk, C.: Development of a music perception test for adult hearing aid users. The South African Journal of communication disorders.58, 19–47 (2011). https://doi.org/10.4102/sajcd.v58i1.38.
26. Looi, V., McDermott, H., McKay, C., Hickson, L.: Music perception of cochlear implant users compared with that of hearing aid users. Ear and Hearing. 29, 421–434 (2008). https://doi.org/10.1097/AUD.0b013e31816a0d0b.
27. Looi, V., McDermott, H., McKay, C., Hickson, L.: The effect of cochlear implantation on music perception by adults with usable pre-operative acous-tic hearing. International Journal of Audiology. 47, 257–268 (2008). https://doi.org/10.1080/14992020801955237.
28. Hockley, N.S., Bahlmann, F., Fulton, B.: Analog to Digital Conversion to Accommodate the Dynamics of Live Music in Hearing Instruments: http://dx.doi.org/10.1177/1084713812471906. 16, 146–158 (2012). https://doi.org/10.1177/1084713812471906.
29. Chasin, M.: Music and Hearing Aids—An Introduction. Trends in Amplification. 16, 136–139 (2012). https://doi.org/10.1177/1084713812468512.
30. Schmidt, M.: Musicians and Hearing Aid Design—Is Your Hearing Instrument Being Overworked? Trends in Amplification. 16, 140–145 (2012). https://doi.org/10.1177/1084713812471586.

31. Zakis, J.A.: Music Perception and Hearing Aids. In: Popelka, G.R., Moore, B.C.J., Fay, R.R., and Popper, A.N. (eds.) Hearing aids. pp. 217–252. Spring-er, Cham (2016). https://doi.org/10.1007/978-3-319-33036-58.
32. Fay, J.P., Perkins, R., Levy, S.C., Nilsson, M., Puria, S.: Preliminary evaluation of a light-based contact hearing device for the hearing impaired. Otology and Neurotology. 34, 912–921 (2013). https://doi.org/10.1097/MAO.0b013e31827de4b1.
33. Ricketts, T.A., Dittberner, A.B., Johnson, E.E.: High-frequency amplification and sound quality in listeners with normal through moderate hearing loss. Journal of Speech, Language, and Hearing Research. 51, 160–172 (2008). https://doi.org/10.1044/1092-4388(2008/012).
34. Moore, B.C.J., Füllgrabe, C., Stone, M.A.: Determination of preferred parameters for multi-channel compression using individually fitted simulated hearing aids and paired comparisons. Ear and Hearing. 32, 556–568 (2011). https://doi.org/10.1097/AUD.0b013e31820b5f4c.
35. Brennan, M.A., McCreery, R., Kopun, J., Hoover, B., Alexander, J., Lewis, D., Stelmachowicz, P.G.: Paired comparisons of non-linear frequency compression, extended bandwidth, and restricted bandwidth hearing aid processing for children and adults with hearing loss. Journal of the American Academy of Audiology. 25, 983–998 (2014). https://doi.org/10.3766/jaaa.25.10.7.
36. Arehart, K.H., Kates, J.M., Anderson, M.C.: Effects of noise, non-linear processing, and linear filtering on perceived music quality. International Journal of Audiology. 50, 177–190 (2011). https://doi.org/10.3109/14992027.2010.539273.
37. Croghan, N.B.H., Arehart, K.H., Kates, J.M.: Music preferences with hearing aids: effects of signal properties, compression settings, and listener characteristics. Ear and hearing. 35, e170–e184 (2014). https://doi.org/10.1097/AUD.0000000000000056.
38. Rhebergen, K.S., Maalderink, T.H., Dreschler, W.A.: Characterizing Speech Intelligibility in Noise after Wide Dynamic Range Compression. Ear and Hearing. 38, 194–204 (2017). https://doi.org/10.1097/AUD.0000000000000369.
39. Groth, J., Søndergaard, M.B.: Disturbance caused by varying propagation delay in non-occluding hearing aid fittings. International Journal of Audiology. 43, 594–599 (2004). https://doi.org/10.1080/14992020400050076.
40. McDermott, H.J.: A technical comparison of digital Frequency-Lowering algorithms available in two current hearing aids. PLoS ONE. 6, e22358 (2011). https://doi.org/10.1371/journal.pone.0022358.
41. Auriemmo, J., Kuk, F., Lau, C., Marshall, S., Thiele, N., Pikora, M., Quick, D., Stenger, P.: Effect of linear frequency transposition on speech recognition and production of school-age children. Journal of the American Acade-my of Audiology. 20, 289–305 (2009). https://doi.org/10.3766/jaaa.20.5.2.
42. Kirchberger, M., Russo, F.A.: Harmonic Frequency Lowering. In: Trends in Hearing (2016). https://doi.org/10.1177/2331216515626131.
43. Salorio-Corbetto, M., Baer, T., Moore, B.C.J.: Comparison of Frequency Transposition and Frequency Compression for People With Extensive Dead Regions in the Cochlea. https://doi.org/10.1177/2331216518822206.
44. Freed, D.J., Soli, S.D.: An objective procedure for evaluation of adaptive anti feedback algorithms in hearing aids. Ear and Hearing. 27, 382–398 (2006). https://doi.org/10.1097/01.aud.0000224173.25770.ac.
45. Spriet, A., Moonen, M., Wouters, J.: Evaluation of feedback reduction techniques in hearing aids based on physical performance measures. The Journal of the Acoustical Society of America. 128, 1245 (2010). https://doi.org/10.1121/1.3458850.
46. Bentler, R., Chiou, L.K.: Digital Noise Reduction: An Overview. Trends in Amplification. 10, 67–82 (2006). https://doi.org/10.1177/1084713806289514.
47. Chasin, M., Hockley, N.S.: Some characteristics of amplified music through hearing aids, (2014). https://doi.org/10.1016/j.heares.2013.07.003.
48. Kim, H.J., Lee, J.H., Shim, H.J.: Effect of Digital Noise Reduction of Hear-ing Aids on Music and Speech Perception. Journal of Audiology and Otology. 24, 180–190 (2020). https://doi.org/10.7874/JAO.2020.00031.

49. Büchler, M., Allegro, S., Launer, S., Dillier, N.: Sound classification in hearing aids inspired by auditory scene analysis. Eurasip Journal on Applied Sig-nal Processing. 2005, 2991–3002 (2005). https://doi.org/10.1155/ASP.2005.2991.
50. Yellamsetty, A., Ozmeral, E.J., Budinsky, R.A., Eddins, D.A.: A Comparison of Environment Classification Among Premium Hearing Instruments. Trends in Hearing. 25, (2021). https://doi.org/10.1177/2331216520980968.
51. Clark, Graeme.: Cochlear Implants: Fundamentals and Applications (Modern Acoustics and Signal Processing). Springer Verlag New York, Inc., New York (2003).
52. Kiefer, J., Hohl, S., Stürzebecher, E., Pfennigdorff, T., Gstöettner, W.: Com-parison of speech recognition with different speech coding strategies (SPEAK, CIS, and ACE) and their relationship to telemetric measures of compound action potentials in the nucleus CI 24M cochlear implant system. International Journal of Audiology. 40, 32–42 (2001). https://doi.org/10.3109/00206090109073098.
53. Koch, D.B., Osberger, M.J., Segel, P., Kessler, D.: HiResolutionTM and con-ventional sound processing in the HiResolutionTM bionic ear: Using appropriate outcome measures to assess speech recognition ability. In: Audiology and NeuroOtology. pp. 214–223. Karger Publishers (2004). https://doi.org/10.1159/000078391.
54. Choi, C.T.M., Hsu, C.H.: Conditions for generating virtual channels in cochlear prosthesis systems. Annals of Biomedical Engineering. 37, 614–624 (2009). https://doi.org/10.1007/s10439-008-9622-9.
55. M. Choi, C.T., Lee, Y.-H.: A Review of Stimulating Strategies for Cochlear Implants. In: Cochlear Implant Research Updates. InTech (2012). https://doi.org/10.5772/47983.
56. El Fata, F., James, C.J., Laborde, M.L., Fraysse, B.: How much residual hearing is "useful" for music perception with cochlear implants? In: Audiology and Neurotology. pp. 14–21. Karger Publishers (2009). https://doi.org/10.1159/000206491.
57. Smith, Z.M., Delgutte, B., Oxenham, A.J.: Chimaeric sounds reveal dichotomies in auditory perception. Nature. 416, 87–90 (2002). https://doi.org/10.1038/416087a.
58. Macherey, O., Delpierre, A.: Perception of musical timbre by cochlear implant listeners: A multidimensional scaling study. Ear and Hearing. 34, 426–436 (2013). https://doi.org/10.1097/AUD.0b013e31827535f8.
59. Arnoldner, C., Riss, D., Brunner, M., Durisin, M., Baumgartner, W.D., Ham-zavi, J.S.: Speech and music perception with the new fine structure speech coding strategy: Pre-liminary results. Acta Oto-Laryngologica. 127, 1298–1303 (2007). https://doi.org/10.1080/00016480701275261.
60. Magnusson, L.: Comparison of the fine structure processing (FSP) strategy and the CIS strategy used in the MED-EL cochlear implant system: Speech intelligibility and music sound quality. International Journal of Audiology. 50, 279–287 (2011). https://doi.org/10.3109/14992027.2010.537378.
61. Looi, V., Winter, P., Anderson, I., Sucher, C.: A music quality rating test battery for cochlear implant users to compare the FSP and HDCIS strategies for music appreciation, (2011). https://doi.org/10.3109/14992027.2011.562246.
62. Peng, T., Kohlberg, G.D., Cellum, I., Mancuso, D., Lalwani, A.K.: Novel web-based music re-engineering software for music enjoyment among cochlear implantees. Otolaryngology - Head and Neck Surgery (United States). 159, P134 (2018). https://doi.org/10.1097/MAO.0000000000003262.
63. Dincer D'Alessandro, H., Ballantyne, D., Portanova, G., Greco, A., Mancini, P.: Temporal coding and music perception in bimodal listeners. Auris Nasus Larynx. (2021). https://doi.org/10.1016/J.ANL.2021.07.002.
64. Chang, Son A; Shin, Sujin; Kim, Sungkeong; Lee, Yeabitna; Lee, Eun Young; Kim, Hanee; Shin, You-Ree; Chun, Y.-M.: Longitudinal music perception performance of postlingual deaf adults with cochlear implants using acoustic and/or electrical stimulation. Phonetics and Speech Sciences. 13, 103–109 (2021). https://doi.org/10.13064/KSSS.2021.13.2.103.
65. Fletcher, M.D., Verschuur, C.A.: Electro Haptic Stimulation: A New Approach for Improving Cochlear-Implant Listening. Frontiers in Neuroscience. 15, 613 (2021). https://doi.org/10.3389/fnins.2021.581414

Music Therapy: A Best Way to Solve Anxiety and Depression in Diabetes Mellitus Patients

Anchana P. Belmon and Jeraldin Auxillia

1 Introduction

A group of disease characterized by a high glucose level causing high impairment in insulin production and insulin action is known as diabetes mellitus. There are mainly two types of diabetes mellitus. They are (i) type 1 diabetes mellitus and (ii) type 2 diabetes mellitus. Type 1 diabetes is a type of diabetes with no insulin production. Lack of insulin resistance causes type 2 diabetes mellitus, which is mainly observed in 90–95% of diabetes mellitus patients. The other comorbidity disorders mainly characterized by retinopathy, neuropathy, and cardiovascular problems are the main symptoms of type 2 diabetes mellitus. Also, the other problems include psychological disorders, anxiety, and even dermatological disorders. The neurochemical changes accompanied with diabetes are the significant reason for depression with adverse health conditions. To avoid mood and mental health problems, an effective methodology used is called music therapy. Beck Anxiety Method and Beck Inventory methods are used as the methods for analyzing music therapy results. Mean, standard deviation, and covariance are used as a measurement for the level of depression and anxiety. Diabetes mellitus and depression together constitute lack of functioning abilities.

A. P. Belmon (✉)
Maria College of Engineering and Technology, Attoor, India

J. Auxillia
St.Xaviers Catholic College of Engineering, Chunkankadai, India

2 Relationship Between Depression and Anxiety

The global prevalent disorder that affects majority of human population is depression.

Depression and anxiety are related to the metabolic disorder in children and adults. HbA1c known as glycosylated hemoglobin is the common factor reflecting metabolic control at the earlier stage of diabetes. According to Maronian et al. [3, 4, 23] based on DSM-IV criteria children and adolescents with diabetes have 28 severe mental disorders than other diabetic patients. They also have poor metabolic control and needs psychiatric consultation. Thus, the patients with higher HbA1c have the poor metabolic control consisting of severe eating disorders, behavior change, and other affective malfunctions as shown in Fig. 1. But the value of HbA1c varies significantly for anxiety conditions.

Depression levels in diabetic patients are due to the non-modifiable diabetic factors, modifiable diabetic factors, and other risk factors. Non-modifiable diabetic factors are due to longtime disease or due to history of diabetes. Modifiable diabetes risk factors cause comorbidity of depression. Modifiable risk factors have a poor value of HbA1c. Other risk factors involve smoking and poor education status.

The book chapter is organized as follows: existing work, music therapy in diabetes patients with two basic methods, results, and discussions.

3 Existing Work

Depression [8, 15] is the main psychological factor affecting 25% of the population. According to the effective study of Anderson et al. [2, 10], depression causes psychological impacts to diabetic patients rather than nondiabetic ones. There is a correlation between diabetes and depression [9, 13, 14] complications. Studies in diabetes patients revealed a clear idea about the depression and pathological self-impacts in the history of diabetic patients by WHO in 2000. The possibility of diabetes with a high intensity of depression is found to be 44% with anxiety and depression symptoms. Timidity, aggression, anxiety, depression in adolescents are

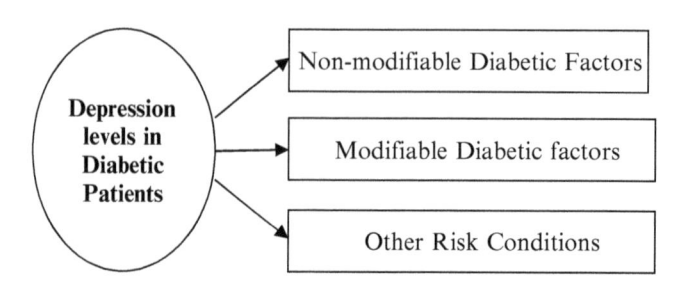

Fig. 1 Depression levels in different diabetic patients

serious problems encountered in their family. According to Conley and Graham [7], many patients encounter problems in anxiety, fear, and troubles. The pharmacologic treatments are available nowadays to decrease the level of anxiety with associated complications. Music therapy and relaxation and patients training are the main nonpharmacological methods. However, feelings, emotions, and human cognitions without speech and language is named as music. Melody and rhythm are the two crucial factors of music for human utilization.

Music as an art is highly intervening and soothing. The effect of music [5] in human body is unbelievable. The format of nonwordly explanations is expanded by means of the music therapy [12, 33, 34]. According to Choi [6], the person's mental, physical, and emotional health are strengthened by music therapy. Loneliness effects [6, 11] can easily reduce in sadist, guilty, and depressed peoples. The positive effects of mood and cognition are the two basic music backgrounds. The music selection accuracy predicts the nature of depression reduction. Anger, sadness, and frustrated effects are smoothened by music rhythm [1, 31, 32]. Tension in old persons can be reduced with a positive spirit of confidence [21]. Studies revealed that [16, 17, 18, 19, 20, 22] music therapy outperforms compared to psychotherapy.

Poor health and worse adhesion causes depression in adult population. Another review findings assess the significance of low-dose music therapy. Less than 20-minute music therapy does not cause any significant impact. However, only [24] medium- and high-dose variants are effective. Musical training causes [25] changes in human brain. Composing and listening music suppresses unrelated emotional connectivities. Music also reviews neuroplasticity [26, 27, 28, 29, 30] in the brain. The underlying positive effects of music are the only reason for neuroplasticity in the brain. Thus, music therapy causes a sense like improved depression, loneliness, and emotional aspects in health.

4 Music Therapy in Diabetic Patients

Music therapy involves mainly two types, namely, passive music therapy and active music therapy [24, 35, 36]. Passive music therapy involves listening to selected songs for half an hour three times a day. The active music therapy includes listening to music at least about 1 hour a day [37, 38, 39, 40] with active discussion and debate on selected songs as a group discussion. The active music therapy includes mime-like performances, and the researcher can question to the physical and emotional experience to experienced persons. Test-based evaluation methods can also be included. The test evaluation methodologies are (Dementia et al. 2012) (i) Beck Anxiety Inventory and (ii) Beck Depression Inventory methods.

4.1 Beck Anxiety Method

Anxiety is an emotional and unpleasant awareness in a feared condition. It is also a feel of discomfort, illness, and anticipated condition. Anxiety as a basis of neurophysical approach can be treated as a feeling of numbness and hypervigilance with increased sweating and heart rate. The biological terms "diaphoresis" and "tachycardia" are associated with higher sweating and heart rate. For an adaptive survival, the mind and body prepare itself for any anxiety treat. When the anxiety persists for a long duration, it leads to lots of medical and psychological disorders to mankind. The Beck Anxiety Inventory methods is a self-evaluated anxiety measurement test. It involves mainly 21 items as in Table 2. The score pattern is described in Table 1.

The entire score is obtained as a sum of 21 particulars. The score between 0–21 indicates low anxiety, 22–35 indicates moderate anxiety, and 36 above indicates a higher potential anxiety. The commonly noticed anxiety symptoms are hot feelings, numbness, unable to relax, nervousness, heart pounding, worst happening fear, etc. The Beck Anxiety Inventory is checked for almost a month to effectively evaluate the performance in persons. A sample Beck Anxiety in Korean adults is in Table 2. High mean score is the significance of anxiety disorder in diabetic patients. High score is also an impact of several panic disorder and phobic neurosis. There are almost similar mean scores for depressed and anxious patients.

4.2 Beck Depression Inventory (BDI)

The second method is the Beck Depression Inventory method. This is a very common and psychological method. This method not depends on the dominancy of societal changes, but it is applicable to all strategies of the society.

The BDI total score evaluate the amount of depression. The emotional, motivational, and cognitive measure of depression is the worldwide BDI score.

The test evaluates about 21 areas of depression and their various aspects. As the Beck Anxiety Inventory, this also consists of test with score ranging from 0 to 3. The entire evaluation score involves values from 0 to 63. The test reported a good reliability score of 0.67 with a concurrency of 0.79. The score variation in different situations is depicted in Table 3.

The pleasurable, energizing, and sad elements are checked in each sessions to evoke the spirit of memories and solace. A sample Beck Depression Inventory

Table 1 Score pattern in Beck Anxiety Inventory

	Not bothered	Mildly bothered	Moderately bothered	Severely bothered
All questions	0	1	2	3

Table 2 Beck Anxiety Method performance evaluation in Korean adults

Sl. no	Criteria	1	2	3	4	5	6	7	8	9	10	11	12	13	14	15	16	17	18	19	20	21
1	Tingling	1																				
2	Hotness	0.37	1																			
3	Ricketiness in legs	0.37	0.38	1																		
4	Inability to relax	0.30	0.33	0.34	1																	
5	Apprehensiveness	0.20	0.28	0.25	0.45	1																
6	Light-headed	0.41	0.43	0.35	0.36	0.30	1															
7	Racing	0.29	0.37	0.36	0.49	0.43	0.37	1														
8	Unsteady	0.27	0.30	0.31	0.51	0.52	0.36	0.46	1													
9	Petrified	0.24	0.27	0.30	0.42	0.64	0.29	0.43	0.60	1												
10	Nervous	0.28	0.36	0.29	0.47	0.42	0.37	0.38	0.58	0.45	1											
11	Suppressing	0.24	0.31	0.36	0.31	0.31	0.31	0.29	0.31	0.36	0.21	1										
12	Hands trembling	0.34	0.32	0.44	0.30	0.29	0.29	0.33	0.27	0.29	0.24	0.31	1									
13	Shaky	0.28	0.33	0.39	0.35	0.37	0.29	0.42	0.39	0.42	0.31	0.33	0.55	1								
14	Fear of losing control	0.22	0.26	0.30	0.27	0.39	0.22	0.26	0.40	0.42	0.38	0.30	0.28	0.39	1							
15	Breathing difficulty	0.23	0.27	0.32	0.26	0.22	0.26	0.34	0.21	0.26	0.22	0.40	0.31	0.29	0.22	1						
16	Die fear	0.19	0.18	0.31	0.20	0.34	0.23	0.26	0.28	0.35	0.25	0.37	0.28	0.26	0.32	0.32	1					
17	Scared	0.18	0.26	0.31	0.31	0.45	0.30	0.35	0.40	0.52	0.32	0.35	0.35	0.40	0.35	0.36	0.51	1				
18	Abdominal discomfort	0.32	0.37	0.32	0.37	0.32	0.42	0.31	0.40	0.32	0.46	0.26	0.29	0.29	0.21	0.32	0.27	0.27	1			
19	Faint	0.32	0.33	0.36	0.31	0.31	0.42	0.26	0.33	0.33	0.26	0.43	0.33	0.36	0.30	0.30	0.34	0.34	0.27	1		
20	Face burning	0.25	0.45	0.28	0.36	0.30	0.33	0.38	0.34	0.30	0.31	0.29	0.30	0.29	0.26	0.19	0.18	0.36	0.32	0.27	1	
21	Sweating	0.33	0.43	0.27	0.33	0.30	0.30	0.35	0.28	0.30	0.30	0.26	0.28	0.30	0.28	0.22	0.23	0.30	0.36	0.28	0.46	1

Table 3 Levels of depression in Beck Depression Inventory

Score	Levels of depression
1–10	Normal
11–16	Mild
17–20	Borderline
21–30	Moderate
31–40	Severe
Over 40	Extreme

Table 4 Beck Depression Inventory sample discussion

Sl. no	Criteria	Mean	Standard deviation
1	Sorrowful	0.04	0.21
2	Melancholy	0.26	0.47
3	Past failure	0.29	0.62
4	Reduced satisfaction	0.16	0.37
5	Feelings of guilty	0.27	0.51
6	Feelings of punishment	0.13	0.56
7	Self-abasement	0.02	0.09
8	Self-reproach	0.13	0.42
9	Suicidal thoughts	0.02	0.17
10	Sobbing	0.07	0.32
11	Anxiety	0.17	0.38
12	Loss of interest	0.06	0.24
13	Irresolution	0.11	0.31
14	Feelings of worthlessness	0.11	0.31
15	Energy loss	0.11	0.31
16	Sleeping pattern change	0.14	0.35
17	Touchiness	0.41	0.78
18	Changes in food	0.07	0.26
19	Concentration	0.11	0.36
20	Tiredness	0.11	0.36
21	Lack of sexual interest	0.29	0.53

is displayed in Table 4. The criteria for mapping the basic human emotions are calculated based on the mean and standard deviation methods.

Beck Depression Inventory measures 21 items, namely, sorrowful, melancholy, past failure, reduced satisfaction, guilty feelings, punishment feelings, self-abasement, energy loss, sobbing, etc.

5 Results and Discussion

Anxiety problems will lead to depression, enuresis, lack of behavioral control, social developmental skills, poor human interactions, etc. Anxiety development in the early childhood stages creates a negative impact in adolescence. Negative thoughts also create a distorted condition in the near future. Simple anxiety measurements like BAI (Beck Anxiety Inventory) gives a psychophysical safety to childhood. Also, it paved a way to manage anxiety for all emotional circumstances. The participation with social activities is reduced by means of depression. It also has a greater impact on the daily activities and results in diseases. Music can be used as a good mood recoverer. Experimental groups are the treatment groups, which are useful for research. The evaluation process is carried on among 50 diabetic patients by using mean and standard deviation measurements with Beck Inventory and Beck Depression Inventory methods. The evaluated results of each control group and the experimental group are discussed in Table 5. The anxiety and depression measures are evaluated in each stages of evaluation, namely, the pre-evaluation, post-evaluation, and follow-up stages.

The research effectively examines all the depression and anxiety measurements in the people. The main hypothesis is the evaluation of music therapeutic effects in the diabetic people (Tables 6).

The results indicated a significant impact on the level of depression in diabetic patients. The sad themes always promote a negative energy, while the desirable rhythms of depletion and melodies provide a strong feeling with a positive energy. An ability to reduce the weakness, lethargy, and depression is converted to a feeling of delight and pleasant. Stress effects and the soul relaxation can be improved in each circumstances by listening music and employing communications. In pre-evaluation stage, the covariance measurement is found to be 0.57 for depression. This is the higher value of severity. The post-evaluation depression value is 0.35, which is a reduced value. In follow-up stage, also the significant impact shows reduced level of depression and anxiety conditions with 0.28 and 0.48, respectively.

6 Conclusion

Music is an effective tool to accelerate the recovery time of patients. The people admitted in the surgical ward have anxiety and stress. The presence of music can effectively change the anxiety subjects with a faster recovery rate. The future work can be extended to find a quick evaluation method for diabetic patients. The music therapy can also be used to strengthen the mental and emotional health of patients with diabetes. The main merits of this current methods are the improved quality of life with reduced hospital charges.

Table 5 Mean and standard deviation measurement in each stages

Measurement strategies	Group	Pre-evaluation stage		Post-evaluation stage		Follow-up stage	
		Mean	Standard deviation	Mean	Standard deviation	Mean	Standard deviation
Depression	Experimental team	31.4	4.17	22.02	4.2	25.2	25.64
	Control team	31.06	3.78	32.12	3.3	33.17	23.65
Anxiety	Experimental team	32.32	3.65	20.45	4.7	23.45	4.45
	Control team	33.86	3.21	33.87	4.48	35.17	4.02

Table 6 Covariance measurement in each stages

Measurement strategies	Pre-evaluation stage Covariance	Post-evaluation stage Covariance	Follow-up stage Covariance
Depression	0.56	0.35	0.28
Anxiety	0.67	0.51	0.48

References

1. American music therapy association, What is music therapy?. Available from: www.musictherapy.org., (2007).
2. Anderson, R.J., Freedland, K.E. Clouse, R.E., Lustman, P.J.: The prevalence of comorbid depression in adults with diabetes: a meta-analysis. Diabetes Care, PMID:11375373, vol. 24(6), 1069–1078. (2001)
3. Automated assessment of COVID-19 reporting and data system and chest CT severity scores in patients suspected of having COVID-19 using artificial intelligence N Lessmann, CI Sánchez, L Beenen, LH Boulogne, M Brink, E Calli, JP ... Radiology, (2021).
4. Bulechek, G.M.1., McCloskey, J.C.: Nursing interventions classification (NIC). Medinfo., Pt 2:1368. PMID:8591448, vol. 8 (1995).
5. Bunt, L., Hoskins, S: The handbook of Music Therapy, New York: Brunner Rutledge (2005).
6. Choi, B.C.: Awareness of music therapy practices and factors influencing specific theoretical approache., J. Music Ther., PMID:18447575, vol. 45 (1), 93–109. (2008).
7. Conley, E.M., LE. Graham, Evaluation of anxiety and fear in adult surgical patients, Nurs. Res., PMID: 5205157, vol. 20(2), 113–22, (1971).
8. Critchley Macdonald, Henson, R.A.: Music and brain: Studies in the Neurology of Music. Heinemann Educational Books (1977).
9. De Groot, M., Anderson, R., Freedland, K.E., Clouse, R.E., Lustman, P.J.: Association of depression and diabetes complications: a Meta analysis, US. Psycho mastic Medicine, PMID:11485116, vol.24, (6), 1069–1078, (2001).
10. De Groot, M., Jacobson, A.M., Samson, J.A., Welch, G.: Glycemic control and major depression in patients with type 1 and type 2 diabetes mellitus. J Psychosom Res., PMID:10404477, vol. 46 (5), 425–335. (1999).
11. Gillam, N.: Evaluated a music workshop which was established in 1995, Research and development officer mental health, (2003).
12. Gold, C., Heldal, T.O., Dahle, T., Wig Tam, T.: Music therapy for schizophrenia of schizophrenia like illnesses. Cochrane librar. https://doi.org/10.1002/14651858.CD004025.pub2, (2005).
13. Goldberg, R.J.: Anxiety reduction by self regulation: theory, practice and evaluation. Ann. Intern. Med., 96, 483–487, (1982).
14. Hosseini, J., Survey of depression rate among diabetic patients with type II of diabetes in the educational therapeutic center of Amol City. M.Sc. Dissertation. Iran University of Medical Sciences, Tehran Psychiatric Institute, 99–101. (Persian)
15. Kaplan Harold I., Sadock, J. Benjamin. Synopsis of Psychiatry-Behavioral Sciences-Clinical Psychiatry, 5th edition. Published by Williams & Wilkins, (1988).
16. Mirshahi, S.: Application of music therapy in physiotherapy. Music therapy blog [Internet]. available at: http://musictherapy.persianblog.com.
17. Riello, R., Frisoni, G.B.: Music therapy in Alzheimer's disease: Is an evidence-base approach possible? Recenti. Progressi. in Medicina., PMID:11413888, vol.92 (5), 317–321, (2001).
18. Schaffer, J.: Music therapy in dementia treatment: Aging well magazine Music therapy in dementia care, Available from: www.agingwellmag.com., (2005).
19. Vaghar Anzabi, M., 1999. Comparison of emotional disorder and self-esteem between diabetic and non-diabetic patients. MSc Dissertation. Tarbiat Moallem University, 1–8. (Persian)

20. WHO: Life skills training courses. Welfare Organization, Deputy of Cultural and Prevention, 122–137, (2000).
21. Sheibani Tazraji, F., Pakdaman sh, Dadkhah, A., Hassanzadeh, M.: The effect of music therapy on depression and loneliness in old people. Iranian Journal of Ageing, vol. 5(16), (2010).
22. Castillo-Pérez, S., Gómez-Pérez, V., Calvillo Velasco, M., Pérez-Campos, E.: Mayoral, M.A. Effects of music therapy on depression compared with psychotherapy. The Arts in Psychotherapy, vol. 37 (5), 387–390, (2010).
23. Maronian, S., Vila, G., Robert, J.J., Mouren-Simeoni, M.C.H.: Troubles DSM-IV, e'quilibre me'tabolique et complications somatiques dans le diabe'te insulino-de'pendant de l'enfant et de l'adolescent. Annales Me'dico Psychologiques, vol. 157, 320–331, (1999).
24. Metzner, S., Verhey, J., Braak, P., Hots, J. Auditory sensitivity in survivors of torture, political violence and flight: An exploratory study on risks and opportunities of music therapy. The Arts in Psychotherapy, vol. 58, 33–41 (2018).
25. Constantin, F.A.: Music therapy explained by the principles of neuroplasticity. Bulletin of the Transilvania University of Braşov vol. 11(60). https://doi.org/10.14510/araj.2017.4113 (2018).
26. Micheal J. Fowler: Microvacular and Macrovascular Complications of Diabetes. Clinical diabetes; vol. 26(2): (2008).
27. Amy C, Weintrob, Daniel Sexton J. Susceptibility of infections in persons with diabetes mellitus. Uptodate (2014).
28. Denise Mann. New Links Seen Between Depression and Diabetes. Web MD. [Online] November 22, 2010. [Cited: December 27, (2014). http://www.webmd.com/depression/news/20101122/newlinks-seen-between-depression-and-diabetes. 4. Does Diabetes Cause Depression? American Diabetes Association ADA. [Online] May 14, 2014. [Cited: December 27, 2014.] http://www.diabetes.org/living-with-diabetes/complications/mentalhealth/depression.html.
29. Mezuk, B., Johnson-Lawrence, V., Lee, H., Rafferty, J.A., Abdou, C.M., Uzogara, E.E.: Is ignorance bliss? Depression, antidepressants, and the diagnosis of prediabetes and type 2 diabetes. Health Psychol., vol. 32: 254–263 (2013).
30. Ghazala Rafique, Furqan Shaikh. Identifying needs and barriers to Diabetes Education in Patients with Diabetes. Journal of Pakistan Medical Association; vol. 56: 347–352 (2006).
31. Gul, Naheed. Knowledge, Attitudes and practices of Type II diabetic patients. Journal of Ayub Medical College, vol. 22: 128–31 (2010).
32. Family Health History and Diabetes. National Diabetes Education Program. [Online] [Cited: December 27, 2014.] http://ndep.nih.gov/am-i-at-risk/family-history/#main.
33. Gebel, Erika. How Diabetes Differs for Men and Women. Diabetes Forecast. [Online] Oct 2011. [Cited: Dec 27, 2014.] http://www. diabetesforecast.org/2011/oct/how-diabetes-differs-for-men-andwomen.html.
34. Genetics of Diabetes. American Diabetes Association ADA. [Online] [Cited: December 27, 2014.] http://www.diabetes.org/diabetes-basics/genetics-of-diabetes.html. 11. Type 2 diabetes runs in the family. Australian Diabetes Council. [Online] [Cited: December 27, 2014.] http://www.australiandiabetescouncil.com/about-diabetes/type-2-diabetesruns-in-families.
35. Ruiz, R.J., Stowe, R.P., Goluszko, E., Clark, M.C., Tan, A.: The relationships among acculturation, body mass index, depression, and interleukin 1-receptor antagonist in Hispanic pregnant women. Depression, vol. 17(2): 338–343, (2007).
36. Healthcare. Pakistan ranked 7th in diabetes prevalence. The Express Tribune. [Online] July 19, 2013. [Cited: 27 December, 2014.] http://tribune.com.pk/story/578880/healthcare-pakistanranked-7th-in-diabetes-prevalence/. 14. Pakistan ranks seventh in Diabetes population. Dawn.com. [Online] December 11, 2011. [Cited: December 27, 2014.] http://www.dawn.com/news/679741/pakistan-ranks-seventh-in-diabetes-population.
37. People suffer from diabetes in Pakistan. Daily Times. [Online] March 6, 2014. [Cited: December 27, 2014.] http://www.dailytimes.com.pk/sindh/06-Mar-2014/7-1m-people-suffer-from-diabetes-inpakistan.

38. Viswanathan, M., Mohan, V., Snehalatha, C., Ramachandran, A.: High prevalence of type 2 (non-insulin dependent) diabetes among offspring of conjugal type 2 diabetic parents in India. Diabetologia, vol. 28, 907–910 (1986).
39. West, K.M.: Diabetes in the tropics: some lessons for western diabetology. In: l'odolsky S, Viswanathan M, eds. Secondary diabetes: the spectrum of the diabetic syndromes. New York: Raven. 249–255 (1980).
40. Wvbenga DR, Pileggi VJ, Dirstine P'H, et al. Direct manual determination of serum total cholesterol with a single stable reagent. Clitn Chem 1970;16:980–4.

Music and Stress During COVID-19 Lockdown: Influence of Locus of Control and Coping Styles on Musical Preferences

Junmoni Borgohain, Rashmi Ranjan Behera, Chirashree Srabani Rath, and Priyadarshi Patnaik

1 Introduction

The ongoing COVID-19 pandemic has incapacitated the entire world. Until now, the pandemic has claimed millions of lives worldwide, and the death toll is rising with each passing hour. The astronomical rise in cases and deaths around the globe initially shattered the global health infrastructure resulting in unprecedented measures like complete lockdown of nations, quarantines, isolations, restriction on movement, and social distancing for months. Countries' economies have been shattered [1], dismantling the social fabric of human cultures by forcing people to adhere to the "new social norms" of social distancing and staying isolated in homes [2]. Enforcement of stern measures by administrations, duration of lockdown, fear of infections/contamination, feelings of loneliness, boredom and frustration, inadequate supplies, and insufficient information [3, 4], along with deaths of loved ones [5, 6], loss of livelihood and jobs, and financial difficulties [4], all taken together, have adversely affected the psychological and mental health of people globally. Even though many studies worldwide suggest that lockdowns have had

J. Borgohain (✉)
Department of Humanities and Social Sciences, Indian Institute of Technology Kharagpur, Kharagpur, India

R. R. Behera · C. S. Rath
Advanced Technology Development Centre, Indian Institute of Technology Kharagpur, Kharagpur, India

P. Patnaik
Department of Humanities and Social Sciences, Indian Institute of Technology Kharagpur, Kharagpur, India

Rekhi Centre of Excellence for the Science of Happiness, Indian Institute of Technology Kharagpur, Kharagpur, India
e-mail: bapi@hss.iitkgp.ac.in

profound effects on everyday life worldwide, most people are psychologically resilient to their impact [7]. People with different coping styles and exhibiting different locus of control (LOC) have found ways and strategies to cope with psychological stress during lockdowns [8–13].

Music has been one of the most effective strategies to cope with psychological stress during COVID-19 lockdowns [11, 14–17]. During the lockdown, people listened to music to feel emotionally better with the situation, to feel comfort, to forget problems, to be energetic, to decrease sad feelings, to relax, to cheer up, to forget concerns, to express feelings, to reduce anxiety, to remember better times, to relieve boredom, to mentally stimulate themselves, and to ward off stressful thoughts [53]. Music has helped people deal with stress and relax, improving overall well-being and life satisfaction, reducing loneliness, and creating a sense of togetherness [15, 18–20]. Previous studies have shown that music reduces stress. However, in unprecedented and challenging times like the COVID-19 period, the ways people deal with stress are influenced by individual traits or locus of control that people possess. There is less information regarding how people's musical tastes are influenced by different coping styles and locus of control. Furthermore, not much is known regarding the musical preferences that people have during the lockdown. Thus, this chapter attempts to identify people's musical preferences with different coping styles and locus of control during COVID-19 lockdown in India. Therefore, the novelty of the work resides in analyzing the individual traits and their use of music in difficult times. In the rest of Sect. 1, we document the study variables and related work. In Sect. 2, the method followed is described, the participants' demographic details, psychological measures used, and how they were recruited. Section 3 provides details of the different statistical tests used to analyze the obtained data. The results obtained and discussion of the results are reported in Sects. 4 and 5, respectively. The final sections of this study, 6 and 7, deal with the limitations and conclusion of the work.

1.1 Coping Styles

A coping style combines conscious and unconscious cognitive and behavioral strategies to master, minimize, or tolerate stress and conflict [21, 22]. Even though numerous coping styles have been identified and studied, coping styles are generally categorized into three basic types: problem-focused (PFC), emotion-focused (EFC), and avoidant coping (AC) styles [23]. While PFC consists of efforts aimed at solving the problem, EFC involves emotional reactions. On the other hand, AC involves activities and cognitions to avoid stressful situations and seek distraction or social diversion [23]. The use of specific coping styles, however, may vary throughout a stressful experience [24], as "goodness of fit" has to be achieved between one's appraisal of the situation and the selected coping strategy to maximize its effectiveness [22]. Studies on coping styles during lockdown from around the globe suggest that people used a range of styles and strategies depending on one's appraisal of the situation to cope with psychological stress and experience well-being [25–28].

1.2 Locus of Control

Locus of control is an individual's belief system regarding the extent of control that he/she has over things that happen to him/her and the factors to which that person attributes success or failure [29]. While individuals with an internal locus of control (ILC) believe life outcomes result from their actions and personal characteristics, individuals with an external locus of control (ELC) believe life outcomes are determined by external forces such as fate, luck, or chance [29]. People who believe they have control over situations and surroundings are less vulnerable to stress in managing one's stress. However, people who have an ELC are more likely to suffer from stress than people who exhibit an ILC [30]. It is all the more relevant to the current pandemic scenario. Although the scope of personal control is low as the spread of the disease depends on the behaviors of other people, individuals exhibiting ILC can engage in preventative behaviors like using masks, washing hands, and maintaining social distance to protect themselves from contracting the disease [31], and adopting strategies to deal with stress. Studies also suggest that people demonstrating ILC reported low levels of psychological distress during lockdown [31, 32].

1.3 Music and Lockdown

Literature is replete with studies suggesting the relevance of music in generating and regulating a wide range of basic and complex emotions [33–36] either through increasing positive emotionality, using negative emotions to achieve certain goals, decreasing negative emotions emotionality, or increasing emotional intensity or arousal [33, 37]. Music activities (listening, singing, playing instruments, and music-making) are also associated with positive outcomes in health and well-being [38]. Music also helps in reducing stress, anxiety, nervousness, tension, and fear [38–40]. It also acts like medicine [41], as many therapists use it to reduce depressive symptoms and pain [42–44] and for motor and neurological rehabilitation [45, 46]. There is little consensus in literature when it comes to the effectiveness of music type on psychological stress. For example, classical and soothing music are assumed to positively affect relaxation and stress reduction than hard or heavy music [47, 48]. At the same time, extreme metal fans experience positive emotions when listening to death metal [49] and extreme music [50]. Also, people experiencing pain tend to choose music that is higher in energy and danceability [51]. Similarly, people often listen to sad when sad, as sad music can help individuals cope with negative emotions [52]. However, literature on sad music is also conflicting, and evidence is mixed on whether sad music is beneficial for listeners [52–54]. Although there is little consensus in the literature on the effectiveness of music type on psychological stress, self-chosen music listening for regulating negative emotions cannot be repudiated [55, 56].

2 Method

2.1 Participants

One hundred thirty-eight Indians participated in the study. Their age ranged from 17 to 35 years (M= 23.7; SD= 5.14); 51 (36.9%) were female; 86 (62.3%) percent were male; and 1 (0.72%) preferred not to reveal his/her gender. Among the participants, 134 reported they worked from home, and four participants worked from the office during the lockdown. Thirty-one of them lived in a joint family, 90 lived in a nuclear family, and 17 resided alone. Of musical activities, 113 reported listening to music, two were composing, 20 were engaged in learning (vocal/instrumental music), and three watched online concerts. Among the occupation categories, 102 participants were students; eighteen were private employees, six were government employees, six worked in multinational corporations, three were businessmen, two were homemakers, and one was a professional practicing privately. 33.8% of participants reported listening to their preferred music mostly during the night, and 31.6% preferred the evening. The time of the day preferred least to listen to music was morning, as agreed by 14.9% of the participants, and 19.6% preferred afternoon. During the first lockdown, 61.5% reported listening to music, whereas 38.4% listened to music during the second lockdown. We also asked them about engagements and activities that kept them busy during the lockdown period. An analysis of the same revealed that 23% relied on OTT[1] (over the top), 11.1% engaged in social interactions with friends over the phone, 10.2% invested their time acquiring new/technical skills, online gaming accounted for 6.84%, and 5.98% involved in academic/career pursuits and practiced indoor games and musical activities. Other engagements included household chores, physical exercise, drawing/painting, working from home, outdoor games, reading books, relief work, and pursuing hobbies.

2.2 Measures Used

Brief-COPE Inventory [57] is an abbreviated version of the COPE (Coping Orientation to Problems Experienced) Inventory. It is a self-report questionnaire

---

[1] OTT stands for "over the top" referring to media services (TV and film content) available through the Internet or cable connection.

with 28 items developed to assess a broad range of coping responses. The scale consists of 14 domains/sub-scales (self-distraction, active coping, denial, substance use, emotional support, instrumental support, behavioral disengagement, venting, positive reframing, planning, humor, acceptance, religion, self-blame) of two items each. Domains such as active coping, instrumental support, positive reframing, and planning fall under PFC. Emotional support, venting, humor, acceptance, religion, and self-blame are subsumed under EFC. Self-distraction, denial, substance use, and behavioral disengagement are characterized by AC. Participants were asked to respond to each item on a four-point Likert scale, indicating what they generally do and feel when they experience lockdown-related stressful events (1 = I have not been doing this at all −4 = I have been doing this a lot). A higher score on each coping style indicates the greater the use of the specific coping strategy. These three coping styles have shown adequate internal structure and consistency [58].

Perceived Stress Scale [59] is a ten-item scale used for measuring the perception of stress. It is a measure of the degree to which situations in one's life are appraised as stressful. The scale was adopted for the context of our study in which we asked them how lockdown induced stress in the lives of the participants (0 = Never - 4 = Very often). The psychometric properties of this scale were found to be superior to those of the 14-item Perceived Stress Scale in populations of college students or workers [60].

The Cantril Ladder [61] is a single-item measure that asks the participants to place themselves on an 11-step ladder. The worst possible life represents the lowest step, and the best possible life represents the top step. In our study, we asked them to evaluate their happiness in their life in general during the pandemic on the 11-step ladder, 0–6 (unhappy), 7–8 (average), and 9-10 (high).

Brief Locus of Control Scale [62] consists of six items in a Likert scale format, three questions for each dimension. Participants were asked to respond to the statements indicating their feelings (1 = Strongly disagree - 5 = Strongly agree). The six-item scale shows a good internal consistency of 0.68, indicating a good homogeneity among the test items measuring this construct.

2.3 Procedure

The survey for this study was conducted online through the Crowdsignal Polldaddy platform. This platform allows the researcher to create online surveys, quizzes, and polls. The participants gave their informed consent before proceeding to complete the survey. Demographic data were collected from the participants, which includes their age, gender, and occupation. They were also asked questions on their musical engagement such as listening, composing, or watching online concerts, hours spent in a week listening to music, lockdown in which music was heard, time of the day preferred to listen to music (morning, afternoon, evening, night) if music helped them cope with COVID-induced stress (1 = Not at all - 5 = To a large extent), and importance of music in their life (1 = Not at all important - 5 = Very important). Psychological scales such as BRIEF-COPE Inventory, Brief Locus of Control Scale, Perceived Stress Scale, and Cantril ladder were used. The list of music genres was made by a group of experts working extensively in the field of musicology. Indian authors developed the 23-music genre list which was also referred to [63]. Genres like Korean pop and Japanese pop were also added to the list. Participants had to rate them on a Likert-type scale (0 = Never listened - 4 = Almost always) to show the extent to which they listened to these genres during the lockdown.

3 Data Analysis

Descriptive statistics (mean, standard deviation, and percentage) were utilized to measure the study's variables. Principal component analysis (PCA) was conducted to analyze the factor structure of the participants' music genres during the lockdown. This analysis method increases the interpretability of large datasets by uncovering dimensions that link together unrelated variables providing insight into the underlying structure of the data. Correlation analysis, denoted by the symbol "r," was also carried out to see the relationship between the study variables: locus of control and coping styles, stress, happiness, use of music, and music preferred during the lockdown.

Inferential statistics such as t-test and ANOVA was also employed. A t-test is used to determine if there is a significant difference between the means of the two groups. It yields a t-score, and a higher t-score denotes a higher difference between and within the groups. A p-value in an inferential statistic measures the probability that an observed difference could have occurred just by random chance. The lower the p-value, the greater the statistical significance of the observed difference. A t-test was employed to examine the difference between the two LOCs (internal and external).

Analysis of variance (ANOVA) test, which analyzes the difference among means of three or more groups, was employed to test whether the means of the three coping styles were different. The "F" in ANOVA stands for the variation between/within the sample means. A higher F value in the test denotes a higher chance of difference between the groups and vice versa. The post hoc test in ANOVA is conducted after a statistical significance is found among the variables. Games-Howell post hoc test was conducted to examine which coping style significantly differed from the others. The data analysis was carried out using the Statistical Package for the Social Sciences (SPSS 20.0) software.

4 Results

4.1 Descriptive Statistics

The mean (M) values and standard deviations (SD) are given in Table 1. The results indicate that the use of ILC was more than ELC. A t-test was done to examine if any significant difference existed between the two traits. It was found that both the traits were significantly different from each other, t (137) = 4.51, $p < 0.001$. The participants' stress levels were average as indicated by the mean (M= 20.91, SD= 5.7) with a happiness level of (M = 6.51; SD = 2.0), indicating average happiness. The means from the two coping styles (EFC, PFC) showed average scores except for AC which was low as 15.26. Results from the ANOVA suggest that a significant difference exists between the three coping styles (EFC, PFC, and AC), F (2, 414) = 252.17, $p < 0.001$. Games-Howell post hoc test was conducted to examine how different coping styles were adopted/preferred by the participants. EFC (M = 27.02) was the most adopted/preferred style followed by PFC (M = 21.72) and AC (M = 15.26) ($p < 0.001$). Participants rated the importance of music in their lives (M = 4.14; SD = 0.83), suggesting that music occupied an important position. The mean hours of music listening in a week was (M = 9.27 hours; SD = 10.29). Among the participants, 76.09% reported listening to music within the range of 0–10 hours, 12.32% within the range of 11–20 hours, 6.52% within the range of 21–30 hours, and 5.07% within the range of 31–40 hours. According to 97.1% of the participants, music helped them cope with stress. At the same time, the extent to which music helped them cope with stress was (M = 3.79, SD = 0.92).

4.2 Relationships Among Locus of Control, Coping Styles, Stress, Happiness, and Music

The correlation analysis showed that ILC was not significantly correlated with any of the variables except music assisting in coping with stress, as shown in Table 2.

Table 1 Means and standard deviations of variables

Variables	Mean	SD
1. Internal locus of control	10.93	1.9
2. External locus of control	9.88	1.9
3. Problem-focused coping	21.72	4.2
4. Emotion-focused coping	27.02	5.2
5. Avoidant coping	15.26	3.4
6. Perceived Stress Scale	20.91	5.7
7. Happiness	6.51	2.0
8. Importance of music	4.14	0.8
9. Music listened in a week (hrs)	9.27	10.2
10. Music helped cope with stress	3.79	0.9

Table 2 Correlation among variables

Variables	1	2	3	4	5	6	7	8	9	10
1. Internal locus of control	1									
2. External locus of control	−0.016	1								
3. Problem-focused coping	0.148	0.056	1							
4. Emotion-focused coping	0.028	0.233**	0.573**	1						
5. Avoidant coping	−0.122	0.267**	0.236**	0.538**	1					
6. Perceived Stress Scale	−0.113	0.275**	0.002	0.229**	0.359**	1				
7. Happiness	0.057	−0.192*	0.087	−0.117	−0.202*	−0.563**	1			
8. Importance of music	0.144	0.098	0.229	258**	0.123	0.104	0.072	1		
9. Music listened in a week (hrs)	0.142	0.036	0.084	0.245**	0.112	0.119	−0.166	0.397**	1	
10. Music helped cope	0.243**	0.109	0.237**	0.309**	0.142	0.049	0.097	0.524**	0.325**	1

Note. *$p < .05$, **$p < .01$

Higher ELC was significantly correlated with EFC (r = 0.233, $p < 0.01$) and AC (r = 0.267, $p < 0.01$). It also held a positive and significant association with stress (r = 0.257, $p < 0.01$) and a negative association with happiness (r = −192, $p < 0.01$). Participants who scored high on PFC had higher correlations with EFC (r = 0.573, $p < 0.01$) and AC (r =0.236, $p < 0.01$). EFC also showed high correlations with AC (r = .538, $p < 0.01$), and an increase in this use of coping style was related to perceived stress (r = .229, $p < 0.01$). Avoidant coping similarly increased perceived stress (r =0.359, $p < 0.01$) and negatively linked with lower happiness (r = −0.202, $p < 0.01$). The table below suggests the increased importance of music in participants' lives, and it showed significant correlations with PFC (r = 0.229; $p < 0.01$) and EFC (r =0.258, $p < 0.01$). Additionally, it heightened spending more time listening to music (r = 0.397, $p < 0.01$) and assisting in coping with stress (r = 0.524, $p < 0.01$). Time (hrs) spent in music listening in a week was strongly linked with EFC (r = 0.245; $p < 0.01$). The greater the EFC style spent time listening to music, it helped them cope with stress (r = 0.309, $p < 0.01$). Similar was the case for both PFC (r = 0.237, $p < 0.01$) and ILC (r = 0.243, $p < 0.01$). It can be summarized that irrespective of copying styles, more listening time was associated with less stress. The other parameters did not show any significant correlations (Table 2).

4.3 Preference of Music

Preference of music was assessed using a 0–4 scale, and it was observed that new film (M = 2.23; SD = 1.16) and old film Bollywood music (M = 2.09; SD = 1.1) were preferably listened to during the lockdown. Regional film music (M = 1.80; SD = 1.1), indie pop (M = 1.78; SD = 1.1), and regional film music (old songs) (M = 1.72; SD = 1.1) and western pop (M = 1.7; SD = 1.3) were also indicated as preferred and were close to mean. Music genres that were least preferred were Korean pop (M = 0.53; SD = 1.03), Japanese pop (M = 0.44; SD = 0.9), heavy metal (M = 0.86; SD = 1.01), and slokas and chants (M = 0.97; SD = 1.07). This shows the preference as well as the popularity of music genres in difficult times (Fig. 1).

4.4 Factor Structure of Music Preferences

Previous studies on music preference have shown a structure/dimensions, and some genres were similarly rated. While understanding the music preferences during the pandemic, the dimensions would be helpful to make assertions. Some earlier studies revealed that music preferences could fall under four dimensions [64]. However, there were arguments regarding inconsistencies in the number of dimensions or the appearance of single musical styles in these dimensions [64–66]. Hence, a principal component analysis (PCA) was conducted to look at the factor structure of the music preferred. Bartlett's test for sphericity was performed to determine if all the variables in the sample were uncorrelated. The same confirmed the uncorrelatedness of the variables (X^2=1112.199, df=171, $p < 0.001$). Kaiser's measure of psychometric sampling adequacy was also analyzed to see whether the correlation matrix could be used for factor analysis. The sampling adequacy was 0.784, indicating the usefulness of the dataset for the PCA. We extracted four dimensions from the analysis with an overall explained variance of 57.3%. Out of the 23 genres, three (Japanese pop, Korean pop, and Western classical music) were eliminated from the analysis as they showed cross-loadings on the dimensions extracted.

The four dimensions from the PCA were a mixture of genres, and they were named after the formal characteristics of music. As shown in Table 3, the first dimension is intense and electronic (I&E): consists of western music with upbeat and a conventional genre like jazz. The second dimension, cultural, emotional, and melodious (CE&M), comprises classical melodies and emotional melodies from the early 1970s and genres belonging to different cultural traditions. The third dimension, Indian contemporary and popular music (IC&P), also forms from the upbeat music of the Indian subcontinent, consisting of fusion and rhythmic elements. The last dimension, devotional music (DM), centers around religious and spiritual themes listened to by the participants, such as slokas/chants and prayer/devotional songs. Our music preference structure is somewhat similar to the

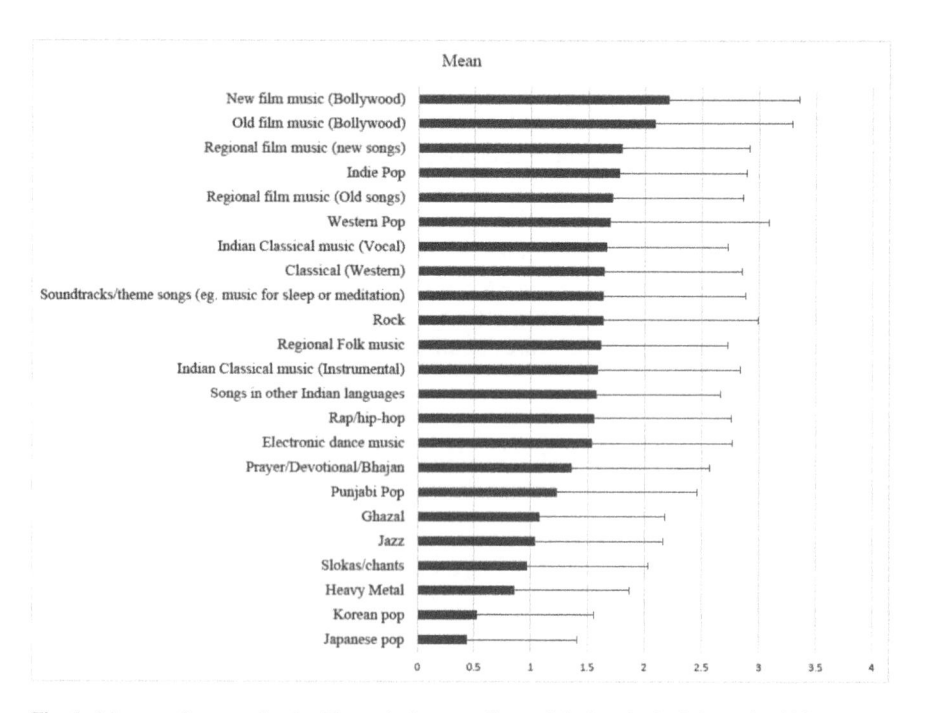

Fig. 1 Mean preferences for the 23 musical genres listened during the lockdown (n=138)

Table 3 Factor loadings of the 23 musical genres on the four dimensions of music preferred during lockdown

Preference dimensions	I&E	CE&M	IC&P	DM
1. Rock	0.79			
2. Heavy metal	0.78			
3. Rap/hip-hop	0.76			
4. EDM	0.76			
5. Western pop	0.73			
6. Jazz	0.68			
7. Indie pop	0.64			
8. Ghazal		0.75		
9. Indian classical music (instrumental)		0.73		
10. Regional film music (old songs)		0.68		
11. Old film music (Bollywood)		0.68		
12. Indian classical music (vocal)		0.62		
13. Regional folk music		0.58		
14. Regional film music (new songs)			0.78	
15. New film music (Bollywood)			0.68	
16. Songs in other Indian languages			0.49	
17. Punjabi pop			0.47	
18. Slokas/chants				0.84
19. Prayer/devotional/bhajan				0.80

Note. I&E = intense and electronic; CE&M = cultural, emotional, and melodious; IC&P= Indian contemporary and popular music; DM = devotional

structure found by an earlier Indian study on music genre preference [53]. However, the last dimension (devotional) loaded separately in our study.

4.5 Music Preferred by Coping Styles and Locus of Control

Results indicated that ELC had a strong correlation with intense and electronic music ($r = 0.195$, $p < 0.022$) and Indian contemporary and popular music ($r = 184$, $p < 0.031$). PFC showed strong correlations with all the dimensions of music genres. In contrast, EFC showed a high correlation with intense and electronic music ($r = 0.262$, $p < 0.002$), cultural emotional and melodious music ($r = 0.170$, $p < 0.046$), and devotional music ($r = 0.312$, $p < 001$). Lastly, AC correlated with intense and electronic music ($r = 0.276$, $p < 0.001$) (Table 4).

Table 4 Table showing correlation between coping styles, locus of control, and music dimensions

LOC and coping styles	I&E	CE&M	IC&P	DM
1. External locus of control	0.195*	0.036	0.184*	−0.004
2. Internal locus of control	0.047	0.148	0.004	−0.049
3. Problem-focused coping	0.197*	0.219**	0.191*	0.251**
4. Emotion-focused coping	0.262**	0.170*	0.102	0.312**
5. Avoidant coping	0.276**	0.031	0.109	0.141

Note. *$p < .05$, **$p < .01$,
I&E = intense and electronic; CE&M = cultural, emotional, and melodious; IC&P= Indian contemporary and popular music; DM = devotional

5 Discussion

The purpose of this study was to investigate if different coping styles and LOC influenced music preference during the COVID-19 lockdown period. It also examined the relationship between the use of music by different coping styles and LOC and if they benefitted from it to cope with stress

5.1 Stress, Coping Styles, and Locus of Control

The values of ILC were high, and PFC was used more than EFC and AC styles. However, higher ILC did not show any significant correlations with either of the coping styles. On the other hand, ELC was correlated with EFC, AC, and stress. It means that the higher the ELC, participants showed increased EFC and AC. These findings support earlier research showing that ELCs see these events or stressors as uncontrollable and hence will engage more in these coping behaviors and increased stress [67]. Externals believe that events and outcomes in their lives are out of their control and attribute them to some other cause. This might prompt them to engage in more EFC and avoid the stressors that came their way as preparedness is less for this type. This finding is supported by studies where participants with a high ELC are indicated to have high levels of stress when facing negative life events [68]. It might also be that people high in this domain might look for less information and develop readiness for events. Even they sometimes fail to look at the problem objectively and set goals to resolve them.

5.2 Music Helped Cope with Stress

The importance of music shown by the participants had links with hours of listening and coping with stress. These findings support the recent research results revealing that individuals use music mostly to combat stressors such as social, work-related, and performance [56]. However, the duration of music listened to in a week was not associated with coping stress. It may be that the participants were browsing different genres of music, and earlier studies had shown that participants had higher stress levels when they reported music listening for less than 5 min [69]. Nonetheless, the greater number of hours was devoted by the EFC style. This finding is consistent with the previous literature showing that it is one of the approach-type strategies used by this coping style to regulate emotions [16].

Participants who scored high in PFC and EFC style and ILC used music to cope with stress. The PFC style is characterized by threat devaluation (deactivating the stressor by directly dealing with it) and positive reappraisal (attempts to change the meaning of the stressor). This style might draw inspiration from music and themes that help them keep engaged with their own lives and eliminate stressors that are harmful to their well-being. The use of music to reduce stress is considered a palliative coping approach in that individuals' internal psychological strategies are enhanced to manage tension and minimize distress [70].

Participants adopted the EFC style mostly observed during the pandemic [71], among the other coping styles. When the stressor remains unchanged, some coping strategies are listening to music or talking on the radio [71]. Emotion-focused coping depends on the environmental aspects of the stressful encounter: if the person can do nothing to change the situation as done by the PFC styles, then it may be effective in emotional regulation and stress reduction [55, 71]. In particular, the COVID-19 pandemic pushed off-limits of individuals where they could not change the source of the problem, or problem-solving mechanisms were ineffective. Hence, EFC was adopted as a viable mechanism to deal with stressful encounters. Additionally, this coping style is renowned for the use of music for emotional regulation [56].

The EFC styles' inclination toward music shows that despite the presence of social presence (kinship: family), music helped them cope with stress. Recent studies also showed that people used music for solitary emotional regulation, whereas people experiencing increased positive emotions used music as a proxy for social interaction [55]. Music paves the way for self-determination and distraction, which may benefit those who scored high on this coping style.

Furthermore, ILC engaged in good behaviors (wearing masks, maintaining social distancing) for their physical well-being [72]. This points to their use of music as a resource for coping with stress. A body of research also suggests that individuals with internal attribution styles may experience greater psychological distress due to feeling little control over the outcome of an event [67]. These significant results show that a pandemic like COVID can strip off personal resources like ILC and induce individuals to use external resources like music. Furthermore, personality variables (LOC) sometimes have weaker effects when the situational variables are strong [73].

The AC style does not show music-related correlations. We believe that people who score high in this domain socially divert themselves by involving in various other activities to avoid the stressors [74].

5.3 Music Preferences

When music preferences were examined closely in the context of use during the lockdown, results revealed that new film music and old film music charted the top. Our findings are supported by a survey study where film music (Bollywood) was one of the most preferred musical genres [75]. 60% listened to the new Bollywood music, 53% listened to old Bollywood songs, and regional film music (oldies) topped the list in consumption.

The nature of new Bollywood music is faster, rhythmic, romantic, and joyful, and old film music that portrays romance and nostalgia has cathartic benefits on the participants. Previous research has also observed that the end goal of music was not to derive positive emotions but instead to ruminate and induce melancholy or nostalgic feelings [76]. Emotions in this genre are portrayed musically in a "monopathic" form or through a narrative or song-dance routine [77]. "Emotional experience is seen not as a secondary, individual state, but as the level at which reception takes place as a social experience" [78]. Overall, this genre heightens the emotional experiences of audiences [79], which is rewarding and often listened to by the participants [80].

5.4 Music Preferred by Locus of Control and Coping Styles

The PFC style showed significant positive correlations with all the music genres. Previous research has found strong associations of this coping style with the personality trait of openness to experience [81]. It includes traits such as creativity, insightfulness, and originality [64, 82]. It has been found in studies that this trait prefers and values different kinds of music, such as folk, ethnic music, rock, and heavy metal. [83, 84]. The explanation for this might be that people who possess this style might be willing to try new approaches in their coping strategies [85].

On the other hand, the EFC style preferred intense and electronic music; cultural, emotional, and melodious music; and devotional music. Studies also show that emotion-focused coping style is closely related to neuroticism, which uses music as a medium for emotion regulation [86, 87]. Emotion-focused coping styles are quite varied, but they all diminish the negative emotions associated with stressors; thus, this coping is action-orientated [88]. Neuroticism correlated positively with emotional use of music, an association explained in terms of the higher emotional sensitivity of neurotics to music with their emotionally stable counterparts [87]. Recent studies on COVID-19 and the pandemic have brought into limelight findings that helped them cope emotionally with positive and negative emotions. It also gave them a spiritual experience and drew inspiration [55]. Avoidant coping styles showed an inclination toward rebellious and intense music.

Internals showed no correlation with the different music genres, and a nonsignificant and negative relationship was found with devotional music. This signifies that people with a high ILC are goal-oriented and individually believe in the outcomes [89]. At the same time, since their perceived stress was low, they did not choose music consistent with their mood [90]. Externals showed a preference for upbeat music, including western as well as Indian. The results showed that stress was significantly high for externals; they might use the choice for upbeat music to augment emotional arousal. Externals perhaps have felt in control and could internalize more the threat of the pandemic, which may lead to negative feelings [91], and used music as a tool for mood regulation.

6 Limitations

The study is not without limitations. Due to lockdown and strict protocols, face-to-face interviews could not be conducted, and an online survey was conducted. Owing to the length of the survey questionnaire, we could not assess the music functions used by different coping styles and locus of control which could have added a new dimension to this study. Furthermore, the emotions and semantics preferred in the songs were also not gauged, which could be focused on later. Apart from that, this study is novel in assessing the use of music in socio-emotional coping during COVID-19 imposed lockdowns.

7 Conclusion

The study's objective was to investigate music and music preferences of different coping styles and LOC. The observations of the study are summarized below:

(i) Correlation analysis between LOC and coping styles showed that a significant relationship between ILC and coping styles was not observed. ELC had positive correlations with EFC and AC style demonstrating higher use of these coping styles during the lockdown by those possessing this trait.
(ii) The mean of the coping styles indicated that most participants used the EFC style during the lockdown.
(iii) This study also highlighted the different music genres preferred during the lockdown. The most preferred genre was new film music (Bollywood). Furthermore, participants with higher ILC, EFC, and PFC utilized music more to cope with stress except for ELC. However, ILC did not show any specific preference to the music dimensions extracted in the analysis. ELC had a strong correlation with intense and electronic music and Indian contemporary and popular music. Participants endorsing high PFC showed a preference for all the dimensions of music genres. Similarly, EFC showed a high correlation with intense and electronic music; cultural, emotional, and melodious music; and devotional music. Finally, the AC style preferred intense and electronic music.

This study has implications of how music can become a tool for socio-emotional management during stressful times. Identification of coping styles can help future researchers to make tailor-made music therapy programs as an intervention to deal with psychological needs. This could also help machine learning experts to make music-recommendation systems based on preferences catering to different coping styles and LOCs.

References

1. Dennis, M.J.: The impact of COVID-19 on the world economy and higher education. Enrollment Management Report 24(9), 3–3 (2020)
2. Behera, R.R., Borgohain, J., Rath, C.S., Patnaik, P.: Well-being of female domestic workers during three months of COVID-19 lockdown: Case study from IIT Kharagpur campus. Indian Journal of Health and Wellbeing 12(1), 83–92 (2021)
3. Brooks, S.K., Webster, R.K., Smith, L.E., Woodland, L., Wessely, S., Greenberg, N., Rubin, G. J.: The psychological impact of quarantine and how to reduce it: rapid review of the evidence. The lancet 395(10227), 912–920 (2020)
4. Pietrabissa, G., Simpson, S.G.: Psychological consequences of social isolation during COVID-19 outbreak. Frontiers in Psychology 11, 2201 (2020)
5. Mohammadi, F., Oshvandi, K., Shamsaei, F., Cheraghi, F., Khodaveisi, M., Bijani, M.: The mental health crises of the families of COVID-19 victims: A qualitative study. BMC Family Practice 22(1), 1–7 (2021)
6. Joaquim, R.M., Pinto, A. L., Guatimosim, R.F., de Paula, J.J., Costa, D.S., Diaz, A.P., Malloy-Diniz, L.F.: Bereavement and psychological distress during COVID-19 pandemics: The impact of death experience on mental health. Current Research in Behavioral Sciences 2, 100019 (2021)
7. Prati, G., Mancini, A.D.: The psychological impact of COVID-19 pandemic lockdowns: A review and meta-analysis of longitudinal studies and natural experiments. Psychological Medicine 51(2), 201–211 (2021)
8. Millar, E.B., Singhal, D., Vijayaraghavan, P., Seshadri, S., Smith, E., Dixon, P., Sharma, A.N.: Health anxiety, coping mechanisms and COVID 19: An Indian community sample at week 1 of lockdown. PLOS ONE 16(4), e0250336 (2021)
9. Chirombe, T., Benza, S., Munetsi, E., Zirima, H.: Coping mechanisms adopted by people during the Covid-19 lockdown in Zimbabwe. Business Excellence and Management 10(1), 33–45 (2020)
10. Klapproth, F., Federkeil, L., Heinschke, F., Jungmann, T.: Teachers' Experiences of Stress and Their Coping Strategies during COVID-19 Induced Distance Teaching. Journal of Pedagogical Research 4(4), 444–452 (2020)
11. Sameer, A.S., Khan, M.A., Nissar, S., Banday, M.Z.: Assessment of mental health and various coping strategies among general population living under imposed COVID-lockdown across world: A cross-sectional study. Ethics, Medicine and Public Health 15, 100571 (2020)
12. Sahni, P.S., Singh, K., Sharma, N., Garg, R.: Yoga an effective strategy for self-management of stress-related problems and wellbeing during COVID19 lockdown: A cross-sectional study. PLOS ONE 16(2), e0245214 (2021)
13. Coulthard, H., Sharps, M., Cunliffe, L., Van den Tol, A.: Eating in the lockdown during the Covid 19 pandemic; self-reported changes in eating behaviour, and associations with BMI, eating style, coping and health anxiety. Appetite 161, 105082 (2021)
14. 50 Vidas, D., Larwood, J.L., Nelson, N.L., Dingle, G. A.: Music Listening as a Strategy for Managing COVID-19 Stress in First-Year University Students. Frontiers in Psychology 12, 647065 (2021)

15. Granot, R., Spitz, D.H., Cherki, B.R., Loui, P., Timmers, R., Schaefer, R.S., Israel, S.: "Help! I Need Somebody": Music as a Global Resource for Obtaining Wellbeing Goals in Times of Crisis. Frontiers in Psychology 12, 1038 (2021)
16. Vajpeyee, M., Tiwari, S., Jain, K., Modi, P., Bhandari, P., Monga, G., Vajpeyee, A.: Yoga and music intervention to reduce depression, anxiety, and stress during COVID-19 outbreak on healthcare workers. International Journal of Social Psychiatry. 00207640211006742 (2021)
17. Ribeiro, F. S., Lessa, J. P. A., Delmolin, G., Santos, F. H.: Music listening in times of COVID-19 outbreak: A Brazilian study. Frontiers in Psychology 12, 1471 (2021)
18. Cabedo-Mas, A., Arriaga-Sanz, C., Moliner-Miravet, L.: Uses and Perceptions of Music in Times of COVID-19: A Spanish Population Survey. Frontiers in Psychology 11, 606180 (2021)
19. Fink, L.K., Warrenburg, L.A., Howlin, C., Randall, W.M., Hansen, N.C., Wald-Fuhrmann, M.: Viral tunes: changes in musical behaviours and interest in corona music predict socio-emotional coping during COVID-19 lockdown. Humanities and Social Sciences Communications 8(1), 1–11 (2021)
20. Krause A.E., Dimmock J., Rebar A.L., Jackson B.: Music listening predicted improved life satisfaction in University students during early stages of the COVID-19 pandemic. Frontiers in Psychology 11, 631033 (2021)
21. Ray, C., Lindop, J., Gibson, S.: The concept of coping. Psychological Medicine 12(2), 385–395 (1982)
22. Lazarus, R.S., Folkman, S.: Stress, appraisal, and coping. NY: Springer. New York (1984)
23. Endler, N.S.: Stress, Anxiety and coping: the multidimensional interaction model. Canadian Psychology/Psychologie Canadienne 38(3), 136–153 (1997)
24. Krypel, M.N., Henderson-King, D.: Stress, coping styles, and optimism: are they related to meaning of education in students' lives?. Social Psychology of Education 13(3), 409–424 (2010)
25. Chinna, K., Sundarasen, S., Khoshaim, H.B., Kamaludin, K., Nurunnabi, M., Baloch, G. M., Memon, Z.: Psychological impact of COVID-19 and lock down measures: An online cross-sectional multicounty study on Asian university students. PLOS ONE 16(8), e0253059 (2021)
26. Ferreira, F.D.O., Lopes-Silva, J.B., Siquara, G.M., Manfroi, E.C., de Freitas, P.M.: Coping in the Covid-19 pandemia: how different resources and strategies can be risk or protective factors to mental health in the Brazilian population. Health Psychology and Behavioral Medicine 9(1), 182–205 (2021)
27. Thai, T.T., Le, P.T.V., Huynh, Q.H.N., Pham, P.T.T., Bui, H.T.H.: Perceived Stress and Coping Strategies During the COVID-19 Pandemic Among Public Health and Preventive Medicine Students in Vietnam. Psychology Research and Behavior Management 14, 795–804 (2021)
28. Lara, R., Fernández-Daza, M., Zabarain-Cogollo, S., Olivencia-Carrión, M.A., Jiménez-Torres, M., Olivencia-Carrión, M. D., Godoy-Izquierdo, D.: Active Coping and Anxiety Symptoms during the COVID-19 Pandemic in Spanish Adults. International Journal of Environmental Research and Public Health 18(16), 8240-8257 (2021)
29. Rotter, J.B.: Generalized expectancies for internal versus external control of reinforcement. Psychological Monographs General and Applied 80(1), 1–28 (1966)
30. Heptinstall, S.T.: Stress Management. In: Everett, T., Dennis, M., Ricketts, E. (eds.) Physiotherapy in Mental Health, pp. 161–187. Butterworth-Heinemann (1995)
31. Alat, P., Das, S.S., Arora, A., Jha, A.K.: Mental health during COVID-19 lockdown in India: Role of psychological capital and internal locus of control. Current Psychology. 1–13 (2021)
32. Flesia, L., Monaro, M., Mazza, C., Fietta, V., Colicino, E., Segatto, B., Roma, P.: Predicting perceived stress related to the Covid-19 outbreak through stable psychological traits and machine learning models. Journal of Clinical Medicine 9(10), 3350 (2020)
33. Thoma, M.V., Ryf, S., Mohiyeddini, C., Ehlert, U., Nater, U.M.: Emotion regulation through listening to music in everyday situations. Cognition & Emotion 26(3), 550–560 (2012)
34. Van Den Tol, A.J.M., Edwards, J.: Listening to sad music in adverse situations: How music selection strategies relate to self-regulatory goals, listening effects, and mood enhancement. Psychology of Music 43(4), 473–494 (2014)

35. Vieillard, S., Harm, J., Bigand, E.: Expressive suppression and enhancement during music-elicited emotions in younger and older adults. Frontiers in Aging Neuroscience 7, 11 (2015)
36. Juslin, P. N.: Musical Emotions Explained: Unlocking the Secrets of Musical Affect. Oxford University Press, USA (2019)
37. Cook, T., Roy, A.R., Welker, K.M.: Music as an emotion regulation strategy: An examination of genres of music and their roles in emotion regulation. Psychology of Music 47(1), 144–154 (2019)
38. de Witte, M., Spruit, A., van Hooren, S., Moonen, X., Stams, G.J.: Effects of music interventions on stress-related outcomes: a systematic review and two meta-analyses. Health Psychology Review 14(2), 294–324 (2020)
39. Thoma, M.V., La Marca, R., Brönnimann, R., Finkel, L., Ehlert, U., Nater, U.M.: The effect of music on the human stress response. PLOS ONE 8(8), e70156 (2013)
40. Seinfeld, S., Bergstrom, I., Pomes, A., Arroyo-Palacios, J., Vico, F., Slater, M., Sanchez-Vives, M.V.: Influence of music on anxiety induced by fear of heights in virtual reality. Frontiers in Psychology 6, 1969 (2016)
41. Sokolova, A.N.: Music as Medicine: The Adyghs' Case. Music Therapy Today 5, 1 (2004)
42. Aalbers, S., Spreen, M., Pattiselanno, K., Verboon, P., Vink, A., van Hooren, S.: Efficacy of emotion-regulating improvisational music therapy to reduce depressive symptoms in young adult students: A multiple-case study design. The Arts in Psychotherapy 71, 101–720 (2020)
43. Nguyen, T. N., Nilsson, S., Hellström, A. L., Bengtson, A.: Music therapy to reduce pain and anxiety in children with cancer undergoing lumbar puncture: a randomized clinical trial. Journal of Pediatric Oncology Nursing 27(3), 146–155 (2010)
44. Bradt, J., Dileo, C.: Music therapy for end-of-life care. Cochrane Database of Systematic Reviews (1), CD007169 (2010)
45. Pacchetti, C., Mancini, F., Aglieri, R., Fundarò, C., Martignoni, E., Nappi, G.: Active music therapy in Parkinson's disease: an integrative method for motor and emotional rehabilitation. Psychosomatic Medicine 62(3), 386–393 (2000)
46. Särkämö, T., Altenmüller, E., Rodríguez-Fornells, A., Peretz, I.: Music, brain, and rehabilitation: emerging therapeutic applications and potential neural mechanisms. Frontiers in Human Neuroscience 10, 103–108 (2016)
47. Burns J. L., Labbé E., Arke B., Capeless K., Cooksey B., Steadman A. et al.: The effects of different types of music on perceived and physiological measures of stress. J. Music Ther. 39, 101–116 (2002)
48. Labbé E., Schmidt N., Babin J., Pharr M.: Coping with stress: the effec-tiveness of different types of music. Appl. Psychophysiol. Biofeedback 32, 163–168 (2007)
49. Thompson W. F., Geeves A. M., Olsen K. N.: Who enjoys listening to vio-lent music and why? Psychol. Popular Media Cult. 8, 218–232 (2018)
50. Sharman, L., & Dingle, G. A.: Extreme metal music and anger pro-cessing. Frontiers in human neuroscience, 9, 272 (2015)
51. Howlin, C., & Rooney, B.: Patients choose music with high energy, dance-ability, and lyrics in analgesic music listening interventions. Psychology of Music, 49(4), 931–944 (2021)
52. Sachs, M. E., Damasio, A., & Habibi, A.: The pleasures of sad music: a systematic review. Frontiers in human neuroscience, 9, 404–416 (2015)
53. van den Tol A. J. M.: The appeal of sad music: a brief overview of current directions in research on motivations for listening to sad music. Arts Psy-chother. 49, 44–49 (2016)
54. Larwood, J. L., & Dingle, G. A.: The effects of emotionally congruent sad music listening in young adults high in rumination. Psychology of Music, 0305735620988793 (2021)
55. Groarke, J. M., & Hogan, M. J.: Listening to self-chosen music regulates induced negative affect for both younger and older adults. PLoS One, 14(6), e0218017 (2019)
56. Giordano, F., Scarlata, E., Baroni, M., Gentile, E., Puntillo, F., Brienza, N., & Gesu-aldo, L.: Receptive music therapy to reduce stress and improve well-being in Italian clinical staff involved in COVID-19 pandemic: A prelimi-nary study. The Arts in psychotherapy, 70, 101688 (2020)

57. Carver, C.S.: You want to measure coping but your protocol's too long: Consider the Brief COPE. International Journal of Behavioral Medicine 4, 92–100 (1997)
58. García, F. E., Barraza-Peña, C. G., Wlodarczyk, A., Alvear-Carrasco, M., Reyes-Reyes, A.: Psychometric properties of the Brief-COPE for the evaluation of coping strategies in the Chilean population. Psicologia: Reflexão e Crítica 31 (2018)
59. Cohen, S., Kamarck, T., Mermelstein, R.: A global measure of perceived stress. Journal of Health and Social Behavior 24, 385–396 (1983)
60. Lee, K.C., Chao, Y.H., Yiin, J.J., Hsieh, H.Y., Dai, W.J., Chao, Y.F.: Evidence that music listening reduces preoperative patients' anxiety. Biological Research for Nursing 14(1), 78–84 (2012)
61. Cantril, H.: The Pattern of Human Concerns. Rutgers University Press, New Brunswick, NJ (1965)
62. Lumpkin, J.R.: Validity of a brief locus of control scale for survey research. Psychological Reports 57(2), 655–659 (1985)
63. Upadhyay, D., Shukla, R., Chakraborty, A.: Factor structure of music preference scale and its relation to personality. Journal of Indian Academy of Applied Psychology 43(1), 104–113 (2016)
64. Rentfrow, P.J., Gosling, S.D.: The do re mi's of everyday life: the structure and personality correlates of music preferences. Journal of Personality and Social Psychology 84(6), 1236 (2003)
65. Delsing, M.J., Ter Bogt, T.F., Engels, R.C., Meeus, W. H.: Adolescents' music preferences and personality characteristics. European Journal of Personality: Published for the European Association of Personality Psychology 22(2), 109–130 (2008)
66. Tekman, H.G., Hortacsu, N.: Aspects of stylistic knowledge: What are different styles like and why do we listen to them? Psychology of Music 30, 28–47 (2002)
67. Scott, S.L., Carper, T. M., Middleton, M., White, R., Renk, K., Grills-Taquechel, A.: Relationships among locus of control, coping behaviors, and levels of worry following exposure to hurricanes. Journal of Loss and Trauma 15(2), 123-137 (2010)
68. Aspelmeier, J.E., Love, M. M., McGill, L. A., Elliott, A. N., Pierce, T. W.: Self-esteem, locus of control, college adjustment, and GPA among first-and continuing-generation students: a moderator model of generational status. Res. High. Educ 53, 755–781 (2012)
69. Linnemann, A., Wenzel, M., Grammes, J., Kubiak, T., Nater, U. M.: Music listening and stress in daily life—a matter of timing. International Journal of Behavioral Medicine 25(2), 223–230 (2018)
70. Sutton, J., De Backer, J.: Music, trauma and silence: The state of the art. The Arts in Psychotherapy 36(2), 75–83 (2009)
71. Ben-Zur, H.: Emotion-focused coping. Encyclopaedia of Personality and Individual Differences 1343–1345 (2020)
72. Kammeyer-Mueller, J.D., Judge, T.A., Scott, B.A.: The role of core self-evaluations in the coping process. Journal of Applied Psychology 94(1), 177 (2009)
73. Pruessner, J. C., Wuethrich, S., Baldwin, M. W.: Stress of self-esteem. Stress Consequences: Mental Neuropsychological and Socioeconomic 53–57 (2010)
74. Parker, J.D., Endler, N.S.: Coping with coping assessment: A critical review. European Journal of Personality 6(5), 321–344 (1992)
75. Indians spend most of the time listening music than rest of the world: Report, *url* : https://www.livemint.com/news/india/indians-spend-more-time-listening-to-music-than-rest-of-the-world-report-1569330576902.html
76. Van Goethem, A., Sloboda, J.: The functions of music for affect regulation. Musicae Scientiae 15(2), 208–228 (2011)
77. Morcom, A.: An understanding between Bollywood and Hollywood? The meaning of Hollywood-style music in Hindi films. British Journal of Ethnomusicology 10(1), 63–84 (2001)

78. Ganti, T.: And yet my heart is still Indian: The Bombay film industry and the (H)Indianization of Bollywood. In: Ginsburg, F., Abu-Laghood, L., Larkin, B. (eds.) Media Worlds: Anthropology on New Terrains, pp. 281300. Berkeley: University of California Press (2002)
79. Rao, S.: "I need an Indian touch": Glocalization and Bollywood films. Journal of International and Intercultural Communication 3(1), 1–19 (2010)
80. Salimpoor, V.N., Benovoy, M., Longo, G., Cooperstock, J.R., Zatorre, R.J.: The rewarding aspects of music listening are related to degree of emotional arousal. PLOS ONE 4(10), e7487 (2009)
81. Afshar, H., Roohafza, H.R., Keshteli, A.H., Mazaheri, M., Feizi, A., Adibi, P.: The association of personality traits and coping styles according to stress level. Journal of research in medical sciences: the official journal of Isfahan University of Medical Sciences 20(4), 353 (2015)
82. Zweigenhaft, R.L.: A do re mi encore: A closer look at the personality correlates of music preferences. Journal of Individual Differences 29 (2008)
83. Bansal, J., Flannery, M.B., & Woolhouse, M.H.: Influence of personality on music-genre exclusivity. Psychology of Music 0305735620953611 (2020)
84. Dollinger, S.J.: Research note: Personality and music preference: Extraversion and excitement seeking or openness to experience? Psychology of Music 21(1), 73–77 (1993)
85. Treacy, A.: Music preferences, and their effect on personality, coping styles and perceived scholastic competence in students (2013)
86. Juslin, P.N., Laukka, P.: Communication of emotions in vocal expression and music performance: Different channels, same code?. Psychological Bulletin 129(5), 770 (2003)
87. Chamorro-Premuzic, T., Swami, V., Cermakova, B.: Individual differences in music consumption are predicted by uses of music and age rather than emotional intelligence, neuroticism, extraversion or openness. Psychology of Music 40(3), 285–300 (2012)
88. Admiraal, W.F., Korthagen, F.A., & Wubbels, T.: Effects of student teachers' coping behaviour. The British Journal of Educational Psychology 70 (Pt 1), 33–52 (2000)
89. Carver, C.S., Scheier, M.F., & Weintraub, J.K.: Assessing coping strategies: a theoretically based approach. Journal of Personality and Social Psychology 56(2), 267 (1989)
90. Mulligan, T.P.: The relationship of music preference and music function with coping in university students. Doctoral dissertation, Oklahoma State University (2009)
91. Sigurvinsdottir, R., Thorisdottir, I.E., Gylfason, H.F.: The impact of COVID-19 on mental health: The role of locus on control and internet use. International Journal of Environmental Research and Public Health 17(19), 6985 (2020)

Biophysics of Brain Plasticity and Its Correlation to Music Learning

Sandipan Talukdar and Subhendu Ghosh

1 Introduction

A musician not only entices us with beauteous and calming melodies, the musician also provides an excellent model system to study plasticity of the human brain. In fact, it is now believed that a musician's brain is an ideal model for studying activity-related plasticity of the brain [1–3]. This pertains to changes in brain function and structure in response to activities performed by a musician. A sound wave traveling in the air finds its way through the ears to the brain, and by interfering with the neural dynamics, it can give rise to structural and functional changes in the brain. It is a gargantuan task to unveil how experience or training causes the changes in the brain. We are still far and far behind from an elaborate knowledge about the mechanism of such plasticity. Training with music comprises of learning how to play an instrument with all the complex patterns involved in it and learning of a myriad of complex musical notes. These involve several components of the sensory and the motor systems of the brain, thus demanding a wide variety of cognitive processes of the higher order. The complexities involved in music training pose manifold challenges, but at the same time, it also provides opportunity to study how sensory and motor systems interact in giving rise to cognition and how different paradigms of training influence the interactions. Music training and its relation to the brain function can broadly be understood in two categories: firstly, how training influences the auditory and sensory-motor systems, and secondly, how the influences of training on auditory and sensory motor systems drive brain plasticity. Again, from the perspective of plasticity, the multi-paradigm nature of music training is of much importance as it enhances plasticity in the brain. The auditory system, which comprises of various parts starting from the ear, cochlea, auditory cortices, and brain stem to other brain

S. Talukdar (✉) · S. Ghosh
Department of Biophysics, University of Delhi South Campus, New Delhi, India

© The Author(s), under exclusive license to Springer Nature Switzerland AG 2023
A. Biswas et al. (eds.), *Advances in Speech and Music Technology*, Signals and Communication Technology, https://doi.org/10.1007/978-3-031-18444-4_14

areas involved in the process of higher auditory cognition, is of critical importance. The auditory system is such a system that is most altered by musical training. Structural and functional changes in response to musical training take place at various levels of the auditory system. For example, changes take place in the brain stem [4], auditory cortices [5], and areas [6] that are involved in higher levels of auditory cognition. Zatorre et al. [7] have reviewed how musical training can be considered as a framework of studying brain plasticity. Exposure to specific music stimuli or pattern can lead to long-term changes in the auditory cortex. The changes in the auditory system can also take different forms depending on the behavioral paradigms that are used for training. For example, classical conditioning, perceptual learning, or stimulus-response learning will bring out different kinds of changes in the auditory system. All these hints toward the remarkable capacity of the brain plasticity in response to music stimuli.

Not only the auditory system that exhibits plasticity in music learning or training but also the motor system is incredibly plastic in this aspect. In learning an instrument like violin or keyboard, a distributed motor network is engaged in the process. A plethora of studies has shown that in acquisition of a new motor skill in case of music, different parts of the motor network contribute deferentially in various phases of learning. Models of learning motor skills have shown [8] that the M1 and the premotor cortices are important parts as far as storing and representing a specific motor sequence is concerned. Again, basal ganglia are more involved in case of stimulus-response association, whereas the cerebellum is involved in the mechanism of error corrections. Interestingly, these features are also found to be fit well in musical training effects. In a study [9] about pianists who are highly trained, anatomical plasticity in motor pathways were observed in white matter. It was also found that the greater the musical practice during childhood, the greater was the integrity in the corticospinal tracts. Most of the auditory and motor changes that have been observed in music learning are based on imaging techniques combined with EEG (electroencephalography) and MEG (magnetoencephalography). Apart from those mentioned above, there are a large swathe of studies representing extraordinary plasticity of the brain arising as a response to music learning. Moreover, music's ability to reorganize the brain has now crept in the realm of clinical applications.

In this chapter, we will focus on the biophysical mechanisms involving in brain plasticity, both at the level of synapses (synaptic plasticity) and single neurons (Intrinsic plasticity). The importance of intrinsic plasticity in learning will be highlighted with the help of experimental evidences. The ways intrinsic plasticity is involved in music learning will also be discussed taking account of experimental evidences. Mostly, the evidences showing the involvement of intrinsic plasticity in music learning have been found in species other than humans, for example, singing birds. This pertains to the fact that the experiments involved in these kinds of studies are highly invasive, making it difficult to use them straightly on human subjects. Nevertheless, the current state of the art in this domain will be discussed considering the information gathered by non- or less invasive biophysical techniques.

2 Brain Plasticity and Learning

Within the realm of physiology, plasticity denotes changeability. With changes in environmental inputs and conditions, the physiology changes in order to adapt itself in a changing environment. When it comes to the brain, plasticity becomes extremely important as it can give rise to changes in behavior, apart from mitigating diseased conditions. Coming straight to humans, we learn and unlearn myriads of tasks and skills throughout our life. Learning and for that matter unlearning are pertained to the extraordinary plasticity ingrained in our brain. In fact, the brain remains plastic up to an advanced age of our life. Brain plasticity or neuroplasticity is an idea having its root at least a century back. Modern neuroscience resourced with sensitive imaging techniques, state-of-the-art molecular equipment, and laboratory techniques has unveiled how neurons, the basic entity of the brain, send out spikes or action potentials and how they can change their activities in response to change in both external and internal inputs.

Synapses, the connections among neurons, form complex networks. The networks are the architecture of memory and learning driven by experience or change in environmental conditions, and they are also the architecture subjected to be reorganized and giving rise to brain plasticity. The idea of neuroplasticity can be dated back to centuries. It was in 1890 when William James, a psychologist and philosopher, postulated [10] the idea that brain and its functions are not fixed entities in his book "*The Principles of Psychology*". James's book was a landmark in the field and quite aptly could suggest that human brain is remarkably capable of reorganizing itself. James's idea remained neglected for quite some time. However, the idea kept on enriching over time. By 1948, a Polish neuroscientist, Jerzy Konorski, first officially coined the term neural plasticity [11]. Konorski suggested that neurons getting activated due to spiking of other neurons in close vicinity together lead to plastic changes in the brain. Konorski's idea of neural plasticity reverberated shortly in the postulates of another psychologist, Donald Hebb. Famous as Hebbian plasticity, Hebb's postulates are considered as one of the founding ideas of the phenomenon known as synaptic plasticity. The synaptic plasticity is the changeability of synaptic strength. The synapses become stronger or weaker in response to stimuli. Hebb's most revered work, the book named "*The Organization of Behavior*" [12], scripted the Hebbian plasticity postulate, which says "When an axon of cell A is near enough to excite cell B and repeatedly or persistently takes part in firing it, some growth process or metabolic change takes place in one or both cells such that A's efficiency, as one of the cells firing B, is increased." This famous para of Hebb's idea is often paraphrased as "Neurons that fire together wire together."

Although the great ideas of neuroplasticity were profoundly postulated by some of the great minds of the time, the brain was believed to be a nonrenewable organ for centuries. It was believed before the great leaders of neuroscience came out with their incredible ideas of neuroplasticity that only within a critical period during childhood connections among the neurons (synapses) are formed and once the critical period is crossed, they remain fixed for the rest of the life. As such, the common belief was that only young brains are plastic. A profound impact of this belief was also reflected in clinical aspect. As the adult brain does not undergo plasticity, if a brain area of an adult is damaged, then there would be no further growth of the neuronal connections and the damaged area would remain damaged permanently. In contrast to this idea, we can take a common and nascent example; we can site the experience of forgetting something that has not been practiced for a long time which is common among many of us. We often tend to forget what we don't practice for a long time and on the other hand practicing makes it easier to learn or remember something. There are molecular and cellular underpinnings of these common aspects. Cornerstones of learning and memory lie in plasticity both at synaptic level and beyond. Besides synaptic plasticity, neurons themselves also possess incredible capacity of changing themselves in response to external stimuli, which is termed as the intrinsic plasticity. The intrinsic plasticity provides neurons to change their propensity of firing action potentials and thus changing their excitability.

2.1 Synaptic Plasticity

Let us have a quick overview of the phenomenon of synaptic plasticity. Any information in the brain is transmitted from one neuron to another through a connection known as the synapse. A synapse contains presynaptic and postsynaptic neurons, separated by the space called the synaptic cleft (see Fig. 1). The presynaptic neuron sends chemical or electrical signals and the postsynaptic neuron receives it. In case of a chemical synapse, the presynaptic neuron releases a chemical known as the neurotransmitter and the postsynaptic neuron's receptors receive it. The neurotransmitters are released from the presynaptic neuron as it fires an action potential in response to stimuli. The action potential travels down the axon of the presynaptic neuron after being initiated at the soma or the cell body. At the end of the axon of the presynaptic neuron, the neurotransmitters are released which is a voltage-dependent phenomenon. The neurotransmitters in turn can cause the postsynaptic neuron to get excited and fire an action potential, provided the process can depolarize the postsynaptic neuron beyond the threshold for firing of action potential.

Notably, a (postsynaptic) neuron can receive synapses from many other presynaptic cells. All the signals combined can make the neuron to fire an action potential. The Hebbian principle saying neurons that fire together wire together can thus be understood in the molecular and cellular underpinnings of forming a synapse. Neuroplasticity denotes the ability of neurons to modify the strength between synapses along with formation of newer ones. Visualizing in this way, neuroplasticity encompasses changes in the strength of mature synapses and the formation and elimination of synapses in both adult and developing brains.

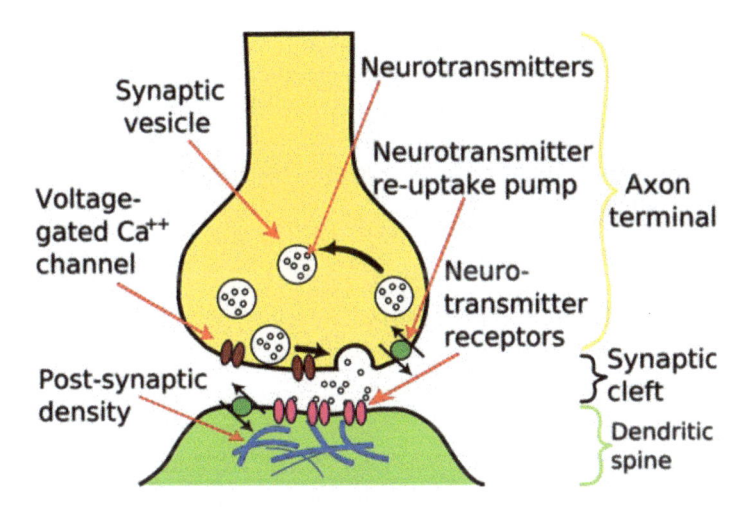

Fig. 1 Chemical synapse between two neurons. Image source Wikimedia commons (with creative common license)

2.2 Intrinsic Plasticity

A synapse involves at least two neurons. The synaptic connections' weakening or strengthening in response to stimuli forms the basis of synaptic plasticity. However, synaptic plasticity is not the only way that the neurons can modify themselves as a way of responding to changing stimuli or environmental conditions. Neurons possess another remarkable level of plasticity which does not require a synapse. This plasticity arises within a neuron and signifies the intrinsic excitability. Such kind of plasticity is termed as the "intrinsic plasticity" of neurons. The intrinsic plasticity, in turn, can also influence synaptic plasticity and vice versa.

The excitability of a neuron primarily denotes its ability to fire action potentials in response to stimuli. To fire an action potential, a neuron has to be stimulated beyond the threshold of firing. The threshold of action potential signifies the voltage that a neuron needs to be depolarized up to. If any stimulus, such as one coming from a synapse, can depolarize a neuron to its firing threshold, then the neuron will generate an action potential. Again, how does a neuron integrate synaptic inputs to fire an action potential is primarily determined by voltage- and calcium-gated ion channels. The number, type, and distribution of the ion channels determine a neuron's firing ability. Intrinsic plasticity is mediated by changes in the ion channels [13], changes in their expression levels or biophysical properties. Speaking broadly, ion channels in the dendritic part of a neuron are responsible for integrating synaptic signals, and ion channels near soma and axon hillock are responsible for the generation of action potential. Experiments, mainly in vitro, suggest that all of these features are plastic, which are subjected to be modified by different neural activities. Although most of the researches on neural plasticity have given main focus on synaptic plasticity, findings on various aspects of intrinsic plasticity have gained pace in the last two decades. Evidences are now plenty which show the important role of intrinsic plasticity in learning and memory as well. The evidences are obtained from studies conducted on both invertebrates and mammals. Zhang and Linden [14] have reviewed some of the best studied evidences about the involvement of intrinsic plasticity in learning and memory.

3 Biophysical Basis of Plasticity

3.1 Synaptic Plasticity

Two of the most widely studied phenomena that are established as evidential representation of synaptic plasticity are the long-term potentiation (LTP) and the long-term depression (LTD). The LTP is the persistent strengthening of synaptic connections, while the LTD signifies the opposite, that is, the long-lasting weakening of synaptic strength. Both LTP and LTD are dependent upon the recent patterns of activity.

LTP: It took almost two decades to get experimental evidence after Hebb's theory first appeared. It was Terje Lømo who first observed the phenomenon of LTP [15] in the hippocampal dentate gyrus of rabbit. Lømo conducted series of experiments on anesthetized rabbits in his attempt to explore the role of hippocampus in producing short-term memory. Lømo's experiments were designed with focus on specific synapses that connect neurons from the perforant pathway to the neurons in the dentate gyrus of the rabbit hippocampus. Lømo's experiments were designed in such a way that presynaptic neurons in the perforant pathway were stimulated while recording the responses from postsynaptic neurons of the dentate gyrus. A single pulse of electrical stimuli applied on the presynaptic neurons caused EPSP (excitatory postsynaptic potential) in the postsynaptic neurons in the dentate gyrus. Lømo could observe a fascinating finding in his experiments; he found that the postsynaptic neurons' response to the single pulse electrical stimuli can be enhanced for a long period of time. However, there is a condition which lies in the nature of the stimuli. The electrical stimuli to the presynaptic neurons need to be of high frequency and applied for a short duration. When such high frequency train of stimuli were applied to the presynaptic neurons, then the EPSPs elicited in the postsynaptic neurons tend to be stronger and long lasting. Here it needs to be mentioned that EPSP is a postsynaptic potential that can make the postsynaptic neurons to fire action potentials. Application of the high frequency stimuli to the presynaptic neurons could result in long-lasting enhancement of the postsynaptic neurons' response, which is what Hebb's theory portrays. The experiments leading to the validation of Hebb's theory of neuroplasticity were possible due to key technological developments of using brain slices [16]. The technique of brain slices combined with the development of the patch clamp technique where it is possible

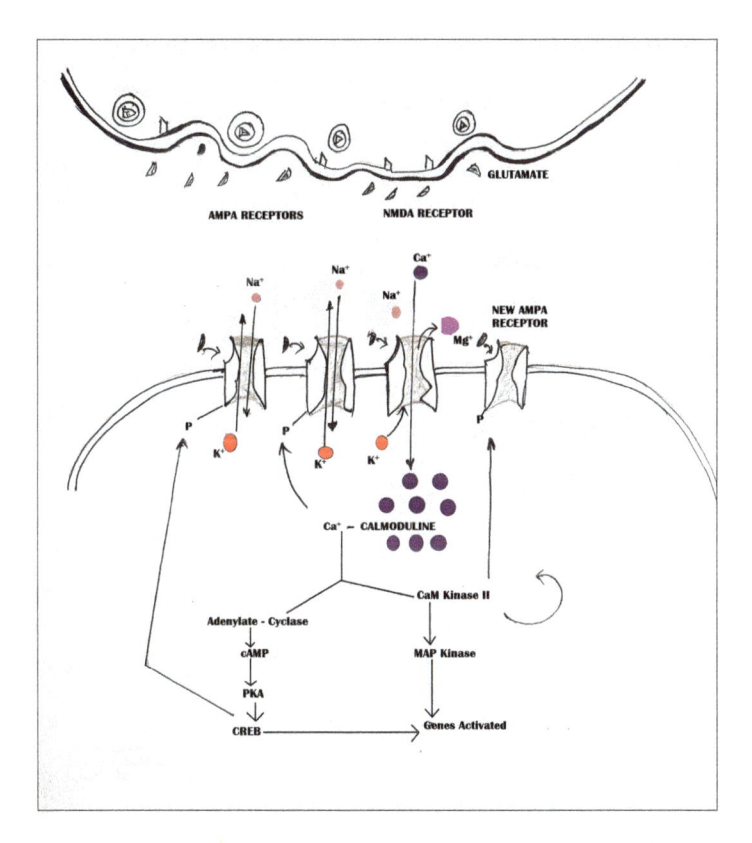

Fig. 2 The LTP mechanism

to conduct intracellular recordings on brain slices has been useful in identifying different plasticity at different synapses across the brain [17].

Well, going deep to the biophysical basis of LTP, we land up in complex interactions among various ions, receptors, and channels. As mentioned earlier, formation of LTP requires high frequency (over 100 Hz) stimuli applied for a short duration (ideally less than a second) in the presynaptic neurons. This procedure causes enough depolarization in the postsynaptic neurons to remove the Mg2+ ions that block the NMDA receptor in the dendrites of the postsynaptic neurons. This results in a large influx of calcium ions to enter (Fig. 2).

Calcium ions are extremely important in cell communications; they can activate many enzymes by means of changing their conformations. Calmodulin is one of such enzymes and it becomes active when four calcium ions bind to it becoming Ca2+/calmodulin. This Ca2+/calmodulin complex is the primary second messenger for LTP. The complex in turn activates other enzymes of importance such as adenylate cyclase and the CaM kinase II. These kinases alter spatial conformations of other molecules by means of phosphorylation, which is adding phosphate ion to a substrate. The adenylate cyclase and cAMP again activate another kinase, the PKA (protein kinase A). The PKA phosphorylates the AMPA (?-amino-3-hydroxy-5-methyl-4-isoxazolepropionic acid) receptors making them remain open for longer time after binding of glutamate. This makes the postsynaptic neuron getting further depolarized and contributing to LTP. It is worth mentioning at this point that glutamate is the neurotransmitter that is released when a neuron fires an action potential. The glutamate released from a presynaptic neuron binds to the AMPA receptor of the postsynaptic neuron. This makes the AMPA receptor to open, resulting in the large influx of sodium ion into the postsynaptic neuron. The sodium ions depolarize the postsynaptic neuron and make it more likely to fire further action potential. Thus, in LTP, the presynaptic and postsynaptic neurons become prone to fire together as they get wired together. This is an example of the NMDA-dependent LTP with a brief overview of the underlying mechanism. However, there are other examples of LTP in other types of synapses with some differences in the mechanism.

In their seminal work [15], Lømo and Bliss performed their experiment in a set of 18 adult rabbits including both sexes. The rabbits were anesthetized with urethane. The experiments used the synapses formed between neurons of the perforant pathway and the neurons from the dentate gyrus. Here, the perforant pathway neurons were the presynaptic neurons and the dentate gyrus neurons acted like postsynaptic neurons. The presynaptic neurons were stimulated with high frequency stimulation of about 4 hertz, and the responses it created in the postsynaptic neurons were recorded.

The experiments found, as expected, that single pulse of electrical stimulation to the presynaptic neurons could generate EPSPs in the postsynaptic neurons. However, when the presynaptic stimulation was of high frequency, then the EPSPs generated were stronger and prolonged as well. Lømo and Bliss reported to have found 14 out of the 18 rabbits showing potentiation for periods ranging from 30 min to 10 h. The prolonged potentiation depended upon the nature of the stimuli applied to the presynaptic neurons. Higher frequency stimuli applied for a shorter time were found to have elicited prolonged potentiation. In the experiment, two paradigms of stimulus input were applied at 10–20/s for 10–15 s and 100/s for 3–4 s.

LTD: The LTD arises as an opposite to LTP and serves to weaken selective synapses. For LTD to take place, presynaptic neuron is stimulated for a longer period of time with low frequency stimuli (nearly 1 Hz) [18]. The LTD depends upon calcium ion to a great extent similar to the case of LTP. Calcium ions' magnitude in the postsynaptic neuron largely determines the direction of a synapse, that is, whether there will be LTP or LTD. In the case of NMDA receptor-dependent LTD, there is a moderate rise in the postsynaptic calcium level, in CA1 hippocampal neurons [19]. The level of calcium below the threshold point leads to LTD, while large influx of the ion leads to formation of LTP. LTD is also associated with phosphatases, the enzymes that dephosphorylate its substrates, just opposite to what the kinases do. A subthreshold level calcium in the postsynaptic neuron leads to the activation of calcium-dependent phosphatases. Apart from hippocampus, LTD can take place in several other brain areas weakening synaptic strength. Both LTP and LTD have significant contributions in learning and memory. These two mechanisms represent the plasticity of synapses in response to external stimuli.

They also represent the phenomenon of brain plasticity with respect to changing nature of environmental stimuli, which is the cornerstone of learning and memory.

3.2 Mechanism of Intrinsic Plasticity and Its Importance in Learning

Acquisition of learning and for that matter memory involves neural correlates that go beyond synaptic plasticity. Evidences have been increasing which relate intrinsic plasticity as having equally important role in learning and memory [14, 20, 21]. Before going in details about how exactly intrinsic plasticity influences learning and memory, it is important to have some basic idea about intrinsic plasticity, especially how neuroscientists detect it and what is the underlying mechanism. As have mentioned earlier, intrinsic plasticity is involved in single neurons. With respect to stimuli, neurons' propensity of firing action potential changes. The figure below shows the classical action potential curve. The figure also signifies the biophysical bases of giving rise to an action potential. When a stimulus (either as a synaptic input or an electrical stimulus in patch clamp) can sufficiently depolarize a neuron, meaning the depolarization is above the threshold of firing, then sodium channels open up. This event leads to large influx of sodium ions and thus further depolarizes the cell. Notably, a typical neuron has a resting potential of around -70 mV and the threshold is about -55 mV. Depolarizing event continues till the membrane potential reaches about 40 mV. At this voltage, sodium channels begin to close and potassium channels start to open. Through the potassium channels, rapid outflux of potassium ions occurs, and in this process, the cell membrane gets back to the polarized state. However, in a typical action potential, potassium channel-mediated repolarization makes the cell hyperpolarized, which is shown as the refractory phase, where the membrane voltage goes up to -90 mV. The refractory phase is called the after-hyperpolarization (AHP) as well. During the AHP phase, no further action potential can be fired by a neuron (Fig. 3).

One of the hallmarks of intrinsic plasticity is the reduction of the AHP in response to repeated stimuli [22]. When subjected to repetitive stimuli, neurons' AHP gradually reduces, and this increases the propensity of firing action potentials and thus increasing the excitability of neurons. One of the first evidences of reduction in AHP depending upon repeated activity came from the experiments of Alkon et al. on the phototaxis of *Hermissenda* [23]. Here they showed in brain slices that *Hermissenda* exposed to repetitive learning paradigm showed reduction in AHP while learning a task. Since then there have been plethora of studies [22, 24–27] which have shown activity-dependent reduction in AHP involved in learning and memory. Moreover, the AHP-mediated intrinsic plasticity can interfere with synapse by lowering the synaptic threshold. Sehgal et al. [28, 29] argued that intrinsic plasticity can act as a metaplasticity mechanism in the formation of memory and

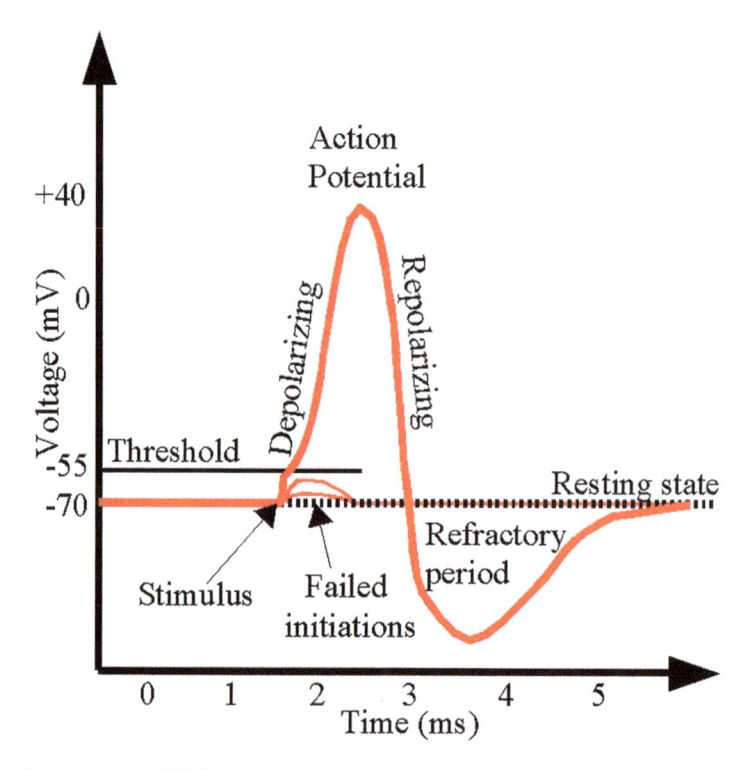

Fig. 3 The action potential. Image source Wikimedia commons

acquisition of learning. The intrinsic plasticity and its contribution to learning have also been elaborated in a recent review [30].

In our previous work [32], we attempted to decipher the mechanism of how neurons give rise to activity-dependent reduction in AHP, which has significant contribution in learning and memory. We adopted the approach of biophysical modeling starting from the Hodgkin-Huxley (HH) model for action potential. It is noteworthy at this point that Hodgkin and Huxley were the first to bring out a model that encompasses the biophysical basis of action potential generation. Their seminal work [31] brought them Nobel Prize in Physiology in 1963. Moreover, the HH model rightly predicted the existence of ion channels on the membrane of neurons. Ion channels were not discovered at that time. They not only predicted the existence of ion channels but also rightly predicted through their biophysical modeling how the ion channels may function. They mentioned about gating variables in the HH model, which are today's ion channels. In our work, we showed that in the context of repetitive stimuli, the threshold of action potential is reduced concomitantly with the reduction in AHP involved in learning. We also brought in a new paradigm in our model, that the potassium reversal potential, which is considered as a constant in the classical HH model, is a variable entity and it modulates the AHP in several types of neurons.

4 Intrinsic Plasticity and Music Learning

Acquiring music skills, whether playing an instrument or learning musical patterns or a combination of both, involves intensive practice, which ought to have neural underpinnings, at synaptic, intrinsic, and network levels. However, deciphering the mechanisms in human subjects is a huge challenge as invasive recording of human brain tissues is a highly ethical issue. In most of the animal studies on learning-related intrinsic plasticity, the animals are subjected to learning experiments, for example, the Morris water maze. After the acquisition of learning was shown in their behavior, slices are made from the brains of the animal subjects, and then recordings were done to enumerate the biophysical changes in the neurons. This pipeline of experiments is not readily possible in human subjects. Nevertheless, there have been instances of studying human brain slices and recording on single neuron. Most of these studies on human subjects are limited to disease conditions, such as epilepsy. Quiroga highlighted single neuron recording in the hippocampal formation in human [33]. In addition to single neuron recording in human brains in diseased conditions, there has also been the use of EEG (electroencephalogram) recording. The EEG recording has seen extensive use in unveiling neural dynamics involved in music perception and learning. Sanyal et al. showed the brain dynamics while experiencing Indian music (raga) through EEG recordings [34]. They showed an arousal activity in the alpha frequency band of EEG when the subjects were exposed to Indian ragas. Again, the alpha and theta band are found to be involved in memory and cognitive performance [35]. There has been a plethora of studies on musical perception and learning conducted with EEG where the underlying neural mechanisms are tried to understand. These studies combined with fMRI (functional magnetic resonance imaging) studies on human brains have brought to us many interesting revealings. However, studies on molecular and cellular patterns involved in neural plasticity in music cognition in human subjects are hugely lacking.

Nevertheless, intrinsic plasticity in song learning and other paradigms of music has been studied through brain slice preparation and electrophysiological recording on other animals. In a study [36] on zebra finch song birds, the role of intrinsic plasticity of the cortical neurons (in the caudal mesopallium area of the zebra finch brain) was deciphered, and interestingly, this plasticity was found to be required for memorizing songs in the bird species. Again, on zebra finch itself, another important study [37] showed that intrinsic properties of neurons could influence network properties and eventually can give rise to behavioral plasticity. Not only on song birds but also in mouse model, researches have been carried out to reveal how learning complex sound pattern changes the intrinsic neuronal properties. In the study [38] by Chen et al., it was found that specific cortical neurons exhibit prolonged firing in the mice that were trained to learn complex sound pattern, which was not discernible in untrained counterparts.

Music learning in humans is also ought to show similar intrinsic neuronal plastic properties. However, single neuron recordings from human brain still remains a difficult task. Nonetheless, with advancement of more sophistications in EEG, fMRI, or other such noninvasive tools, this might get possible in the near future. Being able to do so will enhance our understanding on myriads of aspects of brain plasticity involved in music learning. This will have clinical implications as well, where music has been used in providing therapies in various neurological and cognitive issues. Notably, music has also been found in reducing racial prejudices. In the landmark study [39] conducted by Neto et al. on school-going children in Portugal, children trained with special musical patterns were found to be less prejudiced toward Cape Verdean people.

5 Conclusion

Plasticity is the remarkable feature of the brain involved in learning and memory driven by experiences. The complex biophysical mechanisms involved in both synaptic and intrinsic plasticity have been worked out to a great extent, with new realities pouring in with time. Music is a complex subject to learn and involves cognitive activity of the higher order, which in turn is related to plasticity in the brain. However, how music gives rise to plastic changes in the brain is still elusive for neuroscience community, even after ongoing extensive and active researches across the world. With the development of noninvasive techniques and their ever-sophisticated ways of applying on human subjects, they are expected to bring new lights in understanding how sound waves bring about changes in the neurons, synapses, and brain networks. Efforts in revealing cellular plasticity paradigms, both intrinsic and synaptic, will surely enhance in knowledge building in this direction.

References

1. Jäncke, L., Gaab, N., Wüstenberg, T., Scheich, H., Heinze, H.: Short-term functional plasticity in the human auditory cortex: an fMRI study. Cognitive Brain Research. 12, 479–485 (2001)
2. Münte, T., Altenmüller, E., Jäncke, L.: The musician's brain as a model of neuroplasticity. Nature Reviews Neuroscience 3, 473–478 (2002)
3. Wan, C., Schlaug, G.: Music Making as a Tool for Promoting Brain Plasticity across the Life Span. The Neuroscientist 16, 566–577 (2010)
4. Wong, P., Skoe, E., Russo, N., Dees, T., Kraus, N.: Musical ex-perience shapes human brainstem encoding of linguistic pitch patterns. Nature Neuroscience 10, 420–422 (2007)
5. Bermudez, P., Lerch, J., Evans, A., Zatorre, R.: Neuroanatomi-cal Correlates of Musicianship as Revealed by Cortical Thick-ness and Voxel-Based Morphometry. Cerebral Cortex 19, 1583–1596 (2008)
6. Lappe, C., Herholz, S., Trainor, L., Pantev, C.: Cortical Plastic-ity Induced by Short-Term Unimodal and Multimodal Musical Training. Journal of Neuroscience 19, 28, 9632–9639 (2008)
7. Herholz, S., Zatorre, R.: Musical Training as a Framework for Brain Plasticity: Behavior, Function, and Structure. Neuron 19 76, 486–502 (2012)
8. Penhune, V., Steele, C.: Parallel contributions of cerebellar, striatal and M1 mechanisms to motor sequence learning. Behavioural Brain Research 19 226, 579–591 (2012)

9. Bengtsson, S., Nagy, Z., Skare, S., Forsman, L., Forssberg, H., Ullén, F.: Extensive piano practicing has regionally specific effects on white matter development. Nature Neuroscience 19 8, 1148–1150 (2005)
10. James's, W.: Principles of psychology (1863).
11. Livingston, R. B.: Brain mechanisms in conditioning and learning (1966)
12. Hebb, D.: The organization of behaviour. (2005)
13. Sehgal, M., Song, C., Ehlers, V., Moyer, J.: Learning to learn – Intrinsic plasticity as a metaplasticity mechanism for memory formation. Neurobiology of Learning and Memory 19, 105, 186–199 (2013)
14. Zhang, W., Linden, D.: The other side of the engram: experience-driven changes in neuronal intrinsic excitability. Nature Reviews Neuroscience 19, 4, 885–900 (2003)
15. Bliss, T. V., and Lømo, T.: Long-lasting potentiation of synaptic transmission in the dentate area of the anaesthetized rabbit fol-lowing stimulation of the perforant path. The Journal of physiology 19, 232, 331–356 (1973)
16. Skrede, K., Westgaard, R.: The transverse hippocampal slice: a well-defined cortical structure maintained in vitro. Brain Re-search 19, 35, 589–593 (1971)
17. Mateos-Aparicio, P., Rodríguez-Moreno, A.: The Impact of Studying Brain Plasticity. Frontiers in Cellular Neuroscience 19, 13, (2019)
18. Purves, D., Cabeza, R., Huettel, S. A., LaBar, K. S., Platt, M. L., Woldorff, M. G., and Brannon, E. M.: Cognitive neuroscience. Sunderland: Sinauer Associates, Inc. (2008)
19. Mulkey, R. M., Malenka, R. C.: Mechanisms underlying induction of homosynaptic long-term depression in area CA1 of the hippocampus. Neuron. 19, 9, 967–975 (1992)
20. Debanne, D., Inglebert, Y., Russier, M.: Plasticity of intrinsic neuronal excitability. Current opinion in neurobiology 19, 54, 73–82 (2019)
21. Schaefer, N., Rotermund, C., Blumrich, E. M., Lourenco, M. V., Joshi, P., Hegemann, R. U., Turner, A. J.: The malleable brain: plasticity of neural circuits and behavior–a review from students to students. Journal of neurochemistry 19, 142, 790–811 (2017)
22. Saar, D., Grossman, Y., Barkai, E.: Reduced after hyperpolarization in rat piriform cortex pyramidal neurons is associated with increased learning capability during operant conditioning. European Journal of Neuroscience 19, 10. 1518–1523 (1998)
23. Alkon, D. L., Lederhendler, I., Shoukimas, J. J.: Primary changes of membrane currents during retention of associative learning. Science 19 215, 693–695 (1982)
24. Disterhoft, J. F., Golden, D. T., Read, H. L., Coulter, D. A., Alkon, D. L.: AHP reductions in rabbit hippocampal neurons during conditioning correlate with acquisition of the learned response. Brain research 19 462, 118–125 (1988)
25. Moyer Jr, J. R., Thompson, L. T., Disterhoft, J. F.: Trace eye-blink conditioning increases CA1 excitability in a transient and learning-specific manner. Journal of Neuroscience 19 16. 5536–5546 (1996)
26. Matthews, E. A., Weible, A. P., Shah, S., Disterhoft, J. F.: The BK-mediated fAHP is modulated by learning a hippocampus-dependent task. Proceedings of the National Academy of Sciences 19 105, 15154–15159 (2008)
27. Song, C., Ehlers, V. L., Moyer, J. R.: Trace fear conditioning differentially modulates intrinsic excitability of medial prefrontal cortex–basolateral complex of amygdala projection neurons in infralimbic and prelimbic cortices. Journal of Neuroscience 19 35, 13511–13524 (2015)
28. Sehgal, M., Song, C., Ehlers, V. L., Moyer Jr, J. R.: Learning to learn–intrinsic plasticity as a metaplasticity mechanism for memory formation. Neurobiology of learning and memory 19 105, 186–199 (2013)
29. Song, C., Detert, J. A., Sehgal, M., Moyer Jr, J. R.: Trace fear conditioning enhances synaptic and intrinsic plasticity in rat hippocampus. Journal of neurophysiology 19 107, 3397–3408 (2012)
30. Yousuf, H., Ehlers, V. L., Sehgal, M., Song, C., Moyer Jr, J. R.: Modulation of intrinsic excitability as a function of learning within the fear conditioning circuit. Neurobiology of learning and memory 19 167, 107132 (2020)

31. Hodgkin, A. L., Huxley, A. F.: A quantitative description of membrane current and its application to conduction and excitation in nerve. Bulletin of mathematical biology 19 52, 25–71 (1990)
32. Talukdar, S., Shrivastava, R., Ghosh, S.: Modeling activity-dependent reduction in after hyper-polarization with Hodgkin-Huxley equation of action potential. Biomedical Physics and Engineering Express 19 5, 047001 (2019)
33. Quiroga, R. Q.: Plugging in to human memory: advantages, challenges, and insights from human single-neuron recordings. Cell 19 179, 1015–1032 (2019)
34. Banerjee, A., Sanyal, S., Patranabis, A., Banerjee, K., Guhatha-kurta, T., Sengupta, R., Ghose, P.: Study on brain dynamics by non-linear analysis of music induced EEG signals. Physica A: Statistical Mechanics and its Applications19 444, 110–120 (2016)
35. Klimesch, W.: EEG alpha and theta oscillations reflect cognitive and memory performance: a review and analysis. Brain research reviews 19 29, 169–195 (1999)
36. Chen, A. N., Meliza, C. D.: Experience and sex-dependent intrinsic plasticity in the zebra finch auditory cortex during song memorization. Journal of Neuroscience 19 40, 2047–2055 (2020)
37. Daou, A., Margoliash, D.: Intrinsic neuronal properties represent song and error in zebra finch vocal learning. Nature communications 19 11, 1–17 (2020)
38. Wang, M., Liao, X., Li, R., Liang, S., Ding, R., Li, J., Chen, X.: Single-neuron representation of learned complex sounds in the auditory cortex. Nature communications. 11, 1–14 (2020).
39. Neto, F., Pinto, M. D. C., Mullet, E.: Can music reduce na-tional prejudice? A test of a cross-cultural musical education programme. Psychology of Music 19 47, 747–756 (2019)

Analyzing Emotional Speech and Text: A Special Focus on Bengali Language

Krishanu Majumder and Dipankar Das

1 Introduction

With the advancement of technologies, machines have become intelligent enough to ease several human problems and tasks. This requires input, perception, and processing of information. In present world, the most basic input for communication is speech. Now, researchers are aiming to build machines that are capable of processing such standard inputs in the form of natural language. One of the main objectives is to build machines that will have the same capability of processing natural languages at a level equal or greater to that of a native speaker.

It is always observed that communication involves two-way interaction: one is perception and the other is reaction. While the first one can be achieved in machines with respect to speech recognition, the second one aims to grasp the insights through speech synthesis. Artificial speech synthesis is the technique of producing humanlike speech directly from the machines. These types of system are called text-to-speech models or TTS. These models take textual language representations as input with some other conditional parameters, if required, and turn them into respective humanlike speech.

Text-to-speech can be seen as a part of a bigger system. It is mainly used as a communication channel to convey responses of a system to the end user. Many applications nowadays use TTS as an output medium of their system. We mainly use visual screens as output mediums and the possibilities have not been explored to a great extent. However, with the rising need of intelligent systems, major technology companies are vigorously working to integrate TTS as a part of their products to ease interaction with end users. Some real-world uses are given below.

K. Majumder (✉) · D. Das
Department of Computer Science & Engineering, Jadavpur University, Kolkata, India
e-mail: dipankar.das@jadavpuruniversity.in; http://www.dasdipankar.com/

© The Author(s), under exclusive license to Springer Nature Switzerland AG 2023
A. Biswas et al. (eds.), *Advances in Speech and Music Technology*, Signals and Communication Technology, https://doi.org/10.1007/978-3-031-18444-4_15

- One of the most popular use of TTS can be seen in personal assistants like Apple Siri, Google Assistant, Alexa, etc. which require frequent communications with their end users. This not only simplifies communication but also eases it by allowing the end users to communicate with the system on the go instead of reading it from a fixed output screen.
- TTS plays a very important channel for communication for visually challenged users. These users cannot depend on the screen output and rely on speech output. Screen readers are one such examples where TTS helps to communicate with visually impaired users.
- E-Book Reader, PDF Reader, etc. are some categories of systems that use TTS as the heart of their system.
- Google Maps are another very popular application that uses TTS to give direction and navigational instructions to the user while driving. During driving, it is not possible to always look into the screen for finding routes and it is actually very dangerous. Google Maps avoids this problem by allowing the driver to keep his or her eye on the road and listen to the navigation instructions at the same time.
- We can see the use of TTS in Smart Homes where all the appliances in the house are connected together by preparing a synchronized ecosystem. Here, everything in the house can be controlled with voice commands and in turn, the house can also respond back in its own voice.

Text-to-speech is considered as a generic application which can be integrated with any existing system to establish a natural human-computer interaction medium. Beyond existing systems, TTS can be used with the upcoming new age and future technologies to increase their applicability and incorporate machine intelligence. For instance, self-driving vehicles, autonomous cars, and driving assistant are some of the cutting-edge developments that are planning to have TTS as a part of their systems. Another prospective application implements TTS in our existing ATM machines and other interactive devices in day-to-day life.

Nowadays, man-machine interaction has become very common in this world of emerging technologies. There are a number of state-of-the-art TTS models but most of them lacks natural features and sounds artificial.

The first thing that the existing TTS lacks is the multilingual features. Most of them are trained only for popular languages like English, Chinese, etc. Moreover, very limited work has been carried out on regional languages like Bengali where the resources are limited. Even while working on this topic, we came to know that very limited data is available in these regional languages to carry out any study. This made us realize the need of a proper dataset in Bengali.

Secondly, TTS systems are mainly trained on a single language. This made us think about exploring the possibility of training a TTS on more than one language. Lastly, very limited work has been done for incorporating emotion with synthetic speech. Speech emulating proper contextual emotion can open up new avenues and applications of TTS systems.

The main objective of our present work is to develop a TTS system in Bengali language and incorporate as much naturalistic features as possible. We also intend to explore the emotional aspects in the generated speech and the ways also to enhance it. Lastly, we aim to test our model on bilingual training and find the observations on extending our target language set. We also intend to develop an emotion-specific dataset from scratch and use it to train our model.

2 Literature Survey

A number of interesting works have been going on in the domain of speech synthesis in recent times, and this field is attracting the interest of some of the best scientists in the world. Numerous researches are still going on across the globe to increase the accuracy of the existing systems in natural speech production. Some of the standard and remarked existing models and approaches that are widely accepted are discussed below.

A. Concatenation model [2]: Concatenative approach for text-to-speech is one of the oldest methods of speech synthesis that was proposed in late 1990s. This algorithm uses a very large database for storing pre-recorded voices clips, broken into phoneme level chunks and rearrange them to produce new speeches.

B. Parametric model [3]: Parametric text-to-speech is a statistical model for generation of speech. Though it adds flexibility to the produced speech conditioned on the given input, the generated speech still lacks the natural qualities of human voice as it cannot alter the natural features like pauses, durations, nonspeech sounds, mean energy, etc.

C. Modification of voice features: It was found that emotional speech differs from neutral speech with respect to various features like phoneme duration, pause duration, mean energy level, etc. Researchers at IIIT Hyderabad [4] tried to convert a neutral speech into an anger-oriented speech by modulating the nonuniform duration and other features.

Not only researchers and academicians from universities but also the world leading technology giants Google, Facebook, IBM, Apple, Amazon, Adobe, Baidu, etc. are also conducting their researches in the field of speech synthesis. Some of the state-of-the-art systems came out from these organizations like Adobe's Voco [5] and Facebook's VoiceLoop [6], to name a few. Some state-of-the-art systems are discussed below.

D. WaveNet [7]: Google's DeepMind research group came up with WaveNet that exploits the idea to create amplitude value for one timestep at a time and repeat the process to create a series of amplitudes defining the speech. They took this inspiration when they successfully implemented pixel wise image generation [8, 9] and found that the results were also impressive. They used CNN with dilation to capture the time information over a long period. The residual and skip connections are also used for fast convergence of the training and to propagate the gradient deeper into the network. Though WaveNet is capable of producing high-quality standard speeches, it is time consuming. This problem is solved to a great extent by a better version of the model known as Fast WaveNet [10]. In the upgraded model, the redundant convolutional layers are replaced with convolutional queues from which the sample can be directly drawn from one end and the produced sample is again fed at the end of the queue.

In Parallel WaveNet [11], the researchers tried to train a new WaveNet using a pretrained WaveNet using teacher-student model with a new method a called probability density distillation. The resultant model is claimed to be 20 times faster and supports parallel processing.

WaveGlow [12] is another modification made to WaveNet by researchers from NVIDIA that conditions WaveNet on spectrograms without auto-regression as in Tacotron. This is basically a combined model of GLOW [13] and WaveNet and is claimed to produce fast, efficient, and high-quality audio.

Deep voice [14–16] is speech generative model developed by Chinese technical giant Baidu by using additional conditions to WaveNet. Deep Voice came in three versions. In the first version, the duration and fundamental frequency are predicted from the phonemes, and then they are conditioned on WaveNet to produce waveform. In the second version, duration is predicted first and then it is used to predict the fundamental frequency, and WaveNet is conditioned on them for multi-speaker speech generation. The third installment used an attention-based fully convolutional sequence learning architecture for faster learning with fewer training to create intermediate feature vectors, and final waveforms were created using three different output models, viz., WaveNet, WORLD, and Griffin-Lim.

E. MelNet [17]: Researchers from Facebook came up with MelNet that tries to learn voice synthesis unconditionally from the frequency domain. By doing so, the audio that spans hundreds of thousands of timesteps are scaled down to a few hundred or thousands of timesteps. The constituent frequencies are extracted from the signal. Then the local structures were learned by autoregressive modeling in which one pixel is predicted at a time in time major order (or frequency major order). A spectrogram of low resolution is generated and is continuously unsampled to produce a high-resolution spectrogram. This spectrogram is then converted into time domain to produce naturalistic sound.

F. Tacotron [18]: Tacotron was developed by researchers from Google Brain which also works on the frequency features for voice generation. Their basic architecture trains a neural network to predict spectrogram from encoded text and then sample it to time domain. Since construction of frequency dimension by short-time Fourier transform (STFT) leads to loss of essential information, they introduced a new algorithm called Griffin-Lim reconstruction that can compensate for the loss of phase information during construction of spectrogram. Griffin-Lim algorithm basically tries to find the best suited waveform from the generated spectrogram.

G. Tacotron 2 [19]: Tacotron 2 was a combinational implementation of Tacotron and WaveNet where the WaveNet model is conditioned on the spectrogram produced by the Tacotron to generate waveforms. This new model was reported to have enhanced the capabilities of Tacotron and WaveNet by exploiting their individual capabilities and reduced the reconstruction loss from spectrogram to waveform.

In contrary to the models and architectures, datasets are equally the most valuable component when we plan to develop models using machine learning. The performance of our models solely depends on the datasets used to train the model. There are a number of audiovisual databases for emotion analysis. Some of them are discussed below. Apart from the below mentioned database, HUMAINE database [43] is also an emotional database worth mentioning which contains speeches that are considered to be "pervasive emotion." This is a multimodal database whose speeches are actually taken from nine different emotional databases of different characteristics and different languages.

A. SAVEE [20]: This audiovisual database was developed by the University of Surrey. This is a British English database prepared by four actors aged between 27 and 31 years expressing six different emotions, viz., anger, disgust, fear, happiness, sadness, and surprise. It consists of audio and video recordings of the actors for multimodal systems.

B. RAVDESS [21]: Ryerson Audio-Visual Database for Emotional Speech and Song consists of 7356 files in audio only, video only, and both audio and video. A total of 12 male and 12 female speakers recorded their utterances in 60 trials per speaker for speech and 44 trials per speaker for song in calm, happy, sad, angry, fearful, surprise, and disgust expressions for speech and song contains calm, happy, sad, angry, and fearful emotions.

C. IITKGP-SEHSC [22]: Indian Institute of Technology Kharagpur Simulated Emotion Hindi Speech Corpus is a simulated emotion Hindi speech corpus that was recorded using professional artists from FM radio station in eight emotions, viz., anger, disgust, fear, happy, neutral, sad, sarcastic and surprise.

D. IITKGP-SESC [23]: Indian Institute of Technology Kharagpur Simulated Emotion Speech Corpus is an emotional speech corpus in Telugu similar to IITKGP-SEHSC.

E. MANHOB: MANHOB is a collection of the following three databases developed at the Imperial College London:

- Laughter Database [24]: Here, 12 male and 10 female subjects are recorded in four sessions. In the first session, they were shown funny clips and the reactions were recorded. In the second session, they were asked to give a smile and in the third session, they were asked to give an acted laughter. Finally, they were asked to speak their native language and then English. There are 180 sessions available with a total duration of 3h 49m with 563 instances of laughter, 849 speech utterances, 51 instances of acted laughter, 50 instances of posed smiles, and 167 other vocalizations. It contains audiovisual and thermal recordings.
- MHI-Mimicry Database [25]: Recordings were made of two experiments, viz., a discussion on a political topic and a role-playing game. There are 54 recordings, out of which 34 are of the discussions and 20 are of the role-playing game. Among the participants, 28 were male and 12 female, aged between18 and 40 years.
- HCI Tagging Database [26]: This mainly aims at multimedia content tagging. Thirty subjects were shown small clips. In the first part, a participant was asked

to annotate their own emotive state after each clip on a scale of valence and arousal. In the second part, clips were shown together with a tag at the bottom of the screen, and the subjects are asked if they agree with the tag via a red and a green button.

F. RECOLA [27]—RECOLA is a multimodal database developed at the University of Fribourg containing 9.5 hours of real-time audio, video, electrocardiogram, and electrodermal activity recordings of 46 French-speaking participants who were trying to collaborate to solve a problem.

G. BELFAST [28]—BELFAST was developed at the Queen's University, Belfast, where 650 video clips of naturalistic responses to a series of laboratory-based tasks were recorded from the 82 participants.

H. IEMOCAP [29]—Interactive Emotional Dyadic Motion Capture is a multi-modal audiovisual database collected at the SAIL lab of University of Southern California with motion capture face information, speech, videos, head movement and head angle information, dialog transcriptions, word level, syllable level, and phoneme level alignment as the available modalities. The emotions include anger, happiness, excitement, sadness, frustration, fear, surprise, other, and neutral state spoken by five male and five female actors in English and contain about 12 hours of data.

I. VCTK [30]—Voice Cloning Toolkit database was developed by the University of Edinburgh with the aim of voice cloning. Here, 109 speakers recorded about 400 English sentences in their (different) native accents. The sentences were taken from selected newspapers, the Rainbow Passage, and an elicitation paragraph intended to identify the speaker's accent.

J. Berlin Database of Emotional Speech [31]: This is an emotional speech database in German developed by the University of Berlin, Germany. Ten subjects, five males and five females, are recorded in seven emotions, viz., happy, angry, anxious, fearful, bored, disgusted, and neutral, on ten different texts. In total, it contains more than 5500 utterances.

K. TESS [44]—Toronto Emotional Speech Set was developed by the University of Toronto in which two female speakers (age 26 and 64 years) utter 200 targeted words within a carrier sentence "Say the word ____" in seven different emotions, viz., anger, disgust, fear, happiness, pleasant surprise, sadness, and neutral.

3 Dataset

The first challenge that we faced during this study is the scarcity of open speech dataset in Bengali. We find some speech data in Bengali which are multi-speaker or raw speech without transcription. Such type of data would not have been a good fit for our purpose. Also, we did not find any speech dataset in Bengali that are emotion specific and can be used for emotion analysis. So, we decided to develop our own dataset from scratch according to our needs.

During this study, we have developed a novel emotional database in Bengali consisting of 120 sentences in six different emotions. Each of the emotions contains both emotion-specific sentences and neutral sentences to include variety. Emotion-specific sentences contain emotion-specific words relative to the emotion to reflect the emotion in the sentence and also spoken in that emotion, while the neutral sentences do not have any explicit emotion-specific words but spoken in the said emotion. During the recording, the speaker was suggested to read the given sentences for expressing the particular given emotion. To enhance the emotion trigger, the speaker was asked to imagine a situation relative to the sentence and the emotion and speak the sentence in his natural response.

In doing this, we have assumed that the recorded sentences resemble its natural emotional counterpart. We also believe that the associated emotions are in the right quantity, i.e., the utterances have neither been exaggerated nor been understated with respect to the emotion association. Also, it was assumed that the right emotion is associated with the utterances as was expected from it.

3.1 Data Recording

For the training purpose, we developed a novel database in Bengali language. The database consisted of 120 recordings of short sentences. Twenty sentences were prepared for each of the six emotions out of which 15 are emotion-specific and five are neutral sentences common to all emotions. All the 15 emotion-specific sentences contain emotion-specific words relative to the emotion to trigger the respective emotion in the utterances, while the five neutral sentences do not have any explicit emotion-specific words.

The speaker was a native Bengali male of 24 years of age. The dataset was recorded in six parts for Ekman's six basic emotions [1], viz., happy, sad, fear, disgust, surprise, and anger. Each emotion set is recorded in one go, without any long break, to ensure regularity within the dataset. There was a 2-day gap between the recording sessions of two consecutive emotion-specific sets to ensure diversity and independence between each other. The recordings were made under identical recording circumstances. The voice is recorded at 22 kHz in a silent environment to facilitate noise-free recording. All the recordings were made on a mobile set with Easy Voice Recorder app in Android OS in the same studio-like room. Situation given to speaker prior. While recording the sentences, the speaker was given imaginary situations relative to the given sentence to amplify his emotional responses and to give the utterance a more naturalistic emotional conscience. Table 1. shows examples of sentences belonging to each type of emotion-specific including one of the five neutral sentences common to all the emotions.

Table 1 Examples of emotion-specific sentences used

Emotion	Sentence	English translation	Transliteration
Happy	আজ আমার জন্মদিন	Today is my birthday	Aaj aamar jonmodin
Sad	আজ আমার ব্যাগটা হারিয়ে গেছে	I lost my bag today	Aaj aamar bag ta hariye geche
Fear	আমার হয়তো আর চাকরিটা হবে না	I might not get the job	Aamar hoeto r chakri ta hobe na
Surprise	তুমি বিয়ে করছো	You are getting married	Tumi biye korcho
Disgust	এখানে নংরা ফেলনা	Do not throw garbage here	Ekhane nongra felo na
Angry	তোমার কি মাথা খারাপ	Have you gone mad	Tomar ki matha kharap
Neutral (common)	আমার নাম অমিত	My name is Amit	Amar naam amit

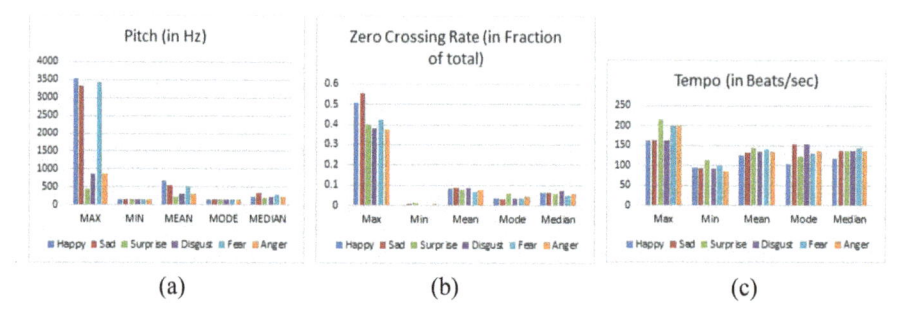

Fig. 1 Quantitative analysis of the recordings. (**a**) Pitch analysis. (**b**) Zero-crossing rate. (**c**) Tempo analysis

3.2 Data Analysis

After the recording, we conducted a visualization and quantitative analysis on the speech data for each emotion class, separately. Some features that we studied are pitch, zero-crossing rate, and tempo. The graphical representation of the findings is given in Fig. 1a–c for better visualization and comparison of the characteristics among different classes.

We also carried out an emotion-wise analysis of the spectrogram and waveform for the six emotions to see the difference in the graph for different emotions. The output spectrogram and waveform of the emotion-specific sentences are given in Figs. 2, 3, 4, 5, 6 and 7.

Lastly, we used the waveform and the spectrogram for the common neutral sentences that are spoken in all the six emotions. This gave us a contrasting comparison among the emotions in terms of waveform and spectrogram. The

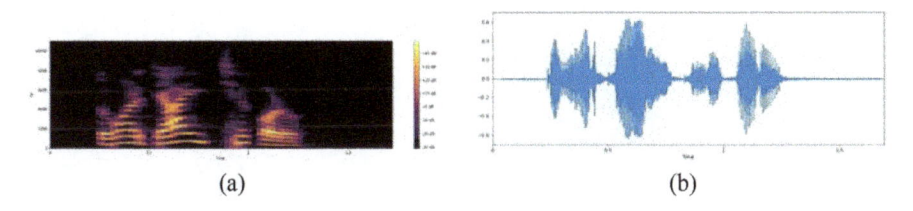

Fig. 2 (**a**) Spectrogram and (**b**) waveform for the sentence "তোমার যা ইচ্ছে কর" in *anger* emotion

Fig. 3 (**a**) Spectrogram and (**b**) waveform for the sentence "আমি তোমার এসব কথা শুনতে রাজি নই" in *disgust* emotion

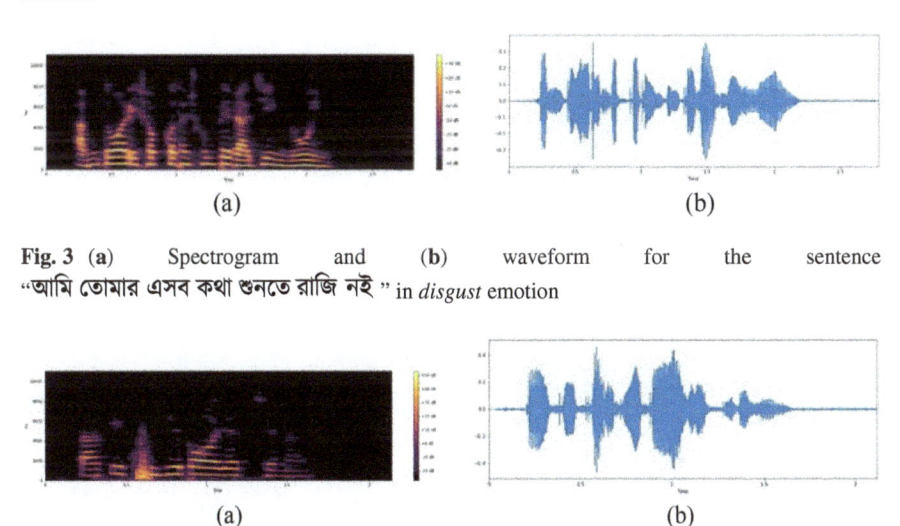

Fig. 4 (**a**) Spectrogram and (**b**) waveform for the sentence "তাকে খুঁজে পাওয়া যাচ্ছে না" in *fear* emotion

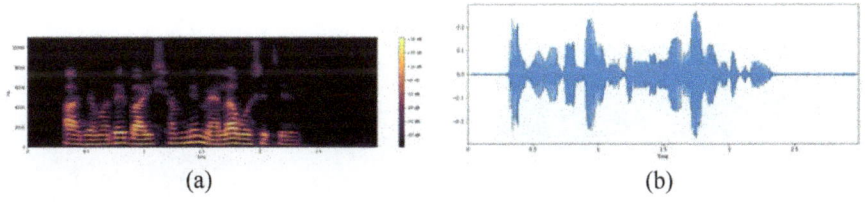

Fig. 5 (**a**) Spectrogram and (**b**) waveform for the sentence "আজ আমাদের বাড়ির সামনে মেলা বসেছে" in *happy* emotion

spectrograms and waveforms of one such neutral sentence are given in Figs. 8, 9, 10, 11, 12 and 13 for all the six emotions.

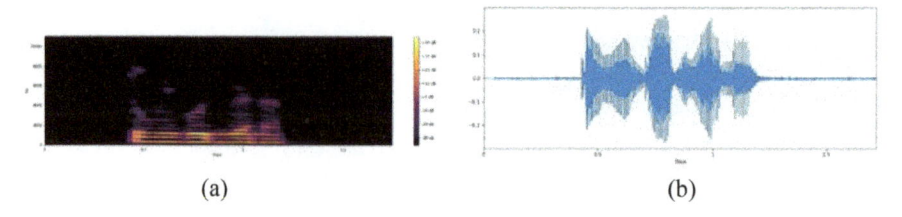

Fig. 6 (**a**) Spectrogram and (**b**) waveform for the sentence "তুমি বুজবে না" in *sad* emotion

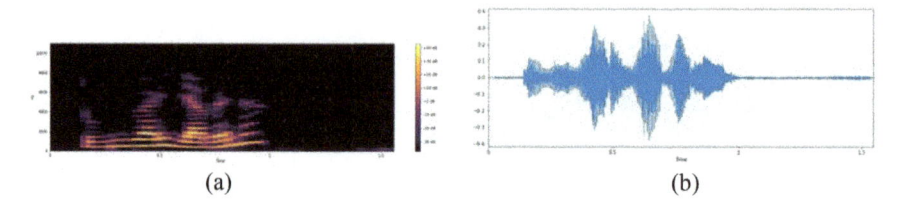

Fig. 7 (**a**) Spectrogram and (**b**) waveform for the sentence "তুমি বাড়ি যাবে না" in *surprise* emotion

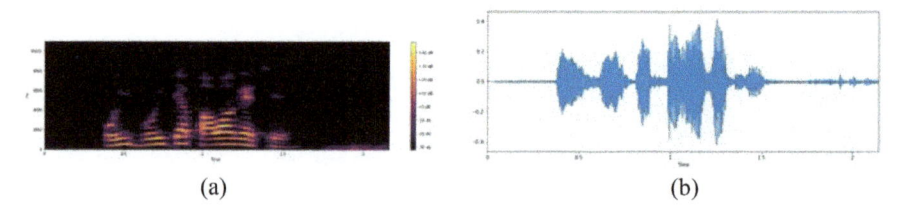

Fig. 8 (**a**) Spectrogram and (**b**) waveform for the emphcommon neutral sentence "ঐ বইটা পাওয়া গেছে" in *anger* emotion

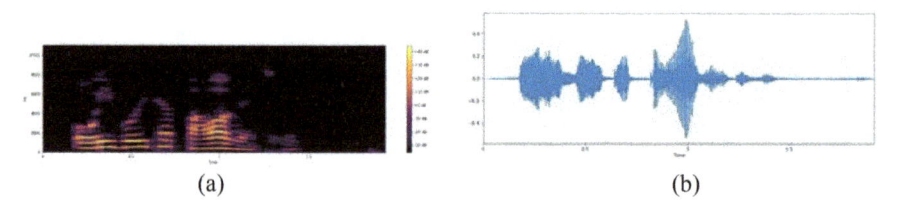

Fig. 9 (**a**) Spectrogram and (**b**) waveform for the emphcommon neutral sentence "ঐ বইটা পাওয়া গেছে" in *disgust* emotion

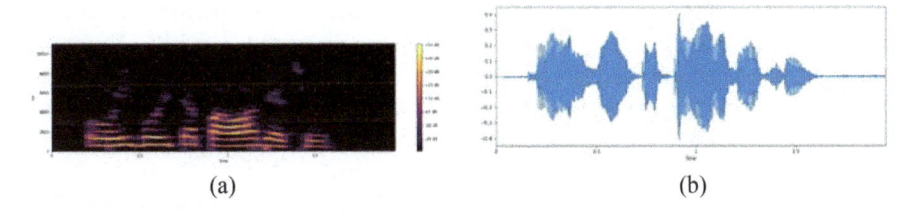

Fig. 10 (**a**) Spectrogram and (**b**) waveform for the emphcommon neutral sentence "ঐ বইটা পাওয়া গেছে" in *fear* emotion

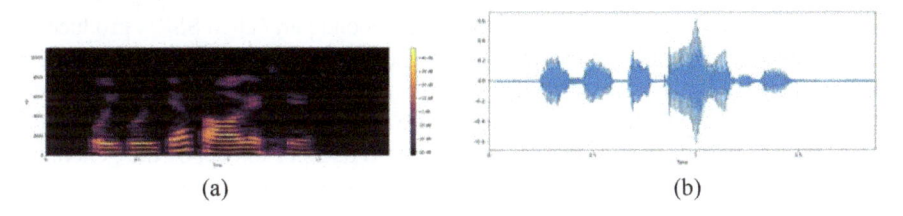

Fig. 11 (**a**) Spectrogram and (**b**) waveform for the emphcommon neutral sentence "ঐ বইটা পাওয়া গেছে" in *happy* emotion

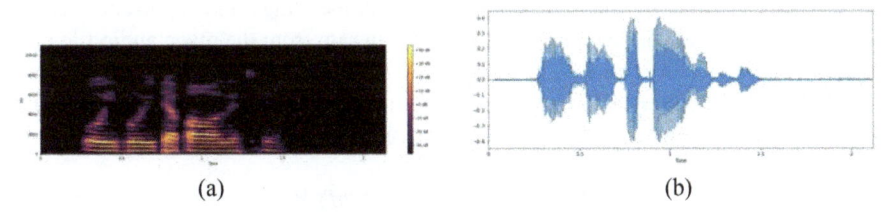

Fig. 12 (**a**) Spectrogram and (**b**) waveform for the emphcommon neutral sentence "ঐ বইটা পাওয়া গেছে" in *sad* emotion

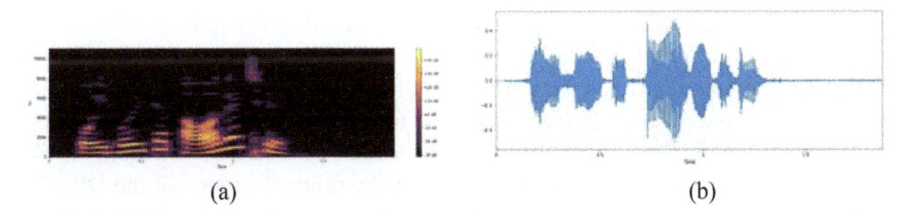

Fig. 13 (**a**) Spectrogram and (**b**) waveform for the emphcommon neutral sentence "ঐ বইটা পাওয়া গেছে" in *surprise* emotion

4 TTS Architecture

4.1 System Framework

We tried to implement a TTS end-to-end model on Bengali language and study its performance on the same. We have chosen convolutional neural networks because they are faster, and recent study has shown that they can outperform traditional sequence learners like RNN or LSTM when implemented carefully. Even the state-of-the-art models like Tacotron or WaveNet also use CNN as their basic architecture. Another important reason is that CNN models are faster to train than sequence models due to its parallelization capability when implemented properly with supporting hardware like GPUs.

For all our experiments, we have considered DCTTS [32] as the baseline. DCTTS is an established state-of-the-art model for English TTS. The proposed network consists mainly of convolutional neural networks due to its speedy trainability. The network is divided into two parts. One part is called Text2Mel which predicts the basic spectrogram from the text encoding. The second part called SSRN produces a high-resolution spectrogram from the basic spectrogram using deconvolution neural nets and upscaling. The Text2Mel module has four subparts, viz., a text encoder to encode the text, an audio encoder to encode the seed spectrogram, a guided attention [33] layer to find the dependencies, and an audio decoder to decode the generated coarse spectrogram. An illustration of the architecture is given in Fig. 14.

We vectorize the text using dictionary positions of the Bengali letters. Then we feed it to a text encoding network which is a stack of CNN with highway connections between layers.

The outputs are two vectors which are text embeddings and represented as K and V. Another network encodes the mel spectrogram from the given audio file into corresponding vectors and it is represented by Q. This K, V, and Q tuples are then fed into an attention layer to find the proper alignment of the encoded text with the encoded spectrogram. Attention is calculated as

$$R = softmax(K^T Q/\sqrt{d}).V \qquad (1)$$

The dot product of K and Q represents the mutual attention of the source and the target vectors. When it is multiplied with V, give the corresponding output as an encoded spectrogram. It is then fed into an audio decoder module which then decodes the tensor into coarse spectrogram. This spectrogram is then fed into SSRN network which then upscales the spectrogram by using deconvolutional neural nets and outputs a higher-resolution spectrogram. This then converted into corresponding waveform using Griffin-Lim algorithm to get the final wave form in .wav format. We have changed the English dictionary as used in the original version of the DCTTS with a Bengali dictionary containing all the characters and numbers from Bengali alphabet set only. Text2Mel and SSRN networks are to be trained separately to get better results.

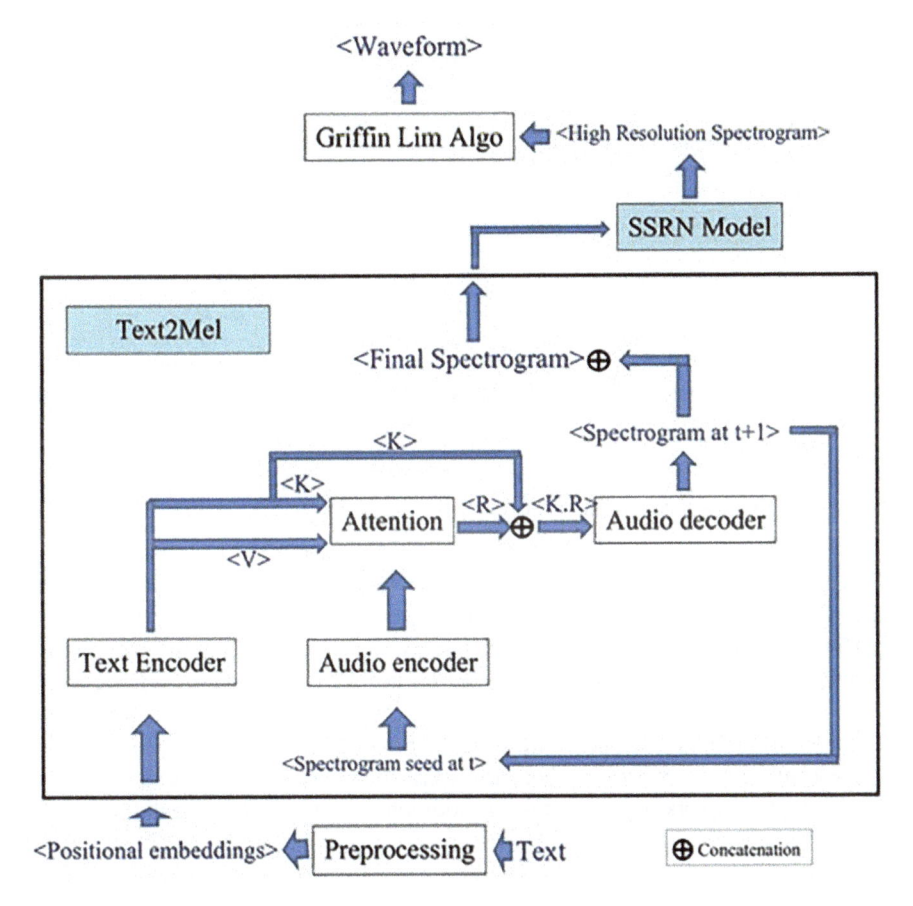

Analyzing Emotional Speech and Text: A Special Focus on Bengali Language   295

Fig. 14 System architecture

4.2 Basic Layout Features

1. Only convolutional neural networks are used to facilitate faster computation.
2. Training is done on two separate networks, one for generating the coarse spectrograms through attention layer, and the other is SSRN network which is trained to upscale the coarse spectrogram.
3. For training the first network, the input was the encoded Bengali text and the target was downscaled spectrogram from the pre-recorded text.
4. For training the SSRN, the input was downscaled spectrogram and the target was full-scaled spectrogram from the pre-recorded text.
5. Guided self-attention is used to locate proper dependencies.
6. Griffin-Lim [34] algorithm is used to convert the spectrograms into waveforms instead of inverse SRFT to reduce approximation error.

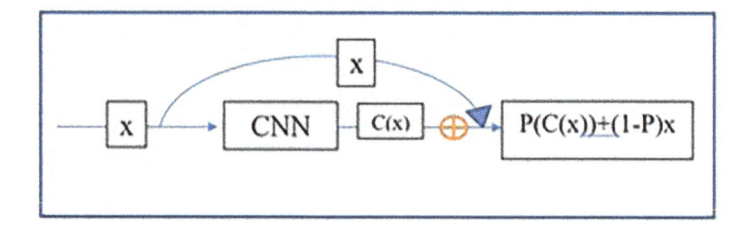

Fig. 15 Highway connection

7. Exclusively tested on Bengali dataset.
8. A new indigenous native Bengali dataset is recorded to facilitate proper training.

4.3 Highway Network [35]

When training deep neural networks, it was observed that increasing the layers might not always give better results. To ensure that the deep network does not perform worse than its shallow counterpart, it was suggested to add bypass connections so that the flow have the option to take the best path and utilize the layers efficiently to give optimal result. If $y = L(x)$ where y is the output of the layer L whose input is x, then y must be the best of x or $L(x)$ whichever gives the better result. If $y = x$, then it chose to bypass the intermediate L. As described in the Highway network paper, we take an additional transform $T(x)$ such that

$$y = T(x) * L(x) + (1 - T(x)) * x \qquad (2)$$

where * means element-wise multiplication operation. This transform T can be thought as the part of the original input that passes through the layer L and implemented with sigmoid function. A high-level illustration of highway connection with CNN layer is shown in Fig. 15.

4.4 Attention

Attention is a very well-known mechanism specifically for encoder-decoder architecture. It was first proposed in the field of machine translation. It works on the fact that while translation, all the output sequence does not depend equally on all the input sequence. Rather, each of the output sequences gives a specific weightage or "attention" to each of the inputs. It is more likely that the present output depends more on its immediate neighbors than the distant neighbors. Attention mechanism is basically based on probability of occurrence of nth term by Markov's assumption.

Mathematically,

$$P(w_1 w_2 n_3 \ldots w_n) \approx \prod_i P(w_i | w_{(i-k)} \ldots w_{(i-1)}) \qquad (3)$$

Here, the input text is embedded into two vectors V and K (values and keys). The spectrogram produced till that point of time is encoded in a vector Q (queries). By multiplying K and Q, we find the mutual dependencies between the text encoding and the spectrogram seed encodings. We then divide it by d where d is the size of that dimension and uses a softmax function to convert them into probabilistic form. The result is then multiplied with the vector V to get the final embedded vector for the spectrogram to be generated. Mathematically,

$$A = softmax(K^T.Q/d) \qquad (4)$$

$$R = A.V \qquad (5)$$

This R is then concatenated with Q and given to audio-decoder as input.

4.5 Text2Mel

Text2Mel basically converts the raw text encoding inputs into coarse low-resolution spectrogram. It contains the following three parts:

(a) Text Encoder: It takes the character embeddings as input and produces two vectors. It contains 14 layers of convolutional neural networks with highway connections to produce a vector which is then split into two equal halves and returned as K and V.
(b) Audio Encoder: It encodes the seed spectrogram into a single vector Q. It consists of 13 layers of convolutional networks with highway connections.
(c) Attention layer: It takes input from text encoder and audio encoder. It produces a single vector which is then concatenated with Q and given to the audio decoder module.
(d) Audio Decoder: It takes a vector as input and produces a coarse spectrogram of low resolution. It uses 11 convolutional layers with highway connection. The expansion of the frequency domain is done by increasing the number of channels in the CNN layer. For training the Text2Mel model, we compute the L1 loss as the difference of the generated spectrogram and the temporally shifted ground truth spectrogram and back-propagated.

4.6 SSRN

SSRN, known as the Spectrogram Super Resolution Network, increases the resolution of the produced coarse spectrogram. It is trained as a different network independent of Text2Mel as proposed in the original paper as it decreases the burden of training. It uses increased channel CNN layers to upscale the frequency domain and deconvolutional neural networks of stride 2 to upscale the temporal domain from T to 4*T. SSRN is a serial combination of convolution and deconvolution neural networks with highway connections.

4.7 Griffin-Lim Algorithm

Spectrograms are constructed from waveforms by using short-term Fourier transform (STFT) to find out the constituent frequencies at the given point of time. Similarly, it is possible to generate waveforms from spectrograms by using inverse STFT. But this has received its own disadvantages. Since this is an approximation formula, the quality gets deteriorated. Thus, instead of using inverse STFT, the proposed model uses Griffin-Lim algorithm to produce the waveform from the generated spectrogram. Griffin-Lim algorithm is a phase reconstruction method used for recreating wave phases from a given spectrogram. It has been used in Tacotron and has proved to be very efficient. This algorithm is expected to recover a consistent, complex valued spectrogram maintaining a given amplitude A. It is calculated with the formula

$$X[m+1] = Pc(PA(X[m])) \qquad (6)$$

$$Pc(X) = GGTX \qquad (7)$$

$$PA(X) = A \bullet X \odot |X| \qquad (8)$$

where X is a complex valued spectrogram updated through iterations, G is the STFT and GT is the pseudo-inverse STFT, A is the amplitude, $\bullet$ is the element-wise multiplication operation, and $\odot$ is the element-wise division operation.

5 Language Independence and English-Bengali TTS Model

5.1 Role of Language in Speech

In earlier days of evolution, people started living in groups called tribes, and each such tribe developed their own language for communication among the tribe.

Language is a cultural as well as a social aspect of human groups. Language gives a sense of oneness within the group and conveys a feeling of attachment. Some widely spoken languages include English, Chinese, German, French, Italian, etc. But there are also languages that might not be as familiar to populations in other continents but are equally important in its place of origin. Some examples of such languages are Hindi, Marathi, Bengali, Tamil, etc. which are some of the well-established Indian languages. To communicate with humans, the machine should be capable of generating the information in a language with which the end user is familiar. But a system is generally devised to be global and cater all types of people. For this, the multilingual capability is a necessary feature for a TTS system. This means that the system must be capable of reproducing the same information in different languages and convey the information in a preferred language of the user.

5.2 Bengali TTS Model

Bengali is subcontinental Asian language mainly native to the regions around Bay of Bengal. Bengali is the national language of Bangladesh. It is also widely spoken in the eastern region of India including the states of West Bengal, Assam, Tripura, Andaman and Nicobar Islands, etc. Due to globalization and international migration, Bengali community also have a significant presence in the Middle East and western part of the world including the USA, Canada, the UK, UAE, etc. Bengali is the fifth most spoken native language and the sixth most spoken language by total number of speakers in the world with more than 250 million Bengali speakers around the globe [42]. According to an UNESCO survey, Bengali was voted the sweetest language in the world.

In this section, we first tried to train the proposed TTS model for Bengali language and study the performance of the same. We have trained the model on the newly developed dataset and reported the output. We did not make much changes in the core architecture except for the preprocessing and postprocessing part.

5.2.1 Preprocessing

As a part of preprocessing, we process the sentence to keep only the Bengali characters and discard all other characters including punctuations and special characters. Initially when the model was trained, we found out it was unable to produce proper speech for big numerical figures in hundreds and thousands as we would read them. So, we mapped all the numerical figures to their word equivalent to compensate for this loss. Next, we made a character-level dictionary and transformed the character sequence into positional encodings according to this dictionary to facilitate computation. We have followed the same preprocessing for all the subsequent experiments.

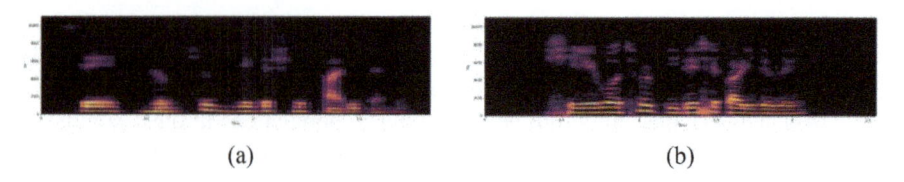

Fig. 16 Spectrogram comparison for the sentence 'সে এবছর বিদেশে পারি দেবে'. (**a**) TTS output. (**b**) Ground truth

5.2.2 Postprocessing

After the speech signals were generated, we tried to increase the quality a bit more. As we know the signals are nothing but amplitude values, we interpolate the value between two consecutive amplitudes by taking their simple average and inserting it in between them. This resulted in a 2× sampling frequency for the same duration of time. We have followed the same postprocessing for all the subsequent experiments.

5.2.3 Experiments and Results

We trained the DCTTS model on our newly developed Bengali dataset for 100 epochs. The two networks, Text2Mel and SSRN, were trained separately as directed in the main paper of DCTTS. We used a personal laptop with AMD A6 processor with clock speed of 2.5 GHz, 1 GB of integrated Graphics Card, 8 GB RAM, 1 TB HDD, and Windows 10 OS. We have used this system for executing all of our experiments including the subsequent experiments. To get quicker results and parallel training, the model was also partially trained on Google Colab [36] and Kaggle [37] to get GPU support. The trained model was able to produce satisfactory results. The spectrogram comparison and waveform comparison of a produced sound and corresponding ground truth are shown in Figs. 16 and 17. It took 334.22 seconds for the model to produce the sound in our local machine.

During initial training stage, the output indicated that the model was not learning the digits or numbers very well. Hence, we had to preprocess the inputs to map all the numbers to their respective word representations and confine the training to Bengali character set only excluding Bengali numbers.

We have experimented on generating sentences with the trained model of variable length, and we subjectively found out that the model works more efficiently on medium-length sentences (>15 characters) than very short sentences. This was probably because the attention layer, which is the heart of the system, was very much dependent on the context which is better available for considerably longer sentences. It was also observed that the quality of the generated speech increases with more training.

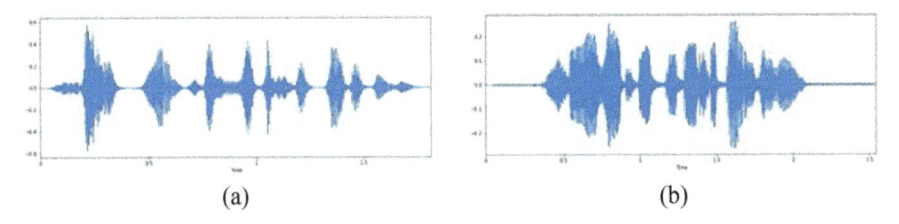

Fig. 17 Waveform comparison for the sentence 'সে এবছর বিদেশে পারি দেবে'. (**a**) TTS output. (**b**) Ground truth

5.3 English-Bengali TTS Model

5.3.1 Language Independence in TTS

In human-computer interaction, language plays a very important role in determining its usability and userbase. The usability will be more if the machine is capable of interacting in multiple languages, hence reaching a larger audience. In this study, we tried to find out if a TTS model can be made multilingual without using multiple language models or language classifiers. If this feature can be incorporated, it will not only make the model multilingual, but also make it lightweight and faster as it will not need any extra classification or separate processing for different language.

5.3.2 Problems in Multilingual Learning and Ambiguity in Pronunciation

Our main motivation for this implementation came about that in TTS; we do not need to find the internal meaning of the text as we do in summarization, translation, or QA systems. Rather, the system just needs to map the letter to its corresponding frequencies to produce the final amplitudes. If each letter produces a specific set of frequencies, the machine can be made to learn the mapping function, and it will be a one-to-one or many-to-one mapping. Problems arise when a single character has different pronunciations and hence different frequencies at different point of time. This needs one-to-many mapping depending on the usage. Fortunately, this case does not arise when working with a one language set as each character in a language dictionary has a specific pronunciation. The problem arises when dealing with multiple languages having common dictionary characters but they differ in pronunciation. For instance, both English and German alphabet set have the character "j" but in German, "j" is pronounced as English "y." This might create an ambiguity during frequency mapping.

During the incorporation of multilingual aspects, we assumed that during transformation from text-to-speech, the phonemes are mapped to their respective amplitudes to produce sound, and if the model can learn this transformation function, it will be able to generate the waveform. When adding more than one language in the training, we assumed that phoneme sequence for each distinct letter is different and there is no ambiguity in the transformation. In simple words, each character will have the same pronunciation every time it is used.

5.3.3 Model Architecture

For experimental purpose, we intend to develop a bilingual system in Bengali and English. For the bilingual TTS, we use the same DCTTS as the baseline system. The only change that we added was a new dictionary to accommodate the English characters alongside the Bengali characters. So, the new dictionary contains English alphabets and Bengali alphabets. Since there are no common characters in English and Bengali alphabet set, there would be no ambiguity in frequency mapping, and hence, it should produce expected results. As a preprocessing part, we have replaced all the English and Bengali numbers to their equivalent wordings to keep it only to the characters as we have seen in the earlier experiment that it produces better results. Then we kept the letters from the dictionary and remove the rest of the outlying characters.

5.3.4 Experiments and Results

From the initial experience with the Bengali model, we preprocessed the inputs to map all the numbers to their respective word representations and confine the training to Bengali and English character set only excluding numbers.

An aggregated dataset was developed by combining our newly developed Bengali dataset and LJ Speech dataset [38]. The DCTTS model was trained on this aggregated dataset for 100 epochs. The two networks, Text2Mel and SSRN, were trained separately as directed in the main paper of DCTTS. The spectrogram comparison and waveform comparison of a produced Bengali speech and corresponding ground truth are shown in Figs. 18 and 19. It took 311.24 seconds for the model to produce the sound in our local machine. The spectrogram comparison and waveform comparison of a produced English speech and corresponding ground truth are shown in Figs. 20 and 21. It took 392.45 seconds for the model to produce the sound in our local machine.

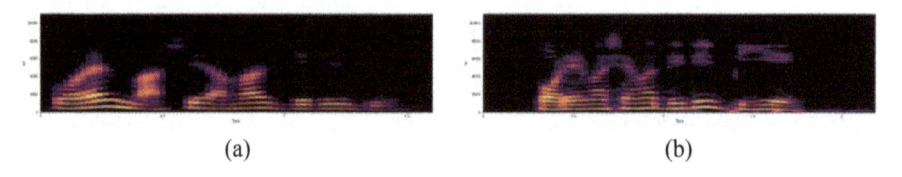

Fig. 18 Spectrogram comparison for the sentence 'পরের মাসে আমার দিদির বিয়ে '. (**a**) TTS output. (**b**) Ground truth

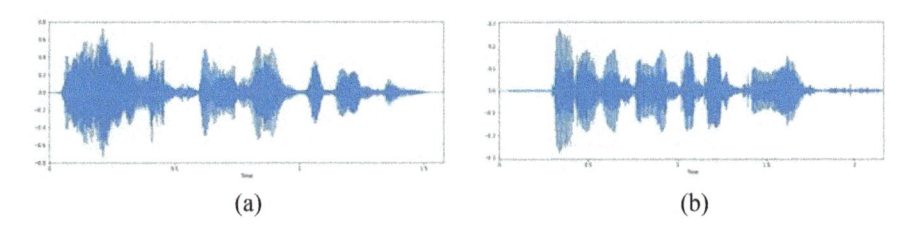

Fig. 19 Waveform comparison for the sentence "পরের মাসে আমার দিদির বিয়ে". (**a**) TTS output. (**b**) Ground truth

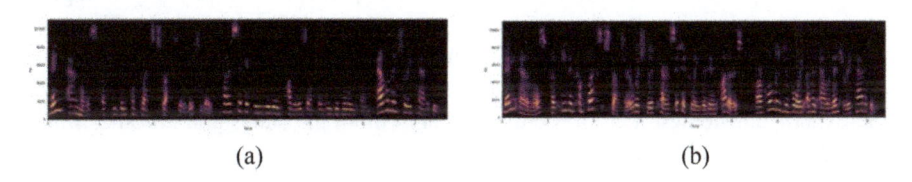

Fig. 20 Spectrogram comparison for the sentence *"Many animals of even complex structure which live parasitically within others are wholly devoid of an alimentary cavity"*. (**a**) TTS output. (**b**) Ground truth

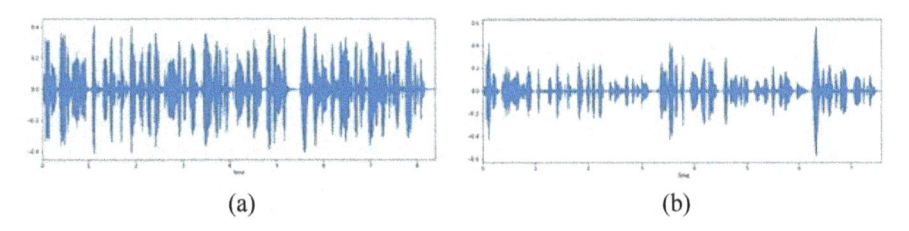

Fig. 21 Waveform comparison for the sentence *"Many animals of even complex structure which live parasitically within others are wholly devoid of an alimentary cavity"*. (**a**) TTS output. (**b**) Ground truth

6 Emotion Incorporation in TTS

6.1 Role of Emotions in Speech

Emotion is an important part of the human communication. It not only gives a special sense to a speech but also adds an extra dimension for conveying information. The intention or the meaning partially depends on the emotions of the context. The same sentence can be spoken in two different emotions giving in two different meanings or intentions. For instance, "My sister is getting married" may have different emotion. If spoken in a happy state, this signifies that the person is excited about the event. But when said in a sad way, it shows the worry of the person which might be regarding separation or anything else related to the event. Though machines have become intelligent enough to produce humanlike speech, it has not been possible yet to incorporate proper emotions aligned with the given context. The addition of emotion to speech synthesis will take the machine intelligence one step closer toward achieving natural humanlike capabilities.

6.2 Types of Emotions

Emotions are state of minds that a person experience and are more subjective that varies from person to person depending on their perception. Many psychologists proposed various classifications of emotions based on different criteria. The most acceptable among them was given by Paul Ekman. Ekman [1] classified emotions in six different classes, viz., happy, sad, anger, disgust, surprise, and fear. Most of the experts considered them to be the primary emotional states and all other states are combination of these states.

6.3 Experiments and Results

In this work, will have tried to find whether the TTS model is capable of capturing and reproducing human emotions through the produced speech. The language was chosen to be English and not Bengali because of the lack of emotional dataset available in Bengali language. From the initial experience with the Bengali model, we preprocessed the inputs to map all the numbers to their respective word representations and confine the training to English character set only excluding numbers. An aggregated dataset was developed by combining happy collections of SAVEE and RAVDESS dataset. The DCTTS model was trained on this aggregated dataset for 100 epochs. The spectrogram comparison and waveform comparison of a produced happy speech and corresponding ground truth are shown in Figs. 22 and 23. It took 258 seconds for the model to produce the sound in our local machine.

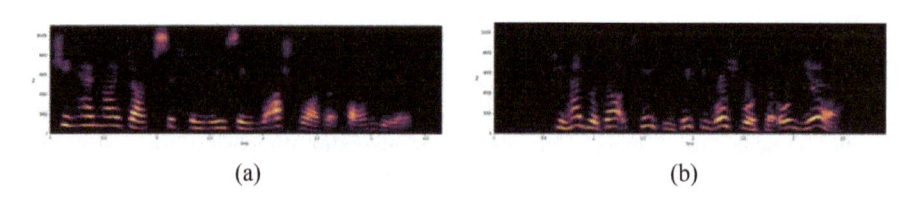

Fig. 22 Spectrogram comparison for the sentence *"She had my dark suit in greasy wash water all year"*. (**a**) TTS output. (**b**) Ground truth

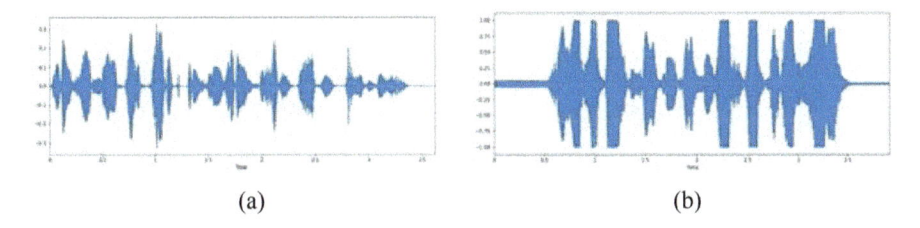

Fig. 23 Waveform comparison for the sentence *"She had my dark suit in greasy wash water all year"*. (**a**) TTS output. (**b**) Ground truth

7 Conclusion and Future Work

Our work mainly focuses on speech synthesis and the possibility of going further than just producing a robotic speech. Not only we tried to explore the speech production capability across different language but also tried to study the incorporation of natural features to it to make the speech more natural and closer to human-produced speech. In this time, we have explored only a part of the larger picture and there is much more to explore in this area.

Though we tried to contribute in terms of a new dataset, this is still not enough for study at a larger scale. As the new-generation researches are data hungry, this dataset can be extended later on into a multi-speaker, multi-gender dataset in Bengali. Also, we noticed that as of now, no open-sourced multimodal database exist in Bengali to study emotion aspects. It will be a great thrust to the future emotion researches if we can contribute in terms of a multimodal database for emotion studies. It will not only be useful for emotion researches in the field of Computer Science and Engineering but also in fields like Psychology and interdisciplinary fields like Linguistics.

In this work, we have done all the experiments with a single model which was already well established. But experimenting with other available models of different categories or coming up with new models might improve the performance in terms of both quality and time. New algorithms like GAN [39], Transformer and Reformer [40], Autoencoders, [41], etc. can be tried in the field of speech generation. Once the best model is found in terms of quality and time, then the experiments of multilingual features and emotion incorporation can be done on the model.

As our experiment of developing bilingual TTS showed satisfactory results with Bengali and English, the next step will be to extend it to more languages. Special care must be taken that the chosen languages do not share any common alphabet in their dictionary or else there might be ambiguity in pronunciation and hence low-quality speech production. For languages that share common alphabet set, multiple models can be trained separately. For that, nonoverlapping disjoint sets of all the target languages must be identified and corresponding models should be trained.

For emotion incorporation, we only experimented with happy emotion. It needs to be done for all the emotional states and results need to be assessed. If we can successfully amplify the emotion associated with the sentence, we can aim to make

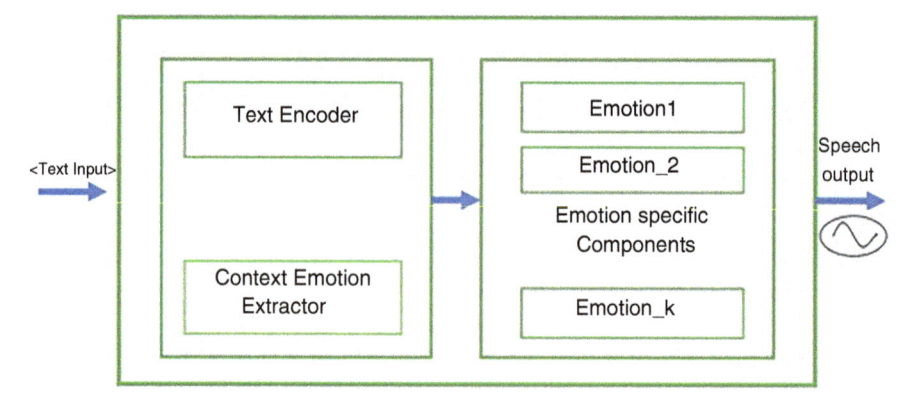

Fig. 24 Hypothetical multi-emotion TTS system architecture

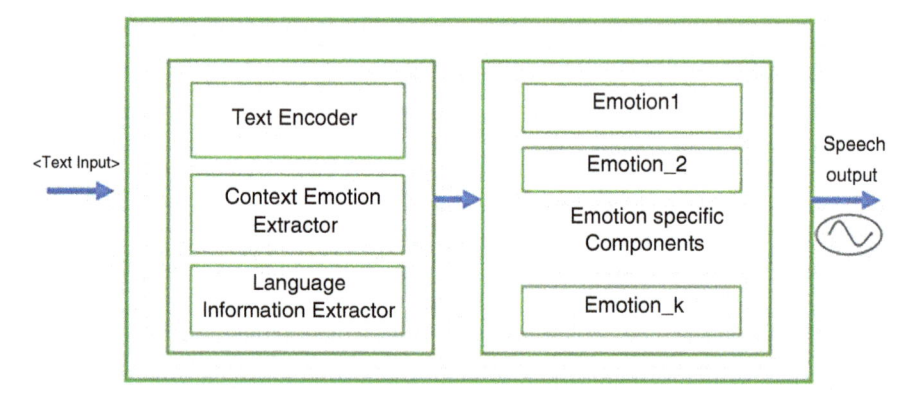

Fig. 25 Hypothetical multi-emotion TTS system architecture

a system that will be capable of identifying and triggering the right emotion for the right sentence depending on the intent. A high-level representation of our proposed idea is shown in Fig. 24.

Once all the above objectives are met, then research needs to be done on how to put all the things together in one box. The ultimate objective will be to build a system that will be multilingual and properly emotional oriented without moving from our main feature, that is, lightweight and time efficient. A high-level representation of our hypothetical concept is shown in Fig. 25.

References

1. P Ekman - Handbook of cognition and emotion, 1999
2. Hunt A. J. and Black A. W. 1996. Unit selection in a concatenative speech synthesis system using a large speech database. In: Proceedings of the Acoustics, Speech, and Signal Processing, 1996. on Conference Proceedings., 1996 IEEE International Conference, Vol. 01

(ICASSP '96). IEEE Computer Society, USA, pp. 373–376. https://doi.org/10.1109/ICASSP.1996.541110
3. Black A. W., et al., "Statistical Parametric Speech Synthesis," 2007. In: IEEE International Conference on Acoustics, Speech and Signal Processing - ICASSP '07, Honolulu, HI, 2007, pp. IV-1229-IV-1232, https://doi.org/10.1109/ICASSP.2007.367298
4. Vuppala, Anil & Kadiri, Sudarsana. (2014). Neutral to Anger Speech Conversion Using Non-Uniform Duration Modification. 9th International Conference on Industrial and Information Systems, ICIIS 2014. https://doi.org/10.1109/ICIINFS.2014.7036614
5. Jin, Zeyu et al. (2017). VoCo: text-based insertion and replacement in audio narration. ACM Transactions on Graphics. Vol. 36. pp. 1-13. https://doi.org/10.1145/3072959.3073702
6. Taigman, Yaniv et al. "VoiceLoop: Voice Fitting and Synthesis via a Phonological Loop."
7. Oord, A.V., et al. (2016). WaveNet: A Generative Model for Raw Audio.
8. Aäron Van Den Oord, et al. 2016. Pixel recurrent neural networks. In: Proceedings of the 33rd International Conference on International Conference on Machine Learning, Vol. 48 (ICML'16). JMLR.org, p. 1747–1756.
9. Aäron van den Oord, et al. 2016. Conditional image generation with PixelCNN decoders. In: Proceedings of the 30th International Conference on Neural Information Processing Systems (NIPS'16). Curran Associates Inc., Red Hook, NY, USA, pp. 4797–4805.
10. Paine, Tom et al. (2016). Fast Wavenet Generation Algorithm.
11. Oord, Aaron et al. (2017). Parallel WaveNet: Fast High-Fidelity Speech Synthesis.
12. R. Prenger, et al., "Waveglow: A Flow-based Generative Network for Speech Synthesis," ICASSP 2019 – 2019. In: IEEE International Conference on Acoustics, Speech and Signal Processing (ICASSP), Brighton, United Kingdom, 2019, pp. 3617-3621. https://doi.org/10.1109/ICASSP.2019.8683143
13. Diederik P Kingma and Prafulla Dhariwal, "Glow: Generative flow with invertible 1x1 convolutions", 2018.
14. Arik, S., et al. (2017). Deep Voice: Real-time Neural Text-to-Speech.
15. Gibiansky, A., et al. (2017). Deep Voice 2: Multi-Speaker Neural Text-to-Speech.
16. Ping, W., et al. (2018). Deep Voice 3: Scaling Text-to-Speech with Convolutional Sequence Learning.
17. Vasquez, Sean & Lewis, Mike. (2019). MelNet: A Generative Model for Audio in the Frequency Domain.
18. Wang, Yuxuan et al. "Tacotron: Towards End-to-End Speech Synthesis." INTERSPEECH (2017).
19. Shen, Jonathan et al. (2018). Natural TTS Synthesis by Conditioning Wavenet on MEL Spectrogram Predictions. pp. 4779-4783. https://doi.org/10.1109/ICASSP.2018.8461368
20. Jackson, Philip & ul Haq, Sana. (2011). Surrey Audio-Visual Expressed Emotion (SAVEE) database.
21. Livingstone SR, Russo FA (2018) The Ryerson Audio-Visual Database of Emotional Speech and Song (RAVDESS): A dynamic, multimodal set of facial and vocal expressions in North American English. PLOS ONE. https://doi.org/10.1371/journal.pone.0196391
22. Koolagudi S. G., et al. "IITKGP-SEHSC: Hindi Speech Corpus for Emotion Analysis," 2011 International Conference on Devices and Communications (ICDeCom), Mesra, 2011, pp. 1-5, https://doi.org/10.1109/ICDECOM.2011.5738540
23. Koolagudi S.G., et al. (2009) IITKGP-SESC: Speech Database for Emotion Analysis. In: Ranka S. et al. (eds) Contemporary Computing. IC3 2009. Communications in Computer and Information Science, vol 40. Springer, Berlin, Heidelberg.
24. Stavros Petridis, et al. 2013. The MAHNOB Laughter database. Image Vision Comput. 31, 2 (February, 2013), p. 186–202. https://doi.org/10.1016/j.imavis.2012.08.014
25. Sanjay Bilakhia, et al. 2015. The MAHNOB Mimicry Database. Pattern Recogn. Lett. 66, C (November 2015), p. 52–61. https://doi.org/10.1016/j.patrec.2015.03.005
26. Wiem, Mimoun & Lachiri, Zied. (2017). Emotion Classification in Arousal Valence Model using MAHNOB-HCI Database. International Journal of Advanced Computer Science and Applications. https://doi.org/10.14569/IJACSA.2017.080344

27. Ringeval F., et al. "Introducing the RECOLA multimodal corpus of remote collaborative and affective interactions," 2013. In: 10th IEEE International Conference and Workshops on Automatic Face and Gesture Recognition (FG), Shanghai, 2013, pp. 1-8, https://doi.org/10.1109/FG.2013.6553805
28. Sneddon, Ian et al. "The Belfast Induced Natural Emotion Database." IEEE Transactions on Affective Computing (2012): vol. 3, p. 32-41.
29. Busso, Carlos et al. (2008). IEMOCAP: Interactive emotional dyadic motion capture database. Language Resources and Evaluation, vol. 42, p. 335-359. https://doi.org/10.1007/s10579-008-9076-6
30. Veaux, Christophe; et al. (2017). CSTR VCTK Corpus: English Multi-speaker Corpus for CSTR Voice Cloning Toolkit, [sound]. University of Edinburgh. The Centre for Speech Technology Research (CSTR). https://doi.org/10.7488/ds/1994
31. Burkhardt, Felix etal. (2005). A database of German emotional speech. In: 9th European Conference on Speech Communication and Technology. vol. 5, p. 1517-1520.
32. Tachibana, Hideyuki et. al. "Efficiently Trainable Text-to-Speech System Based on Deep Convolutional Networks with Guided Attention." 2018. In: IEEE International Conference on Acoustics, Speech and Signal Processing (ICASSP) (2018): p. 4784-4788.
33. Vaswani, Ashish et al. (2017). Attention Is All You Need.
34. Griffin, Daniel W. and Jae S. Lim. "Signal estimation from modified short-time Fourier transform." In: IEEE Transactions on Acoustics, Speech, and Signal Processing vol. 32 (1984): p. 236-243.
35. Srivastava, R. et al. "Highway Networks." (2015)
36. Bisong E. (2019) Google Colaboratory. In: Building Machine Learning and Deep Learning Models on Google Cloud Platform. Apress, Berkeley, CA. https://doi.org/10.1007/978-1-4842-4470-8_7
37. Kaggle https://www.kaggle.com/
38. Ito, K. & Johnson, L. (2017). The LJ Speech Dataset https://keithito.com/LJ-Speech-Dataset
39. Goodfellow, Ian J. et al. "Generative Adversarial Nets." NIPS (2014).
40. Kitaev, Nikita & Kaiser, Łukasz & Levskaya, Anselm. (2020). Reformer: The Efficient Transformer.
41. Liou, C.-Y., et al. (2014). Autoencoder for words. Neurocomputing, vol. 139, pp. 84–96.
42. Wikipedia https://en.wikipedia.org/wiki/Bengali_language
43. Douglas-Cowie, Ellen et al. (2011). The HUMAINE database. https://doi.org/10.1007/978-3-642-15184-2_14
44. Dupuis, K., & Pichora-Fuller, M. K. (2010). Toronto emotional speech set (TESS). Toronto: University of Toronto, Psychology Department

Part IV
Case Studies

Duplicate Detection for for Digital Audio Archive Management: Two Case Studies

Joren Six, Federica Bressan, and Koen Renders

1 Introduction

Many music information retrieval (MIR) technologies have untapped potential. With every passing year, the MIR field presents promising technology and research prototypes [9, 16]. Unfortunately, these academic advances do not translate well to practical use. For digital music archive management especially, MIR techniques are underexploited [5]. We see several reasons for this. One is that MIR technologies are simply not well known by archivists. Another reason is that it is often unclear how MIR technology can be applied in a digital music archive setting. MIR researches see applications as 'self-evident', while a translation step is needed to enthuse end users. A third reason is that considerable effort is needed to transform a potentially promising MIR research prototype into a working, documented, maintained solution for archivists.

In this chapter, we focus on duplicate detection. It is an MIR technology that has matured over the last two decades [8, 11, 23, 26, 29]. It is telling that an overview paper [3] of about 15 years ago is still relevant today. Duplicate detection is regarded as a solved problem and the focus of the MIR community shifted from active research to refinement. A broader impact and application of the technology, however, remains marginal outside big tech. Some of the applications of duplicate detection might not be immediately obvious since it is used indirectly

---

J. Six (✉)
IPEM, Ghent University, Ghent, Belgium
e-mail: joren.six@ugent.be

F. Bressan
Stony Brook University, New York, NY, USA

K. Renders
VRT, Brussel, Belgium

© The Author(s), under exclusive license to Springer Nature Switzerland AG 2023
A. Biswas et al. (eds.), *Advances in Speech and Music Technology*, Signals and Communication Technology, https://doi.org/10.1007/978-3-031-18444-4_16

to complement meta-data, link or merge archives and improve listening experiences and synchronization[24], and it has opportunities for segmentation.

The aim of the chapter is threefold. The first aim is to be explicit about the use of duplicate detection for music archive management. The second aim is to present two case studies using duplicate detection: one determines the amount of unique material in an archive, and the other case study is on reuse of meta-data. The third aim of this chapter is to present and evaluate the improved duplicate detection system that was used for the case studies. Let's first define duplicate detection.

1.1 Duplicate Detection

The definition of duplicate detection in music is less straightforward than it seems at first due to many types of reuse. In [17] there is a distinction made between *exact*, *near* and *far* duplicates. For an exact duplicate, the duplicate contains exactly the same audio as the original. Near duplicates are different only due to technical processing, e.g. by using another lossy encoding format, remastering or compression. Different recordings of the same live concert are also in the near duplicate category. Far duplicates span a whole range of reuse of audio material: samples, loops, instrumental versions, mash-ups, edits, translations and so forth.

Going even further, covers of songs reuse musical material of the original. In all cases, a musical concept is shared between the original and the cover, but only in some cases audio is reused. There is a whole range of different types of covers: live performances, acoustic versions, demo versions, medleys and remixes are only a few examples. A more complete overview can be found in [19]. In this work, we focus on duplicates which contain the same audio material as the original. This includes samples or mash-ups but excludes live versions which do not share audio with the original version, although they might sound similar.

The duplicates of interest share audio material; however, duplicates should still be identifiable even when the audio is slightly changed. Evidently, a match should still be found if volume has changed. Other modifications such as compression, filtering, speed changes, pitch-shifting, time-stretching and similar modifications should be allowed as well. We end up at the following definition:

> Duplicate detection allows to compare an audio fragment to other audio to determine if the fragment is either unique or appears multiple times in the complete set. The comparison should be robust against various modifications.

Assuming a duplicate detection system is available, several applications become possible [17, 21]. Such system is used in the following case studies. After the case studies, a more technical part and evaluation follows which shows the limitations and strengths of the system actually used.

2 Case 1: Duplicates in the VRT Shellac Disc Music Archive

2.1 The VRT Shellac Disc Archive

The VRT shellac disc archive is part of the vast music archive of the Belgian public broadcasting institute. The archive contains popular music, jazz and classical music released between 1920 and 1960 when the public broadcaster was called INR/NIR (Institut National de Radiodiffusion, Nationaal Instituut voor de Radio-Omroep). The total shellac disc archive contains about 100,000 discs of which a selection was digitized. Unique material and material with a strong link to Belgium was prioritized. Basic meta-data is available (title, performer, label) but, notably, a release date is often missing.

The digitized archive contains 15,243 shellac discs. Each disc has a front and backside, which are often not clearly labelled. Both sides of each disc are digitized at 96 Khz/24bit with a chosen EQ curve without further post-processing. The digitization effort was led by Meemoo[1] which also provides long-term storage.

2.2 Determine Unique Material in an Archive

The meta-data suggests that the shellac disc archive contains a significant portion of duplicate material. However, due to nature of meta-data, it is often unclear whether the meta-data describe the exact same audio material or it describes a different rendition of the same song. The main problem is to **conclusively determine the amount of unique audio material in the archive**, or conversely the amount of duplicate material.

There are two opportunities by linking duplicate material. The first deals with sound quality. Some discs are better preserved (and digitized) than others. Linking duplicate material makes it possible to a potentially redirect archivists and listeners to a better preserved duplicate.

A second opportunity relates to meta-data quality. It is of interest to identify when meta-data is inconsistent and how the current meta-data standards for this archive could be improved.

---

[1] From the http://meemoo.be website: 'Meemoo is a non-profit organisation that, with help from the Flemish Government, is committed to supporting the digital archive operations of cultural, media and government organisations'.

2.3 Detecting Duplicates

The archive consists of 30,661 digital files with an average length of 168±5s. Duplicate detection found 11,829 files or 38.6% contain at least 10s of duplicate material. Most are exact duplicates but some are translations (see below).

The meta-data quality is quite high. If the title of a duplicate is compared with the original, 93% match using a fuzzy matching algorithm. To allow slight variations—differences in case use, additional white space, accents and word order—a Sørensen-Dice coefficient is determined between the original and duplicate title. Only if the coefficient is above a threshold the match is accepted. The performer meta-data field fuzzily matches for 83% of the cases: it is sometimes left blank.

One notable meta-data inconsistency results from the fact that the side of each disc was not clearly labelled. The concept of an 'A' side and a 'B' side was not yet established. During digitization and meta-data notation, this is problematic. The cover gives the title of two works but it is unclear on which side the work is located. The meta-data often assigns both titles to each digital file. Since the order is not clear, for a duplicate, the order of the titles can reversed (the sides are switched).

Another more specific finding is that popular orchestral songs were released for multiple markets. The orchestral backing is the same but the sung part is translated, sometimes by the same singer. The practice of dubbing and re-releasing popular hits in another language is much less common now. An example is file *CS-00069022-01* by Jean Walter. One contains *Rêve d'un Soir/Toi Toujours*, the Dutch version is titled *Hou van Mij (Loving You)/Ik Had Een Droom (I Kissed A Dream)*. Note that the order is reversed and that the titles are translated freely. The Dutch version also refers to an English original. Discrimination between a noisy exact duplicate and a translated version is not possible by taking only into account the duplicate detection results.

Meta-data and classical music is generally problematic. Unfortunately, this is also the case for the shellac disc collection, which contains some classical works. It is clear that shoehorning classical composers, performers, soloists, works and parts into a title and performer framework is a source of many of the identified inconsistencies.

One of the opportunities of linking duplicates is shown in Fig. 1. It shows two discs with the same audio material. One disc deteriorated much more than the other. The amount of pops, cracks and hiss on the digitized version of the deteriorated disc makes it hard to listen to. The much better preserved disc resulted in more ear-friendly digital audio. To listen to the noisy and cleaner version, consult the supplementary material.[2] Going one step further is active denoising. There are techniques which use duplicate material to interpolate samples from several sources

---

[2] Supplementary material can be found at http://0110.be/l/duplicates. It contains audio examples, software and raw data on duplicates.

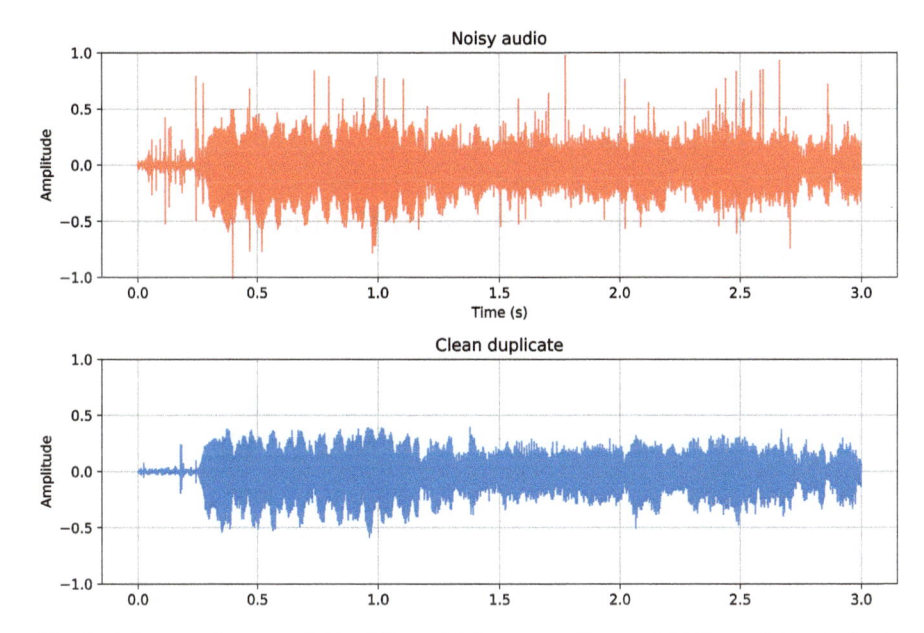

Fig. 1 A fragment from two digitized discs with the same audio material. The one below is clearly less affected by pops and cracks. Identifying and linking duplicates allows listeners to find the best preserved copy

with the aim to reconstruct and denoise gramophone discs [27]. Modern deep-learning techniques to denoise historic recordings [10] also show a lot of promise.

2.4 Some Observations

This case study shows clear advantages of duplicate detection within an archive, especially if the archive is expected to contain many duplicates. It is possible to link audio with others that contain the same recorded material. By post-processing these lists of duplicates, it becomes possible to:

- Identify the amount of unique material within the archive;
- Link low-quality recordings to better preserved duplicates;
- Confront meta-data fields from a recording and its duplicates to get insights into meta-data quality;
- Find surprising links between recordings that share the same recorded material but are clearly different, e.g. a shared orchestral backing of a which the sung part is translated.

The case study also showed a limitation of duplicate detection technology. Discriminating between a near-exact noisy duplicate and a translated version of a song with the same orchestral backing is not possible using only the results of duplicate detection. Additional techniques, meta-data analysis and song-to-text conversion with language recognition, might be employed to solve this automatically.

3 Case 2: Meta-Data Reuse for the IPEM Electronic Music Archive

3.1 The IPEM Music Archive

The Institute for Psychoacoustics and Electronic Music (IPEM) has an archive of tapes originating from 1960 to 1980. IPEM is part of Ghent University in Belgium. To get an idea about the contents of this archive, it helps to sketch a short history of IPEM and the broader context.

After the Second World War, several European broadcasting cooperations started electroacoustic music production studios. In these studios, composers and engineers collaborated to create new instruments and sounds. One of the main drivers behind these investments was the belief that avant-garde music was the logical next step in the western art-music tradition. Two early examples are the 'Colonge Studio für elektronische Musik des Westdeutschen Rundfunks', Germany (1951), and the 'Studio di Fonologia Musicale RAI di Milano', Italy (1955). The insight that avant-garde music studios were not commercially viable warranted institutional backing. In this context, BRT (Belgische Radio en Televisie) and Ghent University ('Rijksuniversiteit Gent' at the time) jointly started IPEM in 1963.

The IPEM production studio was active between 1963 and 1987 [15]. It produced about 450 works of around 100 composers which ended up on about 1000 magnetic tapes [14]. A typical tape cover can be seen in Fig. 2. In the late 1980s, electronic music production became more and more accessible, thanks to the introduction of cheap electronic instruments. The need for institutional backing started to appear anachronistic. After a difficult transition, IPEM established itself as a research centre in systematic musicology and is now part of Ghent University.

3.2 Merging Two Digitized Archives

The archive has been digitized twice. The first digitization campaign was around 2001 [14]. This resulted in a database with high-quality meta-data and set of audio CDs. Unfortunately, the sound cards used at the time (SEK'D ARC 88, 16bit, 48kHz) yielded audio not up to standard for long-term preservation. The choice of writable CDs as long-term storage media was also questionable, due to their limited shelf life. The meta-data, however, that was organized in a relational database is still relevant today.

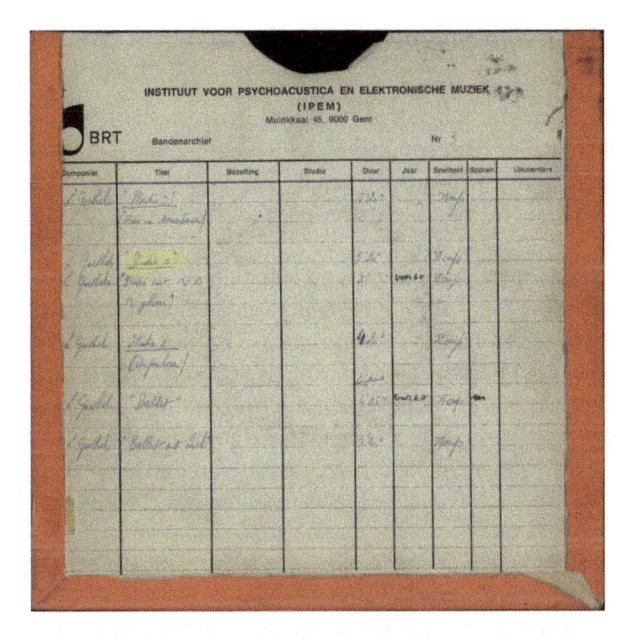

Fig. 2 The cover of a typical magnetic tape in the IPEM archive. The meta-data includes work titles, composers, additional comments and technical meta-data about the tape

The second digitization campaign was organized in 2016 by Meemo. Much better sound cards were available and digitization was done at 96 kHz and 24bit. Long-term storage is done on redundant LTO8 magnetic tapes. However, less effort was spent on detailed meta-data. Complete magnetic tapes were stored in a single audio file together with a list of works on each tape. Each work has a title and a composer. Rough estimates of the duration of the work and additional meta-data (performers, context of the recording) are often missing. Notably, it is unclear where each work starts and stops in the unsegmented digital audio file.

Segmentation of the tapes with avant-garde electroacoustic works is not always trivial. A large part of the tapes do have clear boundaries between works but for others, expert listeners and contextual information is required. This is the case, for example, for works with different parts without a shared sound language. Another example is where silences on tape should be filled by an acoustic instrument during performance (which is not recorded on tape). The meta-data on the duration of the work and the duration of audio on tape can differ widely in such cases.

The problem in the case of the IPEM music archive is to **connect high-quality meta-data and segmentation of a previous digitization to high-quality unsegmented audio of a second digitization.**

3.3 Meta-Data and Segmentation Reuse

With duplicate detection, a link can be found between the low-quality audio and the high-quality audio. The detailed meta-data from the low-quality audio can then be attached to the high-quality audio. The segmentation timestamps for the unsegmented high-quality audio are derived from the recognized parts, as depicted in Fig. 3.

For most works, meta-dating and segmentation is straightforward after a link is found between the low- and high-quality audio archives. There are, however, issues which complicate matters in some cases. The first is the unclear definition of work. Many tapes have been recorded during live concerts. These recordings contain spoken introductions and have an applause at the end. An intro and applause give meaningful context and might be relevant but are not consequently included. A work consisting of many parts might be divided up into its constituent parts or segmented as a single entity. See, for example, IPEM tape number 1076 in the supplementary material.[2]

Another problem is that identical *audio is reused within the archive in different contexts*. Some tapes contain educational material: interviews, talks, lectures and radio shows often contain parts of works but they need not to be segmented as such (see IPEM tape number 1100). Some works also have multiple versions: they share much audio material but have slightly different meta-data. Composers also sampled material created for earlier works in new works: this reuse of material is revealing links between works. However the meta-data and segmentation information should not be simply copied. For example, the works *'Difonium (cadenza, mix)'* and *'Difonium (C, mix)'* by Lucien Goethals share most audio. Many more can be found in the supplementary material.[2]

A third problem is specific to the IPEM archive: different selection criteria have been used for the two digitization campaigns. The material in both collections overlaps for a very large part but both also contain material not present in the other collection. So it is unclear when a link should be found but is lacking due to a flaw in the duplicate detection system. However, the synthetic evaluation below shows a very high reliability of the system used.

To evaluate the automatic segmentation approach, we need two sets of recordings. Both sets need meta-data and segmentation info to check whether the

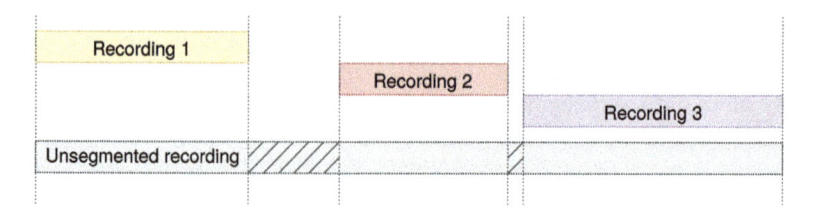

Fig. 3 Three works are found in a long, unsegmented recording. The meta-data of the works can be copied or compared. Segmentation timestamps can be derived from the recognized parts

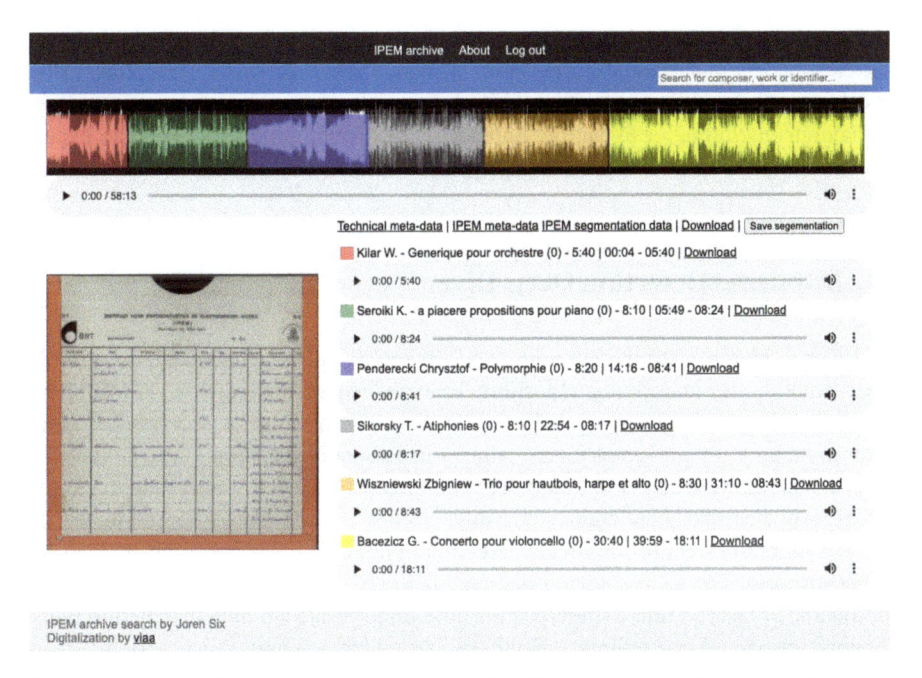

Fig. 4 A web interface to segment tapes manually. The segmentation boundaries (coloured sections above) can be modified on the waveform

matches are correct. Counter-intuitively, we need to manually segment a part of the unsegmented collection simply to be able to evaluate the segmentation timestamps found by duplicate detection. This manual segmentation was done using a custom-build web interface as seen in Fig. 4: the interface allows users to quickly modify segmentation boundaries.

For the 970 tapes in the archive, 2790 segments were annotated. 1158 segmented works were matched with the low-quality archive originating from CDs. Of these 1158, the start and stop location is correct (within 10 seconds) for 348 segments or 30%. The score is mainly due to the consistently higher granularity of segmentation (applause, introduction, parts) for the low-quality audio compared with the annotated segments (works) of the high-quality audio. With the audio correctly linked, a human expert is needed to evaluate whether the higher granularity matches are of interest and may be copied over to generate a definitive segmentation and meta-data set.

3.4 Key Observations: Meta-Data Reuse

The IPEM archive case effectively attaches meta-data and segmentation data from low-quality audio to high-quality audio. For most works, this is a straightforward matter of copying meta-data to the matched high-quality audio. There are, however, caveats which make some cases more difficult: reuse of audio (sampling other works, several similar versions of works) and semantics around segmentation behaviour (including or excluding applause?). It is good practice to keep an expert listener in the loop to verify, confirm or modify meta-data.

4 Duplicate Detection Deep Dive

In order to better grasp the *strengths and limitations* of the technology used in the case studies, the underlying algorithm is explained in this section. An efficient duplicate detection system is able to sort through millions of seconds of audio and come up with a relevant match, almost instantaneously, for a query containing only a handful of seconds of audio. This efficiency and level of accuracy made the previously discussed case studies possible.

The umbrella term for duplicate detection in large archives is known as *acoustic fingerprinting*. The general idea is depicted in Fig. 5. Some feature is extracted from audio and combined into a fingerprint. These fingerprints are then matched to other fingerprints stored in a reference database. If a match is found, it is reported.

An audio feature with attractive properties for acoustic fingerprinting purposes is *local peaks in a spectral representation*. They have been used in many systems [23, 26, 29]. An alternative feature is, for example, energy change in spectral bands [7, 11]. However, there are not many systems which are both freely available and easy to use on larger datasets. One of the few open-source systems is Panako[23], the system used here.

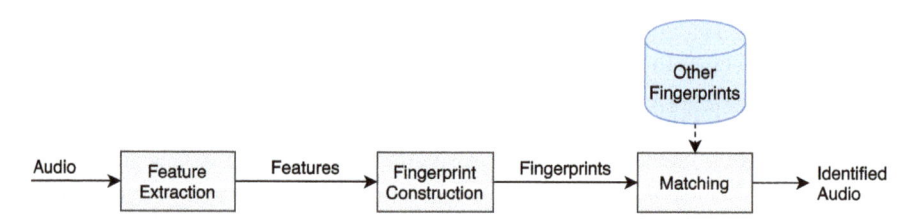

Fig. 5 General acoustic fingerprinting system

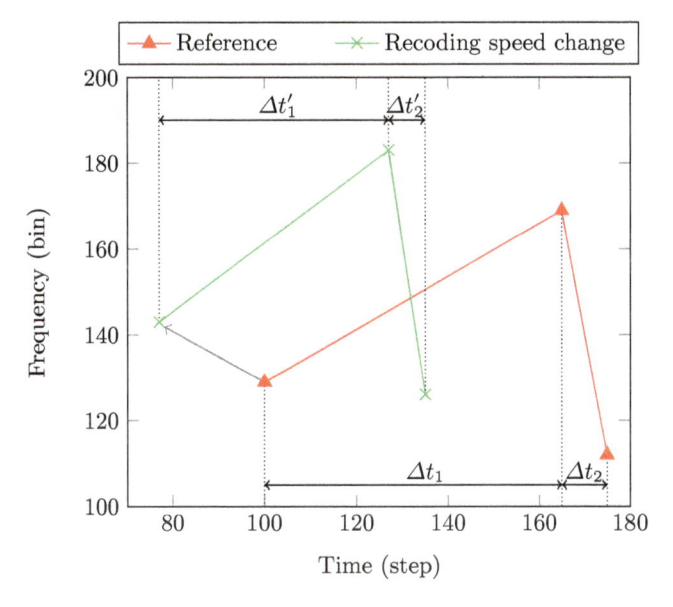

Fig. 6 The effect of speed modification on a fingerprint. The figure shows a single fingerprint extracted from reference audio (—▲—) and the same fingerprint extracted from audio after recording speed modification (—×—)

4.1 Panako: An Acoustic Fingerprinting System

Panako[3] [23] is an acoustic fingerprinting system. It is available under an AGPL license. Panako is based on TarsosDSP [22], a popular Java DSP library. Panako implements a baseline algorithm [29] and the Panako algorithm [23]. The Panako algorithm is able to match queries which are time-stretched, pitch-shifted or sped up with respect to the indexed audio. This is required for monitoring DJ mixes [13, 18, 25] and to match music from analogue media digitized at different speeds. The original Panako paper [23] describes the system in detail.

The key insight used in Panako is that an information of three local peaks in a spectral representation can be combined in a single hash robust against time/pitch modifications. In Fig. 6, each peak has a time t, frequency f and magnitude m component and, for example, $\Delta t_1 / \Delta t_2$ of the reference fingerprint equals $\Delta t'_1 / \Delta t'_2$ after speed-up. Combining only such relative information in hashes allows to match the reference audio with modified audio from the same recorded event even after speed-up, time-stretching or pitch-shifting. More details can be found in the original Panako paper [23].

Panako received updates in 2021 after close inspection of an efficient implementation of a similar algorithm [20] for embedded systems. The underlying

---

[3] http://panako.be: the Panako website (last visited December 19, 2022).

concepts from the original paper still stand but two changes improve the system considerably.[4]

The first change replaces the frequency transform from a classical constant-Q transform [1, 2] to a *constant-Q non-stationary Gabor transform* [12]. The latter is more efficient and allows a finer frequency resolution at equal computational cost. See [28] for a detailed comparison. The spectral peak detection in Panako improves by the use of this finer frequency resolution. In Panako, JGaborator[5] is used: a wrapper around the Gaborator[6] library which implements a constant-Q non-stationary Gabor transform in C++11.

The second change replaces an *exact hash matching* technique by a *near-exact hash matching* approach. Some background helps understand this change. As mentioned before, the first step in the algorithm is a transform a one-dimensional time series into a two-dimensional time/frequency grid. Each bin in this grid has a very short duration and a small frequency dimension. The exact dimensions of these bins are determined by the spectral transform parameters and can be small but they remain discrete. This means that when a query and a match differ by about half the duration of a bin, energy is spread over neighbouring bins. Since time-frequency coordinates of peak magnitude bins are used in fingerprints, off-by-one errors are to be expected, both in time and frequency. Off-by-one errors are handled by the hash in the 2021 version of Panako.

For indexing and matching, a hash is constructed from the components mentioned below. A hash combines fingerprint information into a single integer. The additional information contains an audio identifier used to tally matches. Below, the components are ordered from most to least significant. By only including approximate frequency information and having the time ratio in the least significant bytes, range queries become possible. The last couple of bits can be ignored during a search in ordered hashes: effectively dealing with off-by-one errors.

$$|f_3 - f_2|/4 \, ; \, |f_2 - f_1|/4 \, ; \, \tilde{f}_1$$

$$|t_3 - t_2| > |t_2 - t_1|$$

$$m_3 > m_2 \, ; \, m_3 > m_1 \, ; \, m_1 > m_2$$

$$f_3 > f_1 \, ; \, f_3 > f_2 \, ; \, f_1 > f_2$$

$$(t_2 - t_1)/(t_3 - t_1)$$

This idea of gracefully handling off-by-one errors needs to be reflected in the matching scheme as well. The fingerprints extracted from a query are matched with the index. A list of matching prints is returned and needs to be filtered: hashes

---

[4] Note that this text describes and evaluates Panako as is in the following commit found on GitHub: 6cf936730131d71c94c562a06a1a791e09b4c520.

[5] https://github.com/JorenSix/JGaborator The JGaborator GitHub repository.

[6] http://gaborator.com/ The Gaborator website.

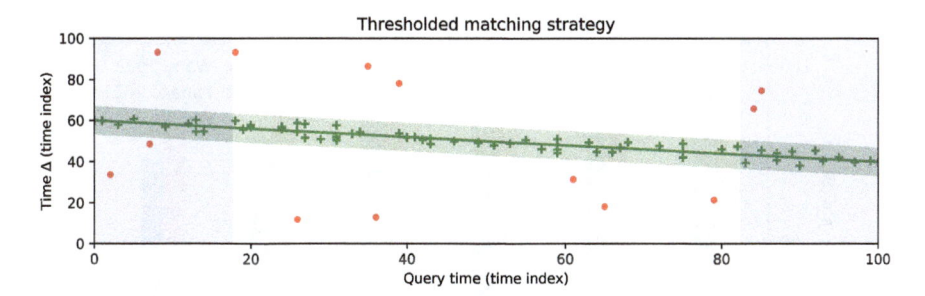

Fig. 7 To discriminate true from false positives, a thresholded matching strategy is used. The first few and last few matches (blue background) are used to calculate a median Δt. Accepted matches (green background) fall in a small range around a linear regression from the first to last median Δt. Some random matches (red dots) are dismissed

might randomly collide or short fragments might match for a very short duration. To filter true positive matches from false positives, the difference in time (Δt) between each reference and query hash. For a true match Δt, is either a constant or changes linearly over time. In the original paper[29], a true positive is only accepted if Δt is a fixed constant. Here, we calculate a linear regression from the first matches (blue in Fig. 7) to the last and allow some small margin in which matches are accepted. In this manner, off-by-one matches and time-stretching/speed-up are supported.

When larger archives are indexed, the characteristics of the key-value store become more and more important. The key-value store stores hashes together with some additional information. A hash combines fingerprint information into a single number. The additional information contains an audio identifier used to tally matches. The 2021 Panako version stores ordered fingerprints using a persistent, compact, high-performance B-Tree[7] [4]. The speed, small storage overhead and performance allow more beneficial trade-offs between query performance: it facilitates storing more fingerprints per second of audio and larger datasets for equal or better query performance.

4.2 Panako Evaluation

To show the strengths and weaknesses of the Panako system, an evaluation is done on the free music archive [6] medium dataset.[8] In total 25,000 music fragments of 30 seconds were used in the evaluation. One fifth of the data set was not indexed to check for true negatives. Readers are encouraged to repeat the evaluation as it is completely reproducible with the evaluation script provided as part of the Panako

---

[7] LMDB: Lightning Memory-Mapped Database Manager, http://lmdb.tech (last visited December 19, 2022).

[8] The data can be downloaded at https://github.com/mdeff/fma (last visited December 19, 2022).

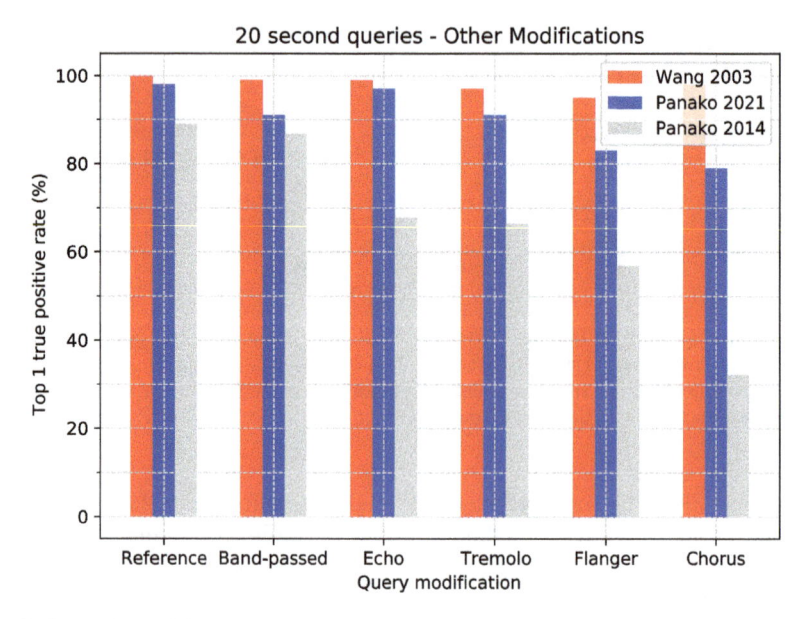

Fig. 8 Comparison of the top one true positive rate after several modifications for 20 second query fragments for Wang [29], Panako [23] and Panako [30]

software distribution. Details on the exact parameters for the modifications can be found there as well.

Panako is evaluated for various modifications (Fig. 8), playback speed modification (Fig. 11), time-stretching Fig. 10 and pitch-shifting Fig. 9. The evaluation follows a straightforward method: a random fragment is selected, a modification is applied and the modified fragment is used to query the index for matches. Only the best match is considered and counted as a true positive, false positive, true negative or false negative. The sensitivity or true positive rate is reported ($TP/(TP + FN)$).

For all modifications in Fig. 8, the query performance from the 2014–2021 version of Panako is clear. The chorus effect impacts the spectrogram the most: the performance increases from about 25–80%. The baseline algorithm (which is also implemented in Panako) is, however, always better but does not support pitch-shift or time-stretch. It shows that there is still headroom for further refinement.

The pitch-shift and time-stretch modifications (Figs. 10 and 9) are calculated from −16% to +16% to reflect a common maximum modification on DJ equipment. Again it is clear that the upgrades drastically improve the performance especially in higher modification factors. The baseline algorithm [29] normally does not support time-stretch modification. The Panako implementation of [29] uses the matching strategy described above which allows time-stretch: Δt is not a fixed constant but is allowed to change linearly.

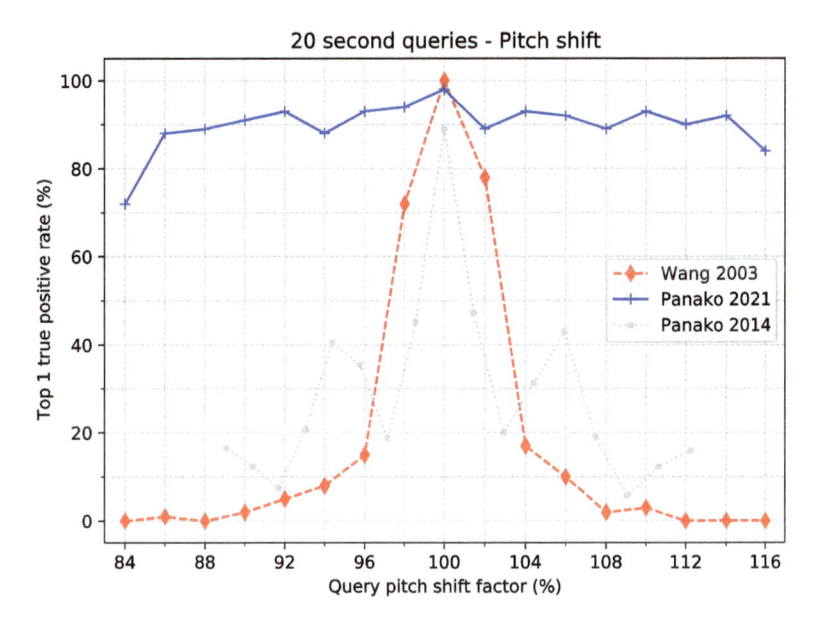

Duplicate Detection for for Digital Audio Archive Management 325

Fig. 9 Comparison of the top one true positive rate after pitch-shifting for 20s query fragments for Wang [29], Panako [23] and Panako [30]

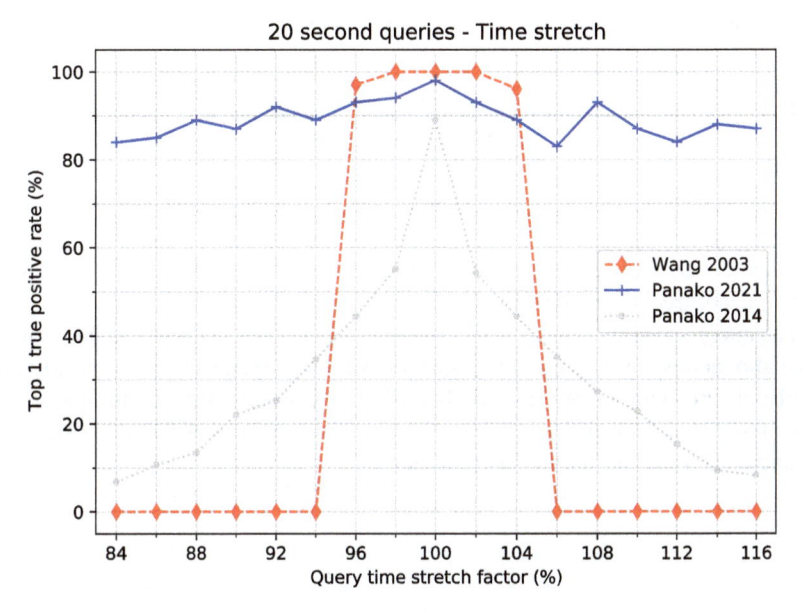

Fig. 10 Comparison of the top one true positive rate after time-stretching for 20s query fragments for Wang [29], Panako [23] and Panako [30]

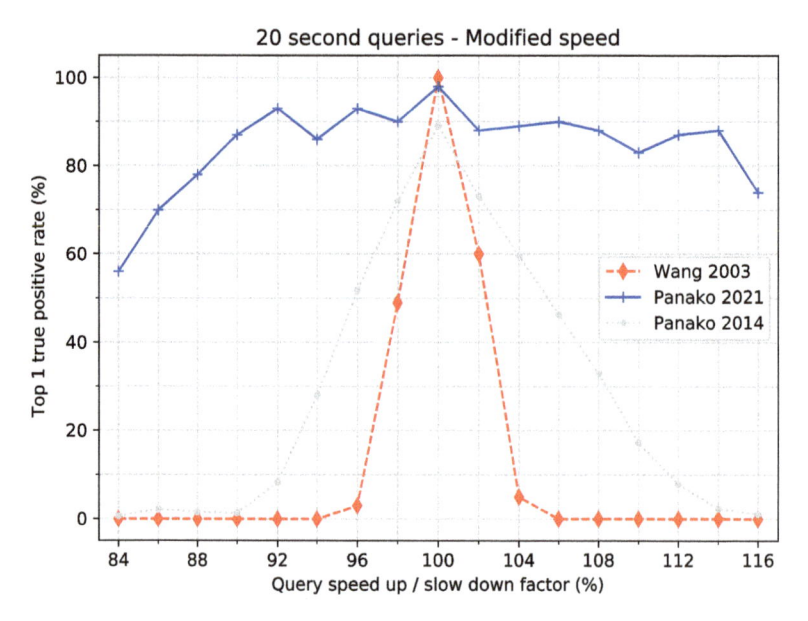

Fig. 11 Comparison of the top one true positive rate after speed-up/slow down for 20s query fragments for Wang [29], Panako [23] and Panako [30]. The audio playback speed is modified from 84 to 116% with respect to the indexed reference audio. If the query is slowed down by 10%, the duration ends up being 22s. For the 2021 Panako algorithm, audio recognition performance suffers (below 80%) when playback speed is changed more than 10%

The speed-up modification (Fig. 11) can be seen as a combination of both time-stretching and pitch-shifting with the same factors. Query performance is, in other words, limited by the time-stretch and pitch-shift performance. For the larger modification factors, the performance drops below 80% but is still much above Panako 2014.[9] More extreme playback speed modification is supported by Sonnleitner and Widmer [26], but their reported query speed is much slower and the system is not freely available.

The query speed of Panako varies with the size of the database and properties of the audio: acoustically dense audio generates more fingerprints. On an Early 2015 MacBook Pro with a 2.9 GHz Dual-Core Intel Core i5, storage and query is 38 times faster than real-time per processor core. This means that 38 seconds of audio is handled in 1 second for each processor core.

---

[9] For Panako, time-stretch and pitch-shift factors do not need to be the same: a fragment pitch-shifted 104% followed by a 92% time-stretch will match the original.

5 Conclusion

In this chapter, a mature MIR technology of duplicate detection was described and put to the test in two case studies. In the first case study, duplicates were detected for a part of the music archive of a public broadcaster. This allowed to identify the unique material in the archive and had additional benefits to check meta-data quality and redirect listeners to higher-quality duplicates. In the second case study, meta-data and segmentation information of low-quality audio was attached to higher-quality duplicates.

The main takeaway from both case studies is that duplicates can be found reliably and easily even in larger archives. For the most part, duplicates are straightforward but some surprising and interesting cases of audio-reuse (sampling, translations, versions) might warrant the need for a human expert in the loop.

The chapter concluded with a technical description and evaluation of Panako—an acoustic fingerprinting system. This was done in order to better understand strengths and weaknesses of the technology. The evaluation of the 2021 version of Panako shows much improved performance over previous versions.

With the case studies, we aimed to directly improve the quality of two collections by showing how duplicate detection technology helps to offer better services to end users. Evidently, we hope to accelerate the adoption of duplicate detection technology by the community of audio archivists and indirectly improve many digital audio archives.

References

1. Judith C Brown. Calculation of a constant q spectral transform. *The Journal of the Acoustical Society of America*, 89(1):425–434, 1991.
2. Judith C Brown and Miller S Puckette. An efficient algorithm for the calculation of a constant q transform. *The Journal of the Acoustical Society of America*, 92(5):2698–2701, 1992.
3. Pedro Cano, Eloi Batlle, Ton Kalker, and Jaap Haitsma. A review of audio fingerprinting. *The Journal of VLSI Signal Processing*, 41:271–284, 2005.
4. Douglas Comer. Ubiquitous b-tree. *ACM Computing Surveys (CSUR)*, 11(2):121–137, 1979.
5. Reinier de Valk, Anja Volk, Andre Holzapfel, Aggelos Pikrakis, Nadine Kroher, and Joren Six. Mirchiving: Challenges and opportunities of connecting mir research and digital music archives. In *Proceedings of the 4th International Workshop on Digital Libraries for Musicology*, DLfM '17, pages 25–28, New York, NY, USA, 2017. ACM.
6. Michaël Defferrard, Kirell Benzi, Pierre Vandergheynst, and Xavier Bresson. FMA: A dataset for music analysis. In *18th International Society for Music Information Retrieval Conference (ISMIR)*, 2017.
7. Dan Ellis, Brian Whitman, and Alastair Porter. Echoprint - an open music identification service. In *Proceedings of the 12th International Symposium on Music Information Retrieval (ISMIR 2011)*, 2011.
8. Sébastien Fenet, Gaël Richard, and Yves Grenier. A Scalable Audio Fingerprint Method with Robustness to Pitch-Shifting. In *Proceedings of the 12th International Symposium on Music Information Retrieval (ISMIR 2011)*, pages 121–126, 2011.

9. Beat Gfeller, Dominik Roblek, Marco Tagliasacchi, and Pen Li. Learning to denoise historical music. In *ISMIR 2020 - 21st International Society for Music Information Retrieval Conference*, 2020.
10. Beat Gfeller, Dominik Roblek, Marco Tagliasacchi, and Pen Li. Learning to denoise historical music. In *ISMIR 2020 - 21st International Society for Music Information Retrieval Conference*, 2020.
11. Jaap Haitsma and Ton Kalker. A highly robust audio fingerprinting system. In *Proceedings of the 3th International Symposium on Music Information Retrieval (ISMIR 2002)*, 2002.
12. Holighaus, Nicki and Dörfler, Monika and Velasco, Gino Angelo and Grill, Thomas. A framework for invertible, real-time constant-q transforms. *IEEE Transactions on Audio, Speech, and Language Processing*, 21(4):775–785, 2012.
13. Taejun Kim, Minsuk Choi, Evan Sacks, Yi-Hsuan Yang, and Juhan Nam. A computational analysis of real-world dj mixes using mix-to-track subsequence alignment. In *21st International Society for Music Information Retrieval Conference (ISMIR), 2020*, 2020.
14. Leman, Marc and Dierickx, Jelle and Martens, Gaëtan. The IPEM-archive conservation and digitalization project. *JOURNAL OF NEW MUSIC RESEARCH*, 30(4):389–393, 2001.
15. Lesaffre, Micheline. 50 years of the Institute for psychoacoustics and electronic music. In Jacobs, Greg, editor, *IPEM Institute for Psychoacoustics and Electronic Music : 50 years of electronic and electroacoustic music at the Ghent University*, volume 004 of *Metaphon*, pages 23–25. Metaphon, 2013.
16. Matija Marolt, Ciril Bohak, Alenka Kavčič, and Matevž Pesek. Automatic segmentation of ethnomusicological field recordings. *Applied Sciences*, 9(3):439, 2019.
17. Nicola Orio. Searching and classifying affinities in a web music collection. In *Italian Research Conference on Digital Libraries*, pages 59–70. Springer, 2016.
18. Diemo Schwarz and Dominique Fourer. Unmixdb: Een dataset voor het ophalen van dj-mixinformatie. In *19th International Symposium on Music Information Retrieval (ISMIR)*, 2018.
19. Joan Serra, Emilia Gómez, and Perfecto Herrera. Audio cover song identification and similarity: background, approaches, evaluation, and beyond. In *Advances in music information retrieval*, pages 307–332. Springer, 2010.
20. Joren Six. Olaf: Overly lightweight acoustic fingerprinting. 2020.
21. Joren Six, Federica Bressan, and Marc Leman. Applications of duplicate detection in music archives: from metadata comparison to storage optimisation. In *Italian Research Conference on Digital Libraries*, pages 101–113. Springer, 2018.
22. Joren Six, Olmo Cornelis, and Marc Leman. TarsosDSP, a real-time audio processing framework in Java. In *Proceedings of the 53rd AES Conference (AES 53rd)*. The Audio Engineering Society, 2014.
23. Joren Six and Marc Leman. Panako - A scalable acoustic fingerprinting system handling time-scale and pitch modification. In *Proceedings of the 15th ISMIR Conference (ISMIR 2014)*, pages 1–6, 2014.
24. Joren Six and Marc Leman. Synchronizing multimodal recordings using audio-to-audio alignment. *Journal on Multimodal User Interfaces*, 9(3):223–229, Sep 2015.
25. Reinhard Sonnleitner, Andreas Arzt, and Gerhard Widmer. Landmark-based audio fingerprinting for dj mix monitoring. In *ISMIR*, pages 185–191, 2016.
26. Reinhard Sonnleitner and Gerhard Widmer. Quad-based Audio Fingerprinting Robust To Time And Frequency Scaling. In *Proceedings of the 17th International Conference on Digital Audio Effects (DAFx-14)*, 2014.
27. Christoph F. Stallmann and Andries P. Engelbrecht. Gramophone noise reconstruction - a comparative study of interpolation algorithms for noise reduction. In *2015 12th International Joint Conference on e-Business and Telecommunications (ICETE)*, volume 05, pages 31–38, 2015.

28. Velasco, Gino Angelo and Holighaus, Nicki and Dörfler, Monika and Grill, Thomas. Constructing an invertible constant-q transform with non-stationary gabor frames. *Proceedings of DAFX11, Paris*, 33, 2011.
29. Avery Li-Chun Wang. An industrial-strength audio search algorithm. In *Proceedings of the 4th International Symposium on Music Information Retrieval (ISMIR 2003)*, pages 7–13, 2003.
30. Six, J.: Panako: a scalable audio search system. Journal of Open Source Software. 7(78), 4554 (2022). The Open Journal. https://doi.org/10.21105/joss.04554

How a Song's Section Order Affects Both 'Refrein' Perception and the Song's Perceived Meaning

Yke Paul Schotanus

1 Introduction

Several musicologists have assumed that specific musical features (such as unexpected notes, tempo and syncopation) can affect the interpretation of song lyrics in a predictable way, i.e. that these features can be used intentionally to further a specific interpretation of those lyrics [1, 4, 5, 8, 21]. However, testing such assumptions experimentally requires the use of various versions of sung stimuli in which only the target feature is different across versions, and recording such stimuli is very hard to do. One would always expect performance-dependent differences to confound the effect of the target features, because it is likely that changing specific musical properties has an effect not only on the listener but also on the performer. Digital technologies, by contrast, can be used to create various versions of musical stimuli without changing the performance. In earlier research, Schotanus [27–32] already has used such digitally manipulated versions of songs and sung sentences in various experiments and has found clear effects of accompanied versus a cappella singing, of out-of-key notes versus in-key notes, of syncopation and of song form on various aspects of song cognition.

In the current study, the effect of formal structure, particularly of song section order and repetition, on the way a song and its lyrics are perceived and interpreted semantically, will be further investigated. Apart from section form (including stanza form), section order also plays an important part in large-scale musical structure, which is assumed to affect both liking and musical meaning and to be of historical relevance [1, 5, 12, 21, 34, 35]; see Schotanus [27] for a brief review. Even in popular music studies, several authors have stressed the significance of musical form and

Y. P. Schotanus (✉)
Institute for Cultural Inquiry, Utrecht University, Utrecht, Netherlands
e-mail: schotschrift.teksten@planet.nl

particularly of repetition and have fought the widespread idea that in popular song musical form is too simple to be investigated [3, 10, 13, 16, 18].

Popular songs are usually classified as examples of one of four categories: strophic songs, AABA songs, verse-chorus songs and verse-chorus-bridge songs [6, 7, 35, 38]. However, it is questionable whether this is doing justice to the variety and the nature of song forms. For example, Schotanus [27] has shown that many songs do not belong to either of those categories and that in analyses of specific songs, the chosen category often does not fit the song's actual form. It is even debatable whether, for example, verse-chorus songs and AABA songs are indeed different categories. Apart from the fact that there are AABA songs in which the AABA part functions as a chorus, there are also songs (e.g. 'Yesterday' by the Beatles) in which the B part, the bridge, is repeated verbatim, as a result of which it may be mistaken for a chorus. What is more, Schotanus, Koops and Reed Edworthy [33] have shown that within a group of straightforward AAA songs such as the Genevan psalms, there are important formal differences between one song and the other which seem to affect song processing. At least they predict psalm popularity.

Therefore, it is necessary to investigate both the relevance of the traditional categorization of popular songs and the possibility that there are other useful ways to approach song form. Schotanus [24, 27], for example, has developed the RAS hypothesis. The RAS hypothesis states that the appreciation and interpretation of a song depends on a set of preference rules for 'song section order', which are based on the assumption that listeners intuitively search for a balance between 'repetition and surprise' (RAS). This hypothesis builds on the cognitive research undertaken by Ollen and Huron [20], Huron [11] and others [15, 22, 23, 36]. According to the RAS hypothesis, a violation of preference rules can cause feelings of tension or boredom which, however, are 'acceptable' to the listener if the lyrics allow for a 'meaningful' interpretation. For example, RAS rule 5 states that late repetitions (either late repetitions of song sections within a song or of melodic phrases within a song section) have a cumulative effect which is only acceptable if it is in line with the lyrical content of the song, or if it is compensated for by other musical features. Partial evidence for RAS rules can be found in several corpus studies [14, 27, 35], and for RAS rule 5 specifically in the Genevan Psalter study [33] and a study concerning Dylan songs [30].

In the current study, the question whether AABA songs and verse-chorus songs are essentially different from each other is investigated by comparing listeners' reactions to different versions of the same songs. Therefore, several experiments were conducted. Two of them, reported on earlier [27], will be summarized below. The last one, an online listening experiment that will be reported here, involved two songs. Both songs consisted of a number of A sections containing at least one refrain line, and one or two bridge sections. Both songs were digitally altered in several alternative versions, one of which had a verse-chorus-like structure with the B sections in a chorus position.

Participants were asked questions about their appreciation for the song, their interpretation of the song and their 'refrein' perception. In Dutch 'the' 'refrein' can refer to either a chorus or a salient refrain line. Therefore, the question which part of the song people think is the 'refrein' is an easy way to determine whether a song is perceived as an AABA song or as a verse-chorus song. For example, if an AABA-like song is something essentially different from a verse-chorus song, the B section is unlikely to be perceived as 'the refrein', even if it is put in a chorus-like position.

However, it is questionable whether B sections in chorus positions will indeed not be perceived as choruses, as both choruses and bridges are supposed to be contrasting song sections. Summach [35] observes that bridges more often show tonal unstability than choruses, but he also observes that such bridge-specific and chorus-specific properties change over time. Moreover, a chorus does not always have a contrastive melody [6]. On the other hand, Van Balen [37] found that choruses in early twentieth-century Dutch popular songs could be retrieved automatically by searching for specific distinctive sound properties. It is, however, undocumented which songs Van Balen used and which song parts were deemed choruses. Consequently, it is unclear whether his research did not involve AABA songs containing highly contrastive B sections.

On the other hand, there may be other song properties that distinguish between bridges and choruses, for example, the nature of the other sections within the song. For example, Summach [35] observes that harmonic tension in A sections in AABA songs tends to be resolved, whereas harmonic tension in verses tends to be left unresolved. Furthermore, in AABA songs, the A sections often contain a refrain line or word (e.g. 'Yesterday') representing the main message of the song, whereas in verse-chorus songs, the main message is assumed to be represented by the chorus [34]. Yet, it is not unusual for verses in verse-chorus songs to contain refrain lines as well (e.g. Radiohead's 'Creep'). The difference may be that listeners will base their interpretations of an AABA song on the refrain line within the A sections, whereas their interpretations of a verse-chorus song are based on the chorus.

If 'refrein' perception would turn out to be dependent on position, either partly or completely, this would call for a cognition-based approach of song form. It would be evidence for both Pattison's assumption that song structure creates certain 'power positions' within the lyrics [21] and that music can be used as a foregrounding device, for example, by accentuating specific song parts [24, 27]. Possibly, the set of RAS rules could be extended with a few preference rules governing 'refrein' perception.

Concerning the existing RAS rules, in one song, RAS rule 5, regarding the effect of late repetitions, was at stake. In one version of that song, there was an accumulation of A sections towards the end of the song. This is a violation of that fifth RAS rule and is assumed to negatively affect the appreciation of this song version.

Although it is not the target issue of this paper, using song versions including sections with verbatim repeated song lyrics, and asking questions about the appreciation and the interpretation of these songs, will make that this study will also contribute to the literature concerning verbatim repetitions of words. It will provide either evidence or counter-evidence to the hypothesis that verbatim repetitions of language are interpreted as more acceptable and more meaningful when presented with music (and in particular when sung) and that repetition can change the meaning of the repeated language at its second occurrence [9, 27]. Apart from that, the study will also contribute to the widespread hypothesis that verbatim repetitions increase liking [19], if only the listeners do not become aware of the fact that their positive feelings are caused by mere repetition [11].

1.1 Previous Experiments

The current experiment was preceded by an online listening experiment reported on in the author's dissertation [27] and a smaller live experiment reported on in a conference poster [25]. In the online experiment, a total of 149 participants, between 15 and 84 years old ($M = 52.58$; $SD = 14.27$), listened to one of four versions of the same song and were asked a few questions about them. The order of the original song was altered in such a way that in one version the B sections were in the middle (the original AABAABAAcoda version), in one version they were at the end (AAAABAABcoda, a verse-chorus-like version) and in one version they occurred in the beginning of the song (ABABAAAAcoda, a version with a cumulation of A sections at the end). Finally, an additional fourth version was created by deleting the last A section from the original version, accentuating the AABA structure (AABAABAcoda).

The hypotheses were that in the original and the fourth version, the content of both A sections and B sections and coda would contribute to the overall interpretation of the song; that in the second version, the contribution of the content of B sections and coda to the overall interpretation would be more prominent; and that in the third version, the content of the A sections would be more influential.

The first, second and fourth versions were hypothesized to be perceived as relatively well formed, in contrast to the third version, which violates the fifth RAS rule. The second version, on the other hand, was hypothesized to be the best structured one in terms of RAS rules, as the first and fourth versions are slightly at odds with RAS rule 5. In the first version, the number of A sections is not decreasing, whereas in the fourth one, the decrease starts relatively late. Finally, the B section was assumed to be mentioned more often as a 'refrein' than the last line of the A section in the second version.

The results were largely in line with the hypotheses. The ABABAAAAc version received the lowest ratings concerning musical quality and lyric quality and showed an A-section-oriented bias in interpretation and refrain perception, whereas the AAAABAABc version received the highest ratings concerning musical and lyric quality and showed a less A-section-oriented bias in interpretation and refrain perception. However, only the latter effect was significant, particularly in comparison with the ABABAAAAc version.

Thus, section order indeed seemed to affect song appreciation, song interpretation and refrain perception in a predictable way. However, the study had several limitations. First, sample size seemed to be too small to receive clear significant results concerning the disapproval of the ABABAAAAc version. Second, the song's formal structure may be too deviant to be perceived as either an AABA or a verse-chorus song, which may have distorted refrain perception anyway. Third, results for the AABAABAc version were difficult to be interpreted, because in hindsight it did not have a clear AABABA pattern, and the deletion of the sixth A creates an unresolved rhyme, because in the original version, the third lines of the fifth and sixth A rhyme with each other. Fourth, after the results were published, the second B section in the AAAABAABc version turned out to be a repetition of the first B section, whereas it was meant to be the second B section of the original. As a consequence, the B section was not only in a chorus position but also repeated verbatim, which may have enhanced the section's appearance as a chorus. What is more, the extra repetition may also have positively affected the appreciation of the song version [11, 19] and may have further enhanced a B-section-oriented interpretation of its meaning. After all, verbatim repetition of words supports processing fluency of those words [19].

In the live experiment [25], 40 participants listened to two AABA songs which were performed as AABABA songs, just by repeating the B section and one of the A sections. After the fourth, the fifth and the sixth song section, an assistant indicated with a hand gesture that a song section was finished, and it was time to answer a few questions. After the fourth section, the only question was whether the song could have been finished by then. After the other two song sections, the same question was asked, followed by the questions whether the last section has been an acceptable addition, and whether it has been a meaningful addition. After listening to the whole song, there were three extra questions: which section would make up the best song ending, was there a 'refrein' and which was the 'refrein' (where participants could choose either the B section or a refrain line taken from the A section)? Most of the answers indicated that both songs are predominantly perceived as AABA(BA) songs and not as verse-chorus songs. However, B sections of both songs were perceived as the 'refrein' by thirty per cent of the participants, and 10 and 16 participants, respectively, even thought that the B sections of song A and B would make up the best song ending. Finally, the fact that most participants judged that either the fifth or the sixth section would make up the best song ending indicates that these song sections were perceived as acceptable and meaningful additions to the song, although they were verbatim repetitions of earlier song sections.

Of course, several participants did not see the assistant's gestures and were not able to identify section divisions, so they did not answer the section-specific questions. Furthermore, it may be hard to answer the question whether a section makes up a good song ending when the song is already going on, and finally, it cannot be excluded that the performers (the author and a guitar player) have influenced the answers through specific accents or flaws in their performance. Nevertheless, the results of this experiment show that the difference between AABA songs and VC songs needs further investigation.

1.2 Current Experiment

As a follow-up to the experiments summarized above, a listening experiment quite similar to the one reported on in the author's dissertation [27] was conducted, in which several of the limitations mentioned above were resolved. This experiment involved two songs.

Song 1 was the same song as the one used in the former online experiment, except that in this case, the AABAABAc variant was not used and the second B section in the AAAABAABc version was indeed the second B part of the original and not a repetition of the first one. Thus, the differences between the ABABAAAAc and the AAAABAABc version could become more clear, and the possible effect of just putting the B sections of the song in 'chorus positions' could be compared with both putting the B sections in chorus positions and replacing the second B section with a verbatim repetition of the first one.

Song 2 was one of the songs used in the live experiment. As this was originally created as an AABA song, it can be considered a straightforward AABA song. Thus, the hypothesis that a B section will be perceived as a chorus if it is placed in a chorus position could be assessed more effectively. Three versions were created: an AABA version, an AABAB version and an AABABA version. The latter was added in order to assess the effect of mere repetition. If the B section would change in a chorus through mere repetition, this would hold in the AABABA version as well. However, if the fact that a song ends with an A section, would turn it into an AABA song, no matter the chorus quality of the B section, the B section will not be perceived as a chorus in such a song version.

Regrettably, funding did not allow for large sample sizes; therefore, the negative effect of late repetition in the ABABAAAAc version of song 1 was again unlikely to be significant. However, if the same pattern would occur, this would at least indicate a certain tendency.

In short, these were the hypotheses at stake:

1. A song section will be perceived as a chorus or a bridge dependent on its position.
2. Chorus perception is different across song versions in which the same song sections occur in different positions.
3. In the AAAABAABc version of song 1 and the AABAB version of song 2, refrain or chorus perception will be more B oriented than in the other versions.
4. The interpretation of a song is different across song versions in which the same song sections occur in different positions.
5. A participants' interpretation of a song is mainly based on the participants' idea which song part is or contains the 'refrein'.
6. The interpretation of the AAAABAABc version of song 1 is more B oriented than the interpretation of the ABABAAAAc version of that song, and the AABAB version of song 2 is more B oriented than both of the other versions of that song.

7. The appreciation for a song is different across song versions in which the same song sections occur in different positions.
8. The interpretation of the AABA version of song 2 is different from the interpretation of the other versions of song 2, although the only textual difference is that parts of the lyrics are repeated.
9. Appreciation for the ABABAAAAc version of song 1 is relatively low.
10. Appreciation for song versions including verbatim repeated song sections is equal to or higher than appreciation for song versions in which all song sections are different from each other.

2 Method

2.1 Participants

A total of 111 participants recruited via Prolific Academic completed the survey and were payed for their work. They were between 18 and 59 years old ($M = 25.96$; $SD = 7.91$), 65 female, 36 male. Most of them (84) were native speakers of Dutch, 17 were not but claimed to be fluent speakers of it. All participants were presented with two songs. Each individual heard one of three versions of each. The version of song 1 was randomly assigned to them first, which was followed by a random version of song 2. After each song, they were asked a few questions about it. Between the songs, they answered a series of questions concerning their musical and literary sophistication, i.e. the complete Gold-MSI questionnaire [2, 17], and 11 items concerning literary sophistication. A principal axis factoring analysis of the latter, the details of which can be found online [26], yielded three factors with an eigenvalue larger than 1, literary activity; 2, passive literary enjoyment; and 3, nonliterary writing activity. Completing the whole survey took about 19 minutes on average.

2.2 Stimuli

The participants all heard one of three versions of the same two songs, all of which can be found online [26]. Both songs were pre-existing cabaret songs in Dutch, composed and sung by the author and accompanied, recorded and digitally altered by Christan Grotenbreg, using a keyboard, connected to ProTools 10 (desktop recording), a Neumann TLM 103 microphone, an Avalon VT 737 SM amplifier and an Apogee Rosetta converter, and in addition, Waves Tune, Renaissance Vox compression and Oxford Eq. voice-treatment software.

The first song, 'Hou'en zo' (Keep it like that), was an AABAABAAcoda song, which was changed and an ABABAAAAcoda song. This could be done without harming the rhetorical logic of the lyrics, because in this song, all A sections mention examples of disasters that did not hit the singer, and in the B sections and the coda, the singer wonders who he should thank for that. The B sections can therefore follow any of the A sections.

Note that the B sections are bridges rather than choruses, as they are longer and more complex than the A sections, have varying lyrics and neither start nor end on the tonic. Moreover, the B sections end with a one-word refrain ('geluk', 'luck'), and have another one-word refrain ('danken', 'say thanks') at the end of the first line, whereas the A sections start with a three-word refrain ('alweer een dag' 'another day') and end with an immediately repeated catch frase catch phrase: 'Hou'en zo' ('Keep it like that'). It is therefore unlikely that the B section will be perceived as a chorus, although there is a clear musical contrast with the A sections because it starts and ends on the dominant, and is partly in a different key (i.e. in B flat minor instead of B flat major).

The second song 'Mijn ogen' ('My eyes') was originally written as an AABA song but has developed over time to an AABABA song in which the second B section is exactly the same as the first one, and the fourth A section is the same as the first one, except for one or two minor changes in the wording. The song was recorded at once as an AABABA song, and after that an AABA and an AABAB version were created by deleting the final sections in such a way that there is a sense of completeness because at least the accompaniment ends on the tonic.

All A sections start with 'Mijn ogen' ('My eyes') and end with a variation on the refrain line 'Maar dan kijk ik met mijn handen naar jou' ('But then I look at you with my hands'). After the fourth A, this refrain line is repeated once. By contrast, the B section is repeated integrally if it occurs twice, which can give it a certain chorus quality. However, it begins and ends on the dominant pitch and is partly in a different key (i.e., E major instead of A minor). In the AABAB version, harmonic tension is resolved by a final A in the base at the moment where in the AABABA version, the last A section begins.

By using pre-existing songs, written by the author, it was possible to work with ecologically valid stimuli which where, nevertheless, very likely to be new to the participants. It also allowed the author to create several alternative versions of them, based on one recording, without copyright issues and such.

2.3 Questionnaire

Apart from the abovementioned general questions concerning, age, gender, musical sophistication and literary experience, there were several song-specific questions. There were two multiple-choice questions per song concerning 'refrein' perception, one multiple-choice question concerning semantic interpretation and a series of Likert-scale questions concerning appreciation and, in the case of song 2, ethic valuation. Finally, there was also a six-item fill-in-the-blank recall test for each song, but this was used only to get an impression of the participants' attitude towards the survey. Therefore, these questions will not be reported on here.

Concerning 'refrein' perception, the participants were asked whether they thought there was a 'refrein' (i.e. a refrain or a chorus), and after that they were asked to choose from several options which part was the 'refrein', or which part they would choose if someone would urge them to indicate a 'refrein' although in the first instance they did not think there was one.

For song 1, the options were as follows: (1) 'Hou'en zo' ('Keep it like that', i.e. the last line of the A section); (2) 'the part that begins with "Alweer een dag" ["another day"] and ends with "Hou'en zo"' (i.e. the entire A section); (3) 'the phrase "Het is geluk"' ('It's all about luck', i.e. the last line of the B section); (4) 'the part about feeling grateful and lucky' (i.e. the entire B section); (5) a combination of 1 and 3; and (6) a combination of 1 and 4.

For song 2, the options were as follows: (1) 'The line "dan kijk ik met mijn handen naar jou"' ('and then I look at you with my hands', i.e. the last line of the A section); (2) 'The part that begins with 'Mijn ogen' ("my eyes") and ends with 'dan kijk ik met mijn handen naar jou"' (i.e. the entire A section); (3) 'The words "Mijn ogen"'; (4) 'The part that begins with "Mijn vingers verkennen..." ("My fingers explore...") and ends with "Ik voel dat je mij echt voelt"' ('I can feel that you really feel me', i.e. the B part); (5) A combination of 1 and 4; and (6) A combination of 3 and 4.

Concerning (semantic) interpretation, the participants were asked to choose one of five or six interpretations of the song's content. For song 1, these were as follows: (1) 'The singer thinks life is full of difficulties and dangers'; (2) 'The singer realizes how fortunate he is, and enjoys this feeling'; (3) 'The singer sees a lot of threats which he hopes to escape'; (4) 'The singer is grateful because he realizes how lucky he is'; and (5) 'The singer is careless, nothing will happen to him'. Interpretation 3 is clearly A-section oriented; interpretation 1 is also more A-section oriented although it does not refer to the repeated catch phrase at the end of it; interpretation 2 is somewhat more B-section oriented; interpretation 4 is clearly B-section oriented; and interpretation 5 is neither B nor A-section oriented; it may occur as a result of overemphasizing the song's ironic tone of voice.

For song 2, the options were as follows: (1) 'The singer assumes that with his hands he can see his partner just as good as with his eyes'; (2) 'The singer thinks bodily contact is at least as important as appearance'; (3) 'The singer thinks beauty can be experienced not only with the eyes but also with tactile sense'; (4) 'The singer is totally immersed in his fantasies about sex with the other'; (5) 'The singer mainly describes how intense contact can be if one does not look but feels'; and (6) 'The singer thinks his partner is so beautiful that he would love to touch her'. Options 1 and 3 are A-section oriented, because they focus on the comparison between looking with eyes and looking with hands; for the same reason, option 2 is predominantly A-section oriented, although it includes the B-section-oriented word 'contact'; option 4 and 5 are clearly B-section oriented as they focus on the action of 'looking with hands'; and option 6 is neither A-section oriented nor B-section oriented as there is no reasoning in it, and it combines the notions of beauty and touching from the A sections with the sense of lust or desire in the B section.

Finally, concerning valuation, there were questions at two instances. First, while listening to the songs, the participants were asked to rate the statement 'I think this is a beautiful song' on a seven-point scale from (1) 'absolutely disagree' to (7) 'totally agree'. Later on, a series of statements to rate on a similar scale followed.

For song 1, the statements were the same as those in the previous experiment [27]: 'The song was cheerful'; 'The song was well structured'; 'The melody was dull'; 'The lyrics were humorous'; 'The lyrics were comprehensible'; 'There were unexpected twists and turns in the song'; and 'I was captivated till the end'.

For song 2, the question about twists and turns was deleted, as it is not a question about appreciation of either lyrics or music, and consequently in the earlier experiment, it did not contribute to one of the factors emerging through a factor analysis on all items. However, three statements were added: 'The tone of voice is light'; 'The song is pornographic'; and 'The song is respectful to women'. These statements were added because the tone of voice in the bridge section is quite erotic and may be perceived as less light, more pornographic and less respectful towards women. In the A sections, the singer only tells his lover that he does not have to look at her with his eyes, because he can also look at her with his hands, but in the B section, he actually describes what looking at her with his hands is like. On the other hand, the B section ends with the line 'en ik voel dat je mij echt voelt' ('and I can feel that you really feel me'), which makes it less male-'gaze' focused.

2.3.1 Analyses

The results were analysed in SPSS using principal axis factoring analyses with oblique rotation (direct oblimin) for the ratings, generalized linear regressions for the factors and both binomial generalized linear regressions and binomial generalized estimating equations for interpretations and 'refrein' perception. In order to run binomial regressions, the multinomial variables representing the choices concerning interpretations and 'refrein' perception were reduced to binomial variables representing or not representing an A-section-oriented bias. In these variables, the value '1' was assigned to all interpretations that were described above as A-oriented interpretations, and all answers referring to 'refrein' candidates from A sections only, and the value '2' to all other options.

3 Results

3.1 'Refrein' Perception

For both songs, the question whether there was a 'refrein' or not was answered significantly different across song versions (*Wald X*(song1; *df* 2) = 7.22, *p* = 0.027; *Wald X*(song2; *df* 2) = 8.92, *p* = 0.012). Binomial generalized linear regressions on this variable showed a significant effect of song version, which is mainly due to the AAAABAABc version of song 1 and the AABA version of song 2. As Table 1 shows, these song versions were remarkably less often thought to have a 'refrein' than the other versions. The differences between the other song versions were marginal, although it is striking that the AABABA version is even more often thought to have a 'refrein' than the AABAB version.

For song 1, the answers to the question which part of the song is the 'refrein' if there must be one are not significantly different across song versions. For song 2, they are. However, a binomial generalized linear regression on the original variable indicating an A-section-oriented bias could not be conducted because of a quasi-complete separation within the data. As Table 2 shows, none of the answers concerning the AABA version involved the B section or a part of it. However, after extending the category of answers involving the B part with answer option 3 (an option which was rarely chosen for all song versions), the effect of song version was still significant (*Wald X*(song2; *df* 1) = 10.28, *p* = 0.006), indicating that the B section is less likely to be perceived as a 'refrein' (i.e. as a chorus) in the AABA version than in the other versions. The difference between the other versions was not significant.

3.2 Interpretation

The question whether section order can change the (semantic) interpretation of a song was assessed by a multiple-choice question in which participants had to choose between several interpretations of the song, some of which were more or less A-section oriented, while others were at least partly based on the content of the B section. The results indicate that the interpretations of song 1 were not significantly different across song versions, whereas those of song 2 are. A binomial generalized linear regression on a variable indicating an A-section-oriented bias or not showed a

Table 1 'Refrein' or not

	Song 1			Song 2		
	AABAABAAc	ABABAAAAc	AAAABAABc	AABA	AABAB	AABABA
Yes	29	20	12	13	20	24
No	12	10	18	22	12	9

Table 2 Numbers of times a 'refrein' candidate is chosen per song version

Song	Song part	Version		
Song 1		AABAABAAc	ABABAAAAc	AAAABAABc
	Last line A section	16	13	15
	A section	18	11	9
	Last line B section	1	0	2
	B section	0	2	0
	Combination or 1 and 3	6	2	3
	Combination or 1 and 4	0	2	1
	Total A	34	24	24
	Total B or A+B	7	6	6
Song 2		AABA	AABAB	AABABA
	Last line A section	16	5	9
	A section	13	9	10
	First words A section	6	5	3
	B section	0	10	6
	Combination or 1 and 4	0	2	2
	Combination or 3 and 4	0	1	3
	Total A	35	19	22
	Total B or A+B	0	13	11

significant effect of song version (*Wald X*(song2; *df* 2) = 8.23, *p* = 0.016), indicating that the interpretation of the AABAB version was significantly more B oriented than the other versions, particularly the AABA version. See Table 3 for more details.

Although the interpretations of song 1 were not significantly different across song versions, those of song 2 were; a binomial regression using generalized estimating equations with song as the within-subject variable, A-section bias in interpretation as the target value and A-section bias in 'refrein' perception as predictor showed that there is a significant relationship between 'refrein' perception and interpretation across songs (*Wald X*(*df* 1) = 11.14, *p* = 0.001).

3.3 Appreciation

For each song, a principal axis factoring analysis was run on the Likert-scale items concerning aesthetic valuation of the songs and their lyrics. For the second song, the additional items related to ethical issues were also included in this factor analysis. For both data sets, both the KMO statistic and the measurements of sampling adequacy (MSA) were above 0.5 (KMO song 1 = 0.80; KMO song 2 + 0.774), the determinants were larger than 0.0001, and Bartlett's test of sphericity was significant. However, in the analysis of the items concerning song 1, the item regarding twists and turns was deleted, because MSA was relatively low, i.e. very

Table 3 Numbers of times an interpretation is chosen per song version

Song	Interpretation	Version		
Song 1		AABAABAA	ABABAAAA	AAAABAAB
	Life full of difficulties and dangers	6	8	4
	Fortunate and happy	7	3	5
	Hope to escape threats	10	5	7
	Grateful because of luck	18	13	13
	Careless	0	1	1
	Total A	16	13	11
	Total A or A+B	25	17	19
Song 2		AABA	AABAB	AABABA
	Hands as good as eyes	4	1	7
	Bodily contact important	4	2	0
	Beauty with tactile sense	18	10	15
	Immersed in fantasies about sex	1	1	0
	Intense contact when feeling	7	12	9
	Beauty raises wish to touch	1	6	2
	Total A	26	13	22
	Total A or A+B	9	19	11

close to 0.5. For song 1, two factors with eigenvalues larger than 1 were retained: positive value (PV) and comprehensible lyrics (CL); for song 2, three of those factors were retained: good song (GS), pornographic song (PS) and good lyrics (GL). As Table 4 shows, there are parallels between PV and PS on the one hand, and CL and GL on the other, but there are striking differences as well, hence the differences in naming.

Although the differences between factor means per song version were not likely to be significant given the sample size, it is still interesting to explore the differences. As Table 5 shows, PV is relatively high for AAAABAABc version compared with the ABABAAAAc one, whereas CL is relatively high for the AAAABAABc version compared with the AABAABAAc version. Furthermore, GS is relatively high for the AABA version of song 2 compared with the AABAB version, PS is relatively high for the AABAB version compared with the AABABA version, and GL is relatively low for the AABAB version. However, none of these effects were significant. Having said that, two covariates did show a significant effect. The Gold-MSI Emotions scale turned out to be a significant predictor CL and GL, and the factor Passive Literary Enjoyment a significant predictor of CL.

Table 4 Factor analyses Likert-scale items, factor loadings and factor specifications

	Song 1		Song 2		
Item/Factor property	PV	CL	GS	PS	GL
Song was beautiful	0.86	0.26	0.96	−0.12	0.32
Song was cheerful	0.57	0.35	0.59	−0.07	0.55
Song was well structured	0.75	0.14	0.74	−0.02	0.54
Melody was dull	−0.64	−0.02	−0.67	−0.11	−0.29
Lyrics were humorous	0.37	0.28	0.29	0.41	0.41
Lyrics were comprehensible	0.25	0.89	0.34	−0.06	0.60
I was captivated till the end	−0.71	0.26	0.78	0.02	0.52
The tone of voice was light			0.56	0.15	0.62
The song was pornographic			−0.05	0.71	0.03
The song was respectful towards women			0.09	−0.40	0.23
Initial eigenvalue	3.25	1.11	3.99	1.48	1.06
Percentage of variance predicted	46.36	15.90	39.72	14.78	10.55
Rotated sum of squared loadings	2.74	1.11	3.36	0.88	2.01

PV = positive value; CL = comprehensible lyrics; GS = good song; PS = pornographic song; GL = good lyrics

Table 5 Factor means per song version

Song	Song version	Mean (SD)		
Song 1		PV	CL	
	AABAABAAc	0.02 (0.93)	−0.10 (1.03)	
	ABABAAAAc	−0.13 (0.96)	0.02 (0.73)	
	AAAABAABc	0.11 (0.92)	0.12 (0.86	
Song 2		GS	PS	GL
	AABA	0.10 (1.09)	0.06 (0.86)	0.06 (0.86)
	AABAB	−0.09 (0.91)	0.15 (0.75)	−0.10 (0.82)
	AABABA	−0.01 (0.95)	−0.21 (0.71)	0.03 (0.84)

PV = positive value; CL = comprehensible lyrics; GS = good song; PS = pornographic song; GL = good lyrics

4 Discussion

In a small two-part listening experiment, a series of hypotheses concerning the effect of section order on 'refrein' perception semantic interpretation of a song and appreciation of a song were tested. The latter, however, was assessed only in an exploratory way.

4.1 'Refrein' Perception

Concerning 'refrein' perception, the results for both songs are in line with the main hypotheses that section order affects 'refrein' perception and, consequently, that the question whether a song part is perceived as a 'chorus' or a 'refrein' is at least partly dependent on its position within the song. The results are also in line with the hypothesis that the B section of song 2 is more likely to be perceived as a chorus in the AABAB version than in the other versions of song 2, although many listeners who have heard this version still think the 'refrein' is in the A sections. This indicates that section order is an important factor in 'refrein' perception, but that, nevertheless, there are essential differences between an AABAB song and a verse-verse-chorus-verse-chorus song.

The results for song 1 are not in line with the third hypothesis. The B section was not more likely to be perceived as a 'chorus' or as a song section containing the 'refrein' in the AAAABAABc version of that song. Conversely, many participants turned out not to recognize a 'refrein' in this version at all. However, these results do show that the AAAABAABc form has made it less likely that the A section is or contains the 'refrein'. The fact that in the second instance the B section was not a good alternative to the A section or parts of it may be due to the fact that the second B section was not a verbatim repetition of the first one. As mentioned before, the second B section of the AAAABAABc version in the first experiment was a verbatim repetition of the first one [27], and in that experiment, the B section or a part of it turned out to be chosen more often as the 'refrein'. Moreover, the fact that in all song versions of song 1 have urged at least some participants to designate the B section as a 'refrein', while this is not the case for the B section in the AABA version of song 2 shows that the mere fact that there is some alternation between A and B sections can 'turn' the B section into the 'refrein', at least for some persons.

These results indicate that both position and verbatim repetition are important features of a 'refrein'. Verbatim repetition of an entire song section can even overrule the more frequent verbatim repetition of a refrain line, although the results for both songs show that also song sections with varying lyrics but including one or two refrain lines or words can be perceived as 'refrein' (in the sense of chorus). This raises the question as to whether AABA songs in which the As and Bs represent separate sections and not parts of a verse or a chorus should be considered chorus-chorus-bridge-chorus songs. However, it seems to be more likely that A sections in AABA songs or related songs such as the AABAABAAc song in this experiment are neither verses nor choruses.

4.2 Interpretation

Concerning interpretation, only the results for song 2 are in line with the hypotheses stated in the introduction. Not only is there a significant effect of section order on the interpretation of the song, but the interpretation of the AABAB version is also more B oriented than the interpretation of the other two versions. What is more, although the differences between the AABA version and the AABABA version are not significantly different concerning A-section bias, the A-section bias tends to be less strong in the AABABA version, and two interpretations are not chosen for that version at all. So, it seems to be the case that this version is interpreted slightly different than the AABA version, although there are no extra lyrics involved. Remarkably, the two interpretations that were not chosen in reaction to the AABABA version were interpretations which, according to the author, are at odds with parts of the lyrics. So, in his eyes, the AABABA version is interpreted not only slightly different but also slightly more correct.

The fact that the results for song 1 are not as predicted does not mean that they are totally at odds with all hypotheses regarding the effect of presentation order on interpretation. In fact, the hypothesis that the interpretation of the AAAABAABc version would be more B-section oriented than the interpretation of the other versions was based on the assumption that the B section of this song version or a part of it was also more likely to be perceived as a 'refrein' of the song in that version. As observed and explained above, this was not the case, so it would have been rather puzzling if the interpretation was more B-section oriented. However, just as the B section was designated to be the 'refrein' in all song versions of song 1 by at least some participants, also several participants have chosen an interpretation of the song which was more or less B-section oriented in reaction to all song versions. In line with that, and with hypothesis 5, a multilevel repeated measure analysis of the relationship between 'refrein' perception and interpretation showed that a participant who thinks the A section or a part of it is the 'refrein' also tends to choose for an A-section-oriented interpretation of the song, whereas participants who think that the B section, or parts of it, can be considered to be a 'refrein', whether or not in combination with the A section (or parts of it) tend to choose for a more or less B-section-oriented interpretation.

Future research could investigate the relationship between 'refrein' perception and interpretation in the data of the previous experiment with song 1 as well. Apart from differences in the judgments concerning the AAAABAABc version, there are also some other remarkable differences between the results of both experiments. Possibly, these differences have something to do with the participants' age (52 on average in the earlier study, versus 26 in the current one). Differences in life experience and musical culture may have caused other interpretations and other ideas of what a 'refrein' is.

4.3 Valuation

As expected, there were no significant differences in aesthetic and ethic evaluation, probably due to sample size. However, as expected, just as in the earlier experiment, appreciation for the ABABAAAc was lower than for the AAAABAABc version, which is in line with RAS rules. Additional research with a much larger sample size is required in order to investigate whether this is indeed an effect of late repetition. Other interesting differences are those concerning GS and PS. The fact that GS is higher but not significantly higher for the AABA version of song 2 than for the other versions is in line with the hypothesis that the verbatim repetitions in the other versions would not decrease the appreciation for these songs, but is at odds with literature suggesting that verbatim repetition of song lyrics would increase liking [19]. As other authors have argued before, it seems likely that the acceptability of repetition is limited, particularly if song lyrics are involved [11, 15, 21, 27]. Therefore, additional research with a larger sample size and more songs is required to investigate to what extent verbatim repetition of song sections is accepted. Finally, the differences in PS are in line with the assumption that the B-section-oriented, AABAB version is perceived as more pornographic and less respectful towards women than the other versions of song 2. If these differences would turn out to be significant in research with a larger sample size, this would show again that song section order affects the interpretation of a song in a predictable way.

4.4 Covariates

As reported in the results section, some of the variables concerning musical and literary sophistication, used as covariates, turned out to be significant predictors of variables indicating appreciation for lyrics, i.e. the factor representing passive literary enjoyment and the Gold-MSI Emotions scale. However, these effects did not affect the effect of song version.

4.5 Limitations

Apart from the fact that sample size may have been too small to detect significant differences in either aesthetic or ethic valuation, sample composition may have affected the results as well. Differences between the results of the earlier experiment with song 1 and the results of the current experiment with the same song cannot be explained by the use of another AAAABAABc version only. Probably, the age of the participants, which was twice as high in the earlier experiment, has caused a different perception of the song.

Another limitation is that the results for the AAAABAABc versions with either a varying or a verbatim repeated B section are difficult to compare because they are in different studies with apparently very different participants. So, in hindsight, it would have been better to include both AAAABAABc versions in this study. For song 2, it would also have been helpful to include a fourth version, i.e. an AABABA version in which the last A section was not exactly the same as the first one (except for a few connectives) but in which it was really a new A section.

Apart from that, the fact that this study involves only two specific songs makes it impossible to detect general rules concerning the effect of song section order. However, as far as it falsifies existing assumptions, its results are of general interest in themselves. And as far as this study develops new hypotheses, it can give direction for further research involving more songs and including a straightforward verse-chorus song with a harmonically stable chorus turned into a VVCVCV song.

Another limitation may be that the effects of formal structure (particularly, of section order and repetition) on 'refrein' perception, interpretation and valuation can, or even should, be assessed in several other ways. Questions can be focused more on emotional meaning [28], can be targeted more on the effect of specific song sections (e.g. on the question whether a repeated song section is perceived as meaningful or not) [25] or can be measured through bodily reactions such as skin conducting and brain potentials [36].

Finally, one may argue that the results of this study are weakened by multiple comparison. In the target sections concerning 'refrein' perception and interpretation, there were seven regressions and only two of them showed a significant effect with a p factor small enough to resist a Bonferroni-like correction through multiplying it by 7. However, several of these analyses were conducted on strictly separated data sets, i.e. data concerning song 1, and data concerning song 2. Moreover, the insignificance of the regression regarding song version-dependent interpretations of song 1 could have been expected, because of the insignificance of a possible section-oriented bias 'refrein' perception. Finally, the results of the different analyses strengthen each other. For example, if 'refrein' perception in song 2 is significant, and the connection between 'refrein' perception and interpretation in both songs is significant, it cannot be the case that the section-oriented bias in song 2 is completely coincidental.

5 Conclusions

The results of a small listening experiment, following an experiment reported on earlier, showed that section order (including the use of verbatim repeated sections) affects 'refrein' perception and (semantic) meaning of a song. Several alternative song versions of two songs were created digitally. The original version of the songs consisted of several A sections containing several refrain lines or words and one or two B sections. Participants were less sure that the refrain lines in the A section or the A section as a whole was the 'refrein' of the song after hearing a song versions in which the B section was put in a chorus position, or in which the B section was repeated verbatim. Moreover, after hearing such a song version, they were less likely to choose an A-section biased interpretation.

An exploratory inspection of the differences in appreciation for the different song version showed some interesting tendencies, which require further investigation using a larger sample size.

These results show that formal structure in popular music is much more complex than is often assumed and cannot simply be categorized in terms of strophic songs, verse-chorus songs, AABA songs and verse-chorus-bridge songs. Refrain lines in A sections and repeated B sections (either verbatim or not) both can be perceived as the 'refrein' of the song and to contain its main message. As a result, the 'refrein' of a song cannot be detected on the basis of strict formal properties. For example, not every section which is repeated verbatim is a chorus. Apparently, the 'refrein' is a song part that is repeated at least once and is perceived as the core of the song semantically. These conclusions are strengthened by the fact that the stimuli were created digitally, avoiding performance-dependent confounding factors.

Acknowledgments This chapter builds on an earlier experiment, reported on in the author's doctoral dissertation [27], which will be summarized within the current chapter. This earlier research was supported by the Netherlands Organisation for Scientific Research (NWO, project number: 023.004.078). The author likes to thank his supervisors Emile Wennekes, Frank Hakemulder and Roel Willems, his participants and those who helped him to find these participants.

References

1. Bourne, J. E., Ashley, R. D.: Listeners reconcile music and lyrics mismatch in song interpretation. Poster presented at ICMPC12 in Thessaloniki. Retrieved from author. (2012)
2. Bouwer, F., Schotanus, Y. P., Sadakata, M., Müllensiefen, D., Schaefer, R.: Measuring musical sophistication in the low countries; validation of a Gold MSI translation in Dutch (in preparation)
3. Braae, N.: Linearity in Popular Song. In N. Braae, K. A. Hansen, (eds.) On Popular Music and Its Unruly Entanglements, pp. 83—99. Palgrave Macmillan, London (2019)
4. Brackett, D.: Interpreting popular music. Cambridge University Press, Cambridge/New York/Melbourne (1995)
5. Burns, L.: Analytic methodologies for rock music: harmony and voice leading strategies in Tori Amos's 'Crucify'. In: W. Everett (ed.) Expression in pop-rock music: a collection of critical and analytical essays (Studies in Contemporary Music and Culture). Garland Publishing, New York/London (2000)
6. Covach, J.: Form in Rock Music: A Primer. In: D. Stein (ed.) Engaging Music: Essays in Music Analysis, pp. 65–76. Oxford University Press, Oxford (2005)
7. Davis, S. The craft of lyric writing. Writers Digest Books, Cincinnati (1985)
8. Everett, W, (ed.): Expression in pop-rock music: a collection of critical and analytical essays (Studies in Contemporary Music and Culture). Garland Publishing, New York/London (2000)
9. Gordon, R. L., Schön, D., Magne, C., Astésano, C., Besson, M.: Words and melody are intertwined in perception of sung words: EEG and behavioral evidence. PLoS One, 5(3), e9889 (2010) https://doi.org/10.1371/journal.pone.0009889
10. Harden, A. C.: Narrativizing recorded popular song. In: N. Braae and K. A. Hansen (eds.) On Popular Music and Its Unruly Entanglements, pp. 39–57. Palgrave Macmillan, London (2019)
11. Huron, D.: A Psychological Approach to Musical Form: The Habituation–Fluency Theory of Repetition. Current Musicology **96**, 7–35. (2013) https://doi.org/10.7916/D8KP81FG
12. Kramer, L.: Music and Poetry: The Nineteenth. Century and After. University of California Press, Berkeley (1984)
13. Lloyd, B.: The form is the message: Bob Dylan and the 1960s. Rock Music Studies **1**(1), 58–76 (2014). https://doi.org/10.1080/19401159.2013.876756

14. Malheiro, R., Panda, R., Gomes, P., and Paiva, R. P.: Emotionally-relevant features for classifications and regression of music-lyrics. IEEE Trans. on Affect. Comput, **9**(2), 240–254. (2016) https://doi.org/10.1109/TAFFC.2016.2598569
15. Margulis, E. H.: On repeat: How music plays the mind. Oxford University Press, Oxford (2014).
16. Middleton, R.: Studying popular music. Open University Press, Milton Keynes/Philadelphia (1995)
17. Müllensiefen, D., Gingras, B., Musil, J., Stewart L.: The Musicality of Non-Musicians: An Index for Assessing Musical Sophistication in the General Population. PLoS One, **9**(2): e89642 (2014) https://doi.org/10.1371/journal.pone.0089642
18. Negus, K.: Bob Dylan. Equinox Publishing Ltd, Sheffield (2008)
19. Nunes, J. C., Ordanini, A., Valesia, F.: The power of repetition: repetitive lyrics in a song increase processing fluency and drive marketing success. J. of Consum. Psychol., **25**(2), 187–199 (2016)
20. Ollen, Joy and Huron, D.: Listener Preferences and Early Repetition in Musical Form, In: S. D. Lipscomb, R. Ashley, R. O. Gjerdingen and P. Webster (eds.) Proceedings of the 8th International Conference on Music Perception and Cognition, pp. 405–407, Casual Productions, Evanston ILL (2004)
21. Pattisson, P.: Writing better lyrics: the essential guide to powerful songwriting. Writer's Digest Books, Cincinnati (2009)
22. Rolison, J., Edworthy, J.: The role of formal structure in liking for popular music. Music Percept., **29**(3), 269–284 (2012) https://doi.org/10.1525/mp.2012.29.3.269
23. Rolison, J., Edworthy, J.: The Whole Song Is Greater Than the Sum of Its Par ts: Local and Structural Features in Music Listening. Psychomusicology, **23**(1), 33–48 (2013) https://doi.org/10.1037/a0032442
24. Schotanus, Y. P.: The musical foregrounding hypothesis: How music influences the perception of sung language. In: J. Ginsborg, A. Lamont, M. Philips, S. Bramley (eds) Proceedings of the Ninth Triennial Conference of the European Society for the Cognitive Sciences of Music, 17–22 August 2015, Manchester, UK (2015)
25. Schotanus, Y. P.: AABA, couplet, refrein: hoe een gebrekkig begrippenapparaat onze kijk op liedjes beperkt (AABA, verse, refrain: how a deficient set of concepts hampers our view on song). Poster presented, May 7th 2016, at the KVNM spring meeting, Leiden. In: Schotanus, Y. P.: "Supplementary materials PhD project Singing as a figure of speech and related publications", https://hdl.handle.net/10411/XRMMFW, DataverseNL, V1. (2020)
26. Schotanus, Y.P.: Supplementary materials for publications concerning three experiments with four songs, https://hdl.handle.net/10411/BZOEEA, DataverseNL Dataverse, V2. (2020)
27. Schotanus, Y. P.: Singing as a figure of speech, music as punctuation: A study into music as a means to support the processing of sung language [doctoral dissertation] Utrecht (2020)
28. Schotanus, Y. P.: Singing and accompaniment support the processing of song lyrics and change the lyrics' meaning. Empirical Musicology Review, **15**(1-2), 18–55 (2020) https://doi.org/10.18061/emr.v15i1-2.6863.
29. Schotanus, Y. P.: Reanalyzing Schotanus 2020: A reaction to Lee's commentary, Empirical Musicology Review, **15**(3-4), 273–278 (2020) https://doi.org/10.18061/emr.v15i3-4.8002
30. Schotanus, Y. P.: Experiencing Dylan: The Effect of Formal Structure and Performance on the Popularity and Interpretation of Two Dylan Songs. In: Gurke, T., Winnett, S. (eds.) Popularity of Words and Music, 39–64. Palgrave Macmillan, Cham (1921) https://doi.org/10.1007/978-3-030-85543-7_3
31. Schotanus, Y. P.: The effect of timing on the singer's tone of voice. Psychomusicology, Advance online publication **31**(3-4), 107–122 (2021). https://doi.org/10.1037/pmu0000278
32. Schotanus, Y. P.: Violations of music syntactic expectations as prosodic cues in sung sentences, Psychol. Music (under review)
33. Schotanus, Y. P., Koops, H. V., Reed Edworthy, J.: Interaction between musical and poetic form affects song popularity: The case of the Genevan Psalter. Psychomusicology, **28**(3): 127–151 (2018) https://doi.org/10.1037/pmu0000216

34. Spyropoulou Leclance, M.: Le refrain dans la chanson Francaise de Bruant a Renaud. Pulim: Limoges (1993)
35. Summach, J.: Form in top-20 rock music, 1955–89 [Doctoral dissertation] UMI 3525244. Proquest, Ann-Arbor MI (2018)
36. Tsai, C.-G., Chen R.-S., and Tsai, T.-S.: The arousing and cathartic effects of popular heartbreak songs as revealed in the physiological responses of listeners. Musicae Scientiae **18**(4): 410–422 (2014). https://doi.org/10.1177/1029864914542671
37. Van Balen, J. M. H.: Audio description and corpus analysis of popular music. [Doctoral dissertation]. Utrecht (2016)
38. Von Appen, R., Frei Hauenschildt, M.: AABA, refrain, chorus, bridge, prechorus: Songformen und ihre historische Entwicklung (Song forms and their historical development). In: Helms, D., Phleps, T. (eds.) Black Box Pop: Analysen populärer Musik. Transcript, Bielefeld, pp. 57–124 (2012)

Musical Influence on Visual Aesthetics: An Exploration on Intermediality from Psychological, Semiotic, and Fractal Approach

Archi Banerjee, Pinaki Gayen, Shankha Sanyal, Sayan Nag, Junmoni Borgohain, Souparno Roy, Priyadarshi Patnaik, and Dipak Ghosh

1 Introduction

Imagine a lightning without the thunder or a river without the ripple or a knock at the door without someone arriving physically! Feels incomplete, right? In the world around us, at any point of time, our eventual and emotional experiences are a post-cognition composite of the data received from all five human senses—auditory and

---

A. Banerjee (✉)
Rekhi Centre of Excellence for the Science of Happiness, IIT Kharagpur, Kharagpur, India

Sir C. V. Raman Centre for Physics and Music, Jadavpur University, Kolkata, India

P. Gayen
Department of Design, Kala Bhavana, Visva Bharati University, Bolpur, India

S. Sanyal
Sir C. V. Raman Centre for Physics and Music, Jadavpur University, Kolkata, India

School of Languages and Linguistics, Jadavpur University, Kolkata, India

S. Nag
Department of Medical Biophysics, University of Toronto, Toronto, ON, Canada

J. Borgohain
Department of Humanities and Social Sciences, IIT Kharagpur, Kharagpur, India

S. Roy
Sir C. V. Raman Centre for Physics and Music, Jadavpur University, Kolkata, India

Department of Physics, Jadavpur University, Kolkata, India

P. Patnaik
Rekhi Centre of Excellence for the Science of Happiness, IIT Kharagpur, Kharagpur, India

Department of Humanities and Social Sciences, IIT Kharagpur, Kharagpur, India

D. Ghosh
Sir C. V. Raman Centre for Physics and Music, Jadavpur University, Kolkata, India

© The Author(s), under exclusive license to Springer Nature Switzerland AG 2023
A. Biswas et al. (eds.), *Advances in Speech and Music Technology*, Signals and Communication Technology, https://doi.org/10.1007/978-3-031-18444-4_18

visual counterparts being the most involved among them. So, it is expected that the emotional experience triggered by only a set of visual or auditory stimuli will be significantly different from when they are experienced together by the audience. The same is applicable for our memories also. Listening to a familiar song can often bring back memories of the ambience where it was first heard, and revisiting a painting in a museum or a home can evoke auditory memories of conversations or an accidental tune from that occasion. Our aesthetic experience being multimodal, these associations also influence the listening and viewing experiences. Such associations, accidental or intentional, fall broadly into three categories—indifferent, compatible, or incompatible. Not all sounds and visuals match with each other—where in some cases, they complement each other (like the ripple sound and the visual of a flowing river) to offer a more fulfilling emotional experience, and in other cases, they may contradict each other (like mourning of people beside funeral pyres burning against a scenic sunset at Varanasi riverbank). Depending on the inherent characteristic features of the concerned audio and visual counterparts, the degree of compatibility between the two varies. The same is applicable in the case of music-visual arts intermediality also. Musical features like tempo, loudness, continuity or visual features like dynamicity, orderliness, crowdedness, etc. primarily determine the nature of emotions evoked through them individually. When they are associated with each other, if the features and consequently the emotional expectation of the music match with that of the visual artwork, they appear as compatible, whereas if the features of the music and the artwork contradict each other, they appear as incompatible to the audience. Where one does not necessarily see any assonance or dissonance between the two can be called an indifferent association. Both music and abstract paintings are very powerful tools capable of evoking a wide range of emotions among the audience, and this work attempts to explore the variations in the total emotional outcomes for a compatible and incompatible combination between the two.

1.1 Musical and Visual Attributes

The primary three attributes which characterize any acoustic signal are its pitch, loudness, and timbre [1]. According to Plack and Oxenham [2], pitch itself is a very important attribute of any auditory stimulus as pitch enhances human ability to perceptually identify different sound sources, depending upon differences in their fundamental frequencies, and also helps us to group together the individual sound components, or harmonics, that arise from the same vibrating source. Although pitch primarily manifests the frequency-related information of the sound signal, it also depends on the sound pressure and the waveform of the stimulus [3]. Frequency of sound is an objective physical quantity, scientifically measured in Hertz (Hz), describing the periodic properties of the signal [4], but pitch is a subjective perception that is measurable only by psychophysical investigation. Most of the sounds that we hear every day, especially the musical ones, are superposition of many different simultaneous frequencies, but the human auditory system combines them into a single percept of one overall pitch [4]. Another most important component of sound is its loudness, which is the manifestation of the intensity or square of amplitude of the vibrating particles, but loudness also has some subjective perceptual component in it and is most commonly measured in dB SPL unit. From a realistic perspective, it is often seen that when the intensity of a natural sound source increases, its frequency also tends to increase, and vice versa. We can observe the same during the acceleration of an engine and human speech [5]. These positive correlations in nature and everyday experience have made people form a mental link between loudness and pitch [5, 6]. Interestingly, Repp [7] reported that during music listening, people expect that the melody lines that rise in frequency will also rise in intensity. A number of studies on music psychology explored a wide range of frequencies, intensities, and timbres and have concluded that changes in any of these perceptual dimensions can influence perceptions of the others [4–6]. The other most important musical features are tempo, pauses, and finally the nature of complexity of the composition. Gabrielsson [8] suggested that musical tempo has major influence on listeners' emotional experiences. Researchers found that audiences have a tendency to relate fast tempo and major modes with joyful or exciting emotions, and slow tempo and minor modes with calmness or sadness [9, 10]. The distribution and lengths of pauses within a musical piece, on the other hand, influence the element of surprise or curiosity in the listeners' minds [11]. Finally, compositional complexity-wise, music of chaotic, loud, and powerful nature are often observed to evoke anger, excitement, anxiety, depression, social isolation, and loneliness [12, 13]. In the domain of visual arts also, a number of previous studies tried to analyze the emotional content related to a particular painting through a number of basic visual features such as line, color, texture, shape, etc. Artists use these features creatively to portray a variety of emotions, e.g., in color—saturation, brightness, and hue are the three properties which play crucial role in determining the valence of emotional expression [14–16]. In a similar manner, the use of lines [14], texture [16], shape, form, and space [17] innovatively in the painting leads to the variation of emotional affect in the observers. Painters use lines to indicate movements and to characterize the changes in feelings, e.g., calmness or relaxation is denoted by horizontal lines, while vertical lines indicate stability, diagonal lines increase tension, and flow is characterized by curvatures [14]. Hough transform is a special technique to describe the measures of static and dynamic lines. Shapes, although having two dimensions, i.e., height and width, are also characterized by features like angularity, roundness, or their simplicity and complexity [17]. Apart from these, there are a number of methods which involve automated emotion recognition from several abstract paintings using a variety of classification features like SVM based on local image statistics [18]. Looking at the associations between auditory and visual features, we observe that sounds can provide visual information to the perceiver about their surroundings by helping them to visualize the position and characteristics of the sound sources. The Doppler effect or Doppler illusion, described by Christian Doppler in 1842, reveals that the pitch of a moving sound source increases as the source approaches the observer, and vice versa. Neuhoff and McBeath [19] suggested that in a similar way, musical frequency can influence the perceived distance between the listener and the sound source. Our everyday experiences also suggest that lower-pitched sounds tend to be linked with larger objects, while soft or low-volume sounds signify that the emitter is distant. These audiovisual correlations which are frequently observed in nature and our everyday experiences often lead the observers to form a mental link between them. These types of mental links were exhaustively explored by the researchers in the domain of consumer cognition and behavior [20]. Recent literature have given further emphasis on investigating such sensory correspondences [21, 22]. Hagtvedt and Brasel [23] explored how in the context of advertising and promotions understanding the effects of musical frequencies on consumers' perceptions and decision-making enhanced the marketers' trading ability. Although aesthetic objects are composites, they have distinctive orientations in terms of the earlier mentioned visual and musical features. In this work, we have chosen a few of them and studied the complementary or contradictory quality among them in order to explore the nature of intermediality between music and visual arts.

1.2 Intermediality Between Music and Visual Arts

Intermediality refers to the translation or connectivity of expressions between two mediums—here music and visual arts. Music can capture and hold our attention naturally, through its rhythmic and melodic tensions [24]. Theories related to psychology of music suggest that music listening creates an associative connection with personal memories and meanings [25]. Music helps us to disconnect from the real world and reconnect to a virtual environment (immersion). From the desire to hear the ancient musical sounds depicted through the poetic ekphrasis on a Greek vase, Keats wrote "heard melodies are sweet ... those unheard are sweeter." In these two lines, Keats expressed the role of imagination in the aesthetic experience of ancient Greek vase-painting [26]. Music-induced visual imageries play a big role in the intermediality of emotions between the two mediums. Juslin and Västjfäll [27] observed that listening to music clip often induces imageries in the audience's mind that can evoke emotions in the process. Cognitive neuroscientists identified this phenomenological experience as the *involuntary musical imagery* (INMI) which happens spontaneously and without conscious control of the audience while she/he is listening to a music piece [28–30]. Musical features can induce episodic memories triggering reminiscing of past incidents, but while experiencing emotions, an audience can also be influenced by incidents and visual imageries not associated with episodic memory. Such kind of visual imageries may be created by similarities in structure [27] or other sociological and cultural associations. Particular musical components, namely, chronological reappearance; predictability in harmonic, rhythmic, and melodic components; and tempo, are effective in stimulating various imageries [31]. Tsang and Schloss [32] identified that color associations can change with the change in musical tempo. Previous research has demonstrated that musical soundtracks can influence the emotional impact, interpretation, and memorization of visual information [33]. On the other hand, Boltz et al. [34] explored the reverse relationship, i.e., whether visual information influences the perception and memory of music and concluded that visual information differentially influenced the perceived emotion of the melody as well as distorted the melody recognition process. In a more recent study, Campos-Bueno et al. [35] investigated whether the affective value (valence and arousal) of paintings can be manipulated by pairing them up with music of opposite affective values in an attempt to neutralize their valence. They observed that when presented simultaneously, music exhibits a superior emotional "power" over painting. Whereas the paintings caused stronger variability in valence and arousal, the music featured stronger effects on valence. In another important work, while studying the quality of experiences, emotionality, and associations evoked by music, paintings, and odors, Herz [36] revealed that the subjective and objective measures of emotional arousal can vary significantly irrespective of any musical type. He concluded that although the quantitative results show that odors caused higher emotional arousal compared to music and paintings, the participants believed that music was able to affect their moods and emotions more than the other two aesthetic sensory stimuli. Gaskill [37] investigated a performance-based art involving color-music, where projections of moving luminous hues on a screen were used to study the association between abstract colors and their cultural meanings. In this context, the names of color-musician Alexander Wallace Rimington and painter Wassily Kandinsky must be mentioned. Cuny et al. [38] investigated the influence of music on e-behavioral intentions of 250 participants about revisiting and recommending a virtual art gallery, which revealed that music promotes e-behavioral intentions. They concluded that emotions and contemplation mediate this intermedial effect. These findings have immense application potential in website designing. Researches on sound design have further revealed that music can play a complementary role on the consumers' response to a product [39]. Additionally, the congruent association between music and visual enhances positive emotions [40] and can help in better memorization of the advertised message [41].

1.3 Background of the Present Study

The earlier review of literature suggest that there have been many attempts in the domain of aesthetics, arts, and marketing research in order to academically understand and commercially use the intermedial relationship between different genres of music and visual arts. However, studies are probably nonexistent that use a comprehensive methodology encompassing multi-approach analysis methods where both the source characteristic features of the music and visual art as well as their individual and combined emotional impact on human mind could be studied from a rigorous scientific quantitative perspective. Another interesting fact is that none of the previous studies used abstract paintings to study this emotional intermediality while viewing them simultaneously with music. Our present study attempts to address all these inadequacies by investigating if (and how) instrumental music integration of complementary and contradictory nature can influence the emotional experience of viewing an abstract painting using feature, audience response, and fractal analysis. The reason behind choosing abstract paintings for this study is their capability of emotion communication only through the basic visual elements (like lines, colors, shapes, orientation, etc.) used in the composition without introducing the context of any objective meaning-making by associating them with realistic objects or events [42]. Thus, through the elimination of semantic dominance, abstract paintings, emotion expression-wise, match well with the abstract emotion evoking nature of instrumental music (also refrained from the semantic influence of lyrics). Hence, a study on the combination of these two will allow one to delve into the intricacies of basic visual and musical features beyond the semantic objectivity and search for the structural origin of intermediality between them.

1.4 Roles of Features, Audience Response, and Fractal Analysis in Exploring Intermediality

Study on intermediality between music and visual arts involves two very strong emotion evoking as well as technically well-explored aesthetic stimuli and demands multi-approach understanding about the details of audiovisual integration from its origination to its perception and cognition. Hence, detailed studies using the available well-established content analysis methodologies, on the music pieces and abstract paintings chosen for this study, are expected to give us significant information about the musical and visual features of the individual paintings and music clips as well as the feature-related similarities, dissimilarities, and compatibility between them. Along with that, extensive audience response surveys can help us to understand the emotional responses of the participants toward the individual paintings and music clips as well as the changes in their emotional response pattern when the two modes are integrated. Earlier researches have revealed that music signals feature a very complex behavior as at every instant components (both in micro- and macroscale) like pitch, timbre, accent, note sequence, melody, etc. are closely linked to each other [43–45]. A number of studies have shown the nonstationary and nonlinear nature of music signals [46, 47], but most of the available linear feature extraction techniques for analyzing time series data (like Fourier transform) are unable to capture the intricate details of the nonlinear fluctuations present in the acoustical waveforms. Hence, a non-deterministic/chaos based approach is needed in understanding the nonlinear and nonstationary music signals [45, 48–50]. Chaos theory, one of the most fascinating discoveries of the twentieth century, is the science of surprises. It is the science of the nonlinear and the unpredictable. The nonlinear technique, which will be used in this work, is detrended fluctuation analysis (DFA) (basically an offshoot of chaos theory) and as a result yields fractal dimension (FD) with which the inherent symmetry scaling present in the complex time series of music can be quantified. Similarly, fractal analysis of paintings was first reported in the seminal paper published in *Nature* by Taylor, Micolich, and Jonas [51], who analyzed a number of the famous artist Pollock's drip paintings. With the help of the evaluated fractal dimension, D, they provided an algorithm with which the authenticity of the paintings could be corroborated. They also showed that the fractal dimension of the paintings created by a particular artist changes over time. Later, there have been a number of criticisms and counter-criticisms regarding this work [52–55], but the fact is, till date, fractal techniques remain one of the most robust methodologies that can be used to quantify the patterns used in a particular painting. The main criticism for Taylor's seminal claim was that "D" was too low to draw any plausible conclusion, while those defending Taylor, including himself, claimed that physical fractals are studied under limited magnification norms. Taylor et al. [55] also studied the interconnection between neurobiology and fractals. Using various biosensors like MRI, EEG, and skin conductance, he established that human stress levels are reduced simply while looking at fractals, and thus the first evidence of physiological measurement of the aesthetic appeal of fractals was proposed. Taylor says that in visual perception, the most elemental abstraction is the deconstruction of an image in terms of just lines and colors. The traditional DFA technique introduced by Peng et al. [56] was extended to the two-dimensional surface by Gu and Zhou [57] to quantify the scaling nature of images. Since then, it has been applied in a number of studies including classification of MRI images [58], for generating music from photographs [59] and also for retina image classification [60]. In this paper, we focused on these basic abstractions of simple and complex paintings and instrumental music clips in the form of various visual and musical features and tried to study their intermediality in the emotional perception of viewers by audience response analysis and source characteristics analysis by calculating the scaling exponents of each abstract painting and instrumental music.

2 Experimental Details

Eight abstract paintings of varying complexity of composition, painted by famous artists like Jackson Pollock, Wassily Kandinsky, and František Kupka, and eight short piano clips (each of 1-minute duration) of different tempo and complexity were chosen for this experiment. Based on their simplicity or complexity of composition, the chosen paintings were named as SP1, SP2, SP3, SP4, CP1, CP2, CP3, and CP4, whereas the chosen music clips were named as SM1, SM2, SM3, SM4, CM1, CM2, CM3, and CM4. All the used abstract paintings and music clips are available at https://bit.ly/3wLV2fp. Figure 1a and b show two of the abstract paintings chosen for this study as sample images. The feature analysis for each of the paintings and music clips was performed to identify the dominant visual and musical features.

Based on the results of feature analysis and review of previous literature, a set of ten musical features (slow-fast, soft-loud, high pitch-low pitch, fragmented-continuous, chaotic-ordered) and ten visual features (static-dynamic, cool-warm,

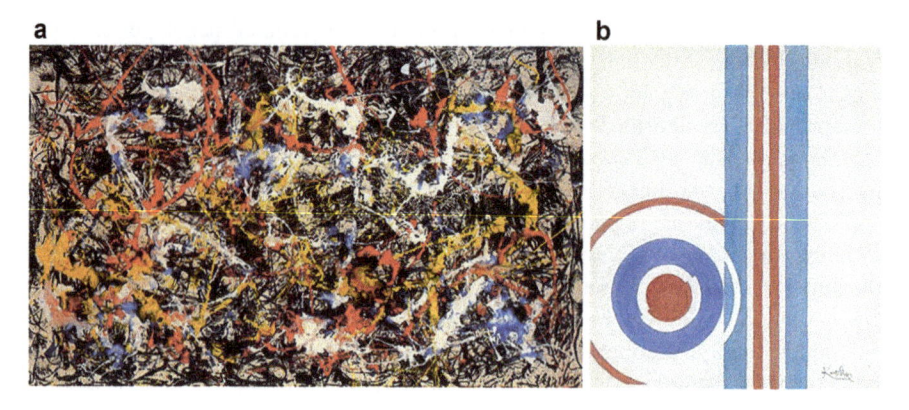

Fig. 1 Two sample abstract paintings with (**a**) complex and (**b**) simple compositions

empty-crowded, fragmented-continuous, chaotic-ordered) were identified. Through an online survey, 45 non-artist (who are untrained in both music and visual arts) participants rated the individual paintings and music clips based on their characteristic features from the given sets of visual and musical features. An audience response survey was also conducted on the same group of 45 participants where they were asked to mark the emotions (from a given set of 11 emotions—happy, sad, anger, fear, calm, tensed, surprise, disgust, romantic, heroic, exciting) for each of the abstract paintings as well as musical clips individually on a five-point Likert scale (where 1 is lowest and 5 is highest intensity of a particular emotion). Following a significant time gap they were asked to mark the emotions for the same abstract paintings again, but this time, different complementary or contradictory music clips were playing simultaneously in the background during the image viewing task. While viewing the paintings along with music, the participants also rated the compatibility between the specific painting and music for each video clip in a similar five-point scale. Finally, an in-depth nonlinear DFA was performed on the images and the music clips to understand their inherent symmetry scaling behaviors, and the resulting scaling exponents were compared to identify a scientific basis for the nature of intermediality between the two mediums.

3 Methodology

For the acoustic source characteristics analysis, the chosen eight piano clips were normalized to 0 dB and digitized at a rate of 44.1 kHz at mono-channel 16-bit format. Using one-dimensional DFA, which is conventionally performed following the algorithm of Peng et al. [56], and following the methodologies used in [49, 61], the DFA scaling exponent or the long-range temporal correlation present in the chosen music clips was calculated. For extracting the scaling exponents corresponding to the eight abstract paintings, a novel 2D-DFA algorithm [62–64] is used here:

3.1 2D-Detrended Fluctuation Analysis

This section describes the steps for computing Hurst exponent using the two-dimensional DFA algorithm for a grayscale image I. The steps are as follows:

(1) The profile $x_{i,j}$ is computed using:

$$x_{i,j} = \sum_{n=1}^{i}\sum_{m=1}^{j} (I_{i,j} - \bar{I}) \qquad (1)$$

where $m = 1, 2, \ldots, M$, $n = 1, 2, \ldots, N$, $I_{n,m} = 0, 1, \ldots, 255$ is the brightness of the pixel at the coordinates (m, n) of the grayscale image and $\bar{I}$ represents the mean value of $I_{n,m}$.

(2) The profile $x_{i,j}$ is divided into square segments of size $s \times s$. We have considered the total number of such small square regions as L_s. The value of s lies in the following range:

$$s_{min} \approx 5 \leq s \leq s_{max} \approx min\{M, N\}/4.$$

We start the algorithm by setting the initial value of s as s_{min}.

(3) Considering a small square region of size $s \times s$ denoted by l, an interpolating curve (local trend) is computed using the formula:

$$G_{i,j}(l, s) = a_l i + b_l j + c_l \qquad (2)$$

for the l_{th} small square region, using a multiple regression procedure. Here, $1 \leq i$, $j \leq s$, and a_l, b_l, and c_l are the coefficients for the l_{th} square which are to be determined using the least squares regression method.

(4) The variance V in the l_{th} small square region is computed for s as follows:

$$V(l, s) = \frac{1}{s^2}\sum_{i=1}^{s}\sum_{j=1}^{s}(x_{i,j} - G_{i,j}(l, s))^2 \qquad (3)$$

This variance should be minimized.

(5) The root mean square $F(s)$ is computed as:

$$F(s) = \left[\frac{1}{L_s}\sum_{l=1}^{L_s} V(l, s)\right]^{1/2} \qquad (4)$$

where L_s denotes the total number of the small square regions of size $s \times s$. After that, the value of s is increased by unity and steps 2 to 5 are repeated until the maximum value of s (i.e., s_{max}) is reached.

(6) The values of $F(s)$ for corresponding values of s are stored. If $x_{i,j}$ has a long-range power-law correlation characteristic, then the fluctuation function $F(s)$ is observed as follows:

$$F(s) \propto s^\alpha \tag{5}$$

where α is the two-dimensional scaling exponent, a self-affinity parameter representing the long-range power-law correlation characteristics of the surface. We fit a least squares regression line in a $\log(F(s))$ vs $\log(s)$ plot to obtain the slope of the line as the value of α.

For the three basic color coordinates red, blue, and green, α has been computed for each chosen painting. The 1D analogue of this technique can be understood as the first step of the method discussed above and used to calculate the 1D DFA scaling exponents for all the chosen music clips. Similar to the 1D DFA scaling exponent, in the case of 2D surfaces also, the α values stay around 0.5 for completely uncorrelated series or white noise and $\alpha > 0.5$ indicates that the data is long-range correlated.

4 Results and Discussions

4.1 Results of Feature Analysis

Abstract Paintings Painting CP1 is a complex abstract composition with different kinds of line patterns, mostly curved, and bright warm colors. The painting has very little empty space and looks crowded. The abstract shapes are depicted asymmetrically, but it has its own harmony. It seems like a very dynamic composition. Painting CP2 also has variety of forms, mostly curved line patterns along with a few spiky straight lines and different color hues. The composition is very complex, crowded, and dynamic. The colors are mostly dark and bright and there are multiple tonal variations which create depth and multiple perspectives. It is an asymmetric composition, but it is rhythmic. Painting CP3 is a complex composition of continuous color strokes. At the background, this painting has black color strokes and over that some brown, yellow, and white color strokes. This painting does not have depth or perspective. It has almost no empty space. Painting CP4 (Fig. 1a) also features a very complex composition with multiple continuous line patterns and colors. There are almost no empty space and no perspective. The colors of this painting are blue, yellow, red, black, white, and brown. The complex distributions of lines and colors have created a dynamic space in this painting. Painting SP1 contains some linear rectangular forms which are dispersed over a white empty space. It does not have any curve line. The discrete geometric forms of this painting are colored with red, deep blue, and light blue and they appear as fixed or static. This painting has a large amount of empty/negative space rather than positive/depicted space. Painting SP2 again contains different discrete linear geometric forms which have created a somewhat dynamic composition. This painting has multiple vanishing points and multiple perspectives. A large amount of this painting is empty, but the emptiness helps to create the depth in this painting. The geometric shapes (rectangles) are colored in yellow ochre, purple, sap green, deep green, deep blue, and light blue. There is no curved line. The empty space has a light grayish pink and light purple tone. Painting SP3 has a well-balanced composition between colors and shapes. The black, light red, and light blue colors have created a contrast composition. There are some discrete rectangular and circular shapes which have been created by straight lines and curve lines. At the background of these colored shapes, there is a large amount of white empty space. Painting SP4 (Fig. 1b) mostly contains straight lines along with some curved lines which creates simple circular and rectangular shapes, mainly colored in light blue and pale red, on the backdrop of a large white empty space.

Instrumental Music Clips Music SM1 is a very soft, slow, simple composition with discrete low-pitched piano strokes being played at equally distributed time gaps in a very ordered manner, with plenty of pauses throughout the clip duration. SM2 is another soft music clip, with a combination of high- and low-pitched discrete piano strokes and chords playing at a slightly higher tempo than SM1, but in a perfectly ordered rhythm cycle. Music SM3 is a composition of discrete single piano strokes being played at gradually lowering pitches in a slow, soft manner with long pauses in between. Music SM4, just like SM1, is a soft, slow composition with discrete combinations of one or two piano strokes being played at equally distributed time gaps in well-defined rhythm with plenty of pauses. Music CM1 is a very-fast-moving, chaotic piano composition with very little pause and sudden fluctuations of loudness and a lot of unexpected pitch sequences and nonharmonic chord progressions. Music CM2, being a typical jazz piano piece, again featured a lot of unexpected pitch combinations with very short pauses, but the loudness fluctuation and tempo were slightly lower than those of CM1. Music CM3, a very-fast-moving piano piece, featured a lot of dramatic fluctuation in loudness, pitch, and rhythmic patterns and very short pauses only at the rhythmic transitions. Music CM4 is a very fast piano piece with a combination of soft and loud harmonic piano pitch sequences and chords along with dramatic pauses and rhythmic changes at a few places. Figure 2a and b show the pitch contours, power spectrum plots, and waveforms of Music SM4 and CM1, respectively.

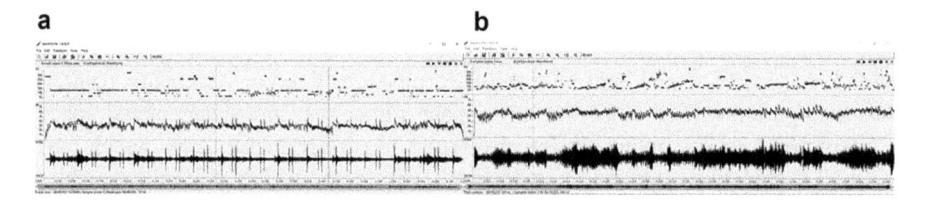

Fig. 2 Pitch contour, power plot, and acoustical waveform of chosen music (**a**) SM4 and (**b**) CM1

4.2 Results of Audience Response Analysis for Exploring the Music-Painting Intermediality

From the results of feature analysis and the review of earlier literature, ten musical features—slow-fast (tempo related), high pitch-low pitch (pitch related), soft-loud (loudness related), fragmented-continuous (pause related), and chaotic-ordered (compositional complexity related)—were chosen for this study. On the other hand, from the visual arts domain, ten visual features like static-dynamic (related to the sense of motion or speed), warm-cool (color related), fragmented-continuous (line related), empty-crowded (negative or empty space related), and chaotic-ordered (compositional complexity related) were chosen. Among these features, some pairs like fragmented-continuous and chaotic-ordered were common for both music and visual arts, while slow-fast tempo of music could be compared indirectly with visual features like static-dynamic. Although scientific theories suggest that most of these features, ideally, will be independent of each other, realistically, a lot of overlapping was expected among many of them because of the subjectivity introduced in each of these parameters which can be measured only through audience response analysis. To study the emotional responses and visual/musical feature perception of the audience, a group of 45 musically and visually untrained participants rated the emotions and the features (from the provided sets of emotions and visual/musical features) corresponding to each individual painting and music clip chosen for this study. Quantitative analysis was performed to calculate the percentage of participants who associated any particular emotion or any specific visual/musical feature with a particular painting/music as well as the weighted average of their rating in terms of intensity of that particular emotion or dominance of that specific feature corresponding to that painting/music. Tables 1 and 2 describe the most associated and highest rated emotions and visual/musical features for the chosen eight abstract paintings and eight music clips, respectively.

Emotion response of individual abstract paintings (Table 1) revealed that most of the participants (75.56%) associated happy emotion with painting SP1, but the highest average intensity (2.87) or intensity was observed for calmness. Paintings SP2, SP3, and SP4 all exhibited calmness as both the most associated and highest rated emotion, though among the three, SP4 was associated with calmness by highest share of participants (86.67%) with highest degree of average intensity (3.38, i.e., between moderate and high). Paintings CP1, CP2, CP3, and CP4 all showed

Table 1 Most associated and highest rated emotions and features corresponding to eight paintings

Painting name	Emotions				Visual features			
	Most associated	% of association	Highest rated	Avg. rating	Most associated	% of association	Highest rated	Avg. rating
SP1	Happy	75.56	Calm	2.87	Static	82.22	Static	3.08
SP2	Calm	78.13	Calm	2.74	Static	77.78	Fragmented	3.27
SP3	Calm	80.00	Calm	3.28	Ordered	91.11	Ordered	3.46
SP4	Calm	86.67	Calm	3.38	Cool	91.11	Ordered	3.70
CP1	Fear	84.44	Exciting	3.57	Dynamic	82.22	Chaotic	4.49
CP2	Exciting	91.11	Exciting	3.41	Chaotic	84.44	Chaotic	4.21
CP3	Tensed	86.67	Tensed	3.64	Chaotic	88.89	Crowded	4.22
CP4	Tensed	80.00	Exciting	3.49	Chaotic	84.44	Chaotic	4.21

Table 2 Most associated and highest rated emotions and features corresponding to eight music clips

Music name	Emotions				Musical features			
	Most associated	% of association	Highest rated	Avg. rating	Most associated	% of association	Highest rated	Avg. rating
SM1	Calm	86.67	Calm	4.15	Soft	86.67	Soft	4.15
SM2	Calm	80.00	Calm	4.08	Soft	86.67	Soft	3.90
SM3	Calm	93.33	Calm	4.55	Ordered	93.33	Ordered	4.38
SM4	Calm	84.44	Calm	4.11	Soft	82.22	Soft	4.38
CM1	Exciting	82.22	Exciting	3.95	Fast	95.56	Fast	4.35
CM2	Exciting	84.44	Exciting	3.66	Fast	88.89	Continuous	3.81
CM3	Happy	82.22	Exciting	4.06	Fast	91.11	Fast	4.17
CM4	Exciting	75.56	Exciting	3.88	Fast	88.89	Fast	4.40

most associations with emotions like fear, exciting, and tensed, with CP2 being associated with exciting emotion by largest proportion of participants (91.11%) among the four. They also showed higher degree of intensity in general compared to SP1, SP2, SP3, and SP4, with CP3 featuring the highest average intensity for tension (3.64) among them. Feature analysis revealed that paintings SP1, SP2, SP3, and SP4 were most associated and highest rated with features like static, ordered, cool, and fragmented with highest associations observed between SP3 and ordered as well as SP4 and cool (91.11%), while dominance rating of specific features revealed that SP4 featured highest average rating for ordered (3.70 or close to high). On the other hand, paintings CP1, CP2, CP3, and CP4 were most associated and highest rated with features like chaotic, dynamic, and crowded with highest associations observed between CP3 and chaotic (88.89%), while dominance rating of specific features revealed that CP1 featured highest average rating for chaotic (4.49, i.e., between high and very high). Similarly, emotion response analysis of individual music clips (Table 2) revealed that music pieces SM1, SM2, SM3, and SM4 all showed most associations as well as highest intensity with calm emotion, with SM3 featuring both highest association (93.33%) and highest average intensity (4.55) among the four. Music CM1, CM2, CM3, and CM4 showed most associations as well as highest intensity with exciting and happy emotions, with CM2 featuring highest association (84.44%) with excitement among the four. In contrast to the case of paintings, music CM1, CM2, CM3, and CM4, in general, showed lower degree of emotional intensity compared to SM1, SM2, SM3, and SM4, with CM3 featuring the highest average intensity for exciting emotion (4.06) among them. Feature analysis revealed that music pieces SM1, SM2, SM3, and SM4 were most associated and highest rated with features like soft and ordered, with highest associations observed between SM3 and ordered (93.33%), while dominance rating of specific features revealed that SM3 featured highest average rating for ordered (4.38) as well as SM4 for soft (4.38). On the other hand, music CM1, CM2, CM3, and CM4 were most associated and highest rated with features like fast and continuous with highest associations observed between CM1 and fast (95.56%), while dominance rating of specific features revealed that CM4 featured highest average rating for fast (4.40, i.e., between high and very high). Combining the findings of Tables 1 and 2, it was observed that soft, slow, smooth music clips SM1, SM2, SM3, and SM4 show great similarity in emotion responses with the static, cool, ordered paintings SP1, SP2, SP3, and SP4. On the other hand, the crowded, chaotic, dynamic paintings CP1, CP2, CP3, and CP4 emotionally resembles with the fast, exciting music CM1, CM2, CM3, and CM4.

Based on these findings, different video clips were made by pairing up each of the chosen eight abstract paintings with two emotionally compatible and two incompatible music pieces, and emotional responses to each video clip were collected from the same 45 participants and again analyzed using the same method of calculating the percentage of association as well as intensity of a specific emotion with a particular video clip. Compatibility between the music and image corresponding to each video was also marked by the participants simultaneously. Figure 3a shows the variations in percentage of participants associating specific

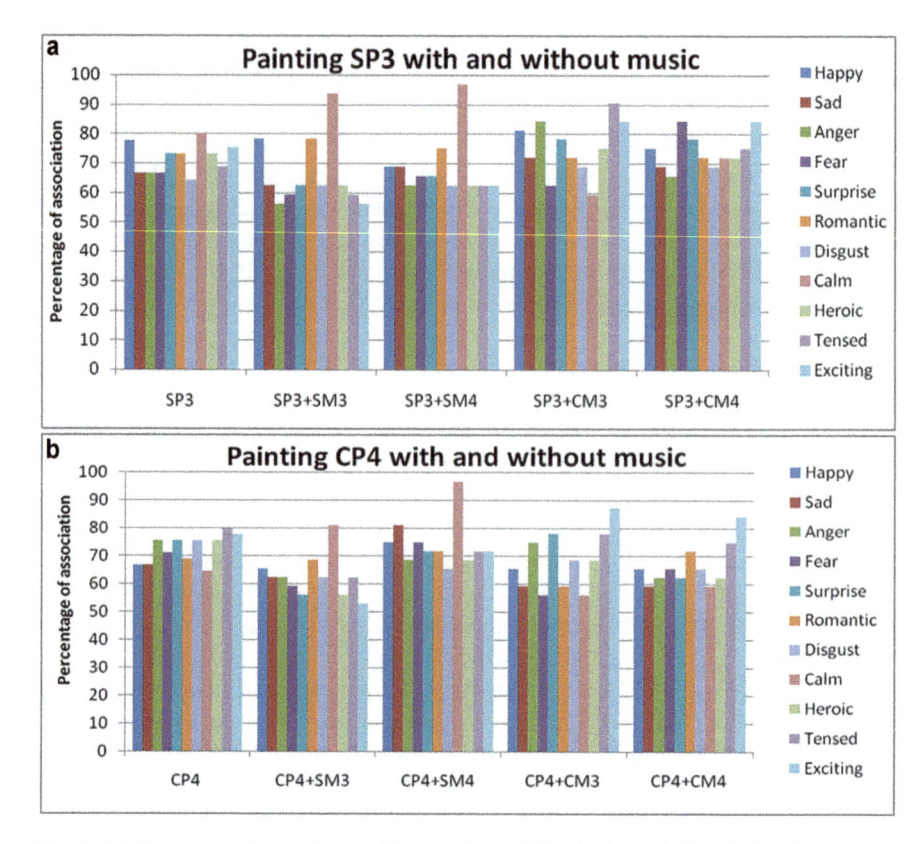

[Figure: bar charts showing (a) Painting SP3 with and without music, and (b) Painting CP4 with and without music, displaying percentage of association for 11 emotions: Happy, Sad, Anger, Fear, Surprise, Romantic, Disgust, Calm, Heroic, Tensed, Exciting]

Fig. 3 (**a**) Percentage of association of 11 emotions while viewing painting CP4 without music, with compatible music (CM3 and CM4), and with incompatible music (SM3 and SM4). (**b**) Percentage of association of 11 emotions while viewing painting SP3 without music, with compatible music (SM3 and SM4), and with incompatible music (CM3 and CM4)

emotions with a particular compatible or incompatible combination of music and painting for SP3. Figure 3b shows the same for painting CP4. (Due to space limitations, only a few combinations could be reported here.)

Figure 3a suggests that while viewing without music, emotions like calm and happy were associated with painting SP3 by most participants. When the same painting was viewed simultaneously with music SM3 (average compatibility 3.09, i.e., moderate) and SM4 (average compatibility 3.25 or between moderate and high), the percentage of association with calm, happy, and romantic emotions increased compared to the without music viewing condition, whereas pairing up with music CM3 (average compatibility 2.25, i.e., close to low) and CM4 (average compatibility 2.19 or low) caused clear decrement in percentage of participants who associated calmness with these combinations while causing an increment in the percentages of association with music embedded emotions like tension, excitement, anger, fear, surprise, etc. A very similar trend was observed in the case of all compatible and incompatible music integrated combinations for paintings SP1, SP2, and SP4.

Figure 3b suggests that when painting CP4 was viewed without music, highest percentage of association was observed for tension, followed by excitement, anger, surprise, etc. When the same painting was viewed simultaneously with fast, dynamic, chaotic music CM3 (average compatibility 3.50, i.e., between moderate and high) and CM4 (average compatibility 3.63 or close to high), the percentage of association with excitement slightly increased, compared to the without music viewing condition, whereas pairing up with slow-soft music SM3 (average compatibility 2.25, i.e., close to low) and SM4 (average compatibility 2.31, i.e., between low and moderate) caused clear decrement in percentage of association with tension and excitement, rather featuring a prominent increment in calmness over all other emotions, introduced by the dominant contribution of music. All compatible and incompatible music combinations with paintings CP1, CP2, and CP3 also exhibited the exact same trend.

Next, looking at the average rating of emotional intensity, we observed a close resemblance with the previous trends. Figure 4a and b show the average emotional intensity ratings of 45 participants for 11 emotions corresponding to different compatible or incompatible combinations of music with painting SP1 and painting SP4, respectively. Figure 5a and b show the same for painting CP4.

Figure 4a suggests that while viewing painting SP1 in without music condition (represented by the deep blue line), the average intensity of calmness was highest among the given 11 emotions, followed by romantic and happy. When viewed simultaneously with music SM1 (average compatibility 3.00, i.e., moderate, red line) and SM2 (average compatibility 3.09 or moderate, violet line), the average

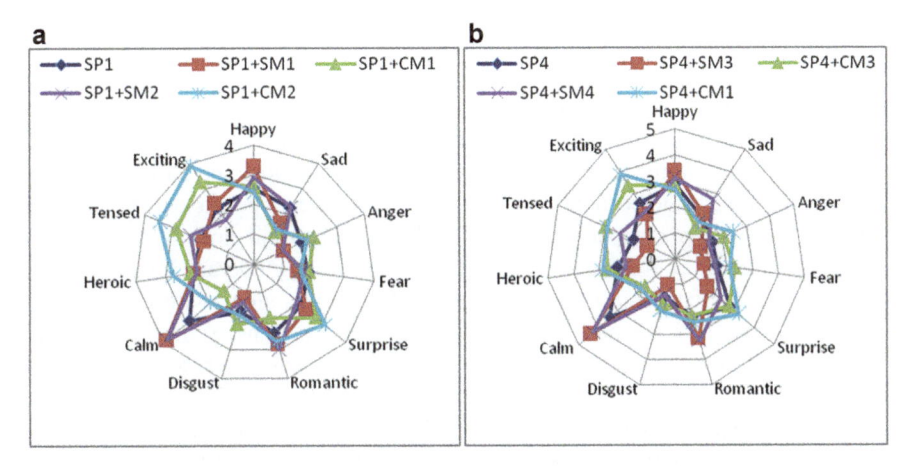

Fig. 4 (**a**) Average intensity rating of 11 emotions while viewing painting SP1 without music, with compatible music (SM1 and SM2), and with incompatible music (CM1 and CM2). (**b**) Average intensity rating of 11 emotions while viewing painting SP4 without music, with compatible music (SM3 and SM4), and with incompatible music (CM1 and CM3)

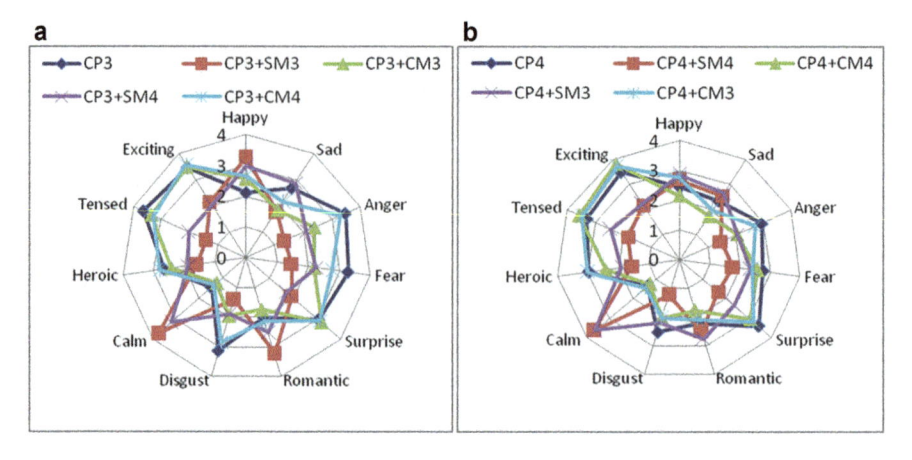

Fig. 5 (a) Average intensity rating of 11 emotions while viewing painting CP3 without music, with compatible music (CM3 and CM4), and with incompatible music (SM3 and SM4). (b) Average intensity rating of 11 emotions while viewing painting CP4 without music, with compatible music (CM3 and CM4), and with incompatible music (SM3 and SM4)

intensity for calm and happy emotions increased compared to without music viewing condition, and a slight decrement in negative emotions like sadness and anger was observed. On the other hand, pairing up with music CM1 (average compatibility 1.94, i.e., close to low, green line) and CM2 (average compatibility 1.78 or between very low and low, light blue line) caused clear decrement in average intensity of calmness along with a significant increment in the average intensities of tension, excitement, heroic, anger, fear, surprise, etc. under the dominant influence of music.

Figure 4b shows that while viewing painting SP4 without music, the average intensity of calmness was highest among the given 11 emotions, followed by surprise and happy. When viewed simultaneously with music SM3 (average compatibility 3.69, i.e., nearly high) and SM4 (average compatibility 3.53 or between moderate and high), the average intensity for calm, romantic, and happy emotions increased compared to without music viewing condition, whereas pairing up with music CM1 (average compatibility 2.66, i.e., between low and moderate, green line) and CM3 (average compatibility 2.41 or between low and moderate, light blue line) caused clear decrement in average intensity of calmness along with a significant increment in the average intensities of music-induced emotions excitement, tension, heroic, anger, fear, surprise, etc. A very similar trend was observed for paintings SP2 and SP3.

Figure 5a suggests that while viewing painting CP3 in without music condition, the average intensity rating of tension was highest among the given 11 emotions, closely followed by excitement, anger, fear, and disgust. When viewed simultaneously with music CM3 (average compatibility 3.31, i.e., moderate) and CM4 (average compatibility 3.66 or close to high), the average intensity for all the inherent emotions of the painting remained unchanged except slight occasional increment in excitement and some decrement in sadness, anger, fear, etc. On the other hand, pairing up with music SM3 (average compatibility 2.28, i.e., low) and SM4 (average compatibility 2.34 or low) caused a complete change in the participants' emotion arousal pattern with significant decrement in average intensity of tension, excitement, heroic emotions, as well as anger, fear, surprise, etc., but the positive valences like calm, romantic, and happy emotions dominate the overall emotional arousal under the influence of calm soft music.

Figure 5b shows that just like painting CP3, while viewing painting CP4 without music, the average intensity of excitement was highest among the given 11 emotions, followed by tension and surprise. When viewed simultaneously with music CM3 and CM4, slight increment in the average intensity of tension and excitement was observed, whereas combinations with music SM3 and SM4 again caused a reversal in the emotional response pattern, with music-induced calmness featuring the highest intensity in both cases. A very similar trend was observed for paintings CP1 and CP2 also. Combining all these results, we can clearly observe that in general, music was showing dominance over the paintings in determining the total emotional outcome. When the compatibility between the music and the visuals was higher, the music helped in projecting the inherent dominant emotions of the paintings, but when the combinations had low compatibility among themselves, the musical emotions mostly overruled the visual contributions both in terms of associating a particular emotion with a video and the perceived intensity of that emotion. Another interesting observation was that in the case of compatible combinations, the influence of calm-soft-slow music on the static-ordered paintings was prominently higher than the influence of fast-chaotic music on dynamic-chaotic paintings.

4.3 Results of Fractal Analysis

1D DFA was used for calculating the monofractal scaling exponent for each of the chosen eight piano music clips, and their values are reported in Fig. 6. An interesting trend was observed in the symmetry scaling behavior of the music clips as for all the chosen four fast, complex, chaotic music clips (CM1, CM2, CM3, CM4), the DFA scaling exponent featured a lower value compared to those for soft, slow, ordered music clips like SM1, SM2, SM3, and SM4, indicating a lower presence of long-range temporal correlations in the case of fast-exciting music compared to slow-calm music. Among the four fast-chaotic music clips, CM4 featured the lowest DFA scaling exponent or long-range temporal correlation, and CM1 featured the highest DFA scaling exponent, whereas among the four soft-slow music clips, SM2 featured the highest long-range temporal correlation, and SM1 featured the lowest value of DFA scaling exponent.

On the other hand, analyzing the chosen eight abstract paintings using 2D DFA technique revealed that in the long-range correlations or the 2D DFA scaling

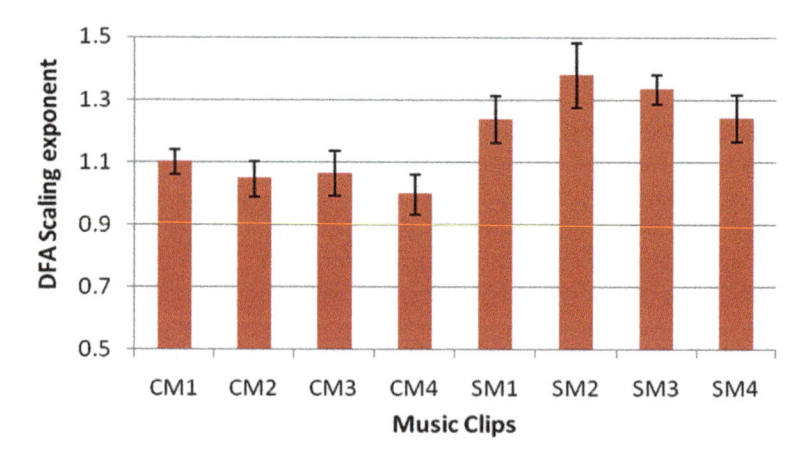

Fig. 6 Variation in DFA scaling exponents for the chosen music clips

exponents along the red, blue, and green color coordinates, no such clear distinctive pattern was observed for complex-chaotic-dynamic paintings and simple-static-ordered paintings. Rather, both types of paintings featured a combination of high, low, and moderate long-range correlations along red, green, and blue coordinates. Among the four complex paintings (Fig. 7a), CP1 featured the highest long-range correlations along all three fundamental color coordinates, whereas CP3 featured the lowest values for the same. Overall, paintings CP1 and CP2 featured higher long-range correlations along RGB axes compared to paintings CP3 and CP4. Among the four static-ordered paintings (Fig. 7b), SP2 featured the highest long-range correlations along all three fundamental color coordinates, whereas SP3 featured the lowest scaling exponent value along red color coordinate and SP4 exhibited the lowest long-range correlations along blue and green color coordinates. Overall, SP2 and SP1 featured higher long-range correlations along RGB compared to SP3 and SP4. But the most interesting observation was that for each of the complex-chaotic-dynamic paintings, the 2D DFA scaling exponents across the three primary color coordinates yielded almost same values, indicating an equivalent presence of long-range correlations along these three color axes, whereas for each of the static-ordered paintings, a visibly higher variation among the long-range correlations along the three color axes was observed.

This analysis gives us a comparative idea about the inner geometry or more precisely, the inherent symmetry scaling nature of the eight chosen music clips and eight chosen abstract paintings, which in the future can help in better understanding of the origin and nature of intermediality between the two mediums from a scientific quantitative approach.

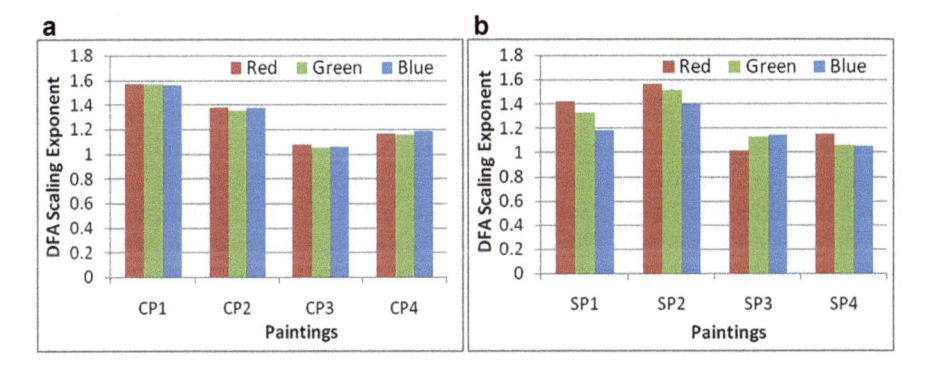

Fig. 7 Variation in the 2D DFA scaling exponents along the RGB coordinates of the chosen (**a**) complex and (**b**) simple abstract paintings

Table 3 Chi-square results for eight paintings with different music integrated combinations

	Chi square	df	N	p value
SP1	57.69*	30	1694	0.0017
SP2	56.98*	30	1708	0.0021
SP3	85.11*	30	1753	3.55E-07
SP4	74.15*	30	1753	0.00001
CP1	58.77*	30	1694	0.0013
CP2	50.24*	30	1751	0.0117
CP3	62.94*	30	1732	0.0004
CP4	69.49*	30	1708	0.00006

*$p < .05$

4.4 Statistical Analysis

To test the statistical significance of our findings, chi-square test of independence was carried out by considering different compatible and incompatible music integrated combinations for each of the chosen eight abstract paintings as independent variables and the participants' emotional responses toward them as dependent variables. The null hypothesis in this case was that the emotional responses of the participants were independent of the type of music associated with a specific abstract painting. Table 3 gives the detailed chi-square results for frequency of emotion association corresponding to each of the chosen eight abstract paintings with compatible and incompatible music combinations.

There were four different musical pair ups for each painting, and corresponding emotional responses from the participants were collected for a set of given 11 emotions; hence, the degrees of freedom were 30 in each case. To test the hypothesis, we wanted to check if the frequency of association with specific emotions corresponding to a particular abstract painting changes under the influence of different types of musical integrations. The results of chi-square test of independence revealed that at the previously decided 95% confidence level or $p < .05$, for painting SP1, when integrated with four different music clips SM1/SM2/CM1/CM2, the frequency of association with 11 different emotions varied significantly [χ^2 (30, N = 1694) = 57.69, p <.05] for different music-painting combinations. Similarly, for all the other abstract paintings also, irrespective of their level of compositional complexity, significant associations were observed between the nature of music integration and the frequency of specific emotion associations at 95% confidence level. These findings again suggest that music has a very strong influence on the overall emotional outcome when paired with an abstract painting and experienced simultaneously by the audience.

5 Conclusion

The objective of this study was to investigate how music of complementary and contradictory nature influences the emotional experience of viewing an abstract painting. For this, eight short piano music clips of different tempo and complexity and eight abstract paintings of varying complexity of composition were chosen. The intermediality of music and visual arts (here, abstract paintings) was explored using a three-way approach of feature analysis, audience response analysis, and nonlinear fractal analysis. The main observations of this study are summarized below.

i. Feature analysis of the individual paintings and music clips suggested that depending upon the variation in line, color, shape usages, and compositional complexities in the chosen eight abstract paintings, ten visual features were identified. Similarly, following the variation in musical components like pitch, loudness, tempo, pause, and complexity of composition, ten musical features were identified from the chosen eight music clips. Comparing the visual and musical features, it was observed that some musical components like fragmented-continuous, chaotic-ordered have direct resemblances in the domain of visual arts also, whereas some other musical features like slow-fast have an indirect correlation with visual features like static-dynamic. Also, there are some features which are unique to the domain of music (soft-loud, high pitch-low pitch) or visual arts (cool-warm, empty-crowded). A match or mismatch in these characteristic features of a music clip and a painting lead to a complementary or a contradictory relationship between the two.

ii. For audience response analysis, these chosen eight paintings and eight music clips were used as stimuli for a group of musically and visually untrained audience (N=45) where they viewed/listened to the paintings or music clips both individually and after a time gap and experienced them simultaneously in different compatible and incompatible combinations. The participants marked their emotional responses corresponding to each painting/music/video clip along with the compatibility between the painting and the music corresponding to each video clip. Results of audience response indicated that while viewing a painting, pairing up with a highly compatible music enhances the arousal of the inherent emotional valence of the painting, but the combination with a contradictory music does the reverse. Rather our findings suggested that music exhibited a more dominating influence on the entire emotional experience of the participants over the visuals when they were presented simultaneously. These results show some agreements with the findings of Campos-Bueno et al. [35]. We also observed a greater influence of soft-calm music on static-ordered paintings compared to that of fast-exciting music on dynamic-chaotic-complex paintings, which suggests that abstract paintings are capable of communicating high arousal emotions like tension, excitement, anger, and fear more strongly than positive emotions like calmness, romantic, or happiness, whereas music can communicate calm and romantic emotions very prominently.

iii. Results of nonlinear fractal analysis using detrended fluctuation analysis suggested that soft, slow, and ordered music clips featured a prominently higher range of long-range temporal correlations compared to the fast, loud, chaotic music clips, whereas in the domain of visual arts, we observed that static, ordered paintings featured a higher variance in the long-range correlations along red, green, and blue coordinates compared to complex, dynamic, chaotic paintings. In other words, a closer look upon the source characteristics of the input music and visual arts stimuli using nonlinear DFA technique could reveal a great deal of information about the inherent symmetry scaling nature or long-range correlation present in the two mediums, which in turn is expected to open a new dimension of scientific research on the origin of intermediality between music and visual arts.

In the future, this study will be further elaborated to investigate the human brain responses toward music-visual arts intermediality using neuro-biosensors (like EEG), and the effects of musical training, visual arts training, and sociocultural and gender bias on the same will be researched in depth. Nevertheless, the unique findings of this study have immense importance in the domain of audiovisual perception and cognition, and their future application in entertainment media (movies, theatre) as well as in advertisement and marketing.

Acknowledgments The first author Archi Banerjee acknowledges the Department of Science and Technology (DST), Govt. of India, for providing the DST CSRI Post Doctoral Fellowship (SR/CSRI/PDF-34/2018) to pursue this research work. Shankha Sanyal acknowledges DST CSRI, Govt. of India, for providing the funds related to this major research project (DST/CSRI/2018/78 (G)).

References

1. Banerjee, A., Sanyal, S., Roy, S., Nag, S., Sengupta, R., & Ghosh, D. A novel study on perception–cognition scenario in music using deterministic and non-deterministic approach. *Physica A: Statistical Mechanics and its Applications, 567*, 125682 (2021)
2. Plack, C. J., & Oxenham, A. J. The Psychophysics of Pitch. *Pitch: neural coding and perception. Plack, C. J., Oxenham, A. J., & Fay, R. R. (Eds.).* Springer Science & Business Media. (Vol. 24), 7–55 (2005)

3. ANSI, A. American National Standard Acoustical Terminology. *ANSI Sl, 11994* (1994)
4. Stainsby, T., & Cross, I. The perception of pitch. *The oxford handbook of music psychology. Hallam, S., Cross, I., & Thaut, M. (Eds.).* Oxford University Press. 47–58 (2009)
5. Mcbeath, M. K., & Neuhoff, J. G. The Doppler effect is not what you think it is: Dramatic pitch change due to dynamic intensity change. *Psychonomic bulletin & review, 9*(2), 306–313 (2002)
6. Neuhoff, J. G., McBeath, M. K., & Wanzie, W. C. Dynamic frequency change influences loudness perception: A central, analytic process. Journal of Experimental Psychology: *Human Perception and Performance, 25*(4), 1050 (1999)
7. Repp, B. H. Detectability of duration and intensity increments in melody tones: A partial connection between music perception and performance. *Perception & Psychophysics, 57*(8), 1217–1232 (1995)
8. Gabrielsson, A. The relationship between musical structure and perceived expression. *The oxford handbook of music psychology. Hallam, S., Cross, I., & Thaut, M. (Eds.).* Oxford University Press. 215–232 (2009)
9. Dalla Bella, S., Peretz, I., Rousseau, L., & Gosselin, N. A developmental study of the affective value of tempo and mode in music. *Cognition, 80*(3), B1-B10 (2001)
10. Gayen, P., Borgohain, J., & Patnaik, P. The Influence of Music on Image Making: An Exploration of Intermediality Between Music Interpretation and Figurative Representation. In *Advances in Speech and Music Technology* (pp. 285–293). Springer, Singapore. https://doi.org/10.1007/978-981-33-6881-1_24 (2021)
11. Lissa, Z. Aesthetic Functions of Silence and Rests in Music. *The Journal of Aesthetics and Art Criticism, 22*(4), 443–454. doi:10.2307/427936 (1964)
12. Shafron, G. R., & Karno, M. P. Heavy metal music and emotional dysphoria among listeners. *Psychology of Popular Media Culture, 2*(2), 74. doi:10.1037/a0031722 (2013)
13. Sharman, L., & Dingle, G. A. Extreme metal music and anger processing. *Frontiers in human neuroscience, 9*, 272 (2015)
14. Machajdik, J.; Hanbury, A. Affective Image Classification Using Features Inspired by Psychology and Art Theory; Association for Computing Machinery: New York, NY, USA; pp. 83–92 (2010)
15. W. Wang, Y. Yu, and S. Jiang. Image retrieval by emotional semantics: A study of emotional space and feature extraction. In IEEE SMC (2006)
16. B. Li, W. Xiong, W. Hu, and X. Ding. Context-aware affective images classification based on bilayer sparse representation. In ACM MM (2012)
17. X. Lu, P. Suryanarayan, R. B. Adams Jr, J. Li, M. G. Newman, and J. Z. Wang. On shape and the computability of emotions. In ACM MM (2012)
18. Yanulevskaya, V.; Uijlings, J.; Bruni, E.; Sartori, A.; Zamboni, E.; Bacci, F.; Melcher, D.; Sebe, N. In the Eye of the Beholder: Employing Statistical Analysis and Eye Tracking for Analyzing Abstract Paintings; Association for Computing Machinery: New York, NY, USA; pp. 349–358 (2012)
19. Neuhoff, J. G., & McBeath, M. K. The Doppler illusion: The influence of dynamic intensity change on perceived pitch. *Journal of Experimental Psychology: Human Perception and Performance, 22*(4), 970 (1996)
20. Krishna, A. An integrative review of sensory marketing: Engaging the senses to affect perception, judgment and behavior. *Journal of consumer psychology, 22*(3), 332–351 (2012)
21. Argo, J. J., Popa, M., & Smith, M. C. The sound of brands. *Journal of Marketing, 74*(4), 97–109 (2010)
22. Sunaga, T., Park, J., & Spence, C. Effects of lightness-location congruency on consumers' purchase decision-making. *Psychology & Marketing, 33*(11), 934–950 (2016)
23. Hagtvedt, H., & Brasel, S. A. Cross-modal communication: sound frequency influences consumer responses to color lightness. *Journal of Marketing Research, 53*(4), 551–562 (2016)
24. Meyer, L. B. *Emotion and meaning in music.* University of chicago Press (2008)
25. Zentner, M., Grandjean, D., & Scherer, K. R. Emotions evoked by the sound of music: characterization, classification, and measurement. *Emotion, 8*(4), 494 (2008)

26. Laferrière, C. Painting with Music: Visualizing Harmonia in Late Archaic Representations of Apollo Kitharōidos. *Greek and Roman Musical Studies, 8*(1), 63–90 (2020)
27. Juslin, P. N., & Västfjäll, D. Emotional responses to music: The need to consider underlying mechanisms. *Behavioral and brain sciences, 31*(5), 559–575 (2008)
28. Farrugia, N., Jakubowski, K., Cusack, R., & Stewart, L. Tunes stuck in your brain: The frequency and affective evaluation of involuntary musical imagery correlate with cortical structure. *Consciousness and cognition, 35*, 66–77 (2015)
29. Zatorre, R. J., Halpern, A. R., Perry, D. W., Meyer, E., & Evans, A. C. Hearing in the mind's ear: a PET investigation of musical imagery and perception. *Journal of cognitive neuroscience, 8*(1), 29–46 (1996)
30. Hubbard, T. L. Auditory imagery: empirical findings. *Psychological bulletin, 136*(2), 302 (2010)
31. McKinney, C. H., & Tims, F. C. Differential effects of selected classical music on the imagery of high versus low imagers: Two studies. *Journal of Music Therapy, 32*(1), 22–45 (1995)
32. Tsang, T., & Schloss, K. B. Associations between color and music are mediated by emotion and influenced by tempo. *The Yale Review of Undergraduate Research in Psychology, 82* (2010)
33. Cohen, A. J. Music as a source of emotion in film. In P. N. Juslin & J. A. Sloboda (Eds.), *Handbook of music and emotion: Theory, research, applications* (pp. 879–908). Oxford University Press (2010)
34. Boltz, M. G., Ebendorf, B., & Field, B. Audiovisual interactions: The impact of visual information on music perception and memory. *Music Perception, 27*(1), 43–59 (2009)
35. Campos-Bueno, J. J., DeJuan-Ayala, O., Montoya, P., & Birbaumer, N. Emotional dimensions of music and painting and their interaction. *The Spanish journal of psychology, 18* (2015)
36. Herz, R. S. An examination of objective and subjective measures of experience associated to odors, music, and paintings. *Empirical Studies of the Arts, 16*(2), 137–152 (1998)
37. Gaskill, N. The Articulate Eye: Color-Music, the Color Sense, and the Language of Abstraction. *Configurations, 25*(4), 475–505 (2017)
38. Cuny, C., Fornerino, M., & Helme-Guizon, A. Can music improve e-behavioral intentions by enhancing consumers' immersion and experience?. *Information & Management, 52*(8), 1025–1034 (2015)
39. van Egmond, R. Emotional experience of frequency modulated sounds: Implications for the design of alarm sounds. *Human factors in design*, 345–356 (2004)
40. MacInnis, D. J., & Park, C. W. The differential role of characteristics of music on high-and low-involvement consumers' processing of ads. *Journal of consumer Research, 18*(2), 161–173 (1991)
41. Heckler, S. E., & Childers, T. L. The role of expectancy and relevancy in memory for verbal and visual information: what is incongruency?. *Journal of consumer research, 18*(4), 475–492 (1992)
42. Brinkmann, H., Commare, L., Leder, H., & Rosenberg, R. Abstract art as a universal language?. *Leonardo, 47*(3), 256–257 (2014)
43. Bigerelle, M., & Iost, A. Fractal dimension and classification of music. *Chaos, Solitons & Fractals, 11*(14), 2179–2192 (2000)
44. Bernardi, A., Bugna, G., & De Poli, G. Musical signal analysis with chaos. Musical Signal Processing, 187–220 (1997)
45. Sanyal, S., Banerjee, A., Patranabis, A., Banerjee, K., Sengupta, R., & Ghosh, D. A study on improvisation in a musical performance using multifractal detrended cross correlation analysis. Physica A: Statistical Mechanics and its Applications, 462, 67–83 (2016)
46. Datta, A. K., Sengupta, R., Banerjee, K., & Ghosh, D. Evaluation of Musical Quality of Tanpura by Non Linear Analysis. In Acoustical Analysis of the Tanpura (pp. 133–149). Springer, Singapore (2019)
47. Bhaduri, S., Bhaduri, A., & Ghosh, D. Acoustical genesis of uniqueness of tanpura-drone signal—Probing with non-statistical fluctuation pattern. Physica A: Statistical Mechanics and its Applications, 551, 124206 (2020)

48. Sengupta, R., Dey, N., Datta, A. K., & Ghosh, D. Assessment of musical quality of tanpura by fractal-dimensional analysis. Fractals, 13(03), 245–252 (2005)
49. Banerjee, A., Sanyal, S., Patranabis, A., Banerjee, K., Guhathakurta, T., Sengupta, R., ...& Ghose, P. Study on brain dynamics by non linear analysis of music induced EEG signals. Physica A: Statistical Mechanics and its Applications, 444, 110–120 (2016)
50. Ghosh, D., Sengupta, R., Sanyal, S., & Banerjee, A. Musicality of human brain through fractal analytics. Springer Singapore (2018)
51. Taylor R.P., Micolich, A.P. and Jonas D. 'Fractal analysis of Pollock's drip paintings', Nature 399, 422 (1999)
52. Jones-Smith, K. and Mathur, H. 'Fractal analysis: Revisiting Pollock's drip paintings', Nature 444 issue, E9-E10 (2006)
53. Jones-Smith, K., Mathur, H. and Krauss, L.M. 'Drip paintings and fractal analysis', Phys. Rev. E 79, 046111 (2009)
54. Coddington J., Elton, J., Rockmore, D. and Wang, Y. 'Multifractal analysis and authentication of Jackson Pollock Paintings', Proc. SPIE 6810, Computer Image Analysis in the Study of Art, 68100F (2008)
55. Taylor, R.P., Spehar, B., Van Donkelaar, P. and Hagerhall, C.M. 'Perceptual and Physiological Responses to Jackson Pollock's Fractals', Frontiers in Human Neuroscience, www.frontiersin.org, 45–57 (2011)
56. Peng, C. K., Buldyrev, S. V., Havlin, S., Simons, M., Stanley, H. E., & Goldberger, A. L. Mosaic organization of DNA nucleotides. *Physical review e, 49*(2), 1685 (1994)
57. Gu, G. F., & Zhou, W. X. Detrended fluctuation analysis for fractals and multifractals in higher dimensions. *Physical Review E, 74*(6), 061104 (2006)
58. Wang, J., Shao, W., & Kim, J. Combining MF-DFA and LSSVM for retina images classification. *Biomedical Signal Processing and Control, 60*, 101943 (2020)
59. Kawakatsu, H. Methods for evaluating pictures and extracting music by 2D DFA and 2D FFT. *Procedia Computer Science, 60*, 834–840 (2015)
60. Wang, J., Shao, W., & Kim, J. Automated classification for brain MRIs based on 2D MF-DFA method. *Fractals, 28*(06), 2050109 (2020)
61. Sanyal, S., Banerjee, A., Basu, M., Nag, S., Ghosh, D., & Karmakar, S. Do musical notes correlate with emotions? A neuro-acoustical study with Indian classical music. In Proceedings of Meetings on Acoustics 179ASA (Vol. 42, No. 1, p. 035005). Acoustical Society of America (2020, December)
62. Nag, S., Sarkar, U., Sanyal, S., Banerjee, A., Roy, S., Karmakar, S., ... & Ghosh, D. A Fractal Approach to Characterize Emotions in Audio and Visual Domain: A Study on Cross-Modal Interaction. arXiv preprint arXiv:2102.06038 (2021)
63. Gayen, P., Banerjee, A., Sanyal, S., Nag, S., Patnaik, P., & Ghosh, D. Influence of "indeterminate music" on visual art: a phenomenological, semiotic and fractal exploration. In *Journal of Physics: Conference Series* (Vol. 1896, No. 1, p. 012021). IOP Publishing (2021, April)
64. Wang, F., Liao, D. W., Li, J. W., & Liao, G. P. Two-dimensional multifractal detrended fluctuation analysis for plant identification. Plant methods, 11(1), 1–11 (2015)

Influence of Musical Acoustics on Graphic Design: An Exploration with Indian Classical Music Album Cover Design

Pinaki Gayen, Archi Banerjee, Shankha Sanyal, Priyadarshi Patnaik, and Dipak Ghosh

1 Introduction

In the field of visual arts, graphic design is a significant area where constant changes and experiments are noticed with changing time. Graphic designers always attempt for making a design more attractive to the people with his/her maximum creative endeavor. On the other side, there are many structural components identified by researchers that are known and accessible to most of the designers. So, automatically it became a competitive field. It is believed that an impactful design work is something that is formal and situation dependent, and a reliable design work can be created with the arrangements of certain design components as per design rules (discovered by design researchers like [1] and [2]). It has been observed that the fundamental strategies of design in the field of applied art have their individual mechanism/rule and components. Such components and rules are shaped by the design field, creation methods, purposes, and resources. Among different

---

P. Gayen (✉)
Department of Design, Kala Bhavana, Visva Bharati University, Bolpur, India

A. Banerjee
Rekhi Centre of Excellence for the Science of Happiness, IIT Kharagpur, Kharagpur, India

Sir C.V. Raman Centre for Physics and Music, Jadavpur University, Kolkata, India

S. Sanyal
School of Languages and Linguistics, Jadavpur University, Kolkata, India

P. Patnaik
Department of Humanities and Social Sciences, IIT Kharagpur, Kharagpur, India

Rekhi Centre of Excellence for the Science of Happiness, IIT Kharagpur, Kharagpur, India

D. Ghosh
Sir C.V. Raman Centre for Physics and Music, Jadavpur University, Kolkata, India

© The Author(s), under exclusive license to Springer Nature Switzerland AG 2023
A. Biswas et al. (eds.), *Advances in Speech and Music Technology*, Signals and Communication Technology, https://doi.org/10.1007/978-3-031-18444-4_19

opinions about design principles, there are main six principles that are widely acknowledged by designers, and the six main graphic design principles are such as visual-harmony/visual-unity, visual-balance, order/hierarchy, proportion/size, visual-dominance/specific-emphasis, as well as visual-similarity/visual-contrast [1, 2]. Similarly, if we talk about music, we would find that musical acoustics as a form of aesthetic expression produced by silence and audible sound for a definite time duration and has features like musical-pitch, musical-rhythm, dynamism, sonic quality, timbre, and musical-texture [3, 4]. The crucial question raised here is that what are the semiotic correspondences found between musical components and design components, as well as what impact would occur if intermedial translation is done between these two mediums. In search of an effective music album cover design of Indian classical instrument music, we attempted these queries. In this regard, Seker [5] analyzed the Western classical music album covers' digital age characteristics by comparing the usage of the main six graphic design principles before and after 2000s. But there is no such study in context of Indian classical music album cover so far. Most of the works on such themes have been done in context of Western music.

Previous studies in this field suggested that musical acoustics and paintings have certain associations. Both evoke visual imageries and emotion share similar kind of aesthetic sensibility. For example, intermedial similarities between musical acoustics and color hues used in paintings are profoundly connected by emotions [6, 7]. Music can effectively impact on the assessment of figurative as well as nonfigurative art [8]. Contrarily, music assessment is significantly influenced by various kinds of visual elements [9]. In this regard, some researchers identified the perception- and interpretation-related correspondences in the artistic construction of musical acoustics and visual arts [10].

Musical acoustics induce visual imageries in the audiences' brain that may evoke emotional experiences during the procedure [11]. Neuroscientists suggested such kind of experience as the "involuntary musical imagery" (INMI), which emerges automatically and with no conscious control of the listener while he/she is listening to a music clip [12–17]. Musical acoustic can evoke periodic memories triggering reminiscing of earlier incidents, although while perceiving emotional experiences, an audience may get regulated with different events as well as visuals that are not associated with the periodic memory. These visuals could be regulated or formed by correspondence in construction [11] or other sociocultural associations. Specific musical features, like sequential repetition, expectedness in musical acoustics, harmonic and rhythmic components, as well as slow tempo, are successful in inducing a variety of visual imageries [18]. In this regard, tempo has significant impact on emotional experience [19]. Many studies have similar observation that listeners have a tendency to associate faster tempo and major modes with happiness emotion and slower tempo and minor modes with sadness emotion [20–24]. It has also been observed that to a great level, color relations depend on musical features. For example, if the music is regulated in the direction of fast or slow, the audiences' perception of color relation will change [25].

1.1 Synesthesia: Music and Visual Arts

Synesthesia is a neurological trait where there is a matching or joining of senses that are generally not connected. Any stimulation in one sense causes an involuntary action in one or the other separate sensors. Synesthesia can occur among any two senses and, in some rare cases, among all the five senses. One such trait – the ability to visualize music – was a part of Indian traditional learning system and is well documented in historical musical texts such as *Raga Sagara* and *Sangeeta Ratnavali*, where we have instances of *dhyanaslokas*. *Dhyanaslokas* are short descriptions of *ragas* or musical modes, which briefly describe the *raga* as a male or female divinity or human form with specific personality traits and associated emotions [26–28]. Interestingly, these *dhyanaslokas*, in Sanskrit and Hindi, were written across four centuries and had a great deal of resemblance – for specific ragas – to one another. Moreover, in visual arts tradition, they were represented in miniature art across North India as *ragamala* paintings (with the *dhyanaloka* inscribed within the paintings). It was believed that at the deepest level of musical understanding, the *ragas* could be visualized as men, women, gods or goddesses possessing certain traits, and these visualizations were documented as *dhyanas* – meditation or contemplation on the embodiment of music as forms.

Experiments with synesthesia in the visual arts date back several centuries further in time than scientific research into the subject. Since the seventeenth century, artists have actually been dealing with the connections between various sensory modalities. The Italian artist Arcimboldo, as an example, explored at the court of Rudoph II in Prague by accompanying the hearing of music from a harpsichord with the sight of corresponding images of strips of colored paper [29]. At the end of the nineteenth century, scientific examinations into audition colorée (shade hearing) began to come to be a substantial area of psychological research. Prior to that time, artists, including painters, dramatists, and also composers, had established a body of knowledge about how to utilize correspondences between the senses [30]. The focus on synesthesia has actually transformed the aesthetic sensibilities and visual arts across the last few centuries. The passion of visual artists in synesthetic sensations and also communications in between aesthetic stimulations from various sense modalities has formed the seeds for new art movements. As an example, ideas about sensory correspondences representing a higher reality by the French poet Charles Baudelaire had a substantial impact on the mystic paintings of the Symbolists. Ideas regarding Gesamtkunstwerk (artworks that incorporate music, picture, dance, and also various other disciplines) by the German composer Richard Wagner encouraged artists like Wassily Kandinsky and also Piet Mondrian, who ended up being crucial believer of abstraction in painting [29, 31]. In more modern-day times, the modern digital arts are indebted to synesthetic experiments right into audiovisual perception on very early computers in the 50s as well as 60s, by artists like Edgar Varèse in France and also James Whitney in the United States [32, 33].

Since no tools were available to reveal synesthetic phenomena to audiences (painters had their combination and artists their musical instruments), these early pioneers of synesthetic art were usually developers too. A great example is the colossal color light organ that Alexander Rimmington constructed at the end of the nineteenth century [34, 35]. As a result of this kind of engineering, synesthetic experiments in the visual arts have not only added new concepts but also concrete tools to the Western society. Scientific study has actually improved these artistic devices in direct and also indirect ways. The manufacture of digital algorithms to convert songs right into images appears to have inspired scientific experimental study. Later, these algorithms located their method into popular culture by their implementations in personal computers and also on the Internet (e.g., in the video clip application of music players). The abovementioned evidences of literature indicate that there is a lack of scientific exploration in context of the impact of musical acoustics on graphic design, particularly how to create Indian classical instrumental music album cover design with the help of this particular genre of music.

1.2 The Present Study

The present study attempted to break the conventional design monotony and to add some new creative strategies in the Indian classical instrumental music album cover design. To explore this, we conducted an experimental study with a group of 30 designers, where two preselected instrumental (*sitar*) Indian classical music clips of contrast emotions (previously rated by the audiences as sad/calm and happy/exciting music) were played at a gap of 1 hour and the designers were asked to create a suitable album cover for each of the music pieces during listening to them cautiously. We used semiotic analysis and Detrended Fluctuation Analysis (fractal analysis) of the chosen music pieces as well as the corresponding cover designs made by the designers to explore the impact of different musical acoustical features on the graphic designs of the music album covers and the nature of intermediality between these two mediums.

It has been observed that musical acoustics has a multifaceted character as various music features are generally entangled at each moment, moreover in the context of Indian classical instrumental musical acoustics where musical intensity, music pitch, timbre, music tempo, and sequential repetition of these musical components create the entire auditory waveform multifaceted as well as it is not possible to investigate with the help of deterministic or linear procedures. Such components bear a resemblance to that of a disorganized and self-similar nonlinear structure. In the case of graphic design also, various image components, namely, line features, color hues, different forms, grouping/synchronization, visual harmony, compositional balance, arrangement order, size/ratio, supremacy/importance, as well as resemblance/dissimilarity, remain present with some kind of intertwined way. Therefore, the graphic art naturally creates some amount of complication. Like musical acoustics, the experimentation on album cover design with simple deterministic and linear procedure formulates a great level of lose regarding some crucial data information; thus, most recent state-of-the-art chaos-based nonlinear equipment should be implicated to understand the entire structure of musical acoustic as well as album cover design. In this regard, DFA (Detrended Fluctuation Analysis) procedure predicts significant amount of connotation, because DFA is able to scientifically calculate (using quantitative methods) the symmetry scaling structures existing within the entrenched geometry of the musical acoustics as well as in the graphic design. Consequently, in the concluding stage, DFA was implicated on the musical acoustics of the two Indian classical instrumental music clips and the corresponding music album cover designs to investigate potential correspondences among these two mediums' symmetry scaling.

2 Objective of the Study

(i) The objective of this study is to identify whether any intermedial correspondence exists between the musical acoustics and the visual representations when graphic designers create album cover designs while listening to Indian classical instrumental music.
(ii) To discover novel methodologies for creating effective and aesthetically appealing album cover designs for Indian classical music

3 Experiment Details

3.1 Participants

Thirty undergraduate students, who were enrolled in the design program (at the Architecture and Regional Planning) of the IIT Kharagpur, took part in the experimentation. They have been taught visual communication in a course. The participants' ages were from 18 to 22 years (18 males and 12 females; mean age 20.5 years, SD 1.04 years). None of the participants reported any previous formal training in music.

3.2 Stimuli Used in the Study

Two Indian instrumental (*Hindustani Classical rāga*) musical acoustics were used in this study. The music clips are (a) *Komal Rishav Asavari* and (b) *Jaunpuri drut*. These two musical acoustics had been rated earlier by 70 audiences as sad-peace (*Komal Rishav Asavari*) and happy-exciting (*Jaunpuri drut*) in a self-report method [36].

3.3 Materials for the Study

The materials used in this experiment are, namely, white papers (12″ × 18″), Camlin color crayons, and 4B pencils.

3.4 Experimental Process

This experimentation took place in separate experiment rooms (two rooms) where 15 respondents had to take seats (maintaining certain distance) in each experiment room. The participating respondents were instructed that they would be listening to an Indian classical instrumental audio piece in a cyclic mode. The audio piece (musical acoustics) may induce some visual imageries as well as emotional experiences in their brains, and they had to represent or depict the musical experiences in the form of a music album cover design of that particular music. The participants were given the necessary equipment (pencil, paper, color, etc.) for the same. A group of 15 participants were engaged in listening to the *Komal Rishav Asavari* music piece, and the other group listen to the *Jaunpuri* music clip. They were given 1 hour to complete the task.

3.5 Analytical Strategy

The obtained album cover designs created by the participants were analyzed using an inter-rater reliability method where four experts were involved in the evaluation process. The analysis was done in terms of semiotic analysis between the musical features and the design attributes, taking into consideration the "six principles" of graphic design, namely, visual-harmony/visual-unity, visual-balance, order/hierarchy, proportion/size, visual-dominance/specific-emphasis, as well as visual-similarity/visual-contrast [1]. After that, DFA (fractal analysis) method was implied to analyze possible correlation and effectiveness of the album cover design of the corresponding music clips.

4 Methodologies

4.1 Analysis of Musical Acoustics

In the case of musical acoustical investigation, the authors (who are expert in the field of music) decided specific six segments (every segment time ~ 5 seconds) from the two Indian classical instrumental music clips. After that, both the musical pieces were standardized to 0 dB as well as digitized at a pace of 44.1 kHz at mono-channel 16-bit form. The pitch contours and intensity contours for both the clips were generated and illustrated with the help of WaveSurfer 1.8.8p4 software.

4.2 DFA Method for Understanding Musical Acoustics and Album Cover Designs

Both the music pieces were analyzed with the help of one-dimensional DFA (Detrended Fluctuation Analysis) method, which measures the degree of self-symmetry inherently fixed in the musical acoustical time sequence. This technique is usually implemented following the algorithm of Peng et al. [37]. For each audio piece, DFA analysis yields a particular specification known as scaling exponent (α), which is an exclusive quantitative parameter of the long-range temporal correspondences present in the acoustical time sequence. The significance of α is almost 0.5 for all uncorrelated sequences, $0 < \alpha < 0.5$ for anticorrelated sequences, $0.5 < \alpha < 1$ for long-range temporal correspondences, as well as $\alpha > 1$ for powerful correlations that are not of power-law form. In the present study, we used the same Detrended Fluctuation Analysis method for extracting the scaling exponent matching to various album cover design. Novel 2D Detrended Fluctuation Analysis algorithm is implemented for this reason following the algorithm as specified in Nag et al. [38]:

4.3 2D DFA Method Details

This section describes the steps for computing Hurst exponent using the two-dimensional DFA algorithm for an image I. The steps are as follows:

1. The profile $x_{i,j}$ is computed using:

$$x_{i,j} = \sum_{n=1}^{i} \sum_{m=1}^{j} \left(I_{i,j} - \overline{I} \right)$$

where $m = 1, 2, \cdots, M$, $n = 1, 2, \cdots, N$, $I_{n,m} = 0, 1, \cdots, 255$ is the brightness of the pixel at the coordinates (m, n) of the image and $\overline{I}$ represents the mean value of $I_{n,m}$.

2. $x_{i,j}$ is divided into small regions of size $s \times s$, where s is set as:

$$s_{\min} \approx 5 \leq s \leq s_{\max} \approx \min\{M, N\}/4.$$

3. An interpolating curve is computed of $x_{i,j}$ using:

$$G_{i,j}(l,s) = a_l i + b_l j + c_l$$

in the lth small square region of size $s \times s$, which can be given by using a multiple regression procedure.

4. The variance in the lth small square region is computed for $s = s_{min}, s_{min} + 1, \ldots, s_{max}$, which is given by:

$$F_{i,j}^2(l,s) = \frac{1}{s^2} \sum_{n=1}^{i+s} \sum_{m=1}^{j+s} (x_{i,j} - G_{i,j}(l,s))^2$$

5. The root mean square $F(s)$ is computed as:

$$F(s) = \left[\frac{1}{L_s} \sum_{l=1}^{L_s} F_{i,j}^2(l,s) \right]^{1/2}$$

where L_s denotes the number of the small square regions of size $s \times s$.

6. If $x_{i,j}$ has a long-range power-law correlation characteristic, then the fluctuation function $F(s)$ is observed as follows:

$$F(s) \propto s^\alpha$$

where α is the two-dimensional scaling exponent, a self-affinity parameter representing the long-range power-law correlation characteristics of the surface. α has been computed for red/blue/green color coordinates.

Similar to the 1D DFA scaling exponent, in case of 2D surfaces also, the measurements are almost similar. That is, when the 2D series is completely uncorrelated (Gaussian or non-Gaussian probability distribution), the calculation of the scaling exponent results to 0.5, also called white noise. If $\alpha < 0.5$, the data is anticorrelated, and if $\alpha > 0.5$, the data is long-range correlated.

5 Results and Discussion

To understand the musical structure, we analyzed the acoustical features of the two music clips in WaveSurfer 1.8.8p4 software. We observed that the *Komal Rishav Asavari* has less oscillation in the pitch contour, and it has certain continuity (see Fig. 1 top section). The intensity contour of the *Komal Rishav Asavari* indicates that it has moderate level of fluctuation in the intensity contour, and it is continuous in nature. The intensity contour of this music also indicates that there are continuous gliding transitions that look like curve (see Fig. 1 middle section). In the case of

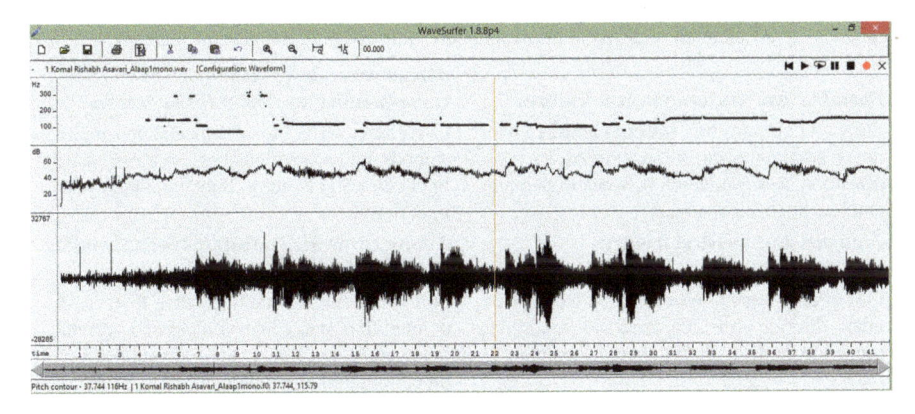

Influence of Musical Acoustics on Graphic Design: An Exploration with Indian... 387

Fig. 1 *Komal Rishav Asavari* musical structure (pitch contour, intensity contour, and musical waveform)

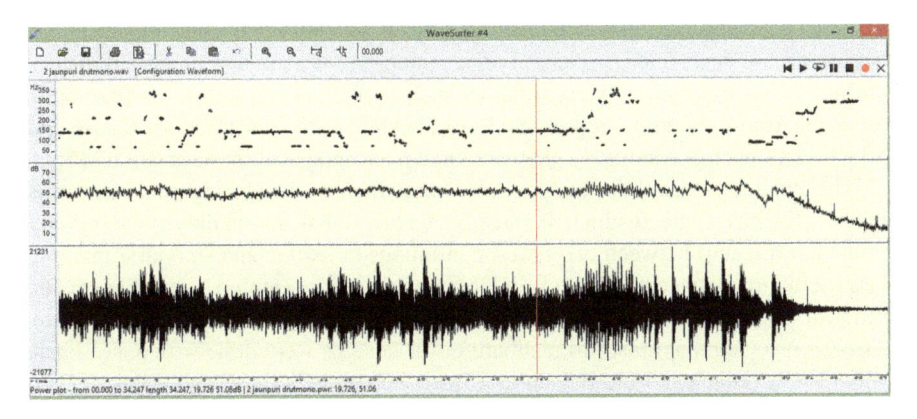

Fig. 2 *Jaunpuri drut* musical structure (pitch contour, intensity contour, and musical waveform)

Jaunpuri drut, we observed that this music clip has high level of fluctuation in the pitch contour, which is fragmented and discontinuous (see Fig. 2 top section). But in the case of intensity contour of the *Jaunpuri drut,* we observed that this music clip has less fluctuation between two gliding transitions, and it is continuous in nature (see Fig. 2 middle section).

5.1 Semiotic Analysis

Semiotic analysis was done on the chosen music pieces (*Komal Rishav Asavari* and *Jaunpuri drut*) and on the corresponding cover designs made by the designers to understand the impact of different musical acoustical features on the graphic designs of the music album covers and the nature of intermediality between these

Table 1 Representation of different musical features through different visual features

Komal Rishav Asavari musical features	Album cover design features corresponding to *Komal Rishav Asavari*
Slow gliding transition (meend) between two notes, continuous transition, smooth transition, less oscillation between high pitch and low pitch, slow tempo	Curve line, wavy line, ornamental design, smooth line, contentious line, dominant use of light color, light yellow, light blue, light pink, light orange
Jaunpuri drut musical features	Album cover design features corresponding to *Jaunpuri drut*
Fast gliding transition (*meend*) between two notes, discrete transition, continuous, jumping transition, sudden oscillation between high pitch and low pitch, fast tempo	Spiky line, fragmented line, curve line, cross, pointed line, shaky line, vertical line, scribble, dominant use of light color, light yellow, light blue, light pink, light orange

two mediums (music and visual representation). As we know that when direct association is observed between different objects and the represented visuals, such depictions are considered as "iconic representation." When the represented visuals are not found to be directly identifiable, such depictions are known as "symbolic representation." When the represented visuals elicit some other actions (cause and effect relationship), commonly, they are known as "indexical representation" [6].

The results of semiotic analysis of *Komal Rishav Asavari* revealed that this music clip has some distinctive musical features that are mainly slow gliding transition (meend) between two notes, continuous transition and smooth transition, less oscillation between high pitch and low pitch, and slow tempo (see Table 1). The semiotic analysis of the album cover designs corresponding to the *Komal Rishav Asavari* revealed that most of the album cover designs were done with curve lines, wavy lines, ornamental motifs, smooth lines, contentious lines, and dominant use of light colors like light yellow, light blue, light pink, and light orange (see Fig. 3a, b). On the other hand, the results of semiotic analysis of *Jaunpuri drut* revealed that this music clip has distinctive features like fast gliding transition (meend) between two notes, discrete transition, continuous, jumping transition, sudden oscillation between high pitch and low pitch, fast tempo, etc. The semiotic analysis of the album cover designs corresponding to the *Jaunpuri drut* revealed that most of the album cover designs were done with spiky lines, fragmented lines, curve lines, cross lines, pointed lines, shaky lines, vertical lines, scribble, dominant use of light color, light yellow, light blue, light pink, light orange, etc. (see Fig. 4a, b).

The semiotic analysis also revealed that while the designers represented their musical experiences through the album covers, they mostly depicted the album covers in three different ways, namely, (1) the designers directly represented the moods and emotions of the music clips in their album cover designs. In this case, the designers dominantly used symbolic representation followed by indexical representation. (2) Some of the designers represented the visual imageries that were evoked by the music clips. This type of album covers have dominant pattern of indexical representation followed by iconic representation. (3) Some designers

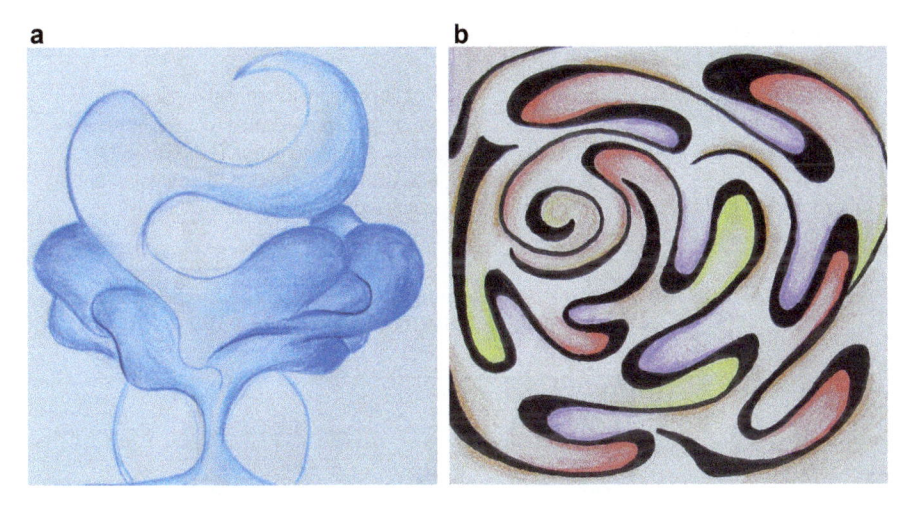

Fig. 3 (**a**) Album cover design for *Komal Rishav Asavari*. (**b**) Album cover design for *Komal Rishav Asavari*

Fig. 4 (**a**) Album cover design for *Jaunpuri drut*. (**b**) Album cover design for *Jaunpuri drut*

represented the musical features in their album cover designs. This type of album covers mostly comes under iconic representation (see Table 2).

During the analysis of the album cover designs, the experts observed that there is significant amount of evidence of the "six principles of design" in the album covers created by the designers, and the components are, namely, visual-harmony/visual-unity, visual-balance, order/hierarchy, proportion/size, visual-dominance/specific-emphasis, as well as visual-similarity/visual-contrast [1].

Table 2 Different types of musical analogies used by the designers

Designers' preferences of depiction of their musical experiences in the album covers	Dominant representational category
Musical moods and emotions	Symbolic representation (dominantly) and indexical representation (subordinately)
Images evoked by the musical acoustics	Indexical representation (dominantly) and iconic representation (subordinately)
Musical features	Iconic representation

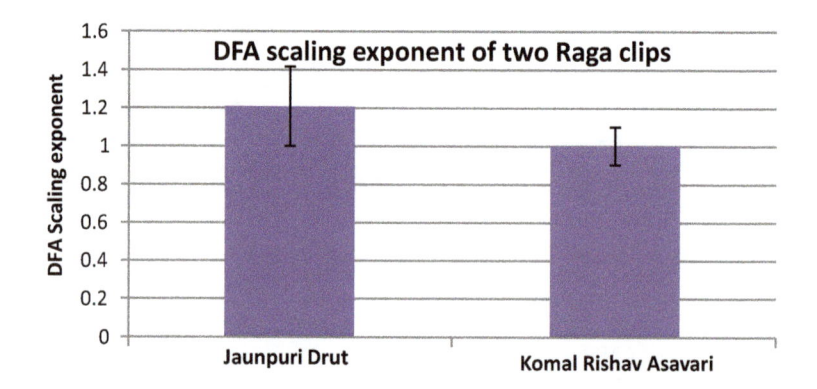

Fig. 5 DFA scaling exponents of the two *Raga* clips – *Komal Rishav Asavari* and *Jaunpuri Drut*

5.2 Fractal Analysis of Music and Comparison with Semiotic

Each of the two chosen *raga* clips of 30 seconds duration were divided into six equal parts, and using the conventional 1D DFA technique, the DFA scaling exponents (or the amount of long-range correlations) for each part of the two input music stimuli or the two Indian *Raga* clips (*Komal Rishav Asavari* and *Jaunpuri drut*) were calculated and the detailed results are reported in Fig. 5.

Figure 5 revealed that when the average of the DFA scaling exponents for the six parts of each *raga* clip was calculated, the fast tempo *Jaunpuri Drut* featured an overall higher scaling exponent or greater long-range correlations compared to the slow-tempo excerpt of *Komal Rishav Asavari*. Another interesting observation was that the variance of DFA scaling exponent between the six parts of the *Jaunpuri Drut* was much higher compared to that of *Komal Rishav Asavari*, which can be attributed to a higher presence of pitch and amplitude change in the fast-tempo *Jaunpuri Drut*. Next, using the two-dimensional extension of the conventional DFA method, that is, the 2D DFA technique, the long-range correlations along the three primary color coordinates – red, blue, and green – were calculated for each of the CD cover designs corresponding to the two *Raga* clips. Figure 6a, b reported the same in details.

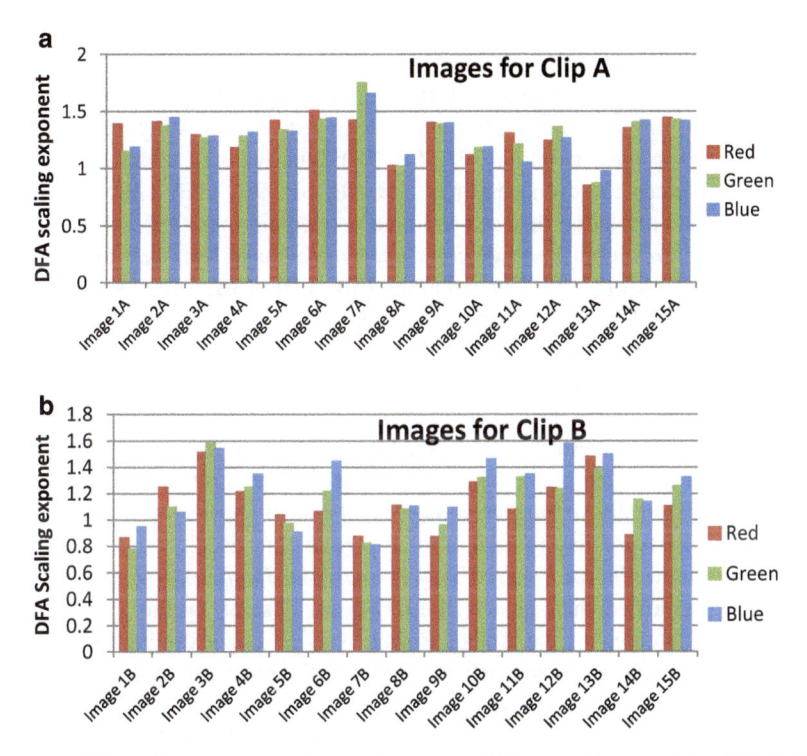

Fig. 6 (**a**) DFA scaling exponents along red, green, and blue coordinates of the 15 individual CD cover designs corresponding to *Komal Rishav Asavari*. (**b**) DFA scaling exponents along red, green, and blue coordinates of the 15 individual CD cover designs corresponding to *Jaunpuri Drut*

The 2D DFA results of the 15 individual CD cover designs created by the 15 participants corresponding to the two chosen *raga* clips revealed that for clip A or *Komal Rishav Asavari*, the "between subject" or individualistic variations in the long-range correlations along the red, blue, and green coordinates of the 15 depicted paintings were much lower compared to those cover design corresponding to *Jaunpuri Drut*. A detailed observation of Fig. 6a suggested that, among the 15 images corresponding to *Komal Rishav Asavari*, the long-range correlation along red color was higher than that of blue and green in four paintings (1A, 5A, 6A, 11A); blue featured a higher correlation than red and green in three paintings (4A, 8A, 13A); green featured higher correlation than red and blue in two paintings (7A, 12A), while the long-range correlations along the three color coordinates or at least two of them were very close to each other in six paintings (2A, 3A, 9A, 10A, 14A, 15A). Among all the images, highest DFA scaling exponent along red was observed in 6A, whereas highest long-range correlations along blue and green were observed in 7A. The lowest symmetry scaling along all three colors were observed in design 13A. Overall, for the CD cover designs related to *Komal Rishav Asavari*, the results hint toward a higher possibility of getting equivalent DFA scaling exponents along red, blue, and green color coordinates compared to the dominance of long-range correlation along one color over other two.

A closer look at Fig. 6b suggested that, among the 15 images corresponding to *Jaunpuri Drut*, the long-range correlation along red color was higher than that of blue and green in three paintings (2B, 5B, 7B); blue featured a higher correlation than red and green in seven paintings (1B, 4B, 6B, 9B, 10B, 12B, 15B); green featured higher correlation than red and blue in one paintings (3B), while the long-range correlations along the three color coordinates or at least two of them were very close to each other in four paintings (8B, 11B, 13B, 14B). Among all the images, highest DFA scaling exponent along red was observed in 3B, whereas highest long-range correlations along blue and green were observed in 12B and 3B, respectively. The lowest symmetry scaling along all three colors were observed in design 7B. Overall, for the CD cover designs related to *Jaunpuri Drut*, the results hint toward a greater possibility of getting higher DFA scaling exponent along blue, compared to the red and green color coordinates. For generalization, we calculated the average DFA scaling exponents along red, blue, and green coordinates for all the 15 paintings corresponding to each of the two *raga* clips used for this study. Figure 7a, b shows the results for *Komal Rishav Asavari* and *Jaunpuri drut,* respectively.

Figure 7a revealed that, indeed, the CD cover designs by the 15 participants created after listening to the *Komal Rishav Asavari* clip exhibited nearly equal amount of average long-range correlations along all three primary, that is, red, blue, and green color coordinates, whereas Fig. 7b yielded a gradually increasing long-range correlations from red to green to blue color coordinates for the 15 designs created corresponding to *Jaunpuri Drut.* Figure 7b also featured a higher margin of errors or higher individualistic variations among participants in the 2D DFA scaling exponent values along red, blue, and green color coordinates.

Combining the findings of the fractal analysis of the input music clips and the output images, we observed that for the fast tempo *Jaunpuri Drut* with higher variance of long-range correlations throughout the clip, the designers were also influenced to create designs with a higher subjective variances and their designs

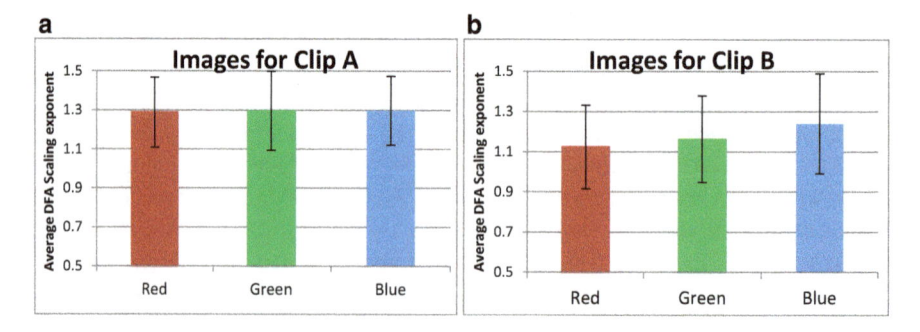

Fig. 7 (**a, b**) Average DFA scaling exponents along red, green, and blue coordinates of the CD cover designs corresponding to (**a**) *Komal Rishav Asavari* and (**b**) *Jaunpuri Drut*

featured a higher long-range correlation value along the blue color coordinate compared to that of red and green, whereas the slow-tempo *Komal Rishav Asavari* with low variance of long-range correlations throughout the clip inspired the designers to create designs that featured lesser subjective variances and an equivalent amount of long-range correlations along all three primary color coordinates of the designs in most cases.

6 Conclusions

Findings revealed that while the designers represented their musical experiences through the album covers, they mostly depicted the album covers in three different ways, namely, (1) the designers directly represented the moods and emotions of the music clips in their album cover designs. In this case, the designers dominantly used symbolic representation followed by indexical representation. (2) Some of the designers represented the visual imageries that were evoked by the music clips. This type of album covers have dominant pattern of indexical representation followed by iconic representation. (3) Some designers represented the musical features in their album cover designs. This type of album covers mostly comes under iconic representation. A comparative quantitative study on the symmetry scaling behavior (using fractal analysis) of the acoustical waveforms of the two music clips as well as the designers' created images also indicated that there was a clear correspondence between musical acoustical features and the depicted visual features in the album cover designs. Moreover, the findings of this study provided us a new innovative approach for creating music album cover design beyond the conventional approaches. In the DFA technique, we observed that the fast-tempo *Jaunpuri Drut* featured an overall higher scaling exponent or greater long-range correlations compared to the slow tempo excerpt of *Komal Rishav Asavari*. Another interesting observation was that the variance of DFA scaling exponent between the six parts of the *Jaunpuri Drut* was much higher compared to that of *Komal Rishav Asavari*, which can be attributed to a higher presence of pitch and amplitude change in the fast-tempo *Jaunpuri Drut*. Overall, in the findings of the fractal analysis of the input music clips and the output images, we observed that for the fast-tempo *Jaunpuri Drut* with higher variance of long-range correlations throughout the clip, the designers were also influenced to create designs with a higher subjective variances and their designs featured a higher long-range correlation value along the blue color coordinate compared to that of red and green, whereas the slow-tempo *Komal Rishav Asavari* with low variance of long-range correlations throughout the clip inspired the designers to create designs that featured lesser subjective variances and an equivalent amount of long-range correlations along all three primary color coordinates of the designs in most cases.

Acknowledgments The authors are thankful to the participants of this study and also thankful to Junmoni Borgohain, Dr. Mainak Ghosh, and Subhajit Das for their cooperation and support.

References

1. Dabner, D., Stewart, S., & Vickress, A. (2017). *Graphic design school: The principles and practice of graphic design*. John Wiley & Sons.
2. Wong, W. (1993). *Principles of form and design*. John Wiley & Sons.
3. Goodall, H. (2013). *The story of music*. Random House.
4. Berry, W. (1987). *Structural functions in music*. Courier Corporation.
5. Seker, C. (2017). New Classics: *The Analysis of Classical Music Album Covers' Digital Age Characteristics*. European Scientific Journal.
6. Gayen, P., Banerjee, A., Sanyal, S., Nag, S., Patnaik, P., & Ghosh, D. (2021, April). Influence of "indeterminate music" on visual art: a phenomenological, semiotic and fractal exploration. In *Journal of Physics: Conference Series* (Vol. 1896, No. 1, p. 012021). IOP Publishing. https://doi.org/10.1088/1742-6596/1896/1/012021
7. Palmer, S. E., Schloss, K. B., Xu, Z., & Prado-León, L. R. (2013). Music–color associations are mediated by emotion. *Proceedings of the National Academy of Sciences*, 110(22), 8836–8841.
8. Limbert, W. M., & Polzella, D. J. (1998). Effects of music on the perception of paintings. *Empirical studies of the arts*, 16(1), 33–39.
9. Parrott, A. C. (1982). Effects of paintings and music, both alone and in combination, on emotional judgments. *Perceptual and Motor Skills*, 54(2), 635–641.
10. Actis-Grosso, R., Lega, C., Zani, A., Daneyko, O., Cattaneo, Z., &Zavagno, D. (2017). Can music be figurative? Exploring the possibility of cross-modal similarities between music and visual arts. *Psihologija*, 50(3), 285–306.
11. Juslin, P. N., &Västfjäll, D. (2008). Emotional responses to music: The need to consider underlying mechanisms. *Behavioral and brain sciences*, 31(5), 559–575.
12. Zatorre, R. J., Halpern, A. R., Perry, D. W., Meyer, E., & Evans, A. C. (1996). Hearing in the mind's ear: a PET investigation of musical imagery and perception. *Journal of cognitive neuroscience*, 8(1), 29–46.
13. Halpern, A. R. (2001). Cerebral substrates of musical imagery. *Annals of the New York Academy of Sciences*, 930(1), 179–192.
14. Liikkanen, L. A. (2008, August). Music in every mind: Commonality of involuntary musical imagery. In *Proceedings of the 10th international conference on music perception and cognition* (pp. 408–412). Sapporo, Japan: ICMPC.
15. Hubbard, T. L. (2010). Auditory imagery: empirical findings. *Psychological bulletin*, 136(2), 302.
16. Herholz, S. C., Halpern, A. R., & Zatorre, R. J. (2012). Neuronal correlates of perception, imagery, and memory for familiar tunes. *Journal of cognitive neuroscience*, 24(6), 1382–1397.
17. Farrugia, N., Jakubowski, K., Cusack, R., & Stewart, L. (2015). Tunes stuck in your brain: The frequency and affective evaluation of involuntary musical imagery correlate with cortical structure. *Consciousness and cognition*, 35, 66–77.
18. McKinney, C. H., &Tims, F. C. (1995). Differential effects of selected classical music on the imagery of high versus low imagers: Two studies. *Journal of Music Therapy*, 32(1), 22–45.
19. Gabrielsson, A. (2009). The relationship between musical structure and perceived expression. In *Oxford handbook of music psychology*.
20. Crowder, R. G. (1984). Perception of the major/minor distinction: I. Historical and theoretical foundations. *Psychomusicology: A Journal of Research in Music Cognition*, 4(1–2), 3.
21. Dalla Bella, S., Peretz, I., Rousseau, L., & Gosselin, N. (2001). A developmental study of the affective value of tempo and mode in music. *Cognition*, 80(3), B1–B10.
22. Gagnon, L., & Peretz, I. (2003). Mode and tempo relative contributions to "happy-sad" judgements in equitone melodies. *Cognition and emotion*, 17(1), 25–40.
23. Hunter, P. G., Schellenberg, E. G., & Schimmack, U. (2010). Feelings and perceptions of happiness and sadness induced by music: Similarities, differences, and mixed emotions. *Psychology of Aesthetics, Creativity, and the Arts*, 4(1), 47.

24. Webster, G. D., & Weir, C. G. (2005). Emotional responses to music: Interactive effects of mode, texture, and tempo. *Motivation and Emotion, 29*(1), 19–39.
25. Tsang, T., & Schloss, K. B. (2010). Associations between color and music are mediated by emotion and influenced by tempo. *The Yale Review of Undergraduate Research in Psychology, 82.*
26. Gangoly, O. C. (2017). *Ragas & Raginis.* Shubhi Publications.
27. Ebeling, K. (1973). *Ragamala painting* (p. 118). Basel: Ravi Kumar.
28. Kaufmann, W. (1968). *The ragas of north India* (No. 1). Indiana University Press, Bloomington.
29. Gage, J. (1999). *Color and meaning: Art, science, and symbolism.* Univ of California Press.
30. Van Campen, C. (2013). Synesthesia in the visual arts. In *Oxford Handbook of Synesthesia.*
31. Düchting, H. (1996). *Color at the Bauhaus: Synthesis and Synaesthesia.* Brothers man.
32. Brougher, K., Strick, J., Wiseman, A., & Zilczer, J. (Eds.). (2005). *Visual Music: Synaesthesia in Art and Music Since 1900: The Museum of Contemporary Art, Los Angeles, [13 February-22 May 2005]: Hirshhorn Museum and Sculpture Garden, Smithsonian Institution, Washington DC, [23 June-11 September 2005].* Thames & Hudson.
33. Daniels, D., & Naumann, S. (Eds.). (2010). *Audiovisuology: An Interdisciplinary Survey of Audiovisual Culture. Compendium.* Walther König.
34. Sidler, N., & Jewanski, J. (2006). Farbe-Licht-Musik, Synästhesie und Farblichtmusik.
35. Peacock, K. (1988). Instruments to perform color-music: Two centuries of technological experimentation. *Leonardo, 21*(4), 397–406.
36. Gayen, P., Borgohain, J., & Patnaik, P. (2021). The Influence of Music on Image Making: An Exploration of Intermediality Between Music Interpretation and Figurative Representation. In *Advances in Speech and Music Technology* (pp. 285–293). Springer, Singapore. https://doi.org/10.1007/978-981-33-6881-1_24
37. Peng, C. K., Buldyrev, S. V., Havlin, S., Simons, M., Stanley, H. E., & Goldberger, A. L. (1994). Mosaic organization of DNA nucleotides. *Physical review e, 49*(2), 1685.
38. Nag, S., Sarkar, U., Sanyal, S., Banerjee, A., Roy, S., Karmakar, S., & Ghosh, D. (2019). A Fractal Approach to Characterize Emotions in Audio and Visual Domain: A Study on Cross-Modal Interaction. *Journal of Image Processing & Pattern Recognition Progress, 6*(3), 1–7.

A Fractal Approach to Characterize Emotions in Audio and Visual Domain: A Study on Cross-Modal Interaction

Shankha Sanyal, Archi Banerjee, Sayan Nag, Souparno Roy, Ranjan Sengupta, and Dipak Ghosh

1 Introduction

Quantification and classification of emotions elicited in humans from different modalities have intrigued researchers from various domains over the last few decades. Most of these studies resort to various psychological parameters to measure the variation of emotional appraisal from two different modalities – say, for example, audio and visual stimulus. A few of them even try to look into how the perceptual strength of emotions expressed in the two modalities differ from one another [1–4]. It is already known that certain emotions (i.e., brief affective states triggered by the appraisal of an event in relation to current goals; [5]) such as awe and wonder [6] are frequently reported in relation to the contemplation

---

S. Sanyal (✉)
Sir C.V. Raman Centre for Physics and Music, Jadavpur University, Kolkata, India

School of Languages and Linguistics, Jadavpur University, Kolkata, India

A. Banerjee
Sir C.V. Raman Centre for Physics and Music, Jadavpur University, Kolkata, India

Department of Humanities and Social Sciences, IIT Kharagpur, Kharagpur, India

Shrutinandan School of Music, Kolkata, India

S. Nag
Sir C.V. Raman Centre for Physics and Music, Jadavpur University, Kolkata, India

Department of Biomedical Physics, University of Toronto, Toronto, ON, Canada

S. Roy
Sir C.V. Raman Centre for Physics and Music, Jadavpur University, Kolkata, India

Department of Physics, Jadavpur University, Kolkata, India

R. Sengupta ·D. Ghosh
Sir C.V. Raman Centre for Physics and Music, Jadavpur University, Kolkata, India

© The Author(s), under exclusive license to Springer Nature Switzerland AG 2023
A. Biswas et al. (eds.), *Advances in Speech and Music Technology*, Signals and Communication Technology, https://doi.org/10.1007/978-3-031-18444-4_20

of artworks. These emotions typically occur when the human brain perceives an object or an event as highly complex and novel, and creates a sense of being in the presence of something greater than oneself [7]. However, it has also been recently emphasized that affective responses to art are more diverse and often include emotions such as sadness [8] and nostalgia [9], which are also experienced in other everyday situations that do not involve contemplation of artworks. If we try to look into more basic features of a painting, i.e., the usage of basic colors (red, green, and blue) along with their offshoots, earlier works [10–17] suggested that warm colors – such as red, yellow, and orange – can spark a variety of emotions ranging from comfort and warmth to hostility and anger. On the other hand, cool colors – such as green, blue, and purple – often spark feelings of calmness as well as sadness. A recent study [18] outlines the effect of variation of hue and saturation on emotional perception. From the emotional ratings, they claim that saturated and bright colors were associated with higher arousal, with the arousal-based activities being significantly affected by the hue of the colors used, increasing from blue and green to red. The ratings of valence were the highest for saturated and bright colors and also depended on the hue. Strezoski et al. [15] provides a new interface – ACE, the Art, Color and Emotion browser, which essentially is a platform for exploring the visual sentiment and emotion in artistic paintings over time. Regarding the association of the two modalities – color and music – with emotion, a number of studies [18–20] look into the cross-modal interaction between the two. While some studies posit that the emotional appraisal corresponding to the two are similar in effect, others suggest differently. Palmer et al. [19] found strong correlations ($0.89 < r < 0.99$) between emotional associations of music with the colors chosen to go with the music in an experiment performed on US and Mexican population, supporting an emotional mediation hypothesis in both cultures. Barbiere et al. [17] report that brighter colors such as yellow, red, green, and blue were usually assigned to the happy songs, and gray was usually assigned to the sad songs concluding that music-color correspondences occur via the underlying emotion common to the two stimuli. However, all these are psychological studies based on analysis of behavioral response data. The correlation of the behavioral response data with the computational data obtained from the mathematical analysis of the source signal data has been attempted in this particular study.

To achieve our goal, we have tried to evaluate long-range temporal correlations present in the three basic color components in paintings and compare them with the emotional appraisal related to those paintings. That music is able to elicit a variety of emotional arousal-based effects in humans has long been known. In recent years, the use of musical stimuli as an important mean of emotional appraisal is being developed with special focus on cross-modal transfer of emotions [21–25]. Most of these studies look into the psychological and cognitive aspects of the musical and visual stimulus, focusing mainly on how the musical priming affects the emotional appraisal corresponding to the visual domain. Very few studies take into consideration the source characteristics of the input audio and visual stimuli [20, 26] mainly due to the lack of robust features to quantify them. The development of the International Affective Picture System (IAPS) has been followed by a similar collection of sounds, the International Affective Digitized Sounds (IADS) [27, 28] – a series of naturally occurring human, nonhuman, animal, and environmental sounds (e.g., bees buzzing; applause, explosions). In two experiments by Bradley and Lang [29], it was shown that valence and arousal ratings of these sounds were comparable to affective pictures from the IAPS. On a physiological level, emotionally arousing sounds elicit large electrodermal activity, which is generally known to be sensitive to the arousal of emotional stimuli.

In this paper, the main aim is to classify the emotional sound and visual stimuli solely from their source characteristics, i.e., the time series generated from the audio signal and the two-dimensional matrix of pixels generated from the affective picture stimulus. The sample data consists of six audio signals of around 15 seconds each and six affective pictures, of which three from each category belongs to positive and negative valence, respectively. The emotional ratings corresponding to the visual and audio stimulus were standardized a priori with the help of different psychological tests and corroborated with standardized measures present in literature. The present work aims to provide a comparative assessment of the results of psychological tests in the perceptual domain with the mathematical quantitative output obtained from the multimodal source signals itself. As a powerful mathematical tool, fractal theory initiated by Mandelbrot [30] has been widely applied to many areas of natural sciences ranging from DNA sequencing [31], economics [32], and heart rate variability [33] to quantifying EEG signals [34–37]. Since the simple iterative algorithm in the fractal theory can generate a variety of complex images, fractal dimension is considered as an effective measure of the complexity of the target object. A robust tool called Detrended Fluctuation Analysis (DFA) have been applied here to calculate the long-range temporal correlations or the Hurst exponent corresponding to the auditory signals [38–40]. A higher value of the scaling exponent (α) implies greater complexity corresponding to the audio signal in question. The 2D analogue of the DFA technique have been applied on the array of pixels corresponding to affective pictures of contrast emotions, which essentially gives the long-range spatial correlations of individual color components. We have utilized the scaling exponent (or the Hurst exponent) obtained from the audio clips and the visual images as a robust parameter to quantify their emotional valence. Thus, we have a single unique scaling exponent corresponding to each 1D audio signal and three scaling exponents corresponding to red/green/blue (RGB) component in each of the visual images. The DFA scaling exponents obtained in this manner have been used as a quantitative classification parameter for identifying emotional cues in auditory and visual domain using the source signals itself. Further, correlation features among the paintings and audio clips have been extracted using the 2D/1D Detrended Cross-correlation (DCCA) technique [41, 42]. This nonlinear correlation technique gives the degree of correlation between the two individual domains (audio and visual) and highlights on the applicability of cross-modal transfer of emotion between the two domains from a mathematical point of view. Using the DFA scaling exponents from the audio and visual clips, Pearson correlation coefficient was also calculated to further elucidate on the point of direct correlation between the two modalities. To conclude, we propose a novel algorithm with which emotional arousal can be classified in cross-modal scenario using only the source audio and visual signals while also attempting a correlation between them. The study is expected to go a long way in research on multimodal interaction of emotional cues across multiple domains. The results and implications have been discussed in detail.

2 Experimental Details

2.1 Choice of Three Pairs of Audio and Visual Stimuli

Six clips of around 15 s each (three clips each belonging to the positive and negative valence) were chosen from the IADS [29] database of musical clips and normalized for acoustic analysis. In a similar manner, six famous paintings were chosen, of which, conventionally, three belonged to the positive and negative valence each. All the clips and paintings chosen were subjected to human response analysis of 50 participants to validate the emotional appraisal of each of the stimulus chosen. The human response test was conducted in standard laboratory setting of Sir C.V. Raman Centre for Physics and Music, Jadavpur University. Table 1 is the template that was provided to the participants for the psychological response test on the paintings and sound clips suggested by Eerola et al. [43, 44]. Tables 2 and 3 give the details of audio clips and paintings that were chosen for analysis and used as a stimulus for the psychological tests. The participants were asked to mark the emotions in a 10-point Likert scale as per their choice, and they were free to rate the clips/images in more than one checkboxes. The audio clips were presented to the participants in a jumbled manner followed by the paintings after an interval of 5 min.

In the following sections, using novel nonlinear methodologies, a correlation study is performed across the cross-modal domains to establish the degree of emotional appraisal that is transferred corresponding to the type of stimulus used.

Table 1 Psychological rating template of the audio clips (along with details) chosen for analysis

Clip no.	Anger	Fear	Happy	Sad
1				
2				
3				
4				
5				
6				

Table 2 Details of the audio clips chosen for analysis

Clip no.	Ground truth	Album name	Piece title	Composer's name
1	Happy	Shallow Grave	The Gardener	John Carmichael Band
2	Happy	Nostradamus	Dawn of Creation	Judas Priest
3	Happy	Shine	The Cranberries: "Salvation"	Jesus Jones
4	Sad	Psycho	Marion	Bernard Herrmann
5	Sad	Big Fish	Return to Spectre	Danny Elfman
6	Sad	Band of Brothers	The Wall	Willie Nelson

Table 3 Details of the paintings chosen for analysis

Image no.	Ground truth	Painting title	Painter
1	Happy	Sunflower	Vincent Van Gogh
2	Happy	Japanese Vase	Vincent Van Gogh
3	Happy	Almond Tree	Vincent Van Gogh
4	Sad	The Tragedy	Pablo Picasso
5	Sad	Starry Night	Vincent Van Gogh
6	Sad	Sailboats at Sunset	Ferdinand du Puigaudeau

3 Methodology

One-dimensional Detrended Fluctuation Analysis (DFA) is conventionally done following the algorithm of Peng et al. (1994) and using the methodologies used in [37] and Sanyal et al. [35, 45]. In this work, for extracting the scaling exponent corresponding to different paintings, we propose a novel 2D DFA algorithm here:

3.1 2D Detrended Fluctuation Analysis

This section describes the steps for computing Hurst exponent using the two-dimensional DFA algorithm for a gray scale image I. The steps are as follows:

The profile x is computed using:

$$x = \sum_{i=1}^{M} \sum_{j=1}^{N} (I_{i,j} - \overline{I}) \qquad (1)$$

where $i = 1, 2, \cdots, M, j = 1, 2, \cdots, N, I_{i,j} \in \{0, 1, \cdots, 255\}$ is the brightness of the pixel at the coordinates (i, j) of the gray scale image, and $\overline{I}$ represents the mean intensity value of the image I. Here, the size of the image I is given by $M \times N$.

The profile x is divided into square segments of size $s \times s$. We have considered the total number of such small square regions as L_s. The value of s lies in the following range:

$$s_{\min} \approx 5 \leq s \leq s_{\max} \approx \min\{M, N\}/4. \qquad (2)$$

We start the algorithm by setting the initial value of s as $s_{\min}$.

Considering a small square region of size $s \times s$ denoted by l, an interpolating curve $G_{i,j}(l, s)$ (local trend) is computed using the formula:

$$G_{i,j}(l, s) = a_l i + b_l j + c_l \qquad (3)$$

for the lth small square region, using a multiple regression procedure. Here, $1 \leq i, j \leq s$ and a_l, b_l and c_l are the coefficients for the lth square that are to be determined using the least-squares regression method.

The variance V in the lth small square region is computed for s as follows:

$$V(l, s) = \frac{1}{s^2} \sum_{i=1}^{s} \sum_{j=1}^{s} \left(x_{i,j} - G_{i,j}(l, s)\right)^2 \qquad (4)$$

This variance should be minimized.

The root mean square $F(s)$ is computed as:

$$F(s) = \left[\frac{1}{L_s} \sum_{l=1}^{L_s} V(l, s)\right]^{1/2} \qquad (5)$$

where L_s denotes the total number of the small square regions of size $s \times s$. After that, the value of s is increased by unity, and steps 2 to 5 are repeated until the maximum value of s (i.e., $s_{\max}$) is reached.

The values of $F(s)$ for corresponding values of s are stored. If $x_{i,j}$ has a long-range power-law correlation characteristic, then the fluctuation function $F(s)$ is observed as follows:

$$F(s) \propto s^{\alpha} \qquad (6)$$

where "$\propto$" means "is proportional to", "α" is the two-dimensional scaling exponent, a self-affinity parameter representing the long-range power-low correlation characteristics of the surface. Thus, $F(s)$ is proportional to s^{α}, and we fit a least-squares regression line in a $\log(F(s))$ vs. $\log(s)$ plot to obtain the slope of the line as the value of α.

The 1D analogue of this technique can be understood as a gross generalization of the method discussed above.

For investigating power-law cross-correlations between different simultaneously recorded time series in the presence of nonstationary signals, 1D Detrended Cross-correlation Analysis (DCCA) [41] has been used in many cases. Here, we generalize it in two-dimensional analogue to extract the degree of correlation present between different paintings. Additionally, 1D DCCA was also applied to assess the cross-correlation parameters corresponding to the audio clips of contrast emotion.

3.2 2D Detrended Cross-Correlation Analysis (DCCA)

This section describes the steps for computing cross-correlation scaling exponent using the two-dimensional DCCA algorithm for two gray scale images A and B. The steps are as follows:

The profiles x and y are computed using:

$$x = \sum_{i=1}^{M} \sum_{j=1}^{N} \left(A_{i,j} - \overline{A}\right)$$

$$y = \sum_{i=1}^{M} \sum_{j=1}^{N} \left(B_{i,j} - \overline{B}\right) \tag{7}$$

where $i = 1, 2, \cdots, M, j = 1, 2, \cdots, N$, $A_{i,j} \in \{0, 1, \cdots, 255\}$, $B_{i,j} \in \{0, 1, \cdots, 255\}$ are the brightness of the pixel at the coordinates (i, j) of the gray scale images A and B, respectively, and $\overline{A}$ and $\overline{B}$ represent the mean intensity values of the images A and B, respectively. Here, the size of each image is given by $M \times N$.

Both $x_{i,j}$ and $y_{i,j}$ are individually divided into small regions of size $s \times s$, where s is set as:

$$s_{min} \approx 5 \leq s \leq s_{max} \approx \min\{M, N\}/4. \tag{8}$$

1. Considering a small square region of size $s \times s$ denoted by l, interpolating curves (local trends) for $x_{i,j}$ and $y_{i,j}$ are given by $Gx_{i,j}$ and $Gy_{i,j}$ that are computed using:

$$\begin{aligned} Gx_{i,j}(l, s) &= ax_l i + bx_l j + cx_l \\ Gy_{i,j}(l, s) &= ay_l i + by_l j + cy_l \end{aligned} \tag{9}$$

in the lth small square region of size $s \times s$, which can be given by using a multiple regression procedure. ax_l, ay_l, bx_l, by_l, cx_l, and cy_l are the coefficients for the lth square, which are to be determined using the least-squares regression method.

The variance in the lth small square region is computed for $s = s_{min}, s_{min} + 1, \cdots, s_{max}$, which is given by:

$$V(l, s) = \frac{1}{s^2} \sum_{i=1}^{s} \sum_{j=1}^{s} \left(x_{i,j} - Gx_{i,j}(l, s)\right) * \left(y_{i,j} - Gy_{i,j}(l, s)\right) \tag{10}$$

The root mean square $F(s)$ is computed as:

$$F(s) = \left[\frac{1}{L_s}\sum_{l=1}^{L_s} V(l,s)\right]^{1/2} \quad (11)$$

where L_s denotes the number of the small square regions of size $s \times s$.

If the profiles are long-range power-law correlated, then the fluctuation function $F(s)$ is observed as follows:

$$F(s) \propto s^{\lambda} \quad (12)$$

Thus, $F(s)$ is proportional to s^{λ} as shown in the above equation with the help of a proportionality sign "$\propto$", and "λ" is the two-dimensional scaling exponent computed from the regression plot of $\log(F(s))$ vs. $\log(s)$. The power-law relation increases as λ increases or decreases. $\lambda = 0.5$ indicate the absence of any cross-correlation, while $\lambda > 0.5$ indicates persistent long-range cross-correlations and $\lambda < 0.5$ indicates anti-persistent cross-correlations, while $\lambda > 1.5$ indicates long-range cross-correlations of power-law form [41, 51, 52].

Thus, using the values of λ for different types of data, we compute the power-law cross-correlations existing in the multimodal data consisting of musical clips and paintings. Figure 1 shows a representative regression plot of $\log(F(s))$ vs. $\log(s)$, where the 2D DFA scaling exponent "α" (derived from Eq. (6)) of Images 1 and 2 is shown alongside their analogous 2D DCCA scaling exponent "λ" (derived from Eq. (12)).

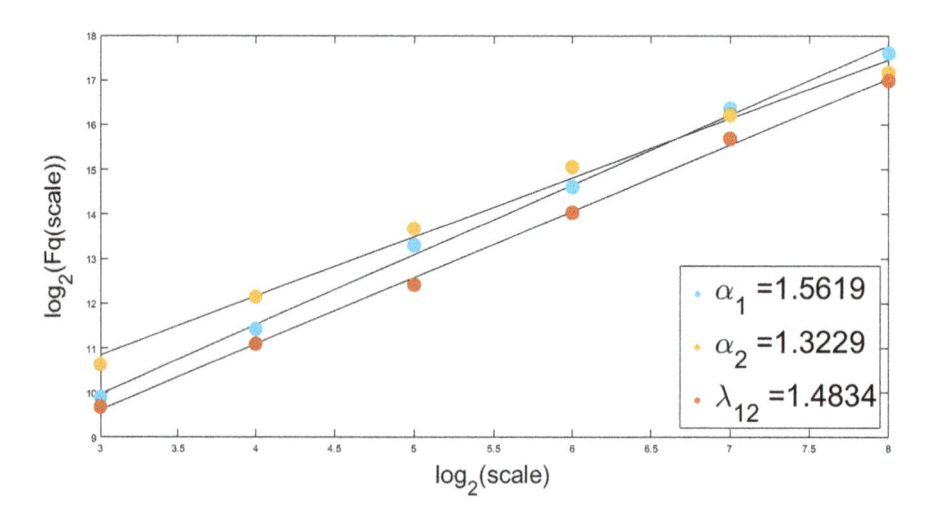

Fig. 1 Regression plot to compare the 2D DFA and DCCA scaling exponent for Images 1 and 2

4 Results and Discussion

Tables 4 and 5 represent the weighted average normalized values of the emotional ratings corresponding to the audio and visual clips used for the human response test of 50 participants. The maximum value for each emotion was 10, while the minimum value was 1.

For classifying the emotional ratings corresponding to the images and clips, we took the help of the 2D circumplex model proposed by Russell [46] (Fig. 2), where emotions are classified in a 2D circular plane along two axes, namely, valence and arousal. These axes are expected to correspond to the internal human representations of emotion.

Using Fig. 2 as our baseline, we have generated radar plots (Fig. 3a–f) comparing the arousal-based effects generated from the clips and images. For this particular study, we had only two target emotions, "Happy" and "Sad." The other two emotional axes, namely, "Anger" and "Sad," were mainly used as distractors chosen randomly. The maximum rating in each emotional axis was 10; hence, the arousal ratings have a maximum rating of 9 corresponding to each emotional axis. The scales have been standardized for all the figures. The emotional valence has been shown using the four axes, and the arousal ratings have been plotted on each of the axis.

As can be seen from Fig. 3a–f generated from human response analysis, the emotional appraisal corresponding to the chosen clips almost matches with that of the standardized measures, i.e., Clips 1–3 and Images 1–3 belong to the positive valence axis, while Clips 4–6 and Images 4–6 belong to the negative valence axis of the Russell's emotional sphere. Corresponding to the cross-modal transfer, the only difference found is in the arousal axis (here, we have represented the arousal axis with the help of ratings 1–7), where it is seen that the arousal-based effects for

Table 4 Psychological ratings of the audio clips chosen for analysis

Clip no.	Anger	Fear	Happy	Sad	Target
1	1.00	1.00	7.33	1.00	Happy
2	1.00	1.00	7.17	1.17	Happy
3	1.00	1.00	7.17	1.00	Happy
4	1.17	1.00	1.00	7.67	Sad
5	1.00	1.33	1.17	7.50	Sad
6	1.00	1.67	1.00	7.50	Sad

Table 5 Psychological ratings of the paintings chosen for analysis

Image no.	Anger	Fear	Happy	Sad	Target
1	0.35	0.20	8.91	0.62	Happy
2	0.10	0.05	8.17	0.13	Happy
3	2.12	0.25	8.29	0.23	Happy
4	1.71	1.33	0.65	6.95	Sad
5	0.09	1.27	2.15	6.88	Sad
6	0.90	0.32	2.75	6.35	Sad

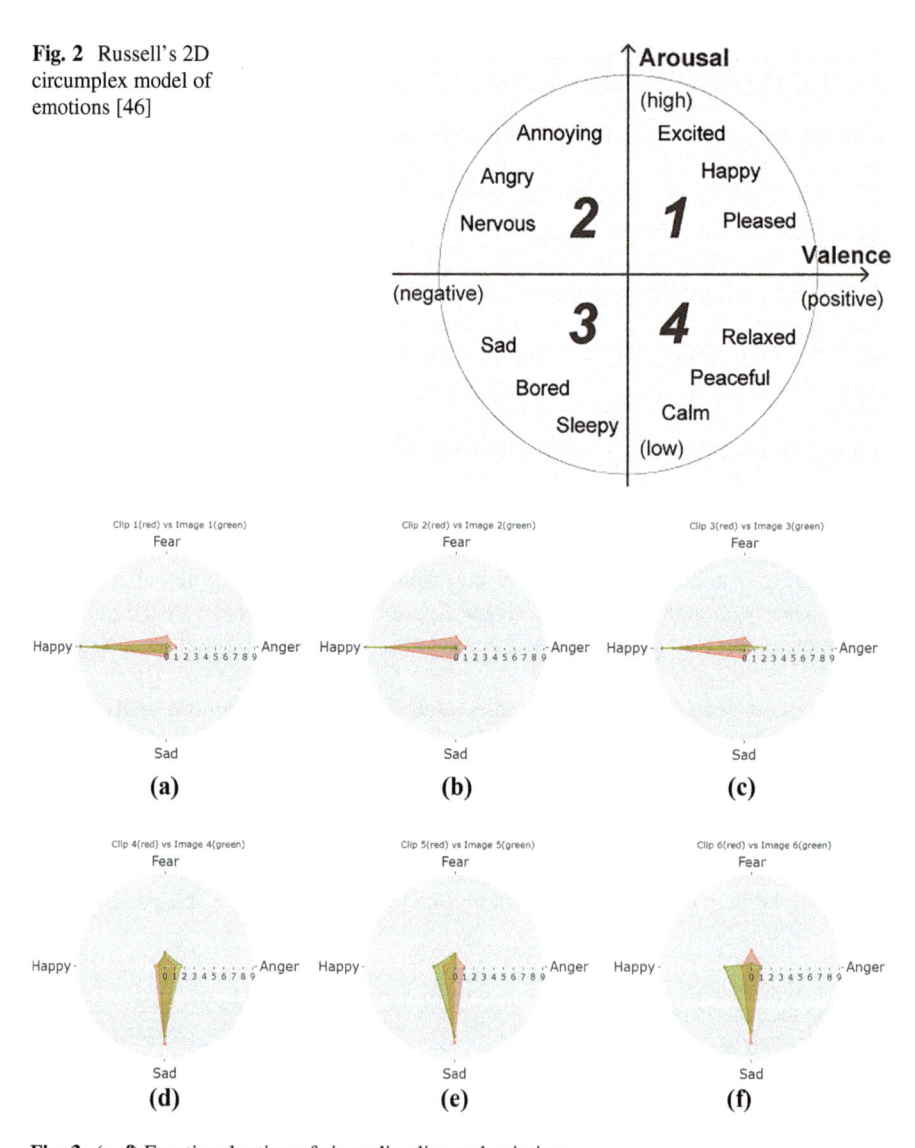

Fig. 2 Russell's 2D circumplex model of emotions [46]

Fig. 3 (**a–f**) Emotional rating of six audio clips and paintings

negative valence images, i.e., Images 4–6 (Fig. 3d–f) are on the lower side compared to that of the audio clips. In this way, we have a standardized measure of each of the stimulus corresponding to both auditory and visual domain using a behavioral response test.

In the next part of our work, DFA exponent was computed for the six audio clips and the six paintings that were put to analysis. In the case of the paintings, α_{red}, α_{green}, and α_{blue} were computed corresponding to the red, green, and blue color component of the painting analyzed. In the following figures, the DFA exponent

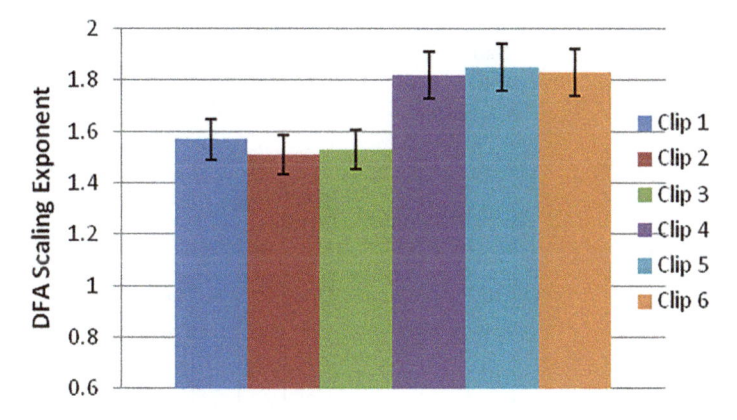

Fig. 4 DFA exponent of audio clips

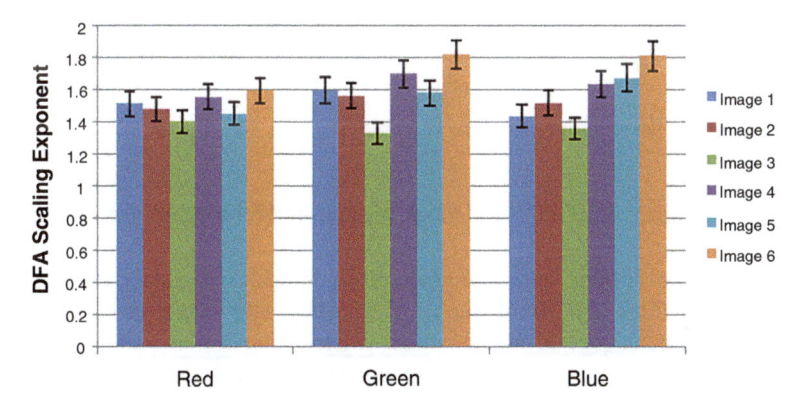

Fig. 5 DFA exponents (color-wise) of visual stimuli

corresponding to each clip and visual stimuli have been plotted. Figure 4 shows the scaling exponents for the six audio clips, which have been classified a priori to positive and negative valence using behavioral measures as described above, while Fig. 5 denotes the scaling exponent for the paintings.

From Fig. 4, it is evident that the scaling exponents of Clips 1 to 3 are lower as compared to the scaling exponents of Clips 4 to 6, i.e., the LRTC present in Clips 1 to 3 are lower than the temporal correlations present in Clips 4 to 6. This can be attributed to various acoustic features of these clips like tempo, rhythm, etc., but the mathematical manifestation here is the decrease/increase in long-range temporal correlations. From Fig. 5, it is evident that the scaling exponents of blue and green colors in Images 4 and 6 are marginally higher than Images 2 and 3. The case is similar for correlations present in each of the three basic colors. Also, an interesting observation is that α_{green} and α_{blue}, i.e., the scaling exponents corresponding to green and blue color, show the maximum increase for Images 4 to 6 (which were classified as evoking sad emotions). Thus, the manifestation of sad emotion can be

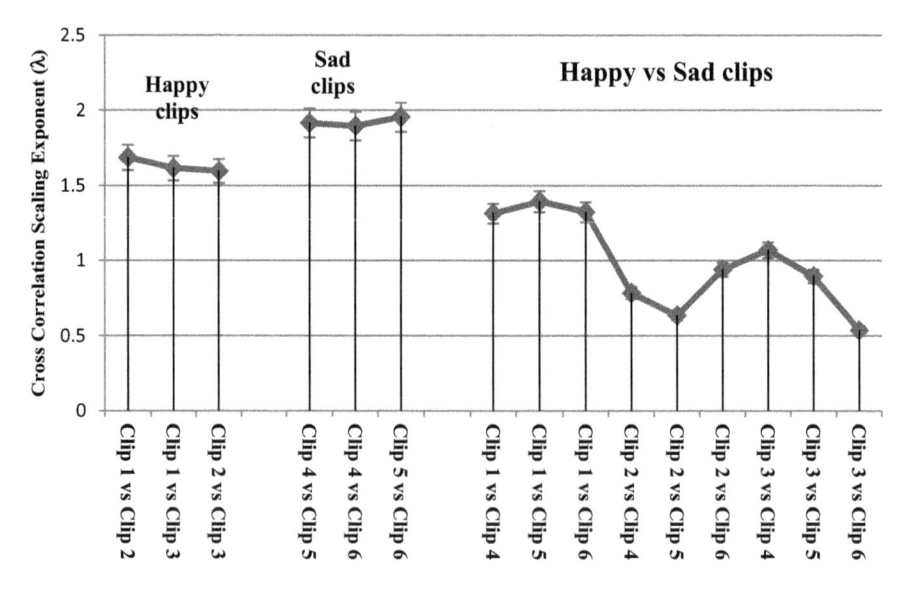

Fig. 6 Cross-correlation scaling exponent (λ) for different clips

attributed to higher order of correlations present in the blue and green color of a painting.

In the next part of our work, the amount of nonlinear cross-correlation existing between the auditory and visual stimuli is evaluated individually using the DCCA (1D and 2D) technique. A higher value of cross-correlation scaling exponent, i.e., $\lambda > 0.5$, indicates strong persistent long-range cross-correlations, while $\lambda > 1.5$ indicates long-range cross-correlations of power-law form [40, 41, 47–52] among the two signals chosen. Figures 6 and 7 represent the values of λ for different combinations of auditory and visual stimuli, respectively.

It is to be noted that in Fig. 7, before calculating the cross-correlation scaling exponent for the visual stimulus, we took an average of the three cross-correlation exponents (i.e., an average of λ_{red}, λ_{green}, and λ_{blue}) obtained from the previous analysis for the simplification of the obtained results.

In Fig. 6, it is seen that the cross-correlation scaling exponent for audio clips belonging to the same valence is on the higher side as compared to the clips belonging to opposite valence. The clips that have been rated as "sad" are the ones that show highest order of long-range cross-correlation, while the clips rated as "happy" also show strong cross-correlation but lower than the "sad" ones. The inter-valence cross-correlations are however much lower than these, while for some pairs, there is almost "no correlation," which is evident from the cases where the value of cross-correlation scaling exponent λ is almost 0.5. From Fig. 7, it is seen that the amount of long-range cross-correlation among Images 4 to 6 is the highest of all the combinations present here, while the correlation among the Clips 1 to 3 is the lowest. Thus, we have an indirect classification of emotional appraisal even

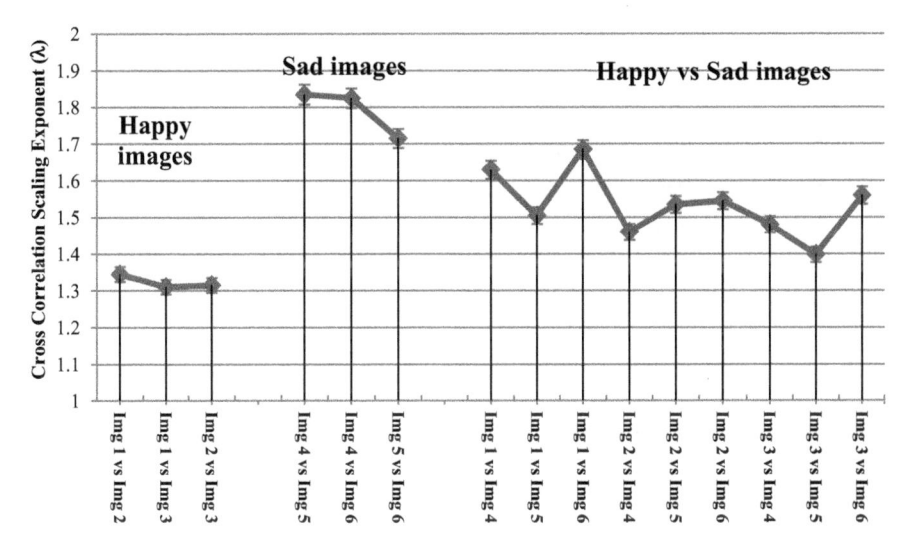

Fig. 7 Cross-correlation scaling exponent (λ) for different images

while performing DCCA also. While the images that have been rated as "sad" have higher order of cross-correlation among them, the "happy" rated images have lower order of long-range cross-correlation. The inter-valence cross-correlation scaling exponents (i.e., the order of cross-correlation among happy and sad images) lie somewhere in between the two cross-correlation values.

5 Conclusions

In this work, we have presented a novel algorithm to automatically classify and compare emotional appraisal from cross-modal stimuli based on the amount of long-range temporal correlations present in the auditory and visual stimulus. This particular methodology takes into consideration the inherent fluctuations, self-similar patterns, and the nonstationary condition present in both audio and visual clips and acts as a mathematical microscope to quantify the same in the form of scaling exponents or degree of correlations. The study provides the following interesting conclusions:

1. For both the auditory and visual stimulus, an averaged DFA scaling exponent of anything greater than 1.5 denotes stimulus belonging to "sad" category. This becomes more evident from Fig. 8 below:
2. The DFA scaling exponent corresponding to blue and green color is highest in the case of "sad" images, while the DFA exponent for "happy" images is high for red color. When we consider the averaged scaling exponent for the paintings,

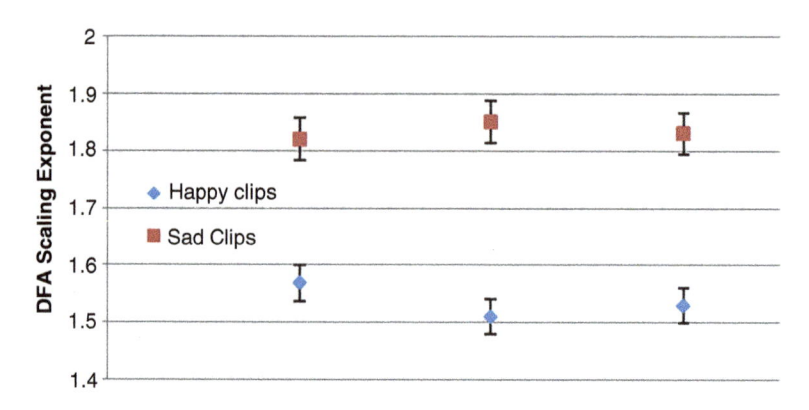

Fig. 8 Clustering of emotions in audio clips with DFA

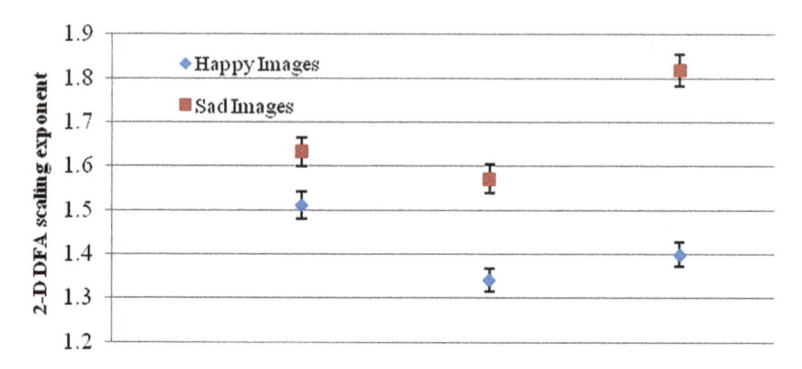

Fig. 9 Clustering of emotions in paintings with 2D DFA

the difference between "happy" and "sad" is not that stark as opposed to that of audio clips, which is evident from Fig. 9:

As is evident from the clustering pattern, the classification precision is higher for the audio clips as compared to that of the images, though no overlap is observed even in the averaged DFA values of visual clips. This result is somewhat similar to what we have already found in the psychological response data, where the intensity of the valence in the case of "sad" paintings was found to be lower as compared to that of the audio clips.

3. The DCCA scaling exponent shows that the amount of cross-correlation is strongest among the sad clips, while the amount of cross-correlation is lowest for the inter-valence clips.
4. From Fig. 7, it is observed that the averaged cross-correlation scaling exponent for happy images is very low (with values lying around "1.3") while that for sad images is considerably high (with values lying around "1.8"). On the other hand, the amount of nonlinear cross-correlation between happy and sad images interestingly lies between the two (with values ranging between "1.3" and "1.8").

Table 6 Pearson coefficient values of DFA exponents from two modalities

Happy (audio) vs. happy (image)	Happy(audio) vs. sad (image)	Sad (audio) vs. happy (image)	Sad (audio) vs. sad (image)
0.987	0.29	−0.493	**0.96**

5. To establish the cross-modality or the amount of emotional transfer from one medium to another, Pearson correlation coefficient is computed from the variation of DFA values belonging to the stimulus from two modalities. The values of Pearson correlation have been reported in Table 6.

From the Pearson coefficient values shown in Table 6, it can be concluded that correlation is highest across two modalities when the clips are from same valence, while in the case of clips from opposite valence, degree of correlation is very low or in some case negative. Thus, an indirect quantitative correlation is obtained between the emotional appraisal of the cross-modal bias of auditory and visual stimuli. Future applications of this novel methodology include emotion quantification from two different domains and also quantification of cross-modal transfer of emotions. This is a pilot study in that direction.

Acknowledgments SS acknowledges the DST CSRI (Govt. of India) for the Research Associateship (**DST/CSRI/2018/78**) and Acoustical Society of America (ASA) to pursue this research. AB acknowledges the Department of Science and Technology (DST), Govt. of India for providing (**SR/CSRI/PDF-34/2018**) the DST CSRI Post-Doctoral Fellowship to pursue this research work.

References

1. Christensen, J. F., Gaigg, S. B., Gomila, A., Oke, P., & Calvo-Merino, B. (2014). Enhancing emotional experiences to dance through music: the role of valence and arousal in the cross-modal bias. *Frontiers in human neuroscience, 8*, 757.
2. Müller, V. I., Habel, U., Derntl, B., Schneider, F., Zilles, K., Turetsky, B. I., & Eickhoff, S. B. (2011). Incongruence effects in crossmodal emotional integration. *Neuroimage, 54*(3), 2257–2266.
3. Grossmann, T., Striano, T., & Friederici, A. D. (2006). Crossmodal integration of emotional information from face and voice in the infant brain. *Developmental Science, 9*(3), 309–315.
4. Scherer, L. D., & Larsen, R. J. (2011). Cross-modal evaluative priming: Emotional sounds influence the processing of emotion words. *Emotion, 11*(1), 203.
5. Scherer, K. R., & Zentner, M. R. (2001). Emotional effects of music: Production rules. *Music and emotion: Theory and research, 16*, 361–392.
6. Zentner, M., Grandjean, D., & Scherer, K. R. (2008). Emotions evoked by the sound of music: characterization, classification, and measurement. *Emotion, 8*(4), 494.
7. Keltner, D., & Haidt, J. (2003). Approaching awe, a moral, spiritual, and aesthetic emotion. *Cognition and emotion, 17*(2), 297–314.
8. Vuoskoski, J. K., & Eerola, T. (2012). Can sad music really make you sad? Indirect measures of affective states induced by music and autobiographical memories. *Psychology of Aesthetics, Creativity, and the Arts, 6*(3), 204.

9. Barrett, F. S., Grimm, K. J., Robins, R. W., Wildschut, T., Sedikides, C., & Janata, P. (2010). Music-evoked nostalgia: Affect, memory, and personality. *Emotion, 10*(3), 390.
10. D'Andrade, R., & Egan, M. (1974). The colors of emotion 1. *American ethnologist, 1*(1), 49–63.
11. Suk, H. J. (2006). Color and Emotion-a study on the affective judgment across media and in relation to visual stimuli.
12. Chai, M. T., Amin, H. U., Izhar, L. I., Saad, M. N. M., Abdul Rahman, M., Malik, A. S., & Tang, T. B. (2019). Exploring EEG effective connectivity network in estimating influence of color on emotion and memory. *Frontiers in neuroinformatics, 13*, 66.
13. Bekhtereva, V., & Müller, M. M. (2017). Bringing color to emotion: The influence of color on attentional bias to briefly presented emotional images. *Cognitive, Affective, & Behavioral Neuroscience, 17*(5), 1028–1047.
14. Gilbert, A. N., Fridlund, A. J., & Lucchina, L. A. (2016). The color of emotion: A metric for implicit color associations. *Food Quality and Preference, 52*, 203–210.
15. Strezoski, G., Shome, A., Bianchi, R., Rao, S., & Worring, M. (2019, October). Ace: Art, color and emotion. In *Proceedings of the 27th ACM International Conference on Multimedia* (pp. 1053–1055).
16. Hanada, M. (2018). Correspondence analysis of color–emotion associations. *Color Research & Application, 43*(2), 224–237.
17. Barbiere, J. M., Vidal, A., & Zellner, D. A. (2007). The color of music: Correspondence through emotion. *Empirical studies of the arts, 25*(2), 193–208.
18. Wilms, L., & Oberfeld, D. (2018). Color and emotion: effects of hue, saturation, and brightness. *Psychological research, 82*(5), 896–914.
19. Palmer, S. E., Schloss, K. B., Xu, Z., & Prado-León, L. R. (2013). Music–color associations are mediated by emotion. *Proceedings of the National Academy of Sciences, 110*(22), 8836–8841.
20. Tsiourti, C., Weiss, A., Wac, K., & Vincze, M. (2019). Multimodal integration of emotional signals from voice, body, and context: Effects of (in) congruence on emotion recognition and attitudes towards robots. *International Journal of Social Robotics, 11*(4), 555–573.
21. Logeswaran, N., & Bhattacharya, J. (2009). Crossmodal transfer of emotion by music. *Neuroscience letters, 455*(2), 129–133.
22. Marin, M. M., Gingras, B., & Bhattacharya, J. (2012). Crossmodal transfer of arousal, but not pleasantness, from the musical to the visual domain. *Emotion, 12*(3), 618.
23. Spence, C. (2020). Assessing the role of emotional mediation in explaining crossmodal correspondences involving musical stimuli. *Multisensory Research, 33*(1), 1–29.
24. Vines, B. W., Krumhansl, C. L., Wanderley, M. M., & Levitin, D. J. (2006). Cross-modal interactions in the perception of musical performance. *Cognition, 101*(1), 80–113.
25. Liang, J., Li, R., & Jin, Q. (2020, October). Semi-supervised multi-modal emotion recognition with cross-modal distribution matching. In *Proceedings of the 28th ACM International Conference on Multimedia* (pp. 2852–2861).
26. Sahoo, S., & Routray, A. (2016, September). Emotion recognition from audio-visual data using rule based decision level fusion. In *2016 IEEE Students' Technology Symposium (TechSym)* (pp. 7–12). IEEE.
27. Stevenson, R. A., & James, T. W. (2008). Affective auditory stimuli: Characterization of the International Affective Digitized Sounds (IADS) by discrete emotional categories. *Behavior research methods, 40*(1), 315–321.
28. Yang, W., Makita, K., Nakao, T., Kanayama, N., Machizawa, M. G., Sasaoka, T., ... & Miyatani, M. (2018). Affective auditory stimulus database: An expanded version of the International Affective Digitized Sounds (IADS-E). *Behavior Research Methods, 50*(4), 1415–1429.
29. Bradley, M. M. & Lang, P. J. (2007). The International Affective Digitized Sounds (2nd Edition; IADS-2): Affective ratings of sounds and instruction manual. Technical report B-3. University of Florida, Gainesville, Fl.
30. Mandelbrot, B. B. (1982). *The fractal geometry of nature* (Vol. 1). New York: WH freeman.
31. Peng, C. K., Buldyrev, S. V., Havlin, S., Simons, M., Stanley, H. E., & Goldberger, A. L. (1994). Mosaic organization of DNA nucleotides. *Physical review e, 49*(2), 1685.

32. Alvarez-Ramirez, J., Alvarez, J., & Rodriguez, E. (2008). Short-term predictability of crude oil markets: a detrended fluctuation analysis approach. *Energy Economics, 30*(5), 2645–2656.
33. Yeh, R. G., Shieh, J. S., Chen, G. Y., & Kuo, C. D. (2009). Detrended fluctuation analysis of short-term heart rate variability in late pregnant women. *Autonomic Neuroscience, 150*(1–2), 122–126.
34. Márton, L. F., Brassai, S. T., Bakó, L., & Losonczi, L. (2014). Detrended fluctuation analysis of EEG signals. *Procedia Technology, 12*, 125–132.
35. Sanyal, S., Banerjee, A., Pratihar, R., Maity, A. K., Dey, S., Agrawal, V., ... & Ghosh, D. (2015, September). Detrended Fluctuation and Power Spectral Analysis of alpha and delta EEG brain rhythms to study music elicited emotion. In *2015 International Conference on Signal Processing, Computing and Control (ISPCC)* (pp. 205–210). IEEE.
36. Sengupta, S., Biswas, S., Nag, S., Sanyal, S., Banerjee, A., Sengupta, R., & Ghosh, D. (2017, February). Emotion specification from musical stimuli: An EEG study with AFA and DFA. In *2017 4th International Conference on Signal Processing and Integrated Networks (SPIN)* (pp. 596–600). IEEE.
37. Banerjee, A., Sanyal, S., Patranabis, A., Banerjee, K., Guhathakurta, T., Sengupta, R., ... & Ghose, P. (2016). Study on brain dynamics by non linear analysis of music induced EEG signals. *Physica A: Statistical Mechanics and its Applications, 444*, 110–120.
38. Banerjee, A., Sanyal, S., Roy, S., Nag, S., Sengupta, R., & Ghosh, D. (2021). A novel study on perception–cognition scenario in music using deterministic and non-deterministic approach. *Physica A: Statistical Mechanics and its Applications, 567*, 125682.
39. Sanyal, S., Banerjee, A., Nag, S., Sarkar, U., Roy, S., Sengupta, R., & Ghosh, D. (2021). Tagore and neuroscience: A non-linear multifractal study to encapsulate the evolution of Tagore songs over a century. *Entertainment Computing, 37*, 100367.
40. Sanyal, S., Nag, S., Banerjee, A., Sengupta, R., & Ghosh, D. (2019). Music of brain and music on brain: a novel EEG sonification approach. *Cognitive neurodynamics, 13*(1), 13–31.
41. Podobnik, B., & Stanley, H. E. (2008). Detrended cross-correlation analysis: a new method for analyzing two nonstationary time series. *Physical review letters, 100*(8), 084102.
42. Cao, G., Cao, J., Xu, L., & He, L. (2014). Detrended cross-correlation analysis approach for assessing asymmetric multifractal detrended cross-correlations and their application to the Chinese financial market. Physica A: Statistical Mechanics and its Applications, 393, 460–469.
43. Eerola, T., & Vuoskoski, J. K. (2011). A comparison of the discrete and dimensional models of emotion in music. *Psychology of Music, 39*(1), 18–49.
44. Eerola, T., Lartillot, O., & Toiviainen, P. (2009, October). Prediction of Multidimensional Emotional Ratings in Music from Audio Using Multivariate Regression Models. In *ISMIR* (pp. 621–626).
45. Sanyal, S., Banerjee, A., Basu, M., Nag, S., Ghosh, D., & Karmakar, S. (2020, December). Do musical notes correlate with emotions? A neuro-acoustical study with Indian classical music. In *Proceedings of Meetings on Acoustics 179ASA* (Vol. 42, No. 1, p. 035005). Acoustical Society of America.
46. Russell, J. A. (1980). A circumplex model of affect. *Journal of personality and social psychology, 39*(6), 1161.
47. Dutta, S., Ghosh, D., & Samanta, S. (2014). Multifractal detrended cross-correlation analysis of gold price and SENSEX. Physica A: Statistical Mechanics and its Applications, 413, 195–204.
48. Ghosh, D., Chakraborty, S., & Samanta, S. (2019). Study of translational effect in Tagore's Gitanjali using Chaos based Multifractal analysis technique. Physica A: Statistical Mechanics and its Applications, 523, 1343–1354.
49. Roy, S., Banerjee, A., Roy, C., Nag, S., Sanyal, S., Sengupta, R., & Ghosh, D. (2021). Brain response to color stimuli: an EEG study with nonlinear approach. Cognitive Neurodynamics, 1–31.
50. Sanyal, S., Banerjee, A., Patranabis, A., Banerjee, K., Sengupta, R., & Ghosh, D. (2016). A study on improvisation in a musical performance using multifractal detrended cross correlation analysis. Physica A: Statistical Mechanics and its Applications, 462, 67–83.

51. Zhou, W. X. (2008). Multifractal detrended cross-correlation analysis for two nonstationary signals. *Physical Review E, 77*(6), 066211.
52. Shadkhoo, S., & Jafari, G. R. (2009). Multifractal detrended cross-correlation analysis of temporal and spatial seismic data. The European Physical Journal B, 72(4), 679–683.

Inharmonic Frequency Analysis of Tabla Strokes in North Indian Classical Music

Shambhavi Shivraj Shete and Saurabh Harish Deshmukh

1 Introduction

Analyses of the sound of music and the frequencies associated with the musical notes have been always a topic of great interest amongst the related researchers. Typically, the sound is measured by its amplitude (dB), frequency (Hz) and duration (sec). There exist a fourth dimension to the sound which itself is a multidimensional called timbre [1]. This attribute of the sound is the one that differentiates the sound of the same amplitude, frequency and duration. For example, the human ear can clearly distinguish the difference between sound produced by a flute and a violin playing C_5 notes of the same amplitude, frequency and duration. Although Indian classical music is melodious in contrast with Western music which is harmonious, the musical instruments that are used exhibit harmonious behaviour. It is interesting to analyse the sound production pattern of Indian musical instruments such as Tanpura, Sitar, or percussion instruments like Tabla. The musical instruments are broadly classified as string, woodwind, brass and percussion. These instruments operate on the principle of acoustic resonance. An acoustic resonance occurs when a musical instrument amplifies its sound frequency that matches with one of its natural frequencies of vibration [2].

In recent years with the advancements in technology, and development in various signal processing techniques, analysis of sound originated from various musical instruments has become easy to some extent. An essential component of any music is its rhythm and musical instruments producing rhythmic structure to assist the vocalists or instrumentalists. Tabla is one of the majorly used accompanying rhythm instruments for North Indian classical music (NICM).

S. S. Shete (✉) · S. H. Deshmukh (✉)
Maharashtra Institute of Technology, Aurangabad, India

Some pitched musical instruments exhibit the harmonic behaviour of overtones. Harmonics are the integer multiple of the fundamental frequency of a vibrating object [3]. These harmonics are higher in frequency and lesser in magnitude than that of the fundamental frequency. When a metallic string, fixed between two points keeping sufficient tension, is plucked in the middle, several harmonics are generated along with the fundamental frequency. If F is the fundamental frequency, then the first harmonic is at $2 \times F$ frequency value. Similarly, all other upcoming harmonics will be $3 \times F$, $4 \times F$, $5 \times F$ and so on. When a note A_4 is played at 440 Hz, it has the harmonics 880 Hz, 1320 Hz, 1760 Hz and so on. The musical instruments based on metallic strings or air columns are harmonic [4].

There also exist some musical instruments that have a high degree of inharmonicity to produce a musical tone. Inharmonicity is the degree to which the overtones differ from their harmonic overtone locations [5]. In such kinds of instruments, the peaks are not in the multiple integer ratios. When a note is played, the fundamental frequency attains the highest magnitude of energy, and the overtone signal energies exhibit diminishing behaviour. The overtones are the resonant frequencies that are not whole multiples of the fundamental frequency [6]. The overtones are produced from the musical instruments Sitar and Tanpura creating a sympathetic resonance of the sound. When the Western string instrument violin is plucked, by applying the different pressure on the bow, the overtones are generated [7].

The sound produced from the circular membrane of percussion instruments also emits overtones. According to hyper-physics, the overtones produced by circular drum membrane are 1, 1.59, 2.14, 2.30, 2.65 and 2.92 multiples of the fundamental frequency [8]. The overtones of a drum membrane are not harmonic. This is due to the physical quality of a drum that makes the drum inharmonic. A vibrating circular membrane has an extra dimension that generates symmetric and asymmetric vibrations [9]. In Timpani, inharmonic frequencies are produced with asymmetric vibrations with integer multiples of 1, 1.35, 1.67, 1.99, 2.30 and 2.61 [10]. The North Indian classical music percussion instrument Tabla is a single pitch musical instrument. The Tabla membrane has a paste applied and dried on its surface, called Syahi (ink). Due to the gradually reducing thickness of the ink from the centre towards the edges, the instrument produces inharmonic overtones.

In this research, we have analysed the overtone patterns of basic Tabla strokes and correlated the overtone frequencies with the musical notes. This research could be useful to the Tabla players and musicians to design, tune and play the Tabla instrument as per the requirements of the musical notes of the Raga of North Indian classical music [11]. The research document is organized section-wise in the following way. Section 2 elaborates the research conducted so far and focuses on various basic Tabla strokes produced by left, right and both the Tabla drums simultaneously. The Tabla stroke mode generation and progression with respect to nodal diameter and nodal circle produced by Tabla drum membrane is explained in Sect. 3. Section 4 explores the spectral analysis of eight overtones of each basic Tabla strokes. Finally, Sect. 5 summarizes the results obtained from the spectral analysis of the Tabla strokes and their corresponding musical notes.

2 Literature Review

The Tabla is a single-sided drum that contains two drums: left and right drums. The left drum also called Dayan is made up of steel or metal alloy. The drum is spherical. This drum is used to emit base frequencies. Traditionally, in NICM the left drum instrument is not tuned with any musical note. The Tabla player ensures that the skin of the drum is not too loose nor too tight. If the skin is left loose, the left drum stroke frequencies are very low and do not produce any harmonic overtone. If the drum skin is too tight, then instead of giving a moderate bass frequency, the drum produces multiple out-of-order frequency mixtures.

The left drum is played by resting 1/3rd of the hand over the membrane, and the strokes are played by using the first or middle finger keeping the rest of the hand touching the surface of the membrane. Most of the time, the Tabla player rests his hand and wrist on the open surface of the membrane. The wrist and the fingers are used such that the wrist touches the ink at one end and fingers hit the drum at the exact opposite side of the ink.

The right drum Tabla is made up of wood, usually teak wood. The size of this drum is smaller than the left drum. Its shape is spherical and narrow towards the open end. The membrane is laced and tightened with animal skin hoops and is tightened at the bottom of the drum [12]. The membrane of both the drums is made up of goatskin with three layers. The first layer and third layer cover the entire area of the open end of both the drums. The middle layer skin is cut with exact dimensions of the ink that would be applied on the outermost layer. With this kind of construction of the membrane, due to the ink weight, the membrane vibrates to a larger extent than that of a drum without membrane ink. The location of the ink over the left drum is asymmetric to its centre and symmetric for the right drum, applied in circular form.

The ink is a mixture of iron particles, rice floor, catechu and gum. It is a fine art to make the paste of these mixtures with consistency and strength. Ink is the most important component for determining the tone of the Tabla. After applying the ink, a polished stone of basalt is used to rub the ink repeatedly until it is spread with reducing thickness. A drum without any such ink when stroked, the sound waves travel from the centre of the drum towards its rim and return towards the centre reinforcing and cancelling each other to form standing waves [13]. Due to harmonic generation and phase cancellation, the sound becomes noisy. When a Tabla player hits on the Tabla instrument that has ink, the sound is converted into a variety of sound that generates melody. Also, the instrument generates modulations in it [14].

The Tabla is tuned with a specially designed steel hammer. The claw side is used to hammer on the dowels and the face side to hammer on the rim of the surface of the Tabla membrane. A high pitch clear sound is audible from the Tabla membrane when stretched. So, the tightening of the surface is done by hammering over the rim and dowels. To reduce the tension over the membrane, the hammer is used in opposite direction. The circular-shaped cotton rings are kept below each drum to prevent vibrations generated by Tabla strokes from reaching to the ground and back-propagating. It is also used to incline both the drums towards each other to give comfort to the Tabla player to produce the strokes.

The art and science of Tabla is not a written theory, but the knowledge is percolated orally from the teacher to the disciples since the inception of this instrument. Therefore, the knowledge, the abbreviations, the terminologies, stroke production techniques, etc. differed over the years. For example, strokes 'Na and Ta' are the same in some traditions, and in others, the stroke production method makes them two different syllables.

Tabla strokes are classified as strokes originated from the left drum and from the right drum and strokes produced by simultaneously hitting both the drums. To analyse the frequency overtones originated from these drums, it is essential to understand the stroke production technique for each drum. Small variation in the location of a finger makes huge changes in frequency overtone production. The following section elaborates Tabla stroke production method.

2.1 Right Drum Strokes

2.1.1 Stroke Na

The Tabla stroke Na is produced using the forefinger on the right drum. The stroke is produced by keeping the ring finger on the diameter of the ink. The ring finger is placed over the membrane in such a way that it does not completely damp the overtones. This finger plays a prominent role in generating harmonic sound out of the stroke Na. By touching the ring finger lightly, the Tabla instrument is hit by the forefinger over the rim producing stroke Na.

2.1.2 Stroke Ta

Ta stroke is produced similar to the stroke Na except with the difference that the forefinger strikes the Tabla membrane in the space between the ink boundary and rim boundary called maidan. The ring finger position remains the same as that of Na.

2.1.3 Stroke Te

The Te stroke is produced by hitting the ink using the forefinger and keeping it on the ink tightly. This damps the resonating sound of the stroke. In some traditions, the stroke Te is also produced using the ring finger and middle finger by striking the ink and keeping the fingers tightly on the ink.

2.1.4 Stroke Tun

Stroke Tun is produced, similar to the stroke Te, except that the forefinger striking the ink is immediately lifted to keep the entire membrane resonating.

2.2 Left Drum Strokes

2.2.1 Stroke Ga

Ga stroke is played by using the middle finger at the centre of the left drum membrane holding the wrist down. This is a resonant stroke and creates open vibrations. Various pitch frequencies could be heard for different Ga strokes. Majorly the wrist pressure applied over the membrane decides the pitch frequency of stroke Ga.

2.2.2 Stroke Ka

Ka stroke is played by striking the left-hand palm sharply over the left drum membrane and resting it there itself. Usually, the palm covers the entire region of ink.

2.3 Both the Drum Strokes

2.3.1 Stroke Dha

The stroke Dha is produced by playing both Ga (on the left drum) and Na (on the right drum) simultaneously. This stroke gives a loud and strong sound. Dha is produced in various ways. In Delhi style, it is played by simultaneously hitting both the strokes Ga and Na. In Purbi style Dha, the right stroke Na is played loudly than the left drum stroke Ga [15].

2.3.2 Stroke Dhin

The left drum stroke Ga and right drum stroke Ta when hit together, a stroke Dhin is produced. The stroke Dhin is an open stroke in which fingers of both hands are lifted immediately after striking the membrane. This generates a ringing sound from the drum.

2.3.3 Stroke Tin

The Tabla stroke Tin is produced by keeping the left drum stroke membrane covered by the entire left-hand palm, producing stroke Ka, and simultaneously hitting the right drum using the forefinger to produce stroke, Tun.

3 Tabla Strokes and Mode Progression

Alike other membrane percussion instruments, Tabla, Mridangam, Pakhawaj, Dholki, etc. come from a family of percussion instruments in which a special ink is applied on the right drum membrane. In the research 'The Indian Musical Drum', the author has experimentally proved the mode progression of Tabla strokes [16]. The author claims that the loading of the drum-head greatly increases the energy of vibration. The loaded membrane also is responsible for the emission of sustained tone. The marginal leather ring influences the duration of the tone. It acts as a damper for higher overtones without disturbing the lower tones.

Table 1 shows Tabla membrane vibration modes considered for overtone analysis and frequency mapping with musical notes. The modes of a circular membrane of Indian drum Tabla are labelled as ND and NC where ND is the number of nodal diameters and NC is the number of nodal circles. In mode of the vibration, the nodal diameter and nodal circles are the stable regions and they do not participate in any oscillation [17]. F0 corresponds to the fundamental frequency of the Tabla stroke exhibiting (0,1) mode, where 0 being the nodal diameter and 1 the nodal circle produced at the rim representing no nodal diameters but one circular edge. The sound produced from (0,1) mode radiates equally in all directions towards the outer surface of the membrane which is called a mono-pole source. The membrane transfers its vibration energy into sound energy where vibrations are removed away. This mode does not contribute to the quality of the tone if the duration of the note is short. This mode is represented by the fundamental frequency of the sound.

Table 1 Mode progression of Tabla membrane [18]

Frequency	Modes	Multiplier
F0	0,1	1
F1	1,1	1.59
F2	2,1	2.14
F3	0,2	2.30
F4	3,1	2.65
F5	1,2	2.98
F6	4,1	3.16
F7	2,2	3.50
F8	0,3	3.60

Similarly, the mode (1,1) represents 1 nodal diameter and 1 nodal circle. The frequency of (1,1) mode is 1.59 times the fundamental frequency. The sound produced from this mode oscillates in two directions representing a dipole source. The half membrane is pushed up, and the half membrane is pushed down. This mode contributes to the musical sound or pitch because it takes a longer duration to decay. When Tabla is hit at the centre or the outer surface, the sound has a definite pitch and is sustained for some time. In this way remaining all the modes could be analysed with respect to the number of nodal diameters and nodal circles and corresponding frequency multiplier to calculate the frequency of each overtone.

The frequency multipliers are the constants used to calculate the frequency of the overtone. These overtone frequencies are mapped with piano scale frequencies. We have tried to correlate the relationship between fundamental musical notes produced by the Tabla instrument and the contribution of various other musical notes generated from the harmonics following the fundamental frequency to give different timbral perception for each basic Tabla stroke. A correlation of overtones produced by resonating string instruments such as Tanpura is made here. Although the harmonics produced by each of the Tanpura instrument strings has been termed as 'Shruti'. We could not find any term similar to Shruti for overtones produced by the Tabla instrument. One of the basic reasons behind this could be that the membrane skin does not vibrate to the extent to which a string resonates. Even though we have identified the harmonic structure of the Tabla strokes, which may not be distinctly audible like Tanpura, it could be sensed through the timbral structure of the Tabla stroke.

4 Experimental Setup

The audio database is recorded and preprocessed in studio environments. A professional Tabla instrument with standard size and shape was tuned to the scale C5. The Ga stroke frequency was kept near to E1. The trained Tabla player was asked to play nine basic Tabla strokes with a metronome tempo of 80 bits per min. This ensured a sufficient gap between two successive strokes of the same type. Tabla player was asked to play each basic Tabla stroke repeatedly for about 3 min.

In this way, nine Tabla stroke audio loops were recorded. Individual Tabla stroke audio files were manually generated using audio editing software. Similarly, some basic rhythms from NICM were recorded with a tempo of 80 bpm containing a different combination of basic stroke. The final audio dataset was prepared by taking five samples from the former recording and five samples from the latter recording. Thus, a total of ten audio samples per Tabla strokes are separated. This procedure was repeated for all nine basic Tabla strokes. Total 90 audio files of duration 1–2 sec with sampling frequency 44,100 Hz, 16-bit pulse code modulation, .wav format constitutes the final database.

Each audio signal was analysed in the time and frequency domain. Fast Fourier transform was applied to extract the frequency spectrum. The frequency component

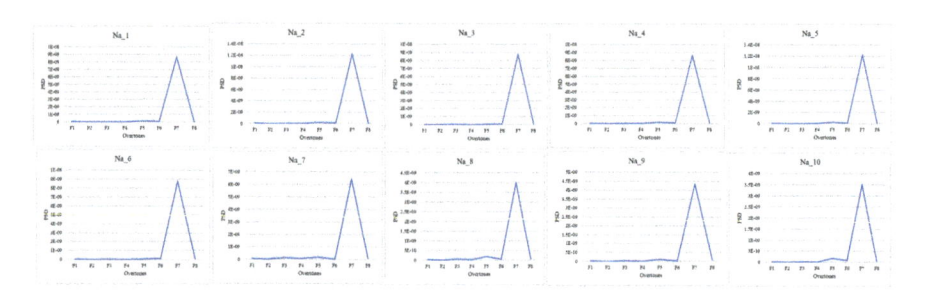

Fig. 1 Overtone frequency analysis of stroke 'Na'

with the highest power spectral density (PSD) is treated as fundamental frequency F0. By using the multipliers, overtone frequencies F1 to F8 were calculated. Based on the higher power spectral density values of these overtones, prominent overtones were separated, and corresponding musical notes of these frequencies were mapped. These high-value overtones contributed to the timbral tonal quality of the stroke. A common analysis of all the ten samples of each stroke with respect to their overtone pattern was made. It is found that except for very few negligible noisy audio files, the behaviour of overtones of all the samples of a single stroke was found to be similar. Thus, it could be inferred that for all basic Tabla strokes, the pattern of the inharmonic overtones is similar, and for each stroke, the amplitude of the overtones also exhibits similar values. Figure 1 shows a graphical representation of eight overtones of stroke Na for ten audio samples.

There exist different principles of music notations. Typically, the frequency for equal-tempered scale A_4 is 444 Hz; however, there exist various tuning choices such as 432 Hz, 434 Hz, 436 Hz, etc. The Shruti that is discovered may not coincide with the exact musical note, but for simplicity of the understanding, we have considered the nearby musical note for each harmonic overtone. Many times, these frequencies may minutely vary based on the exact speed of sound in the environment. Also, the Tabla player may strike the Tabla membrane with different pressure each time hitting the same stroke. Thus, the mode progression slightly differs due to minute variations in fundamental frequency. Although all the audio samples are recorded in the studio environment, Tabla is not an electronic instrument to reproduce each stroke the same way but an acoustic instrument.

5 Results and Discussions

Frequency analysis of 90 audio excerpts containing ten strokes of each basic Tabla stroke is analysed here. The purpose of considering multiple audio excerpts of the same Tabla stroke is two-folded: first, to generalize the system, and the results' multiple instances would clarify the signal analysis in generalized form, and second, due to human error of the Tabla player to produce the same strokes multiple times including (a) slight change of the locations where the stroke is hit and (b) the pressure difference applied on the membrane of the drum each time the stroke is played and see simultaneous hitting of both the drums may vary each time. The following section may have first analysed the harmonic structure and Shruti (overtones) produced by right drum strokes, left drum strokes and both the drum strokes simultaneously.

5.1 Right Drum Strokes

5.1.1 Stroke Na

Na is one of the frequently used Tabla strokes of the right drum. When multiple samples of Na were analysed in the time and frequency domain, it is observed that the single stroke Na gives the fundamental frequency of 528 Hz. This corresponds to the C_5 (Sa) musical note at which the Tabla was initially tuned. After fundamental frequency, the partial modes up to 8th overtones are generated; however, the signal energy and sustain duration is very low as compared to their signal energy and sustain duration of the fundamental frequency, making them aurally less significant. The position of ring finger and its light pressure over the membrane suppress vibrations produced by (0,1), (1,1), (2,1), (0,2), (3,1), (1,2) and (4,1) modes. On the other hand, the seventh overtone produced by mode (2,2) shows significant energy over the other overtones. This can be observed from the magnitude of the power spectral densities of the frequencies produced by the overtones as shown in Fig. 2.

Fig. 2 Overtone frequency analysis of stroke 'Na'

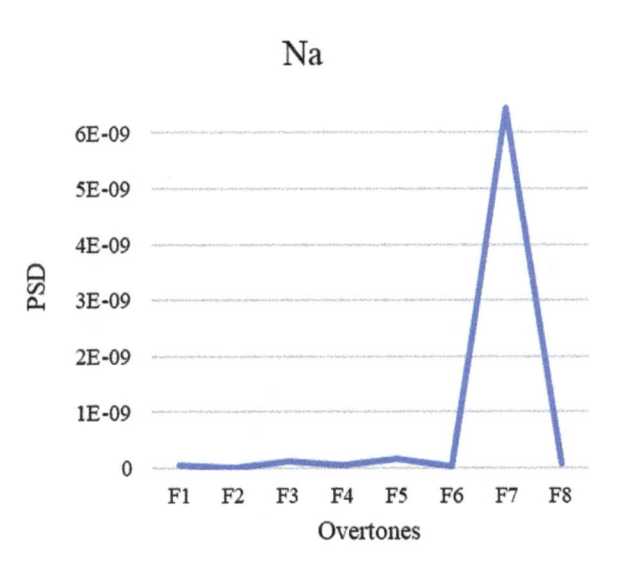

Table 2 Shruti generated from Tabla stroke 'Na'

Overtone number	Frequency (Hz)	Western musical note	Indian musical note
F1	840	$G^\#_6$	Komal Dhaiwat
F3	1242	$D^\#_6$	Komal Gandhar
F5	1573	G_6	Pancham
F7	1848	$A^\#_6$	Komal Nishaad

The seventh harmonic along with other low-energy harmonics and fundamental frequency generates timber of the sound produced by stroke Na. It is observed that the first harmonic F1 is produced due to 1 nodal diameter and 1 nodal circle produced on the surface of the membrane of the Tabla drum. The third harmonic F3 is produced due to (0,2) mode. The fifth harmonic F5 with mode (1,2) is an additive combination of harmonics F1 and F3.

The strongest overtone F7 with mode (2,2) is a combination of mode (2,1) (F2) and mode (0,2) (F3) exhibiting the highest energy density. Mode (2,2) produces a frequency of 1848 Hz that corresponds to the musical note $A^\#_6$. It is also observed that the odd harmonics 1, 3, 5 and 7 possess more spectral energy than even number of harmonics. Table 2 shows the frequency chart of the odd number of overtones generated by stroke Na and their corresponding Western and Indian music notation (in the Hindi language).

5.1.2 Stroke Ta

Out of majorly performed open Tabla strokes, Ta plays a vital role in the foundation of any rhythmic cycle or rhythm. Out of ten Tabla strokes containing the stroke Ta, eight strokes showed similar harmonic structure, and two strokes showed slight variation in the behaviour of the overtones. The fundamental frequency obtained for all these strokes was found to be 528 Hz producing musical note C_5. A Tabla player uses Ta stroke to ensure that the Tabla membrane is equally tightened from all over the surface to produce the fundamental note at which the Tabla is tuned.

Figure 3 shows a graphical representation of all the overtones and their power spectral density values. The mode (0,1) of the stroke Ta is obtained at fundamental frequency F0 and subsequent harmonic overtones are calculated. From the power spectral density obtained for each of these harmonic overtones, it is observed that the magnitude of signal energy at overtones F1, F4 and F7 is higher than the rest of all harmonic overtones. F1 harmonic overtone excites mode (1,1) with 1 nodal diameter and 1 nodal circle. Similarly, the F4 and F7 harmonic overtones excite modes (3,1) and (2,2), respectively.

The stroke Ta is produced at three locations based on the tradition of the Tabla player. Some Tabla players produce Ta by exposing the forefinger half over the outer rim edge and half over the maidan (area between ink and outer rim). Few Tabla players play only at the maidan, and some Tabla players play stroke Ta such that

Fig. 3 Overtone frequency analysis of stroke 'Ta'

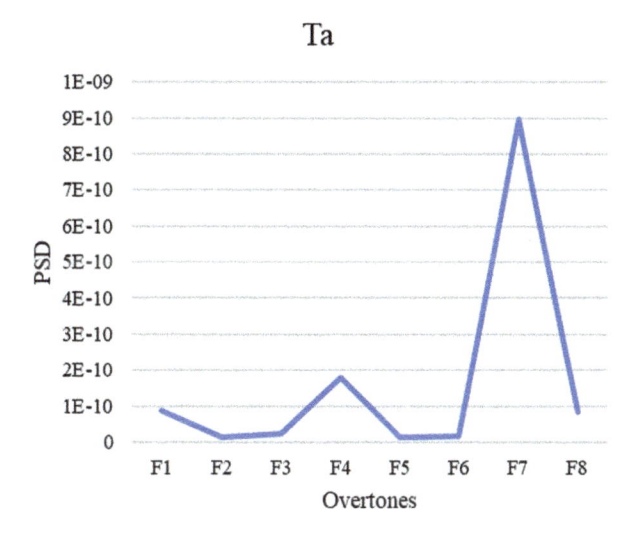

Ta

Table 3 Shruti generated from Tabla stroke 'Ta'

Overtone number	Frequency (Hz)	Western musical note	Indian musical note
F1	840	$G^\#_6$	Komal Dhaiwat
F4	1399	F_6	Madhyam
F7	1848	$A^\#_6$	Komal Nishad

the finger touches the outer area of the ink to damp its future overtones. The Tabla strokes that we have used here for analysis are played by the Tabla player who plays Tabla stroke Ta on the maidan. The two Tabla strokes of the same type exhibit little variation in their harmonic frequency pattern due to the error introduced by the Tabla player while playing the two strokes. The Tabla player may unknowingly hit the Tabla membrane with little different pressure and may change the location of the membrane where the figure is hitting at. Table 3 shows the Shruti that is generated from overtones F1, F4 and F7.

5.1.3 Stroke Te

The fundamental frequency of the stroke Te was obtained as 50 Hz. It has the highest amplitude with musical note G_1. Although the amplitude of the 8th harmonic is next highest to the fundamental frequency amplitude, it cannot be separately heard. The reason behind this is the way with which the stroke is produced than other strokes. While generating the stroke Te, it is played by hitting the finger on the centre of the ink and abruptly stopping the vibrations. The attack value of the stroke generated reaches its maximum value as soon as the Tabla is hit; however, decay, sustain and release values of the stroke are abruptly made equal to 0 by holding the finger tightly on the surface of the membrane. Due to this, multiple overtones are generated with

Te

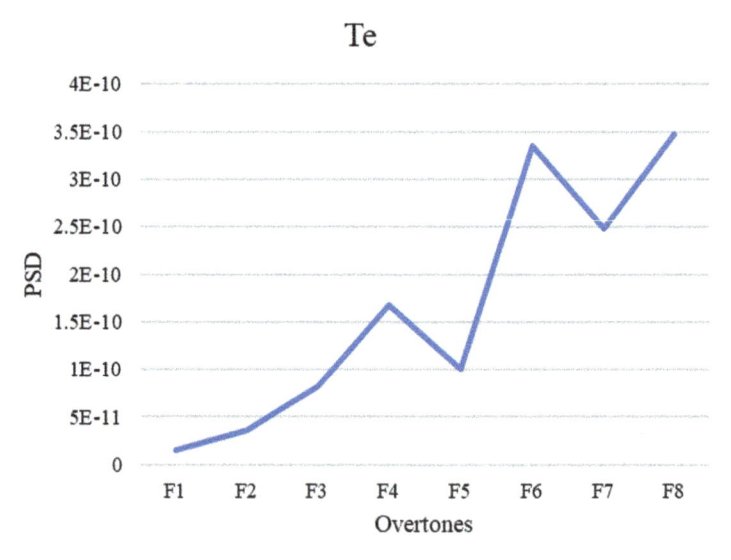

Fig. 4 Overtone frequency analysis of stroke 'Te'

Table 4 Shruti generated from Tabla stroke 'Te'

Overtone number	Frequency (Hz)	Western musical note	Indian musical note
F4	132	C_3	Shadj
F6	158	$D^{\#}_3$	Komal Gandhar
F8	180	F_3	Madhyam

prominent amplitudes that are not audible because of their short-living existence in the time domain.

As shown in Fig. 4, the harmonic overtones F4, F6 and F8 show high power spectral density. The F4 harmonic is generated through mode (3,1) where 3 nodal diameters and 1 nodal circle are produced. Similarly, the harmonics F6 and F8 are generated through modes (4,1) and (0,3), respectively. The rest of all harmonics are damped producing spectral power near to zero. This is because the Tabla player lightly holds his forefinger after striking the stroke 'Te'. The pressure of the figure is not that high to damp all the modes of progression but is sufficient to allow the F4, F6 and F8 in-harmonic overtones. Table 4 shows Shruti generated from Tabla stroke Te.

5.1.4 Stroke Tun

The Tabla stroke Tun is produced by hitting the forefinger over the centre of the ink. The ink is uniformly distributed from the centre of the membrane. The thickness of this ink is diluted towards the circumference of the membrane. Due to this structure of the Tabla membrane surface, a very important observation is made. When the

Fig. 5 Overtone frequency analysis of stroke 'Tun'

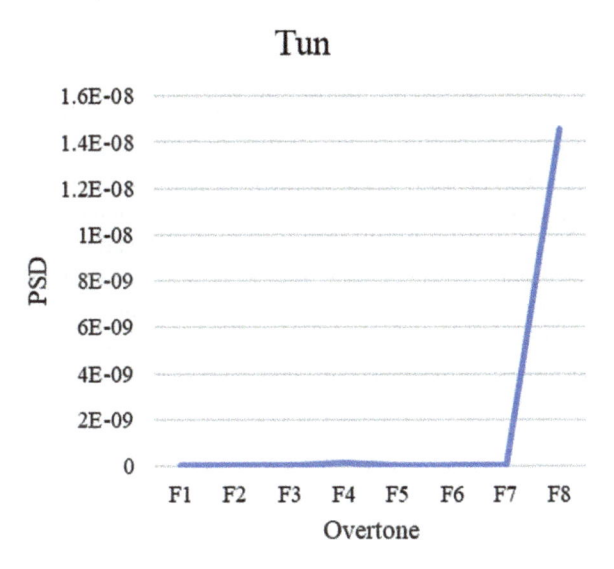

Table 5 Shruti generated from Tabla stroke 'Tun'

Overtone number	Frequency (Hz)	Western musical note	Indian musical note
F8	1044	C_6	Shadj

Tabla instrument is played with an open stroke by immediately lifting the striking finger, it keeps the Tabla membrane resonating. It is observed that the musical note octave for the Tabla stroke is one minus the octave in which the Tabla is originally tuned. It is also important to note that the strokes Na and Ta when they are played by keeping the ring finger lightly touched to the outer surface of the ink, it produces a stroke with a musical note of the fifth octave, and the open resonating membrane produces stroke musical note of exactly one lower octave. The 'Tun' stroke with the fundamental frequency having the highest amplitude was found at frequency 290 Hz exhibiting musical note D_4.

The primary power spectrum density of all the overtones except F8 is found to be near 0 as shown in Fig. 5. The ADSR analysis of stroke 'Tun' shows that sustain time of the audio signal is very high as compared with any other basic Tabla stroke. This causes suppressed overtones from F1 to F7. However, as soon as the signal releases it sustains, the prominent overtone F8 could be heard as shown in Table 5. The 8th harmonic of stroke 'Tun' F8 has a mode (0,3) showing 0 nodal diameters and 3 nodal circles. This generates the frequency of the overtones as 1044 Hz with musical note C_6. Tun is the only Tabla stroke played on the right drum without keeping any other finger being touched to the membrane of the drum.

5.2 Left Drum Strokes

In the construction of the Tabla instrument, the left drum membrane is asymmetric due to its location of ink, while the right drum Tabla membrane is symmetric. The mode progression of the drum stroke with respect to overtone is inharmonic [18]. We have extended the mode progression of the membrane of the symmetric drum to the membrane of the asymmetric drum. The reason behind this is although the left drum membrane is asymmetric in contrast with the right drum, the Tabla player most of the time rests his arm over the space while playing a stroke. Thus, it gives a similar progression of overtones as the right drum. Also, in contrast with other Western percussion instruments in Tabla, combined strokes of both the drums are also treated as a type of stroke. To analyse such type of stroke, the overtone production must be considered the same for both the drums.

5.2.1 Stroke Ga

The left drum has ink applied asymmetrically in contrast with the right drum Tabla. There exist many ways with which the left drum ink is positioned while playing the drum. In all positions, the wrist and the forefingers cover the ink such that the wrist is placed at one end of the ink and the forefinger hits the drum at the other end. For stroke Ga, the wrist is placed lightly touching the skin of the membrane such that it does not attenuate all the overtones but also does not give freedom to the membrane to resonate completely. The fundamental frequency obtained for the stroke Ga is 85 Hz. From the frequency spectrum of the Tabla stroke Ga shown in Fig. 6, it is observed that the second overtone F2, fifth overtone F5 and seventh overtone F7 possess maximum power spectral densities.

Fig. 6 Overtone frequency analysis of stroke 'Ga'

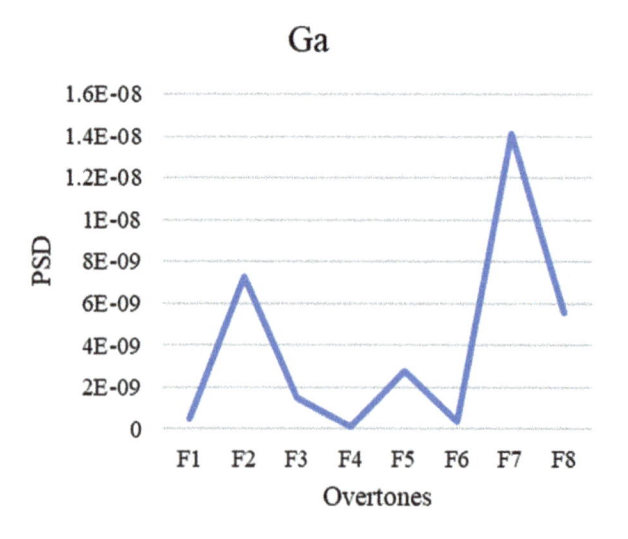

Table 6 Shruti generated from Tabla stroke 'Ga'

Overtone number	Frequency (Hz)	Western musical note	Indian musical note
F2	182	F_3	Madhyam
F5	158	B_3	Nishaad
F7	180	D_4	Rishabh

In NICM, the musical note produced by the left drum is negligible. The purpose of the left drum is just to produce rhythmic bass. However, we have tried to standardize the frequency scale with which the left drum has to be tuned and the overtone structure originated from the left drum for its basic strokes. The modes of vibration for F2, F5 and F7 are the same as the modes prescribed for the Tabla drum. However, an in-depth analysis of the location of the concentrated ink and its effect on the production of its overtones is yet to be analysed and proved. The modes that are excited out of these overtones are F2 (2,1), F5 (1,2) and F7 (2,2) exhibiting frequencies, and the musical note is as shown in Table 6.

As mentioned earlier, on the membranes of both the Tabla drums, an ink is applied and dried. This ink at one side produces inharmonic overtones and to the other side when damped with specific figure at specific location can suppress the power spectral density of the inharmonic overtones that are generated. Thus, in Fig. 6, the inharmonic overtones F3, F4 and F6 produce almost zero PSD, while F2, F5 and F7 show higher range of PSD values.

5.2.2 Stroke Ka

The stroke Ka is a closed stroke played on the left drum. The frequency spectral analysis of Ka shows weird spectral values for different frequencies calculated based on the fundamental frequency for ten audio samples. The left drum was originally tuned to frequency 41 Hz (Gandhaar), and the fundamental frequency for the stroke is obtained as 25 Hz (Pancham). Though the fundamental frequency was near 25 Hz, the remaining overtones exhibit haphazard values for each audio sample of the Tabla stroke.

As shown in Fig. 7, the harmonic overtones F1, F4 and F6 show high power spectral density. The first overtone of stroke Ka produces 1 nodal diameter and 1 nodal circle giving F1 (1,1) mode. This corresponds to frequency 40 Hz with musical note E_1. This overtone could be brightly heard along with fundamental frequency. The next overtone F4 produces 3 nodal diameters and 1 nodal circle giving frequency 66 Hz and corresponding musical note C_2. Similarly, for F6 overtones, the frequency produced is 79 Hz with musical note $D^{\#}_2$ as shown in Table 7.

Ka

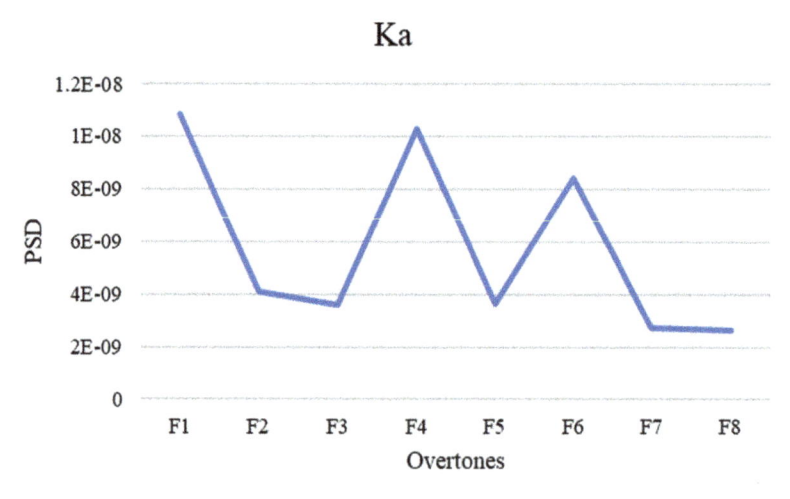

Fig. 7 Overtone frequency analysis of stroke 'Ka'

Table 7 Shruti generated from Tabla stroke 'Ka'

Overtone number	Frequency (Hz)	Western musical note	Indian musical note
F1	40	E_1	Gandhaar
F4	66	C_2	Shadj
F6	79	$D^{\#}_2$	Komal Gandhaar

5.3 Both Drum Strokes

The strokes that are originated from both the drums simultaneously should be analysed for mode progression as per individual left and right basic drum strokes. Mode progression from these individual drum strokes will be similar as explained in the above sections. In this section, we would prefer to analyse the homophonic sound generated as a combined effect of sound from the drums. Hence, in this section, only the fundamental frequencies obtained from combined stroke and upcoming overtones are discussed which are calculated from the frequency multipliers declared for drum membrane.

5.3.1 Stroke Dha

Stroke Dha is obtained by producing Ga stroke on the left drum and Na stroke on the right drum simultaneously and keeping fingers hitting the right drum touched to the rim. The combined effect of the Ga and Na is analysed here. From the spectral analysis of the combined stroke Dha, it is observed that each time Dha is produced, it exhibits a different overtone structure. The factors that cause this change in each stroke produced are as follows. While playing Na on the right drum, the forefinger

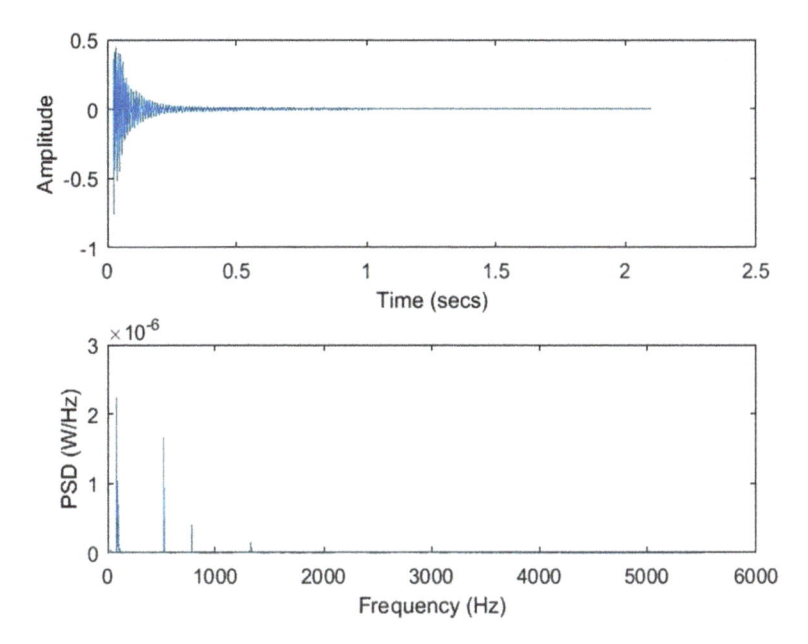

Fig. 8 Time and frequency domain analysis of stroke 'Dha'

is kept stuck on the outer rim. The pressure applied to keep the finger stuck on the drum membrane causes the upcoming overtones to be suppressed. Different pressure applied by the forefinger produces different overtone behaviour of the stroke. It is practically impossible even for a professional Tabla player to maintain the same pressure each time the same stroke is played. The Tabla strokes Ga and Na are open and closed strokes, respectively. Thus, from the frequency spectrum, it is found that it contains frequencies of the stroke Ga and Na. It can be observed that the 7th overtone of Na is prominent as compared with other overtone energies. The same behaviour is observed in the frequency spectrum of the stroke Dha. The 7th overtone frequency of the stroke Dha is found to be prominent over other overtones present. To the normal human ear, the difference of time instance where left drum and right drum stroke played is unnoticeable.

In the spectral analysis of the Tabla stroke Dha, two separate frequencies are visible prominently with high-power spectral density values as shown in Fig. 8. The first spike in the figure shows the fundamental frequency of the left drum (85 Hz) which represents the presence of left drum stroke Ga. The second spike in the figure shows the fundamental frequency of the right-hand Tabla drum (528 Hz) representing the presence of right drum stroke Na. The order of occurrence of left and right drum stroke frequency may reverse based on the stroke that was played slightly earlier than the other in the time domain. This comparison shows that both the strokes are played on both drums simultaneously, and their frequency spectrum for the combined effect of the stroke is analysed. With respect to the signal energy,

Fig. 9 Overtone frequency analysis of stroke 'Dha'

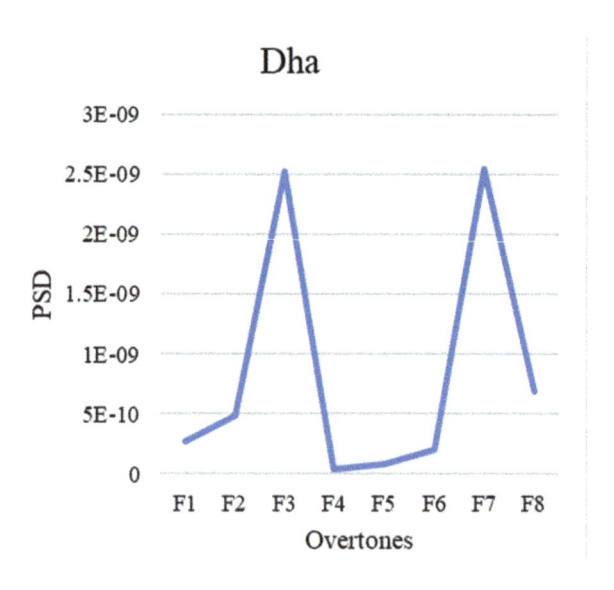

Table 8 Shruti generated from Tabla stroke 'Dha'

Overtone number	Frequency (Hz)	Western musical note	Indian musical note
F3	193	G_3	Pancham
F7	294	D_4	Rishabh

the right-hand drum stroke Na and left-hand stroke Ga are merged to produce a homophonic texture of the sound combinedly called Dha.

The overtone analysis of stroke Dha is as shown in Fig. 9. In the combined stroke where frequencies from both the drums are merged, the overtone structure shows that F3 and F7 contain maximum power spectral density values. The musical notes that are generated from these overtones are F3 (193 Hz) musical note—(G_3)—and for F7 (294 Hz) musical note, (D_4) as shown in Table 8.

5.3.2 Stroke Dhin

The Tabla stroke Dhin is one of the major strokes used in repetitive cycles of some rhythms. The stroke Dhin is produced by hitting and lifting fingers from both the drum membranes simultaneously. The combined effect of the stroke Ga produced on the left drum and Ta produced on the right drum is analysed here. From the analysis of the combined audio signal frequencies and overtones, it is found that the stroke shows variations in its harmonic structure. The reasons behind these variations are three-folded. (A) The pressure applied on the drum membrane by the fingers is different each time the drums are hit. (B) Many times, the Tabla player unknowingly hits the Tabla membrane at slightly different locations for each stroke. (C) Ga and Ta both are played in open mode. It is also important to note that both the strokes

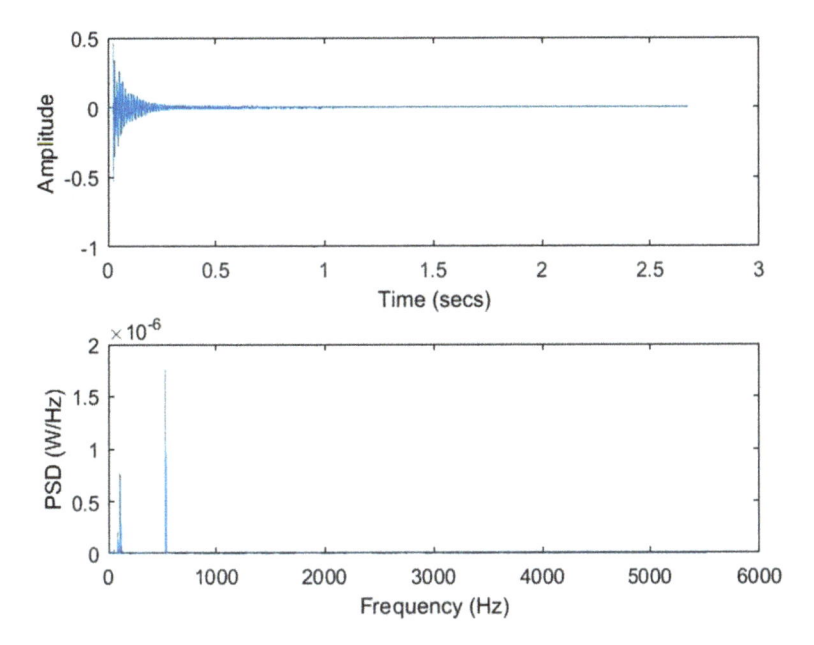

Fig. 10 Time and frequency domain analysis of stroke 'Dhin'

from the left and right drum should be produced at the same instance of time. This difference is unnoticeable by the human ear and causes no effect on the overall sound of the stroke but could be visualized clearly in the frequency spectrum.

With respect to the signal analysis and instance of a time when the stroke is hit, it is found that very rarely both the fingers hit both the drums simultaneously. At times, the left drum is hit earlier than the right drum giving a different harmonic structure for the combined sound. While sometimes the right drum stroke is played earlier than the left drum stroke. In all the cases, the former fundamental frequency and its overtones dominate the later drum stroke fundamental frequency and its overtones.

As shown in Fig. 10, the time domain signal appears to be like the rest of the strokes; however, in the frequency domain, two prominent frequencies are visible with high-power spectral density values. Since the stroke Dhin is a combination of strokes originated from left and right drum simultaneously, there exist two fundamental frequencies, viz. from left drum stroke and right drum stroke. It is observed that for stroke Ga, the overtones F3 and F7 hold nonzero signal energies, while the rest of the overtones hold negligible signal energy.

Figure 11 shows frequency overtones generated by the stroke Dhin. For the stroke Dhin, the curse of the order of left and right drum hit causes change in the spectrogram for different instances of the stroke. Similar behaviour is observed for stroke Dha. This comparison shows that, although both the strokes (Dha and Dhin) are played on both drums (left and right) simultaneously, the pattern of overtone structure for the combined stroke is different for each stroke instance.

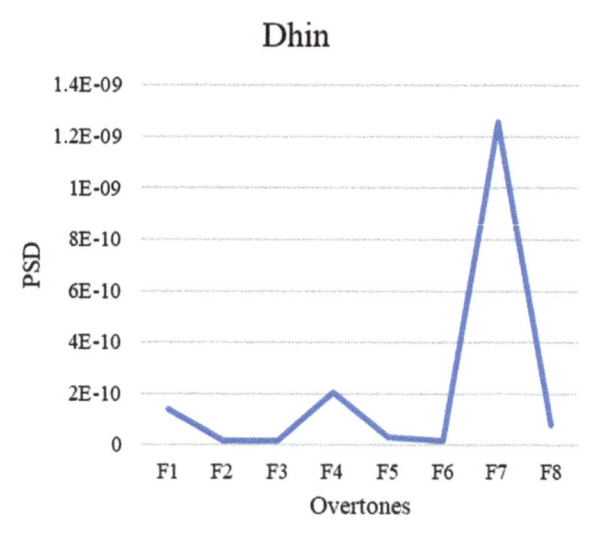

Fig. 11 Overtone frequency analysis of stroke 'Dhin'

Table 9 Shruti generated from Tabla stroke 'Dhin'

Overtone number	Frequency (Hz)	Western musical note	Indian musical note
F1	842	$G^{\#}_5$	Komal Dhaiwat
F4	1404	F_6	Madhyam
F7	1855	$A^{\#}_6$	Komal Nishaad

For stroke Ta, frequency overtones F1, F4 and F7 contain the maximum energy of the signal as shown in Table 9. Some of the left drum overtones and right drum overtones overlap with each other. For the combined signal energy for the stroke Dhin, overtones F1 as 842 Hz, F4 as 1404 Hz and F7 as 1855 Hz have the highest signal energies generating musical note $G^{\#}_5$, F_6 and $A^{\#}_6$, respectively.

5.3.3 Stroke Tin

Tin is the stroke played on the left and right drums simultaneously. It is similar to the stroke Dhin except that on the left drum instead of Ga stroke Ka is played. The right drum stroke Ta, when combined with stroke Ka, produces a resonating stroke Tin. In this stroke, the left drum stroke is closed, and the right drum stroke is open.

As shown in Fig. 12 in the frequency domain, the left drum Tabla stroke Ka has a very low frequency (~25 Hz) G_0, and the right drum Tabla stroke Ta has a fundamental frequency (~528 Hz) C_5. Since the stroke Ka is a closed stroke, its overtones are damped immediately, thus producing very low values of decay, sustain and release. Thus, the overtones are negligible. On the other hand, the right drum stroke Ta, being open stroke, produces its overtones independently. The signal energy of the left drum is dominated by the signal energy of the right drum. The location of the left-hand palm striking the left drum membrane makes different kinds

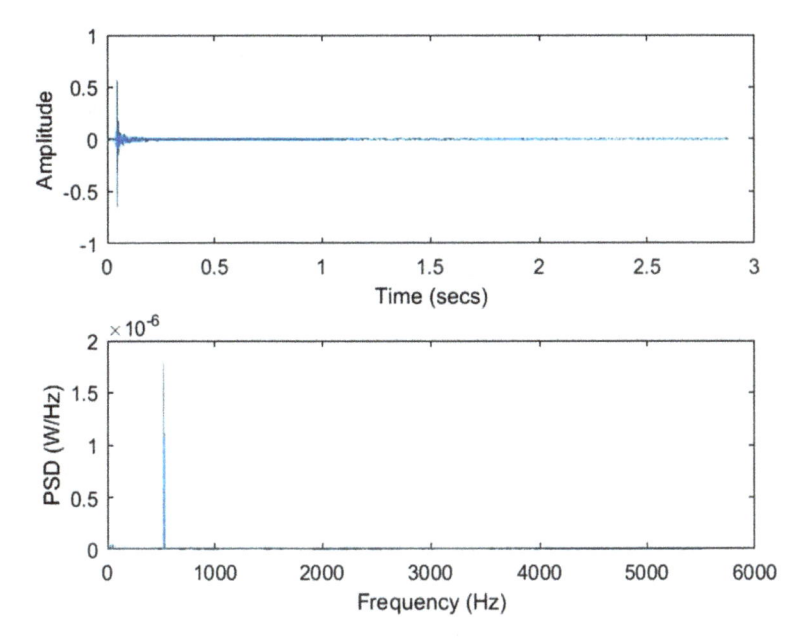

Fig. 12 Time and frequency domain analysis of stroke 'Tin'

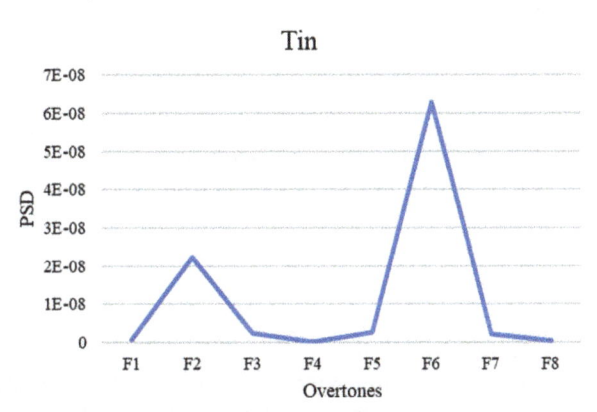

Fig. 13 Overtone frequency analysis of stroke 'Tin'

of overtones. If the fingers of the left hand are striking the membrane on ink, then the overtone structure slightly differs. However, in all cases, the power spectral density of all the overtones is so small that it could be neglected.

Figure 13 shows overtone analysis of stroke Tin. It is observed that the combined stroke produces fundamental frequency F0 as 529 Hz (C_5). The prominent overtones that are generated are at overtone F2 as 1134 Hz and F6 as 1675 Hz as shown in Table 10.

The percussion instrument sound wave propagation theory works on non-integer multipliers to the fundamental frequency to produce overtones. Due to the presence of ink over the membrane of the Tabla instrument, the overtones produced by the

Table 10 Shruti generated from Tabla stroke 'Tin'

Overtone number	Frequency (Hz)	Western musical note	Indian musical note
F2	1134	$C^{\#}_6$	Komal Rishabh
F6	1675	$G^{\#}_6$	Komal Dhaiwat

membrane are found out using nodal diameters and nodal circles. For each stroke, the overtones are calculated and analysed concerning the magnitude of their power spectral density. A combined analysis of all the ten samples of each of the nine basic strokes is done to find out the pattern of the power spectral strength of all the overtones across all the samples. Some important observations and conclusions are as follows:

- It is observed that a slight change in the location, where the finger hits the membrane, does not affect greatly to the fundamental frequency; however, it affects the energy of the overtones. The Tabla player may apply different pressures and decide to lift the finger each time when he plays the same stroke. This causes a substantial difference in the overtone energies causing the negligible yet observable change in the timbre of the sound.
- If the right drum Tabla is tuned to the C_x scale where C is the musical note and x is the octave, then stroke 'Tun' produces fundamental frequency with note $D_{(x-1)}$.
- It is observed that all the basic Tabla strokes exhibit high-power spectral density magnitude at the 7th overtone (F_7).
- The mode (1,1) is responsible for ringing sound and contributes to the musical sound for percussion instruments such as timpani, which does not have a layer of ink on the membrane. The application of the ink on the surface of the membrane makes the Tabla instrument produce strokes that have larger sustain values than other instruments.
- It is observed that mode (2,2) where 2 nodal diameters and 2 nodal circles give larger sustain sound than other modes has a very high-power spectral density as compared to other Tabla strokes. Mode (2,2) takes a long time to decay and does not transfer the vibration energy into radiated sound energy. Thus, this mode rings for a while and contributes to the sound or pitch of the stroke.
- Prominent overtones of the left drum have very little magnitude and frequency, so they contribute negligibly to the timbre of the sound.
- The right drum stroke produces overtone which majorly contains sharp (#) notes such as $G_\#$, $D_\#$, $A_\#$, etc. The Tabla instrument used in this research is a standard Tabla set used professionally. However, the Tabla instrument is not scientifically designed. For example, the Tabla player is unaware of the ingredient types of the ink and their percentage of the components used. Thus, the instrument player only focuses on fundamental frequencies and neglects the overtones.
- When a Tabla player tunes the right-hand drum with musical note C5, then the seventh overtone (F_7) could be musical note (G_6). It is to be mentioned here that

Table 11 Summary of nine basic Tabla strokes and their corresponding Shruti generated

Sr. no.	Stroke	Drum source	Fundamental frequency	Musical note	Prominent overtones	Musical notes
1	Na	Right	528	C_5	F1, F3, F5, F7	$G^\#_6, D^\#_6, G_6, A^\#_6$
2	Ta	Right	528	C_5	F1, F4, F7	$G^\#_6, F_6, A^\#_6$
3	Te	Right	50	G_1	F4, F6, F8	$C_3, D^\#_3, F_3$
4	Tun	Right	290	D_4	F8	C_6
5	Ga	Left	85	E_2	F2, F5, F7	F_3, B_3, D_4
6	Ka	Left	25	G_0	F1, F4, F6	$E_1, C_2, D^\#_2$
7	Dha (Na+Ga)	Both	84	E_2	F3, F7	G_3, D_4
8	Dhin (Ta+Ga)	Both	529	C_5	F1, F4, F7	$G^\#_5, F_6, A^\#_6$
9	Tin (Ta+Ka)	Both	529	C_5	F2, F6	$C^\#_6, G^\#_6$

the experiment is carried using only one set of Tabla instruments tuned at the (C_5) scale.

- Table 11 shows nine basic Tabla strokes with their fundamental frequencies, prominent overtones and corresponding Western and Indian musical notes. The mode progression and analysis of overtone structure of Tabla strokes originated from both the drums can only be analysed by individually observing the right drum stroke and left drum stroke.
- The Tabla sound contains the homophonic mixture of both the drums together; hence, this research is kept limited to the theory of circular membrane mode propagation as described by C. V. Raman [18]. Here, the constant multipliers are used to calculate the overtones irrespective of the actual harmonic component energy of the Tabla stroke.
- The frequency in which the right drum is tuned is usually the fundamental frequency of the stroke originated from the right drum and both the drums simultaneously.

6 Conclusion

North Indian classical music Tabla instrument is one of the majorly used accompanying rhythm instruments, which is also played as a solo performance. For different vocal and musical instrument performances, the Tabla plays important role in providing the base rhythm and tempo. The Tabla instrument is called an 'Eksuri' (single musical note) instrument. However, being an acoustic musical instrument,

Tabla produces various inharmonic musical notes which contribute to the timbral sound quality of the strokes originated from the instrument. In contrast with the other Western rhythm membrane instruments, the Tabla instrument is a special of its kind. The specialty of the Tabla instrument is in the ink applied over the surface of the membrane.

A standard Tabla instrument with tuning scale C_5 (528 Hz) is used here for analysis. The audio database consists of studio-recorded and pre-processed audio excerpts containing ten samples of nine basic Tabla strokes with sampling frequency 44,100 Hz, 16-bit pulse code modulation .wav format. The time domain sound signal is analysed in the frequency domain using fast Fourier transform, and the power spectral density of each frequency component present in the audio signal is analysed. Based on the source of the sound, the basic Tabla strokes are categorized into three classes, namely, strokes from the left drum, right drum and both the drums simultaneously.

Almost negligible research has been done so far towards the sound produced through the Tabla instrument for the timbral attribute and frequency components present in the Tabla stroke sound other than the fundamental frequency. Most of the Tabla instrument players are found to be unfamiliar with the technical aspects of Tabla instrument construction. From generations together, the Tabla manufacturers are illiterate and nontechnical, who hardly understand the frequency overtones originated from a Tabla stroke. The art of Tabla making is solely based on the Tabla maker's capability to understand the timbral quality of the Tabla sound. Similar to Tanpura makers, Tabla instrument manufacturing art is percolated from one generation to the next generation. Due to the presence of the ink over the membrane of the Tabla instrument, and due to the weight of the ink, the Tabla stroke produces inharmonic overtones. These overtones are expressed in the form of nodal diameter and nodal circles generated over the surface of the membrane of the Tabla instrument when struck by bare fingers at some pre-decided locations. Based on the location of the striking finger, different Tabla strokes are generated. Each Tabla stroke uniquely generates a mixture of a different inharmonic component in the form of overtones. These overtones are short-lived and hence usually neglected by the musicians. Although the short-lived inharmonic components present in the Tabla stroke produce low-power spectral density values, their presence contributes to a great extent towards the timbral aspects of the Tabla strokes.

This research contributes towards the standardization of the manufacturing of the Tabla instrument. Each Tabla stroke exhibits different inharmonic overtones. These overtones are mapped onto the Indian and Western musical scales. Thus, for the exact production of the desired musical note from each Tabla stroke, different parameters could be set up during the manufacturing of the Tabla instrument. The various parameters which could be calibrated in advance include the amount of ink applied over the membrane, the density of the ink, the highest thickness at the centre of the membrane, number of ink layers to be applied from the centre towards the circumference of the membrane and the difference between the thickness of each layer concerning the thickness at the centre. These all parameters are very crucial and contribute to a great extent towards the musical notes generated from each basic

Tabla stroke. The human Tabla player may not be able to apply same pressure while producing same strokes repeatedly or may shift the location of the Tabla stroke produced on the surface of the membrane; however, these errors would not matter and could be neglected as far as the generation of basic Tabla strokes and their corresponding musical notes produced due to the presence of inharmonic overtone.

References

1. Park, Tae Hong (2004). Towards Automatic Musical Instrument Timbre Recognition. Ph.D. Thesis, Candidacy: Princeton University.
2. Wikipedia. [Online] April 3, 2021. [Cited: May 01, 2021.] Acoustic Resonance. https://en.wikipedia.org/wiki/Acoustic_resonance
3. Wikipedia. [Online] April 9, 2021. [Cited: May 2, 2021.] Harmonic. https://en.wikipedia.org/wiki/Harmonic
4. Gough, C (2007). Musical Accoustics. New York, NY: Springer.
5. Schneider, A and Frieler, K. (2008). Perception of Harmonic and Inharmonic Sounds: Results from Ear Models, International Symposium on Computer Music Modeling and Retrieval, CMMR 2008. Computer Music Modeling and Retrieval. Genesis of Meaning in Sound and Music. Vol. 5493, pp. 18–44.
6. Gerhard, Albersheim (1948). Overtones and Music Theory. University of California Press, Sept 1948, Bulletin of the American Musicological Society, pp. 69–71.
7. Overholt, Dan (2005). The overtone violin: A new computer music instrument. May 26–28, 2005. New Interfaces for Musical Expression, NIME-05.
8. Classic, Merlot. [Online] January 2006. http://hyperphysics.phy-astr.gsu.edu/hbase/Music/cirmem.html
9. Mehta, Parikshit (2009). Vibrations of Thin Plate with Piezoelectric Actuator: Theory and Experiments. Clemson University, TigerPrints.
10. Nave, R. Preferred Timpani Modes. [Online] http://hyperphysics.phy-astr.gsu.edu/hbase/Music/cirmem.html#c3
11. https://en.wikipedia.Raga. [Online] April 16, 2021. https://en.wikipedia.org/wiki/Raga
12. Tabla. [Online] April 29, 2021. https://en.wikipedia.org/wiki/Tabla
13. Physics Textbook for Class XI, National Council of Educational Research and Training, New Delhi, 2006.
14. Baral, Bibhudutta and Manasa, K. Tabala Making - Varanasi. Digital Learning Environment for Design www.dsource.in
15. Courtney, David and Courtney, Chandrakantha (1998). Basic Strokes and Bols. [Online] https://chandrakantha.com/tablasite/bsicbols.htm
16. Raman, C. V, and Kumar, S. (January 1920). Musical Drums with Harmonic Overtones. Nature. Vol. 104, pp. 500–500.
17. Russell, Daniel A. Acoustics and Vibration Animations. [Online] January 21, 1998. https://www.acs.psu.edu/drussell/Demos/MembraneCircle/Circle.html
18. Raman, C V (1934). The Indian Musical Drums, Proc. Indian Acad. Sci. (Math. Sci.), Vol. 1, pp. 179–188.

Index

A
Acoustic fingerprinting, xi, 320–323, 327
Acoustics, 4, 8, 9, 54, 61, 75, 109, 141, 176, 205, 217, 218, 220, 223, 228, 229, 232, 312, 317, 354, 360, 379–393, 400, 406, 415, 422, 437
AHP reduction, 277, 278
AI-based music, 25
Album cover design, xii, 380, 382–385, 387–389, 393
Audience response, 358–360, 364–371, 374
Audio toolbox, vii
Audio/visual stimuli, xii, 398–401, 406, 407, 411

B
Bengali, x, xi, 283–306
Bias, vi, vii, 41, 52, 61, 62, 71–86, 111, 124, 143, 151, 154, 163, 165, 334, 340–342, 346, 348, 375, 411
Brain plasticity, ix, x, 269–280
Bridge, 35, 332, 333, 336, 338, 340

C
Case studies, v, xi, 311–327
Circular membrane vibration modes, 420, 437
Classifier, viii, 85, 96, 99, 105, 106, 108–110, 138–140, 143, 144, 148–155, 176–181, 184, 186, 188, 193–198, 202, 203, 208, 209, 211, 229–230, 301
Cochlear implant, 218, 222, 230–233, ix

Collaborative approach, 54–55
Comorbidity, 237, 238
Computational analysis, 113, 114
Content-based, vi, 52–54, 56, 60, 65, 72, 75–77, 79, 80, 91, 141
Context, vi, 36, 52, 53, 56–60, 62–64, 75–78, 86, 95, 96, 102, 108, 143, 145, 194, 196, 197, 253, 262, 278, 300, 303, 304, 316–318, 356–358, 380, 382
Convolutional neural network (CNN), viii, 18, 27, 106, 127–129, 134, 160, 163–164, 169–173, 176–184, 186–188, 190, 191, 194, 195, 198–202, 206, 211, 285, 294, 296–298
Coping style, x, 249–264
Cross-modal valence, 398–400, 405, 406, 408, 410, 411

D
Data augmentation, viii, 15, 20, 177, 180, 195, 199, 203–204
Deep learning, vi, viii, ix, x, 3, 4, 13–18, 20, 25–46, 57, 63, 109, 114, 124, 125, 139, 148, 159–173, 176, 177, 181, 193, 196, 198–205, 210, 211
Deep neural networks (DNNs), vi, vii, viii, 13, 14, 18, 26, 121–134, 138, 193–211, 296
Diabetes mellitus, ix, 237–245
Digital signal processing, 218, 220, 223, 225–230
Duplicate detection, xi, 311–327

E
Emotion, 3, 37, 52, 72, 94, 137, 193, 232, 239, 251, 284, 343, 354, 380, 397

F
Feature analysis, 359, 362–364, 367, 374
Feature extraction, vi, vii, 4–8, 13, 20, 52, 93, 95–97, 104–108, 137–155, 161, 162, 177, 184, 195–198, 358, 376
Fractal analysis, xii, 358–359, 371–375, 384, 390–393

G
Generative adversarial networks (GANs), 26, 27, 29, 33, 35, 36, 38, 43, 195, 196, 199, 203, 210, 305

H
Hearing aids, ix, 122, 217, 218, 220–233
Hodgkin-Huxley (HH) model, 278
Homophonic sound, 430, 432
Hurst exponent, xiii, 361, 385, 399, 401
Hybrid approach, vi, 52, 56–57

I
Indian classical instrumental music, 380, 382–384
Indian classical music, v, vii, viii, xii, 91–114, 159–173, 259, 379–393, 415–439
Intermediality, xi, xii, 353–375, 382, 387
Intrinsic plasticity, 270, 272–274, 277–280

K
k-nearest neighbours (k-NN), viii, 150, 151, 175–191

L
Locus of control, 249–264
Long-term depression (LTD), 274, 276
Long-term potentiation (LTP), 274–276
Lyrics, 153, 206, 232, 331–334, 337, 338, 340, 342–347, 358

M
Machine learning (ML), v, vii, 3, 13, 34, 52, 55, 65, 72, 76, 77, 80, 81, 85, 124, 137–155, 160–162, 166, 176, 182, 193, 194, 196–198, 200, 202, 205, 208, 264, 287
MATLAB, vii, 123, 125
MIR applications, 311
Movie music recommendation, 193
Multi-class classification, 163, 165
Music, v–xiii, 25–46, 51–65, 71–86, 91–114, 125, 137–147, 152, 154, 159–173, 175–191, 193–211, 217–233, 237–245, 249–264, 269–280, 311–314, 316–321, 323, 327, 331–334, 337, 338, 340, 346, 347, 349, 353–375, 379–393, 398, 400, 415–439
Musical instrument recognition, 175, 177
Musical notes, 218, 222, 269, 416, 417, 420–427, 429, 430, 432, 434, 436–439
Music archives, xi, 311–317, 323, 327
Music cognition, 279
Music emotion recognition, viii, 85, 194, 197–204
Music generation, 26, 28, 30, 31, 33, 34, 36–46
Music information retrieval (MIR), xi, v, viii, xi,, 25, 73, 76, 81, 92–94, 100, 159, 175, 176, 180, 196, 311, 327
Music perception, ix, x, 217–233, 279
Music preference, x, 61, 254, 256–260, 262–263
Music processing, vii, ix, 222, 225, 228
Music recommendation, v, vi, vii, x,, 51–65, 138, 175, 264
Music recommender systems, 71–86, vi
Music therapy, ix, 237–245

N
Neural networks, vii, viii, x, 4, 26–28, 96, 108, 122, 124–127, 151, 160, 166, 177, 178, 286
Non-superstar artists, 72, 75, 81, 82, 84–86
North Indian Classical Music, xiii, 415–439

P
Peak signal to noise ratio (PSNR), vii, 122, 123, 125, 130, 132, 134
Popularity bias, 52, 61, 62, 84

R
Recurrent neural network (RNN), viii, 33, 35, 38, 42, 163, 165–173, 179, 294
Refrain, xi, 332–336, 338, 339, 345, 348, 349, 358
Regression, 33, 44, 151, 184, 194, 195, 286, 323, 340–342, 348, 361, 362, 386, 402–404
Repetition, 40, 113, 331–336, 345, 347, 348, 380, 382

S
Semiotic analysis, xii, 382, 384, 387–390
Signal processing, 123, 144, 160, 173, 176, 196, 217–233, 415, ix, v, vii
Signal to noise ratio (SNR), vii, 15, 122–125, 129, 130, 132, 134
Social influence, 355
Sound archive management, xi, 311–327
Speaker recognition, v, vi, 3–20
Spectro–temporal processing, 5
Speech, 3, 96, 121, 165, 176, 195, 217, 239, 355
Speech processing, v, vi
Stimuli creation, 337
Synaptic plasticity, 270–274, 277

T
Tabla, xi, xiii, 109, 159, 415–439
Text-dependent, 4, 5, 8, 9, 15, 20
Text-to-speech (TTS), x, xi, 283–285, 293, 294, 298–306
Tonic identification, vii, 93, 95–99, 105–107, 109, 114, 160
Transfer learning, ix, 26, 36, 37, 44, 45, 195, 196, 199, 201–203, 208–211

U
User orientation, 52, 58–60

V
Video soundtracks, vii, viii, 137–155
Visual arts, 354–358, 364, 374, 375, 379–382

W
Wiener filtering technique, vii, 121–134

X
XGboostÿ, viii, 151, 160–163, 166–170, 172, 173

Milton Keynes UK
Ingram Content Group UK Ltd.
UKHW020640100124
435787UK00002B/5